De Gruyter Studies in Mathematics 44

Alex Degtyarev

Topology of Algebraic Curves

An Approach via Dessins d'Enfants

De Gruyter

Mathematics Subject Classification 2010: Primary: 14H30, 14H50, 14J27, 14P25;
Secondary: 20F36, 11F06, 05C90, 14H57.

ISBN 978-3-11-025591-1
e-ISBN 978-3-11-025842-4
ISSN 0179-0986

Library of Congress Cataloging-in-Publication Data

A CIP catalog record for this book has been applied for at the Library of Congress.

Bibliographic information published by the Deutsche Nationalbibliothek

The Deutsche Nationalbibliothek lists this publication in the Deutsche Nationalbibliografie; detailed bibliographic data are available in the Internet at http://dnb.dnb.de.

Typesetting: Da-TeX Gerd Blumenstein, Leipzig, www.da-tex.de
Printing and binding: Hubert & Co. GmbH & Co. KG, Göttingen
∞ Printed on acid-free paper
Printed in Germany

www.degruyter.com

I dedicate this book to my wife Ayşe Bulut:
it is her constant support, patience, and
understanding that make my work successful.

Preface

The purpose of this monograph is to summarize, unify, and extend a number of interrelated results that were published or submitted during the last five years in a series of papers by the author, partially in collaboration with Ilia Itenberg, Viatcheslav Kharlamov, and Nermin Salepci. As often happens in a long research project, the principal ideas have evolved and a few minor mistakes have been discovered, calling for a new, more comprehensive and unified approach to the earlier papers. Furthermore, as the work is still in progress, I am representing older results in a more complete and general form. The monograph also contains several newer results that have never appeared elsewhere. Thus, we complete the analysis of the metabelian invariants of a trigonal curve (see Chapter 6), compute the fundamental groups of all, not necessarily maximizing, irreducible simple sextics with a triple point (see Chapter 8; a few new sextics with finite nonabelian fundamental group have been discovered), and establish the quasi-simplicity of most ribbon curves, including M-curves (see Section 10.3.2). The latest achievement is the complete understanding of simple monodromy factorizations of length two in the modular group (see Section 10.2.2; joint work with N. Salepci). In spite of its apparent simplicity, our description of such factorizations has interesting applications to the topology of real trigonal curves and real Lefschetz fibrations; they are discussed in Section 10.3. There also are a few other advances in the study of monodromy factorizations (see, *e.g.*, Section 10.2.3), but in general the situation still remains unclear and the problem seems wild.

The dominant theme of the book is the very fruitful close relation between three classes of objects:

- elliptic surfaces and trigonal curves in ruled surfaces, see Chapter 3,
- skeletons (certain bipartite ribbon graphs), see Chapter 1, and
- subgroups of the modular group $\Gamma := PSL(2, \mathbb{Z})$, see Chapter 2.

(Slightly different versions of skeletons appeared in the literature under a number of names, the most well known being *dessins d'enfants* and quilts.) When restricted to appropriate subclasses, this relation becomes bijective, providing an intuitive combinatorial and topological framework for the study of trigonal curves, on the one hand, and of subgroups of Γ, on the other.

Undoubtedly, both *dessins d'enfants* and the modular group are amongst the most popular objects of modern mathematics; both are extensively covered in the literature. Thus, the modular group and its subgroups play a central rôle in the theory of modular forms, Moonshine theory, some aspects of number theory and hyperbolic geometry.

Dessins d'enfants, apart from Grothendieck's original idea [83] connecting them (*via* Belyĭ maps) to the absolute Galois group $\mathrm{Gal}(\bar{\mathbb{Q}}/\mathbb{Q})$, are related (*via* moduli spaces of curves and the Gromov–Witten theory) to topological field theories and integrable partial differential equations. Unfortunately, all of these fascinating topics are beyond the scope of this book. Our primary concern is a straightforward application of dessins to topology of trigonal curves and plane curves with deep singularities: the monodromy of such a curve can be computed, in a purely combinatorial way, in terms of its dessin; as a consequence, the monodromy group is a subgroup of Γ of genus zero and, using the well developed theory of such subgroups, we obtain numerous restrictions on the fundamental group of the curve and its more subtle geometric properties. This idea is summarized in Speculation 5.90 in Chapter 5.

Principal results

Originally, my interest in *dessins d'enfants* was motivated by our work (joint with I. Itenberg and V. Kharlamov) on real trigonal curves and real elliptic surfaces and by my attempts to compute the fundamental groups of plane sextics. These two classes of objects remain the principal geometric applications of the theory. Here is a brief account of the most important statements found in the text.

Plane sextics. We use skeletons to classify irreducible plane sextics with a triple singular point (including non-simple ones) and compute their fundamental groups, see Theorems 7.45, 8.1, and 8.2. Thanks to works by J.-G. Yang [166], I. Shimada [148], and the author [46], the classification of simple plane sextics is close to its completion. This classification relies on the global Torelli theorem for $K3$-surfaces and is not quite constructive; the 'visualization' of sextics, necessary for the detailed study of their geometry, remains an open problem. In the presence of a triple point, this problem is solved by means of the skeletons.

A brief survey of the known results concerning plane sextics and their fundamental groups is given in Section 7.2.1 and Section 7.2.3.

Another application in this direction is the classification up to deformation and the computation of the fundamental groups of singular plane quintics, see Theorems 7.49 and 7.50. This result is old, but its complete proof has never been published.

Universal trigonal curves and metabelian invariants. The monodromy group of a non-isotrivial trigonal curve over a rational base is a subgroup of genus zero, and any subgroup $H \subset \Gamma$ of genus zero is realized by a certain *universal curve*, from which any other curve with the monodromy group subconjugate to H is induced, see Section 5.3.1 and Corollary 5.88. These facts impose strong restrictions on the fundamental group and relate the latter to some geometric properties of the curve. (I expect that there should be a reasonable description of all finite quotients of such groups.) As a first step, we obtain universal (independent of the singularities) bounds on the Alexander module of a trigonal curve, see Theorems 6.1 and 6.16. Then, as an illustration of Speculation 5.90 (2), we establish a version of Oka's conjecture for trigonal curves, see Theorem 6.10, and classify the so-called *Z-splitting sections* of such curves, see Theorem 6.15: any Z-splitting section is induced from a certain universal one. This statement may have further implications to the study of tetragonal curves, hence plane sextics with $\mathbf{A}$ type singular points only.

Monodromy factorizations. A long standing question, related to the study of the topology of algebraic and pseudo-holomorphic curves, is whether a simple factorization of a given monodromy at infinity is unique up to Hurwitz equivalence. We answer this question in the negative and show that the problem is much wilder than it might seem: in the group as simple as $\mathbb{B}_3$, the number of nonequivalent factorizations may grow exponentially in length, see Theorem 10.20. On the other hand, we give a complete classification of Γ-valued factorizations of length two, see Theorems 10.27, 10.30, and 10.32 (joint with N. Salepci). As a by-product, we show that any maximal real elliptic Lefschetz fibration is algebraic, see Theorem 10.88. With elliptic surfaces in mind, we also introduce a new invariant of monodromy factorizations, the so-called *transcendental lattice*, and study its properties, see Section 10.2.3.

As another application, we show that for extremal elliptic surfaces (see Section 10.3.1) and for a certain class of real trigonal curves, including M-curves (see Theorem 10.73), the topological and equisingular deformation classifications are equivalent. (Extremal elliptic surfaces are defined over algebraic number fields, and both classifications are also equivalent to the analytic classification in this case.)

Zariski k-plets. We construct a few examples of exponentially large (with respect to appropriate discrete invariants) collections of nonequivalent objects sharing the same combinatorial data. The objects are: extremal elliptic surfaces (see Theorem 9.30), irreducible trigonal curves (the ramification loci of the surfaces above), real trigonal curves (see Example 10.85), and real Lefschetz fibrations (see Example 10.87). All of the examples are essentially based on Theorem 10.20; thus, in each case, the objects differ *topologically*, constituting the so-called *Zariski k-plets*. The trigonal curves also share such commonly used invariants as the fundamental group and transcendental lattice, see Addendum 9.36 and Theorem 9.31.

The transcendental lattice. The j-invariant of an extremal elliptic surface is given by its skeleton. We show that the homological invariant, hence the surface itself, can be encoded by an orientation of the skeleton, see Theorem 9.1, and develop a simple algorithm computing the lattice of transcendental cycles and the Mordell–Weil group of the surface in terms of its oriented skeleton, see Theorem 9.6. More generally, the transcendental lattice and the torsion of the Mordell–Weil group of an arbitrary (not necessarily extremal) elliptic surface can be computed in terms of its homological invariant, regarded as a monodromy factorization, see Corollary 9.26. This algorithm leads us to the definition of transcendental lattice as an invariant of factorizations, see Section 10.2.3, and motivates a topological approach to the study of its arithmetic properties such as the discriminant form and parity.

Contents at a glance

The principal concepts are introduced in Part I: we discuss bipartite ribbon graphs and their relation to the subgroups of (appropriate quotients of) the free group $\mathbb{F}_2$, see Chapter 1, the modular group Γ and closely related braid group $\mathbb{B}_3$, see Chapter 2, and trigonal curves and elliptic surfaces, both complex and real, and their topological counterpart, the so-called *Lefschetz fibrations*, see Chapter 3. For the reader's convenience, I have also included some background material that a topologist may not be familiar with and reproduced the proofs (or at least ideas of the proofs) of a few known statements, which are either difficult to find in the literature or closely related to the main subject. A separate section in Chapter 1 deals with *pseudo-trees*, which are an important special class of skeletons used later on in the construction of various exponentially large examples and in the study of simple monodromy factorizations.

In Chapter 4, we follow [60] and describe the (equivariant) equisingular deformation classes of (real) trigonal curves in terms of *dessins* – certain overdecorated embedded graphs which must be considered up to a number of moves and which can be rather difficult to handle. It turns out that, under some additional extremality assumptions, dessins can be replaced with skeletons, *i.q.* subgroups of the modular group, making their study feasible. In the real settings, this correspondence between skeletons and deformation classes of curves is made precise in Section 10.3.2.

In Chapter 5, we recall the notion of *braid monodromy*, adjusted to the particular case of curves in ruled surfaces, and the Zariski–van Kampen theorem, computing the fundamental group of such a curve in terms of its monodromy group. The principal result here is a purely combinatorial computation of the braid monodromy of a trigonal curve in terms of its dessin/skeleton (see Section 5.2) and, as an upshot, a strong restriction on the monodromy group of a trigonal curve and the notion of *universal curve*. The two latter lead us to Speculation 5.90, which is copiously illustrated in Chapter 6.

Part II deals with the geometric applications, both old and new. Here, the chapter names are self-explanatory. We compute and study the fundamental groups of trigonal curves and related plane curves (see chapters 6, 7, and 8), discuss the transcendental lattice of an extremal elliptic surface and work out a particular series of examples (see Chapter 9), and make a few steps towards the understanding of Γ-valued monodromy factorizations and their applications to the topology of trigonal curves, elliptic surfaces, and Lefschetz fibrations (see Chapter 10; for completeness, a few more or less classical results concerning the free groups $\mathbb{F}_n$, symmetric groups $\mathbb{S}_n$, and other braid groups $\mathbb{B}_n$ are also cited here).

Appendices collect the material that would not fit elsewhere. Appendix A contains a few assorted statements concerning integral lattices and quotient groups, especially the so-called *Zariski quotients*, which appear as the fundamental groups of algebraic curves. In Appendix B, for comparison and as a very simple application, we discuss bigonal (hyperelliptic) curves in Hirzebruch surfaces. Appendix C is a listing of the `GAP` code that handles technical details of some proofs. (Shorter *ad hoc* listings are included into the main text; all `GAP` files are available for download.) Appendix D is a glossary: we fix the notation and give a brief explanation of a few terms, with the selection based upon the author's own background and personal preferences.

Reading this book

Every effort has been made to produce a text as self-contained and cross-referenced as possible, so that it can be read starting at any point with only a very minimal background from the reader.

As usual, the end of a proof is marked with a $\square$. Some statements are marked with a $\lhd$, which means that either the statement is trivial (*e.g.*, most corollaries) or its proof has already been explained. If a statement is marked with a $\rhd$, possibly followed by a list of references, its proof is omitted and the reader is directed to the literature. In most cases, the source is cited at the header.

We use the commonly accepted symbol $:=$ as a shortcut for 'is defined as'.

Most symbols typeset in a special font (bold, Gothic, calligraphic, *etc.*) represent objects or classes of objects introduced somewhere in the book; they should be found in Section D.2. A brief explanation of other more or less common terms, notations, and concepts used throughout the whole text is given in Section D.1.

Acknowledgements

I would like to thank my colleagues Norbert A'Campo, Mouadh Akriche, Igor Dolgachev, Ergün Elçin, Sergey Finashin, Alexander Klyachko, Anton Klyachko, Anatoly Libgober, Viatcheslav Nikulin, Mutsuo Oka, Stepan Orevkov, Ichiro Shimada, Muhammed Uludağ, and Özgün Ünlü, with whom I had numerous discussions while working on this project and earlier papers and who kindly coped with my ignorance in their fields of expertise.

My special gratitude goes to my co-authors Ilia Itenberg, Viatcheslav Kharlamov, and Nermin Salepci, who generously shared many ideas during our work on joint projects. Some of these ideas underlie whole sections of the book.

This book would never have appeared had it been not for Michael Efroimsky, who encouraged me and finally convinced me to undertake this tremendous task.

I had a chance to represent a few selected topics in a mini-course given at *Faculté des Sciences de Bizerte*, and I wish to thank the audience for their hospitality, patience, and valuable suggestions that helped me improve the clarity of the exposition.

Modern research is unthinkable without software and `Internet`. I would like to mention and express my gratitude to:

- founders, maintainers, and contributors of Wikipedia, the free encyclopædia,
- Donald Knuth, the person who created TEX and made the art of mathematical typesetting accessible to the general public,
- Alexander Simonic, the developer of WinEdt, a text editor that makes the joy of TEX truly joyful,
- the creators of `GAP` [76], a symbolic computation package without which some of the results of this book could not have been obtained in finite time,
- the creators of GLE, a software package that helps one replace frustrating mouse based picture drawing with the excitement of writing code and chasing bugs.

The final version of the manuscript was prepared during my sabbatical stay at *l'Instutut des Hautes Études Scientifiques*. I wish to extend my gratitude to this institution and its friendly staff for their hospitality and excellent working conditions.

Ankara, December 2011 Alex Degtyarev
Bilkent University

Contents

Part I

Skeletons and dessins

Chapter 1

Graphs

1.1 Graphs and trees

In this introductory section, we merely fix terminology and notation; most proofs are omitted. We make no attempt to develop a complete combinatorial theory of graphs (this subject is extensively covered in the literature, see, *e.g.*, [82]), nor do we try to adhere consistently to the combinatorial approach. Our main purpose is to make an easy transition to Definition 1.14 in Section 1.2.1, upon which graphs remain geometric objects and are treated using topology and geometric intuition rather than formal combinatorics.

1.1.1 Graphs

Geometrically, a graph is a locally finite CW-complex of dimension one. Thus, we allow infinite graphs; we do however require that the valency of each vertex must be finite. Combinatorially, a graph can be defined as follows.

Definition 1.1. A *graph* is a collection $\mathfrak{G} = (V, E, \varrho, \iota)$, where V and E are sets, possibly infinite, $\varrho\colon E \to E$ is a free involution, and $\iota\colon E \to V$ is a map with the preimage of each point finite.

The elements of the sets $\operatorname{Vtx}\mathfrak{G} := V$ and $\operatorname{End}\mathfrak{G} := E$ are called the *vertices* and *edge ends* of $\mathfrak{G}$, respectively, and $\iota\colon E \to V$ is called the *incidence map*. The *star* of a vertex $v \in V$ is the pull-back $\iota^{-1}(v)$; its cardinality $\deg v := \operatorname{Card}\iota^{-1}(v)$ is called the *valency* or *degree* of v.

An *edge* of a graph $\mathfrak{G}$ is an orbit of ϱ; thus, ϱ is the map sending an edge end to the opposite end of the same edge. A *directed edge* is an ordered orbit of ϱ. For a directed edge $e = (e_0, e_1)$, we let $e(0) := \iota(e_0)$, $e(1) := \iota(e_1)$, and $-e = (e_1, e_0)$. Since each orbit consists of two elements, one can identify a directed edge $e = (e_0, e_1)$ with its end $e_0 \in E$. An *orientation* of $\mathfrak{G}$ is a choice of a direction of each edge of $\mathfrak{G}$; in other words, it is an equivariant map $\mathfrak{o}\colon \operatorname{End}\mathfrak{G} \to \mathbb{Z}_2$, $\mathfrak{o} \circ \varrho(e) = \mathfrak{o}(e) + 1 \bmod 2$. A pair $(\mathfrak{G}, \mathfrak{o})$ is called a *directed graph*. By convention, $\mathfrak{o}(e_0) = 0$ for an edge (e_0, e_1) directed according to the orientation $\mathfrak{o}$.

Remark 1.2. In the literature, one can find a great deal of combinatorial definitions of graphs. Usually, they are defined as a pair of sets, vertices and edges, related in an appropriate way. Our definition in terms of edge ends rather than edges is intended to avoid complications caused by loops, *i.e.*, edges with both ends attached to the same vertex. In the presence of loops, the star of a vertex should be regarded as a set of edge ends rather than edges, and one needs two distinctive ends of each edge to define an orientation. In fact, each graph has a canonical bipartite subdivision, see Remark 1.6 below, where edge ends become edges and the definition becomes more conventional. Alternatively, an edge end can be regarded as a directed edge, as explained in the previous paragraph. For this reason, we will freely use all commonly accepted intuitive terminology related to graphs; some of the notions are formally (re-)introduced further in this chapter.

A *morphism* between two graphs $\mathfrak{G}_i = (V_i, E_i, \varrho_i, \iota_i)$, $i = 1, 2$, is a pair of maps $\varphi\colon V_1 \to V_2$ and $\psi\colon E_1 \to E_2$ preserving the graph structure, *i.e.*, such that $\psi \circ \varrho_1 = \varrho_2 \circ \psi$ and $\psi \circ \iota_1 = \iota_2 \circ \varphi$. A morphism of directed graphs $(\mathfrak{G}_i, \mathfrak{o}_i)$ must commute with the orientation, *i.e.*, $\mathfrak{o}_2 \circ \psi = \mathfrak{o}_1$. A *subgraph* of a graph $\mathfrak{G} = (V, E, \varrho, \iota)$ is a pair (V', E') of subsets, $V' \subset V$, $E' \subset E$, which is a graph with respect to the restricted maps, *i.e.*, such that $\varrho(E') \subset E'$ and $\iota(E') \subset V'$. A subgraph $\mathfrak{G}'$ is called *induced* if $E' = \iota^{-1}(V') \cap \varrho(\iota^{-1}(V'))$; in other words, an edge of $\mathfrak{G}$ belongs to $\mathfrak{G}'$ if and only if so do its both ends. An induced subgraph is said to be *spanned* by its vertices. The inclusion $\mathfrak{G}' \hookrightarrow \mathfrak{G}$ of a subgraph $\mathfrak{G}'$ is a morphism.

We freely apply to graphs notions and terms used for *CW*-complexes; thus, we will speak about connected graphs, connected components, the fundamental group (*cf.* Theorem 1.4 below), *etc.* Sometimes, to avoid confusion, the *CW*-complex representing a graph $\mathfrak{G}$ is referred to as the *geometric realization* of $\mathfrak{G}$ and is denoted by $|\mathfrak{G}|$. Formally, the geometric realization of $\mathfrak{G} = (V, E, \varrho, \iota)$ can be constructed as follows: Direct each edge and let $|\mathfrak{G}|$ be the quotient of the union $((E/\varrho) \times I) \sqcup V$ by the equivalence relation $(e, 0) \sim e(0)$, $(e, 1) \sim e(1)$ for each edge $e \in E/\varrho$. In this construction, each vertex v is represented by a 0-cell and each directed edge e is represented by a closed 1-cell $e \times I$, which comes with a *characteristic map* $\phi_e\colon e \times I \to |\mathfrak{G}|$.

Note that a connected graph is at most countable.

Definition 1.3. A *walk* in a graph $\mathfrak{G}$ is a finite sequence $\gamma = (v; e_1, \dots, e_k)$, $k \geqslant 0$, where v is a vertex and e_i are directed edges of $\mathfrak{G}$, such that $v = e_1(0)$ (if $k > 0$) and $e_i(1) = e_{i+1}(0)$ for all $i = 1, \dots, k-1$. The *initial point* of γ is$\gamma(0) := v$; the *terminal point* is defined *via* $\gamma(1) := e_k(1)$ if $k > 0$ or $\gamma(1) := v$ if $k = 0$. A walk γ is called *closed* if $\gamma(0) = \gamma(1)$.

Note that, if $(\mathfrak{G}, \mathfrak{o})$ is a directed graph, the edges constituting a walk do *not* need to be directed according to $\mathfrak{o}$. If this *is* the case, one speaks about a *directed* walk.

We reserve the term *path* for a topological path in the geometric realization of the graph. A *combinatorial path* is a walk $(v; e_1, \dots, e_k)$ such that no vertex appears more than once in the sequence $v = e_1(0), \dots, e_k(0)$. A closed combinatorial path is called a *cycle*. Similar to walks, one can speak about *directed* paths and cycles.

The *geometric realization* of a walk $\gamma = (v; e_1, \dots, e_k)$ is the path $|\gamma|\colon I \to |\mathfrak{G}|$ defined as follows: if $k = 0$, then $|\gamma|\colon t \mapsto v$ is constant; otherwise, $|\gamma| = \phi_{e_1} \cdot \ldots \cdot \phi_{e_k}$ is the product of the characteristic maps of its edges, regarded as paths.

The *product* $\gamma' \cdot \gamma''$ of two walks $\gamma' = (v'; e'_i)$ and $\gamma'' = (v''; e''_j)$ is defined whenever $\gamma'(1) = \gamma''(0)$ and is the walk obtained by the concatenation $(v'; e'_i, e''_j)$. The *inverse* of a walk $\gamma = (v; e_1, \dots, e_k)$ is the walk $\gamma^{-1} := (\gamma(1); -e_k, \dots, -e_1)$. One has

$$\gamma' \cdot \gamma''(0) = \gamma'(0), \quad \gamma' \cdot \gamma''(1) = \gamma''(1); \qquad \gamma^{-1}(0) = \gamma(1), \quad \gamma^{-1}(1) = \gamma(0).$$

The product of any two closed walks starting at the same vertex v is always well defined and is again a closed walk starting at v; the inverse of a closed walk starting at v is also a closed walk starting at v.

An *elementary contraction* of a walk γ is the walk obtained from γ by removing a consecutive pair of the form $e, -e$ from the sequence of edges of γ. Two walks are said to be *equivalent* if they are related by a finite sequence of elementary contractions and inverse operations.

The following statement is almost straightforward.

Theorem 1.4. *With respect to the product, the set of the equivalence classes of all closed walks at a fixed vertex v of a graph $\mathfrak{G}$ forms a group, called the* (combinatorial) fundamental group *of $\mathfrak{G}$ and denoted by $\pi_1(\mathfrak{G}, v)$. The geometric realization map $\gamma \mapsto |\gamma|$ induces an isomorphism $\pi_1(\mathfrak{G}, v) = \pi_1(|\mathfrak{G}|, v)$.* ▷ [82]

The equivalence class of a closed walk γ at v is denoted by $[\gamma] \in \pi_1(\mathfrak{G}, v)$. Any walk γ in $\mathfrak{G}$ defines the *translation isomorphism*

$$T_\gamma\colon \pi_1(\mathfrak{G}, \gamma(0)) \to \pi_1(\mathfrak{G}, \gamma(1));$$

it is given by $[\alpha] \mapsto [\gamma^{-1} \cdot \alpha \cdot \gamma]$. It follows that, as an abstract group, $\pi_1(\mathfrak{G}, v)$ depends only on the connected component containing v. Thus, if $\mathfrak{G}$ is connected, one can speak about the abstract group $\pi_1(\mathfrak{G})$, which is well defined up to isomorphism.

Bipartite graphs

A *bipartite graph* is a graph $\mathfrak{G}$ equipped with a partition of its vertices into two classes, *i.e.*, a map $p\colon \operatorname{Vtx}\mathfrak{G} \to \mathbb{Z}_2$, such that the composition $p \circ \varrho\colon \operatorname{End}\mathfrak{G} \to \mathbb{Z}_2$ is an orientation. The vertices in the pull-backs $p^{-1}(0)$ and $p^{-1}(1)$ are called $\bullet$ (black) and $\circ$ (white), respectively. For a 'variable' type of vertices we commonly use the symbol $*$, which takes values in the set $\{\bullet, \circ\}$.

Alternatively, a bipartite graph can be defined combinatorially in terms of its edges rather than edge ends, as follows.

Definition 1.5. A *bipartite graph* is a collection $\mathfrak{G} = (V_\bullet, V_\circ, \mathcal{E}, \iota_\bullet, \iota_\circ)$, where $V_\bullet$, $V_\circ$, and $\mathcal{E}$ are sets, possibly infinite, and $\iota_\bullet \colon \mathcal{E} \to V_\bullet$ and $\iota_\circ \colon \mathcal{E} \to V_\circ$ are maps such that the preimage of each point is finite.

The elements of the sets $\operatorname{Vtx}_\bullet \mathfrak{G} := V_\bullet$, $\operatorname{Vtx}_\circ \mathfrak{G} := V_\circ$, and $\operatorname{Edg} \mathfrak{G} := \mathcal{E}$ are called, respectively, the $\bullet$- and $\circ$-*vertices* of $\mathfrak{G}$ and its *edges*. The *star* of a vertex and its *valency* are defined as in the case of ordinary graphs, using $\iota_\bullet$ and $\iota_\circ$ instead of ι: the star of a $*$-vertex v is the set $\iota_*^{-1}(v) \subset \mathcal{E}$, $* = \bullet, \circ$, and its valency is the cardinality $\deg v := \operatorname{Card} \iota_*^{-1}(v)$.

For the formal passage from Definition 1.5 to 1.1, one should let $V = V_\bullet \sqcup V_\circ$ and $E = \mathcal{E} \times \{\bullet, \circ\}$ and define $\varrho \colon (e, \bullet) \mapsto (e, \circ) \mapsto (e, \bullet)$ and $\iota \colon (e, *) \mapsto \iota_*(e)$ for $* = \bullet, \circ$.

Note that each bipartite graph $\mathfrak{G}$ is canonically directed: by convention, we let $e(0) = \iota_\bullet(e)$ and $e(1) = \iota_\circ(e)$ for an edge $e \in \mathcal{E}$.

Remark 1.6. An ordinary graph $\mathfrak{G} = (V, E, \varrho, \iota)$ can be regarded as a bipartite graph: one lets $V_\bullet = V$, $V_\circ = E/\varrho$, $\mathcal{E} = E$, $\iota_\bullet = \iota$ and takes for $\iota_\circ$ the natural projection $E \to E/\varrho$. We call $\bar{\mathfrak{G}} = (V_\bullet, V_\circ, \mathcal{E}, \iota_\bullet, \iota_\circ)$ the *bipartite subdivision* of $\mathfrak{G}$; the edges of $\bar{\mathfrak{G}}$ represent *directed* edges of $\mathfrak{G}$. Intuitively, when passing from $\mathfrak{G}$ to $\bar{\mathfrak{G}}$, one declares the vertices of $\mathfrak{G}$ black and inserts a bivalent $\circ$-vertex in the middle of each edge of $\mathfrak{G}$. Conversely, a bipartite graph $\bar{\mathfrak{G}}$ is the bipartite subdivision of an ordinary graph if and only if all $\circ$-vertices of $\bar{\mathfrak{G}}$ are bivalent. In view of this correspondence, and since bipartite graphs are our primary concern, we introduce most notions for bipartite graphs only, referring to Definition 1.5 or Definition 1.14 below.

Convention 1.7. In the drawings of bipartite graphs we routinely omit bivalent $\circ$-vertices, assuming that such a vertex is to be inserted at the center of each edge connecting two $\bullet$-vertices.

1.1.2 Trees

A *tree* as a simply connected graph, *i.e.*, a connected graph $\mathfrak{T}$ with $\pi_1(\mathfrak{T}) = 0$. Due to Theorem 1.4 and the Whitehead theorem [163, 164], a graph $\mathfrak{T}$ is a tree if and only if its geometric realization $|\mathfrak{T}|$ is contractible. Yet another equivalent definition is that a tree is a connected graph without cycles.

As a straightforward application of Zorn's lemma, any connected graph $\mathfrak{G}$ contains a maximal tree; in fact, a subtree $\mathfrak{T} \subset \mathfrak{G}$ is maximal if and only if it contains all vertices of $\mathfrak{G}$. Since $|\mathfrak{T}|$ is contractible, the quotient projection $|\mathfrak{G}| \to |\mathfrak{G}|/|\mathfrak{T}|$ is a homotopy equivalence. On the other hand, the quotient $|\mathfrak{G}|/|\mathfrak{T}|$ is a *CW*-complex of dimension one and with a single 0-cell $|\mathfrak{T}|/|\mathfrak{T}|$, hence a wedge of circles. By the Seifert–van Kampen theorem, the fundamental group $\pi_1(\mathfrak{G}) = \pi_1(|\mathfrak{G}|/|\mathfrak{T}|)$ is the free group generated by the circles constituting the wedge.

Corollary 1.8. *For a connected graph $\mathfrak{G}$, the geometric realization $|\mathfrak{G}|$ is homotopy equivalent to a wedge of circles and the group $\pi_1(\mathfrak{G})$ is free of rank $1 - \chi(|\mathfrak{G}|)$.* $\triangleleft$

As a consequence of Corollary 1.8, one obtains a simple proof of the *Nielsen–Schreier theorem* and *Schreier index formula*. Purely algebraic proofs of these statements take several pages, see, *e.g.*, [114].

Theorem 1.9 (Nielsen–Schreier). *A subgroup H of a finitely generated free group G is free. If the index $n := [G : H]$ is finite, then H is of finite rank* $\operatorname{rk} H = n(\operatorname{rk} G - 1) + 1$.

Proof. (This proof was discovered by Max Dehn, see [115].) One has $G = \pi_1(W)$, where W is the wedge of $\operatorname{rk} G$ circles, regarded as a graph. A subgroup $H \subset G$ defines a topological covering $\tilde{W} \to W$, which is also a graph. Hence, H is free. If the degree n of the covering is finite, counting cells yields $\chi(\tilde{W}) = n\chi(W)$, which gives the index formula. □

Given a tree $\mathfrak{T}$, any two distinct vertices u, v of $\mathfrak{T}$ can be connected by a unique path γ. Define the *vertex distance* $\operatorname{dist}(u, v)$ as the number of edges constituting γ, and extend this definition to $\operatorname{dist}(u, u) = 0$ for any vertex u. It is immediate that dist is indeed a metric on the set of vertices of $\mathfrak{T}$.

1.1.3 Dynkin diagrams

Dynkin diagrams are a very special family of graphs that keep appearing in various unrelated areas of mathematics. Our interest is due to the fact that Dynkin diagrams classify simple singularities, singular elliptic fibers, and definite lattices generated by vectors of square (-2). In this graph related chapter, we only list the so-called *simply laced* Dynkin diagrams, both elliptic and affine, and discuss their relation to definite lattices. Other applications and a detailed description of the corresponding root systems can be found in [27].

Simply laced Dynkin diagrams constitute two infinite series, $\mathbf{A}_p$, $p \geqslant 1$, and $\mathbf{D}_q$, $q \geqslant 4$, and three exceptional classes $\mathbf{E}_6$, $\mathbf{E}_7$, and $\mathbf{E}_8$, see Figure 1.1. Each diagram appears in two forms, *elliptic* and *affine*; the latter is usually marked with a tilde.

For a characterization (one of the many) of the simply laced Dynkin diagram, let us assign a lattice $\mathbf{L}_\mathfrak{G}$ to each finite loop free graph $\mathfrak{G}$. As a free abelian group, let $\mathbf{L}_\mathfrak{G} := \bigoplus \mathbb{Z}v$, the summation running over all vertices $v \in \operatorname{Vtx} \mathfrak{G}$; the square v^2 of each generator $v \in \operatorname{Vtx} \mathfrak{G}$ equals -2, and the product $u \cdot v$, $u \neq v$, of two distinct generators is the number of edges of the graph connecting u and v.

Proposition 1.10. *For a connected finite loop free graph $\mathfrak{G}$, the lattice $\mathbf{L}_\mathfrak{G}$ is negative definite (semidefinite) of and only if $\mathfrak{G}$ is a simply laced Dynkin diagram (respectively, affine Dynkin diagram). In the affine case, the kernel* $\ker \mathbf{L}_\mathfrak{G}$ *is generated by the vector $z := \sum n_v v$, where n_v are the coefficients shown in Figure* 1.1.

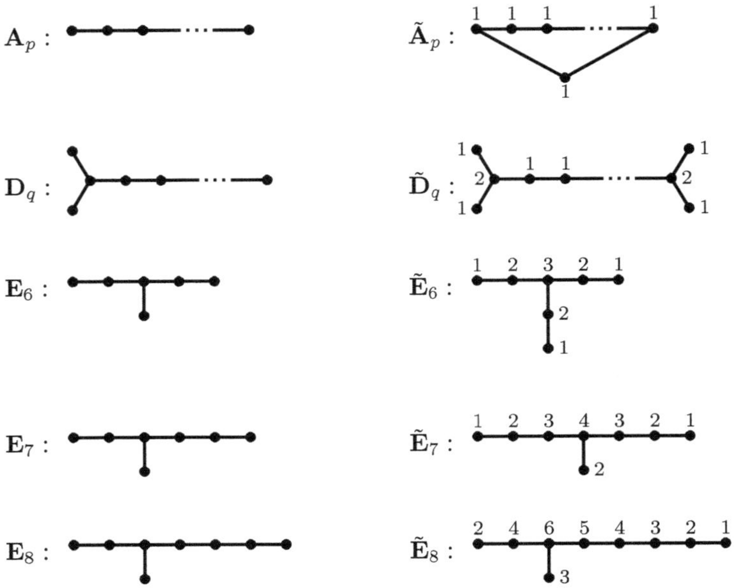

Figure 1.1. Simply laced Dynkin diagrams.

Proof. The lattices $\mathbf{L}_{\mathfrak{G}}$ associated with (affine) Dynkin diagrams are indeed (semi-)definite and the kernels $\ker \mathbf{L}_{\mathfrak{G}}$ are as stated. Hence, the affine Dynkin diagrams are extremal in the sense that adding an extra vertex would produce an element of positive square: for such a vertex u one would have $u \cdot z \geqslant 1$ and hence $(u + 2z)^2 \geqslant 2$. Thus, affine Dynkin diagram exhaust the semidefinite case, and it remains to notice that elliptic Dynkin diagrams are all graphs that do not contain an affine one. □

Convention 1.11. The lattice $\mathbf{L}_{\mathfrak{G}}$ associated with a Dynkin diagram $\mathfrak{G}$ is denoted by the same symbol as $\mathfrak{G}$ (see also Convention A.12). Sometimes, we extend this notation by letting $\mathbf{A}_0 := \mathbf{D}_0 := 0$, $\mathbf{D}_1 := \mathbb{Z}\mathbf{x}$, $\mathbf{x}^2 = -4$, $\mathbf{D}_2 := 2\mathbf{A}_1$, and $\mathbf{D}_3 := \mathbf{A}_3$. In each assignment, the right hand side of the expression exposes certain behavior expected from its left hand side. For example, the following proposition, which gives an explicit description of the $\mathbf{A}$ and $\mathbf{D}$ type lattices, extends to these generalizations.

Proposition 1.12 (see [27]). *The lattices $\mathbf{A}_p$ and $\mathbf{D}_p$ can be defined as follows:*

$$\mathbf{A}_{p-1} = (\mathbf{v}_1 + \cdots + \mathbf{v}_p)^\perp \subset \mathbf{B}_p, \qquad \mathbf{D}_p = \{x \in \mathbf{B}_p \mid x^2 = 0 \bmod 2\}.$$

In other words, $\mathbf{A}_{p-1}$ is the orthogonal complement of the 'minimal' characteristic vector $\mathbf{v} := \mathbf{v}_1 + \cdots + \mathbf{v}_p \in \mathbf{B}_p$, and $\mathbf{D}_p$ is its orthogonal complement modulo 2 or, equivalently, $\mathbf{D}_p$ is the maximal even sublattice of $\mathbf{B}_p$.

Proof. The orthogonal complement $(\mathbf{v}_1+\cdots+\mathbf{v}_p)^\perp \subset \mathbf{B}_p$ is generated by the vectors $\mathbf{v}_i-\mathbf{v}_{i+1}$, $i=1,\ldots,p-1$, and these vectors form a canonical basis for $\mathbf{A}_{p-1}$. For $\mathbf{D}_p$, observe that $\mathbf{B}_p$ is odd and, hence, its maximal even sublattice is of index 2. On the other hand, the vectors $-\mathbf{v}_1-\mathbf{v}_2$ and $\mathbf{v}_i-\mathbf{v}_{i+1}$, $i=1,\ldots,p-1$, do generate an even sublattice of index 2, and these vectors form a canonical basis for $\mathbf{D}_p$. □

1.2 Skeletons

The notion of skeleton is the principal concept used throughout the book: we will employ skeletons to classify trigonal curves, compute their fundamental groups, construct examples, *etc*. Crucial is Definition 1.14, which establishes a correspondence between skeletons and subgroups of certain groups.

1.2.1 Ribbon graphs

Recall that a *cyclic order* on a finite set X is a transitive $\mathbb{Z}$-action on X: given $x \in X$, the elements $\operatorname{succ} x := x \uparrow (+1)$ and $\operatorname{pred} x := x \uparrow (-1)$ are the *immediate successor* and *predecessor* of x, respectively.

A *ribbon graph* is a graph equipped with a cyclic order on the star of each vertex. A typical example is a graph $\mathfrak{G}$ with a proper immersion $|\mathfrak{G}| \hookrightarrow S$ to an oriented surface S: the cyclic order on the star of a vertex v is given by the counterclockwise rotation (with respect to the given orientation of S) about v.

Convention 1.13. In the figures, we use the standard convention, drawing ribbon graphs in the plane, possibly with self-intersections, and assuming the 'blackboard thickening', *i.e.*, the cyclic order induced from the plane.

The *bipartite subdivision* of a ribbon graph is again a ribbon graph, as the star of a bivalent $\circ$-vertex has a unique cyclic order. For bipartite graphs, the definition of a ribbon graph structure can be combined with Definition 1.5 and simplified.

Definition 1.14. A *bipartite ribbon graph* is a triple $\mathfrak{S} = (\mathcal{E}, \operatorname{nx}, \operatorname{op})$, where $\mathcal{E}$ is a set and $\operatorname{nx}, \operatorname{op}\colon \mathcal{E} \to \mathcal{E}$ are two automorphisms of $\mathcal{E}$ with finite orbits. A connected bipartite ribbon graph is also called an (abstract) *skeleton*. A pair $(\mathfrak{S}, e)$, where $\mathfrak{S}$ is a skeleton and $e \in \mathcal{E}$, is called a *marked skeleton*.

The sets of $\bullet$- and $\circ$-vertices of a bipartite ribbon graph are defined as the orbit sets $V_\bullet = \mathcal{E}/\operatorname{nx}$ and $V_\circ = \mathcal{E}/\operatorname{op}$, respectively, the incidence maps $\iota_\bullet$ and $\iota_\circ$ being the canonical projections. In most applications, $\circ$-vertices are at most bivalent and can be regarded as edges, see Convention 1.7, whereas $\bullet$-vertices are usually trivalent. Thus, the notation 'op' and 'nx' stands for 'opposite' and 'next' (with respect to the cyclic order), respectively. We will also use the uniform notation $\varrho_\bullet := \operatorname{nx}$ and $\varrho_\circ := \operatorname{op}$.

The valency of a vertex is the cardinality of the corresponding orbit. Consider a pair $\mathbf{t} := (t_\bullet, t_\circ) \in \mathbb{N}^2$. A skeleton $\mathfrak{S}$ is said to be a *skeleton of type* $\mathbf{t}$, or a $\mathbf{t}$-*skeleton*, if the valency of each $*$-vertex of $\mathfrak{S}$ divides t_*, $* = \bullet, \circ$. If the valency of each $*$-vertex equals t_*, then $\mathfrak{S}$ is called *regular*, or $\mathbf{t}$-*regular*.

A *morphism* between two bipartite ribbon graphs $\mathfrak{S}_i = (\mathcal{E}_i, \mathrm{nx}_i, \mathrm{op}_i)$, $i = 1, 2$, is a map $\varphi\colon \mathcal{E}_1 \to \mathcal{E}_2$ commuting with nx and op, *i.e.*, such that $\varphi \circ \mathrm{nx}_1 = \mathrm{nx}_2 \circ \varphi$ and $\varphi \circ \mathrm{op}_1 = \mathrm{op}_2 \circ \varphi$. A morphism of marked skeletons is required, in addition, to take the distinguished edge e_1 of $\mathfrak{S}_1$ to the distinguished edge e_2 of $\mathfrak{S}_2$. The group of automorphisms of a skeleton $\mathfrak{S}$ is denoted by $\operatorname{Aut}\mathfrak{S}$. We regard Aut as a subgroup of the permutation group $\mathbb{S}(\operatorname{Edg}\mathfrak{S})$.

Warning 1.15. Certainly, any morphism of (bipartite) ribbon graphs induces, in a canonical way, a morphism of the underlying (bipartite) plain graphs. However, the converse is not true. For example, any subgraph $\mathfrak{S}' \subset \mathfrak{S}$ of a bipartite ribbon graph inherits a bipartite ribbon graph structure (*e.g.*, from the minimal supporting surface $\operatorname{Supp}\mathfrak{S}$); however, the inclusion $\mathfrak{S}' \hookrightarrow \mathfrak{S}$ is *not* a morphism unless $\mathfrak{S}'$ is a union of whole components of $\mathfrak{S}$.

Ribbon graphs as $\mathbb{G}$-sets

Let $\mathbb{G} := \langle \mathbf{x}, \mathbf{y} \rangle$ be the free group on two generators. Throughout this chapter, we reserve the notation $\mathbf{x}$, $\mathbf{y}$ for the chosen pair of generators of $\mathbb{G}$, sometimes using the uniform notation $\mathbf{g}_\bullet := \mathbf{x}$ and $\mathbf{g}_\circ := \mathbf{y}$ adapted to the treatment of bipartite graphs.

According to Definition 1.14, a bipartite ribbon graph $\mathfrak{S}$ is merely a set $\mathcal{E}$ with an action of $\mathbb{G}$ such that both $\mathbf{x}$ and $\mathbf{y}$ have finite orbits. For a technical reason, we extend nx and op to a *right* $\mathbb{G}$-action $\mathcal{E} \times \mathbb{G} \to \mathcal{E}$, denoted by $(e, g) \mapsto e \uparrow g$; by definition, $\mathbf{x}$ acts *via* nx and $\mathbf{y}$ acts *via* op. The graph is connected if and only if this action is transitive; thus, a marked skeleton $(\mathfrak{S}, e)$ can be identified with the quotient set $H\backslash\mathbb{G}$, where $H = \operatorname{stab}_{\mathbb{G}} e$ is the stabilizer of e in $\mathbb{G}$. Furthermore, a morphism of bipartite ribbon graphs is merely a morphism of $\mathbb{G}$-sets. In particular, a morphism φ of skeletons is uniquely determined by the image of any edge of the source and has a well defined *degree* $\deg\varphi := \operatorname{Card}\varphi^{-1}(e)$, where e is any edge of the target.

A finite bipartite ribbon graph $\mathfrak{S} = (\mathcal{E}, \mathrm{nx}, \mathrm{op})$ can be represented in `GAP` [76] by a pair of permutations $\mathrm{nx}, \mathrm{op} \in \mathbb{S}(\mathcal{E})$. A few simple functions manipulating skeletons and computing various invariants are shown in Listing C.2 and explained further in this chapter. Thus, functions "`Black`", "`Blacks`", and "`BlackNo`" compute, respectively, the set of $\bullet$-vertices, numbers of $\bullet$-vertices by valencies, and the total number of $\bullet$-vertices; the `White*` counterparts of these functions provide the same information about the $\circ$-vertices. The minimal, in the obvious sense, type $\mathbf{t}$ of a skeleton $\mathfrak{S}$ is given by "`Type`", and "`IsRegularSkeleton`" decides whether $\mathfrak{S}$ is $\mathbf{t}$-regular. The connected components of a bipartite ribbon graph $\mathfrak{S}$ are handled by "`IsConnected`", "`Components`", "`Component`", and "`ExtractComponent`".

For the obvious reasons, in GAP, instead of the infinite group $\mathbb{G}$ we deal with the finite subgroup of $\mathbb{S}(\mathcal{E})$ generated by nx and op. The stabilizer $\operatorname{stab}(e)$, regarded as a subgroup of $\mathbb{S}(\mathcal{E})$, is given by `"stab"`.

If $\mathfrak{S}$ is a skeleton, the stabilizers of all edges of $\mathfrak{S}$ form a whole conjugacy class of subgroups; we call this class the *stabilizer of* $\mathfrak{S}$ and denote it by $\operatorname{Stab}_{\mathbb{G}} \mathfrak{S}$. The requirement that all orbits of both $\mathbf{x}$ and $\mathbf{y}$ must be finite translates into the following condition on the representatives $H \in \operatorname{Stab}_{\mathbb{G}} \mathfrak{S}$:

$(*)$ for each $g \in \mathbb{G}$, there is a pair $(r, s) \in \mathbb{N}^2$ such that $\mathbf{x}^r, \mathbf{y}^s \in g^{-1}Hg$.

Next few statements are immediate consequence of elementary theory of discrete group actions, see, *e.g.*, [90].

Theorem 1.16. *The functors* $(\mathfrak{S}, e) \mapsto \operatorname{stab}_{\mathbb{G}} e$ *and* $H \mapsto (H\backslash\mathbb{G}, H\backslash H)$ *establish an equivalence between the category of marked skeletons and morphisms and that of subgroups* $H \subset \mathbb{G}$ *satisfying* $(*)$ *and inclusions. A skeleton* $\mathfrak{S} = H\backslash\mathbb{G}$ *is of type* $\mathbf{t} \in \mathbb{N}^2$ *if and only if* H *contains the normal subgroup generated by* $\mathbf{x}^{t_\bullet}$ *and* $\mathbf{y}^{t_\circ}$. ◁

Given a type $\mathbf{t} \in \mathbb{N}^2$, denote $\mathbb{G}_{\mathbf{t}} = \langle \mathbf{x}, \mathbf{y} \mid \mathbf{x}^{t_\bullet} = \mathbf{y}^{t_\circ} = 1 \rangle \cong \mathbb{Z}_{t_\bullet} * \mathbb{Z}_{t_\circ}$. There is a canonical epimorphism $\mathbb{G} \twoheadrightarrow \mathbb{G}_{\mathbf{t}}$. Clearly, the $\mathbb{G}$-action on the set of edges of a skeleton of type $\mathbf{t}$ factors through $\mathbb{G}_{\mathbf{t}}$; we use the notation $\operatorname{stab}_{\mathbf{t}}$ and $\operatorname{Stab}_{\mathbf{t}}$ for the stabilizers regarded as (conjugacy classes of) subgroups of $\mathbb{G}_{\mathbf{t}}$.

Note that any subgroup of $\mathbb{G}_{\mathbf{t}}$ lifts to a subgroup $H \subset \mathbb{G}$ satisfying $(*)$.

Corollary 1.17. *Given a type* $\mathbf{t} \in \mathbb{N}^2$, *the functors* $(\mathfrak{S}, e) \mapsto \operatorname{stab}_{\mathbf{t}} e \subset \mathbb{G}_{\mathbf{t}}$ *and* $H \mapsto (H\backslash\mathbb{G}_{\mathbf{t}}, H\backslash H)$ *establish an equivalence between the category of marked* $\mathbf{t}$-*skeletons and morphisms and that of subgroups* $H \subset \mathbb{G}_{\mathbf{t}}$ *and inclusions.* ◁

Corollary 1.18. *The maps* $\mathfrak{S} \mapsto \operatorname{Stab}_{\mathbb{G}} \mathfrak{S}$, $[\![H]\!] \mapsto H\backslash\mathbb{G}$ *establish a canonical one-to-one correspondence between the set of isomorphism classes of skeletons and that of conjugacy classes of subgroups* $H \subset \mathbb{G}$ *satisfying* $(*)$. ◁

Corollary 1.19. *Given a type* $\mathbf{t} \in \mathbb{N}^2$, *the maps* $\mathfrak{S} \mapsto \operatorname{Stab}_{\mathbf{t}} \mathfrak{S}$ *and* $[\![H]\!] \mapsto H\backslash\mathbb{G}_{\mathbf{t}}$ *establish a canonical one-to-one correspondence between the set of isomorphism classes of* $\mathbf{t}$-*skeletons and that of conjugacy classes of subgroups* $H \subset \mathbb{G}_{\mathbf{t}}$. ◁

If a skeleton $\mathfrak{S}$ is fixed, the isomorphism classes of marked skeletons $(\mathfrak{S}, e)$ are in a one-to-one correspondence with the orbits of $\operatorname{Aut} \mathfrak{S}$.

Corollary 1.20. *The conjugacy class* $[\![H]\!]$ *of a subgroup* $H \subset \mathbb{G}$ *satisfying* $(*)$ *is in a canonical bijection with the set of orbits of the group* $\operatorname{Aut}(H\backslash\mathbb{G})$. *In particular,* $\operatorname{Aut}(H\backslash\mathbb{G})$ *is transitive if and only if* H *is normal in* $\mathbb{G}$. ◁

Corollary 1.21. *Given a subgroup* $H \subset \mathbb{G}$ *satisfying* $(*)$, *there is a canonical group isomorphism* $\operatorname{Aut}(H\backslash\mathbb{G}) = \mathcal{N}_{\mathbb{G}}(H)/H$.

Proof. Any $\mathbb{G}$-endomorphism of the homogeneous $\mathbb{G}$-set $H\backslash\mathbb{G}$ is the multiplication $Hx \mapsto Hgx$ by a fixed element $g \in \mathbb{G}$ (such that Hg is the image of the coset $H\backslash H$ under the endomorphism). This multiplication map is well defined if and only if $g^{-1}Hg \subset H$, *i.e.*, if $g \in \mathcal{N}_{\mathbb{G}}(H)$; it is the identity if and only if $g \in H$. □

Corollary 1.22. *A subgroup $H \subset \mathbb{G}$ satisfying $(*)$ is normal if and only if the group* $\operatorname{Aut}(H\backslash\mathbb{G})$ *acts transitively on the set of edges of the skeleton $H\backslash\mathbb{G}$.* ◁

Corollary 1.21 is used in function "`Aut`", see Listing C.2, computing the automorphism group $\operatorname{Aut}\mathfrak{S}$.

1.2.2 Regions

A *region* in a bipartite ribbon graph $\mathfrak{S} = (\mathcal{E}, \mathrm{nx}, \mathrm{op})$ is an orbit of the element $\mathbf{xy} \in \mathbb{G}$. Each region is a $\mathbb{Z}$-set: we let $1 \in \mathbb{Z}$ act *via* $(\mathbf{xy})^{-1}$. The cardinality of a region R is called its *width* $\operatorname{wd} R$. (In the arithmetical theory of the modular group, instead of regions one speaks about cusps and cusp widths; this, and the fact that the term 'degree' is way too overused, explains our terminology.) A region R of finite width n is also referred to as an *n-gon* or an *n-gonal region*, the 'corners' being the $\bullet$-ends of the edges constituting R. The set of regions of $\mathfrak{S}$ is denoted by $\operatorname{Reg}\mathfrak{S}$. We also use the notation $\operatorname{Reg}_w \mathfrak{S}$ for the set of all regions of a given width $w \in \bar{\mathbb{N}}$ and $\operatorname{Reg}_0 \mathfrak{S}$ for the set of all regions of finite width. For a finite skeleton $\mathfrak{S}$, the set $\operatorname{Reg}\mathfrak{S}$ is computed by function "`Reg`", see Listing C.2; the numbers of regions by sizes are given by "`Regs`", and the total number of regions is "`RegNo`".

With the modular group in mind, see page 45, an element of $\mathbb{G}$ conjugate to a power $(\mathbf{xy})^n$, $n \in \mathbb{N}$, is called *parabolic* or *unipotent* of width n. The following statement is a paraphrase of the definition. (For a subgroup $H \subset \mathbb{G}$, a parabolic element $g^{-1}(\mathbf{xy})^n g \in H$ is called *minimal* if $g^{-1}(\mathbf{xy})^m g \notin H$ for $0 < m < n$.)

Lemma 1.23. *For a subgroup $H \subset \mathbb{G}$, the H-conjugacy classes of minimal parabolic elements of H are in a canonical one-to-one correspondence with the finite regions of the skeleton $H\backslash\mathbb{G}$. The width of a minimal parabolic element of H is equal to the width of the corresponding region.* ◁

We do not define the *boundary* ∂R of a region R as a formal object, but we do use it as an intuitive concept. Thus, we say that the $\bullet$- and $\circ$-ends of all edges $e \in R$ belong to ∂R, and so do all (directed) edges of the form e and $-e \uparrow \mathbf{x}$, $e \in R$. If the ribbon graph structure is induced from a proper embedding $|\mathfrak{S}| \hookrightarrow S$ to an oriented surface S, see Section 1.2.1, regions in the sense of our definition are the boundary components (properly understood) of the connected components of the complement $S \smallsetminus |\mathfrak{S}|$ (or rather the cut of S along $|\mathfrak{S}|$), and the above convention concerning the boundary of a region agrees with the common sense. For example, in Figure 1.2, the boundary of $R = \operatorname{reg} e_0$ is formed by the edges $e_i := e_0 \uparrow (\mathbf{xy})^{-i}$, $i \in \mathbb{Z}$, that constitute R, as well as the edges $e_i' := e_i \uparrow \mathbf{y}^{-1} = e_{i+1} \uparrow \mathbf{x}$, which *a priori* do

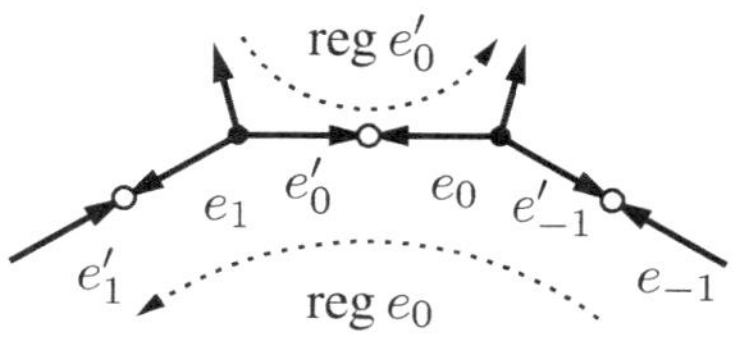

Figure 1.2. A region of a skeleton.

not belong to R. (It may happen, though, that, *e.g.*, reg e_0 and reg e_0' is the same region.)

Each edge e of $\mathfrak{S}$ belongs to a single region, denoted by reg e. If $n := \mathrm{wd}(\mathrm{reg}\, e)$ is finite, the *geometric realization* $|\mathrm{reg}\, e|$ is the closed unit disk $\bar{D}^2 \subset \mathbb{C}$ equipped with the *attaching map* $\psi_e \colon \partial|\mathrm{reg}\, e| \to |\mathfrak{S}|$ defined *via* $\exp(\pi i t/n) \mapsto \phi_{e_s}(t-s)$ for $s \leqslant t \leqslant s+1$, $s \in \mathbb{Z}$, where $e_s = e \uparrow (\mathbf{xy})^{-k}$ if $s = 2k$ is even and $e_s = -e_{s+1} \uparrow \mathbf{x}$ if s is odd. If $\mathrm{wd}(\mathrm{reg}\, e) = \infty$, then $|\mathrm{reg}\, e|$ is the closed upper half plane $\bar{\mathfrak{H}} := \{z \in \mathbb{C} \mid \mathrm{Im}\, z \geqslant 0\}$ with the attaching map $\psi_e \colon \partial|\mathrm{reg}\, e| = \mathbb{R} \to |\mathfrak{S}|$ defined *via* $t \mapsto \phi_{e_s}(t-s)$ for $s \leqslant t \leqslant s+1$, $s \in \mathbb{Z}$, where the edges e_s are the same as above.

Minimal supporting surface

As explained in Section 1.2.1, a typical example of ribbon graphs is that of graphs properly embedded to an oriented surface. It turns out that any ribbon graph structure can be defined in this way, using a certain canonical minimal surface, which is obtained by 'patching' each region of the graph with a disk (or, more precisely, the geometric realization of the region).

Definition 1.24. The *minimal supporting surface* $\mathrm{Supp}\,\mathfrak{S}$ of a bipartite ribbon graph $\mathfrak{S}$ is the quotient of the disjoint union $|\mathfrak{S}| \sqcup \bigsqcup_{e \in \mathcal{R}} |\mathrm{reg}\, e|$, where $\mathcal{R} \subset \mathrm{Edg}\,\mathfrak{S}$ is a set of representatives, one for each region $R \in \mathrm{Reg}\,\mathfrak{S}$, by the equivalence relation $z \sim \psi_e(z)$ for all $e \in \mathcal{R}$ and $z \in \partial|\mathrm{reg}\, e|$.

For a *finite* region $R \in \mathrm{Reg}\,\mathfrak{S}$, the image $c_R \in \mathrm{Supp}\,\mathfrak{S}$ of the origin $0 \in \bar{D}^2$ is called the *center* of R. The *open supporting surface* of $\mathfrak{S}$ is the difference $\mathrm{Supp}^\circ\,\mathfrak{S} := \mathrm{Supp}\,\mathfrak{S} \smallsetminus \bigcup c_R$, the union running over all finite regions of $\mathfrak{S}$. (In the notation, the superscript $\circ$ is not related to $\circ$-vertices; it rather reflects the concept of 'openness'.)

Proposition 1.25. *The minimal supporting surface* $\mathrm{Supp}\,\mathfrak{S}$ *defined above is indeed an oriented surface; up to a canonical homeomorphism, it is independent of the choice of the set* $\mathcal{R}$ *of representatives. The inclusion* $|\mathfrak{S}| \subset \mathrm{Supp}^\circ\,\mathfrak{S}$ *is a proper embedding with the following properties:*

1. *the inclusion* $|\mathfrak{G}| \subset \operatorname{Supp}\mathfrak{G}$ *induces the original ribbon graph structure on* $\mathfrak{G}$;
2. *there is a canonical strict deformation retraction* $\operatorname{Supp}^\circ \mathfrak{G} \to |\mathfrak{G}|$;
3. *the complement* $\operatorname{Supp}\mathfrak{G} \smallsetminus |\mathfrak{G}|$ *is a union of open disks, one disk for each region.*

Proof. Each edge e appears in the boundary of exactly two regions (not necessarily distinct): once in $\partial \operatorname{reg} e$ and once in $\partial \operatorname{reg}(e \uparrow \mathbf{y})$, and the two occurrences are used in the construction of the maps ψ with the opposite orientations, see Figure 1.2. Hence, $\operatorname{Supp}\mathfrak{G}$ is a locally Euclidean space and the complex orientations of the geometric realizations of all regions match to define an orientation of $\operatorname{Supp}\mathfrak{G}$.

If edges e and $e' = e \uparrow (\mathbf{xy})^r$ are in the same region, the pairs $(|\operatorname{reg} e|, \psi_e)$ and $(|\operatorname{reg} e'|, \psi_{e'})$ are canonically homeomorphic: if the width $n := \operatorname{wd}(\operatorname{reg} e)$ is finite, the homeomorphism is the rotation through $2\pi r/n$ about the origin; if $n = \infty$, it is the translation by $2r$ along the real axis. This observation proves the independence of $\operatorname{Supp}\mathfrak{G}$ of $\mathcal{R}$.

For Item 2, a homotopy h_t between the retraction h_0 and identity $\mathrm{id} = h_1$ is constructed separately on each 'cell' $\bar{D}^2 \smallsetminus 0$ or $\bar{\mathfrak{H}}$ constituting $\operatorname{Supp}^\circ \mathfrak{G}$. On a compact cell $\bar{D}^2$, take for h the radial homotopy $h_t(z) = tz + (1-t)z/|z|$; on an unbounded cell $\bar{\mathfrak{H}}$, let $h_t(z) = tz + (1-t)\operatorname{Re} z$.

All other statements follow immediately from the definitions. □

If all regions are finite, the construction of $\operatorname{Supp}\mathfrak{G}$ represents this surface as a *CW*-complex of dimension two, regions being the 2-cells. If $\mathfrak{G}$ is finite, this complex is finite, hence compact. Each connected component of $\mathfrak{G}$ is embedded into its own component of $\operatorname{Supp}\mathfrak{G}$. Thus, if $\mathfrak{G}$ is a finite skeleton, $\operatorname{Supp}\mathfrak{G}$ is homeomorphic to a sphere with $g := \frac{1}{2} b_1(\operatorname{Supp}\mathfrak{G})$ handles. The genus g of $\operatorname{Supp}\mathfrak{G}$ is called the *genus* of $\mathfrak{G}$. Computing the Euler characteristic $\chi(\operatorname{Supp}\mathfrak{G})$, one has

$$2 - 2g = \operatorname{Card}(\mathcal{E}/\mathbf{x}) + \operatorname{Card}(\mathcal{E}/\mathbf{y}) + \operatorname{Card}(\mathcal{E}/\mathbf{xy}) - \operatorname{Card}\mathcal{E}, \qquad (1.26)$$

where $\mathcal{E} = \operatorname{Edg}\mathfrak{G}$. The genus of $\mathfrak{G}$ is computed by "`Genus`", see Listing C.2.

More generally, one can define the *patching* of one or several regions of $\mathfrak{G}$ as the quotient S of the disjoint union $|\mathfrak{G}| \sqcup \bigsqcup_{e \in \mathcal{R}'} |\operatorname{reg} e|$ by the identification $z \sim \psi_e(z)$ for all $e \in \mathcal{R}'$ and $z \in \partial|\operatorname{reg} e|$, where $\mathcal{R}' \subset \operatorname{Edg}\mathfrak{G}$ is a set of representatives, one for each region to be patched. As above, up to canonical homeomorphism the result is independent of the choice of the representatives and the inclusion $|\mathfrak{G}| \hookrightarrow S$ is a proper embedding. The minimal supporting surface of a ribbon graph $\mathfrak{G}$ is obtained by patching all regions of $\mathfrak{G}$.

(Un-)ramified morphisms

Let $\mathfrak{G}_i = (\mathcal{E}_i, \mathrm{nx}_i, \mathrm{op}_i)$, $i = 1, 2$, be two bipartite ribbon graphs and $\varphi\colon \mathcal{E}_1 \to \mathcal{E}_2$ a morphism. Clearly, φ takes each vertex (region) of $\mathfrak{G}_1$ onto a vertex (respectively, region) of $\mathfrak{G}_2$. Moreover, the restriction of φ to each vertex v or region R of $\mathfrak{G}_1$ is a morphism of $\mathbb{Z}$-sets, thus inducing an epimorphism $\mathbb{Z} \twoheadrightarrow \mathbb{Z}$. The order of the kernel of this epimorphism is called the *ramification index* of v or R, respectively. It belongs to $\bar{\mathbb{N}}$ and equals $\deg v/\deg\varphi(v)$ or $\operatorname{wd} R/\operatorname{wd}\varphi(R)$, respectively, where we let $\infty/\infty = 1$.

Definition 1.27. A morphism $\varphi\colon \mathcal{E}_1 \to \mathcal{E}_2$ of bipartite ribbon graphs $\mathfrak{S}_i$, $i = 1, 2$, is said to be *unramified* at a vertex v or region R or $\mathfrak{S}_1$ if the corresponding ramification index equals one. If all ramification indices are equal to one, then φ is said to be *unramified*; otherwise, it is *ramified.*

Any morphism $\varphi\colon \mathcal{E}_1 \to \mathcal{E}_2$ of graphs extends to a map $\tilde{\varphi}\colon \operatorname{Supp}\mathfrak{S}_1 \to \operatorname{Supp}\mathfrak{S}_2$ of their minimal supporting surfaces. To make this extension canonical, for a pair of regions R_1, $R_2 = \varphi(R_1)$ we use the following maps of their geometric realizations:

- $\bar{D}^2 \to \bar{D}^2$, $z \mapsto z^r$, if $\operatorname{wd} R_1 = r \operatorname{wd} R_2 < \infty$,
- $\bar{\mathfrak{H}} \to \bar{D}^2$, $z \mapsto \exp(2\pi i n z)$, if $\operatorname{wd} R_1 = \infty$ and $\operatorname{wd} R_2 = n < \infty$, and
- $\bar{\mathfrak{H}} \to \bar{\mathfrak{H}}$, $z \mapsto z$, if $\operatorname{wd} R_1 = \operatorname{wd} R_2 = \infty$.

The map $\tilde{\varphi}$ thus constructed is a ramified covering, with the extra convention that the exponential function $\bar{\mathfrak{H}} \to \bar{D}^2$ has $0 \in \bar{D}^2$ as a branch point of infinite ramification index. The extension $\tilde{\varphi}$ is (un-)ramified if and only if so is the original morphism φ. Denoting by $r(\,\cdot\,)$ the ramification index under φ, the extension $\tilde{\varphi}$ has ramification index $r(v)$ at each vertex $v \in \operatorname{Vtx}\mathfrak{S}_1$ and ramification index $r(R)$ at the center c_R of each finite region $R \in \operatorname{Reg}_0\mathfrak{S}_1$.

As a consequence, for a morphism $\varphi\colon \mathfrak{S}_1 \to \mathfrak{S}_2$ of finite skeletons, one has the following *Riemann–Hurwitz formula* (see, *e.g.*, [86])

$$2 - 2g_1 = (2 - 2g_2)\deg\varphi - \sum_{v \in \operatorname{Vtx}\mathfrak{S}_1} [r(v) - 1] - \sum_{R \in \operatorname{Reg}\mathfrak{S}_1} [r(R) - 1], \tag{1.28}$$

where g_i is the genus of $\mathfrak{S}_i$, $i = 1, 2$, and r stands for the ramification index. Of course, (1.28) can as well be derived directly from (1.26).

The following corollary of (1.28) restates well known properties of surfaces.

Corollary 1.29. *Let* $\varphi\colon \mathfrak{S}_1 \to \mathfrak{S}_2$ *be a morphism of finite skeletons, and let* g_i *be the genus of* $\mathfrak{S}_i$, $i = 1, 2$. *Then:*

1. *one has* $g_2 \geqslant g_1$; *if* $g_1 = g_2$, *then either* $g_2 \leqslant 1$ *or* $\deg\varphi = 1$;
2. *if* $g_2 = 0$ *and* $\deg\varphi > 1$, *then* φ *is ramified*;
3. *if* $g_2 = 1$, *then* $g_1 > 1$ *if and only if* φ *is ramified*; *otherwise* $g_1 = 1$;
4. *if* $g_1 = 1$, *then* $g_2 = 0$ *if and only if* φ *is ramified*; *otherwise* $g_2 = 1$. ◁

Product of skeletons

Let $\mathfrak{S}_i = (\mathcal{E}_i, \operatorname{nx}_i, \operatorname{op}_i)$, $i = 1, 2$, be two skeletons. The *product* $\mathfrak{S}_1 \times \mathfrak{S}_2$ is the set $\mathcal{E} := \mathcal{E}_1 \times \mathcal{E}_2$ with the diagonal action $\operatorname{nx} := \operatorname{nx}_1 \times \operatorname{nx}_2$, $\operatorname{op} := \operatorname{op}_1 \times \operatorname{op}_2$. Clearly, the

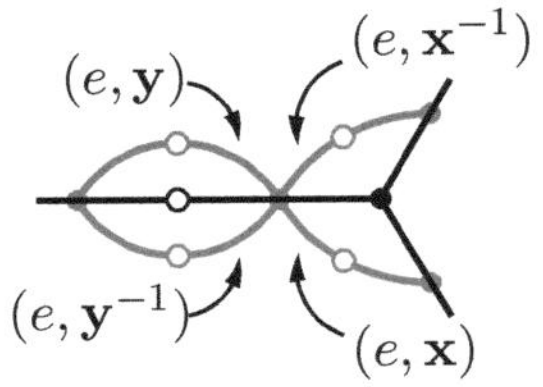

Figure 1.3. The inflation $\bar{\mathfrak{S}}$ (grey) of a skeleton $\mathfrak{S}$ (black).

geometric realization $|\mathfrak{S}_1 \times \mathfrak{S}_2|$ is the fibered product $|\mathfrak{S}_1| \times_{|\mathfrak{S}_0|} |\mathfrak{S}_2|$, and the minimal supporting surface $\operatorname{Supp}(\mathfrak{S}_1 \times \mathfrak{S}_2)$ is the fibered product $\operatorname{Supp}\mathfrak{S}_1 \times_{\operatorname{Supp}\mathfrak{S}_0} \operatorname{Supp}\mathfrak{S}_2$, where $\mathfrak{S}_0 := \bullet\!\!-\!\!\circ$ is the terminal object of the category of bipartite ribbon graphs.

The product of two skeletons does not need to be connected. Fixing a marking e_i of $\mathfrak{S}_i$ and identifying $\mathfrak{S}_i = H_i \backslash \mathbb{G}$, where $H_i := \operatorname{stab}_{\mathbb{G}} e_i$, $i = 1, 2$, one has the following simple relation, often referred to as the *double coset formula*:

$$(H_1 \backslash \mathbb{G}) \times (H_2 \backslash \mathbb{G}) = \coprod_{g \in H_2 \backslash \mathbb{G} / H_1} (H_1 \cap g^{-1} H_2 g) \backslash \mathbb{G}.$$

It follows that the product $\mathfrak{S}_1 \times \mathfrak{S}_2$ provides a convenient geometric way to list all essentially distinct intersections of subgroups conjugate to H_1 and H_2. Furthermore, the graph of any morphism $\varphi\colon \mathfrak{S}_1 \to \mathfrak{S}_2$ is a subobject (in the category theoretical sense, *cf.* Warning 1.15) of $\mathfrak{S}_1 \times \mathfrak{S}_2$ and its projections to $\mathfrak{S}_1$ and $\mathfrak{S}_2$ are morphisms. Hence, one has the following simple statement.

Proposition 1.30. *Given two finite skeletons $\mathfrak{S}_1$, $\mathfrak{S}_2$, the map $\varphi \mapsto$ graph of φ establishes a bijection between the set of morphisms $\varphi\colon \mathfrak{S}_1 \to \mathfrak{S}_2$ and that of connected components of $\mathfrak{S}_1 \times \mathfrak{S}_2$ of cardinality equal to* $\operatorname{Card}\mathfrak{S}_1$. *In particular, $\mathfrak{S}_1 \cong \mathfrak{S}_2$ if and only if* $\operatorname{Card}\mathfrak{S}_1 = \operatorname{Card}\mathfrak{S}_2$ *and $\mathfrak{S}_1 \times \mathfrak{S}_2$ has a component of cardinality* $\operatorname{Card}\mathfrak{S}_1$. ◁

Proposition 1.30 is used in functions "`Morphisms`" and "`IsIsomorphic`" in Listing C.2; the product of two skeletons is given by "`Prod`".

1.2.3 The fundamental group

The *inflation* of a bipartite ribbon graph $\mathfrak{S} = (\mathcal{E}, \varrho_\bullet, \varrho_\circ)$ is the ordinary ribbon graph $\operatorname{Inf}\mathfrak{S}$ whose bipartite subdivision is $\bar{\mathfrak{S}} = (\bar{\mathcal{E}}, \bar{\mathrm{nx}}, \bar{\mathrm{op}})$, where $\bar{\mathcal{E}} = \mathcal{E} \times \{\mathbf{x}^{\pm 1}, \mathbf{y}^{\pm 1}\}$ and

$$\bar{\mathrm{nx}}\colon (e, \mathbf{g}_*^\alpha) \mapsto (e, \mathbf{g}_{\alpha(*)}^{-\alpha}), \qquad \bar{\mathrm{op}}\colon (e, \mathbf{g}_*^\alpha) \mapsto (\varrho_*^\alpha(e), \mathbf{g}_*^{-\alpha}).$$

Here $e \in \mathcal{E}$, $* \in \{\bullet, \circ\}$, $\alpha = \pm 1$, and the multiplicative group $\{\pm 1\}$ is identified with the permutation group $\mathbb{S}\{\bullet, \circ\}$. Geometrically, we do inflate each vertex v of $\mathfrak{S}$, replacing it with the boundary of a small disk surrounding v in $\operatorname{Supp}\mathfrak{S}$, and contract each edge of $\mathfrak{S}$ afterwards. The vertex and four edges emerging from a single edge $e \in \mathcal{E}$ are shown in Figure 1.3.

The vertices of $\operatorname{Inf}\mathfrak{S}$ (corresponding to the $\bullet$-vertices of $\bar{\mathfrak{S}}$) are naturally identified with the edges of $\mathfrak{S}$; the valency of each vertex is four. The *directed* edges of the inflation $\operatorname{Inf}\mathfrak{S}$ (respectively, edges of $\bar{\mathfrak{S}}$) are labeled by pairs $(e, \mathbf{g}_*^{\pm 1})$, where $e \in \mathcal{E}$ and $* = \bullet, \circ$. For such an edge (e, g), one has $(e, g)(0) = e$ (as a vertex of $\operatorname{Inf}\mathfrak{S}$), $(e, g)(1) = e \uparrow g$, and $-(e, g) = (e \uparrow g, g^{-1})$. There are two kinds of regions of $\bar{\mathfrak{S}}$:

- A region $R = \operatorname{reg} e$, $e \in \mathcal{E}$, gives rise to a region $\operatorname{def} R := \operatorname{reg}(e, \mathbf{y}^{-1}) = \operatorname{reg}(e \uparrow \mathbf{x}, \mathbf{x}^{-1})$, called the *deflation* of R (*cf.* Lemma 1.31 below); one has $\operatorname{wd}(\operatorname{def} R) = 2 \operatorname{wd} R$.
- A vertex $v = \iota_*(e)$, $e \in \mathcal{E}$, $* = \bullet, \circ$, gives rise to a region $\operatorname{inf} v := \operatorname{reg}(e, \mathbf{g}_*)$, called the *inflation* of v; one has $\operatorname{wd}(\operatorname{inf} v) = \deg v$.

The regions $\operatorname{def} R$ and $\operatorname{inf} v$ do not depend on the chosen representations $R = \operatorname{reg} e$ and $v = \iota_*(e)$, as one has $(e \uparrow \mathbf{xy}, \mathbf{y}^{-1}) = (e \uparrow \mathbf{x}, \mathbf{x}^{-1}) \uparrow \mathbf{xy} = (e, \mathbf{y}^{-1}) \uparrow (\mathbf{xy})^2$ and $(e \uparrow \mathbf{g}_*^{-1}, \mathbf{g}_*) = (e, \mathbf{g}_*) \uparrow \mathbf{xy}$.

Lemma 1.31. *There is a canonical embedding* $|\operatorname{Inf}\mathfrak{S}| \hookrightarrow \operatorname{Supp}\mathfrak{S}$, *extending to a homeomorphism* $h\colon \operatorname{Supp}(\operatorname{Inf}\mathfrak{S}) \to \operatorname{Supp}\mathfrak{S}$. *such that*

- $h(|\operatorname{def} R|) \subset |R|$ *for each region* R *of* $\mathfrak{S}$,
- $h(c_{\operatorname{def} R}) = c_R$ *for each finite region* R *of* $\mathfrak{S}$, *and*
- $h(c_{\operatorname{inf} v}) = v$ *for each vertex* v *of* $\mathfrak{S}$.

(Here, c_R *stands for the center of region* R*.)*

Proof. One can regard the open disk D^2 and half plane $\mathfrak{H}$ as the hyperbolic plane, with the edges of $\mathfrak{S}$ representing certain segments in the absolute, and connect the centers of consecutive segments by geodesics. These geodesics map to the edges of $\operatorname{Inf}\mathfrak{S}$. (This construction does not quite work for a monogonal region, as the two geodesics would coincide. In this special case, some other pair of curves connecting the centers of the two edges should be chosen and fixed.)

The extension of the map to the 2-cells of $\operatorname{Supp}(\operatorname{Inf}\mathfrak{S})$ is straightforward. The closed disk $\bar{D}^2$ (half plane $\bar{\mathfrak{H}}$) representing a region of the form $\operatorname{def} R$ shrinks radially (respectively, vertically) onto the ideal polygon bounded by the above geodesics. The disk $\bar{D}^2$ representing a region R' of the form $\operatorname{inf} v$ is divided into $\deg v$ sectors, according to the edges of $\operatorname{Inf}\mathfrak{S}$ in the boundary $\partial\bar{D}^2$, and the sector over an edge e' is mapped homeomorphically onto the outer region of the hyperbolic plane bounded by the geodesic representing e'. □

Chains

We use the concept of inflation to introduce the notion of *chain*, which is to replace that of walk in the combinatorial study of the fundamental group of a skeleton. The advantage of this approach is the fact that closed chains can almost be regarded as elements of the stabilizer Stab $\mathfrak{S}$, thus establishing a relation between the groups Stab $\mathfrak{S}$ and $\pi_1(\mathfrak{S})$. A more substantial argument in favor of chains is given after Theorem 1.42.

Definition 1.32. A *chain* in a bipartite ribbon graph $\mathfrak{S}$ is a walk in its inflation Inf $\mathfrak{S}$. Alternatively, a chain is a pair (e, w), where $e \in \operatorname{Edg} \mathfrak{S}$ is an edge of $\mathfrak{S}$ and w is a word in the alphabet $\{\mathbf{x}, \mathbf{x}^{-1}, \mathbf{y}, \mathbf{y}^{-1}\}$. The *evaluation homomorphism* is the map val sending a chain (e, w) to the image of w in $\mathbb{G}$.

Due to the identification $\operatorname{Vtx}(\operatorname{Inf} \mathfrak{S}) = \operatorname{Edg} \mathfrak{S}$ and the description of directed edges of Inf $\mathfrak{S}$ given above, the two definitions of chains are indeed equivalent. In a walk $(e; (e_i, g_i))$, $i = 1, \ldots, k$, one must have $e_1 = e$ and $e_{i+1} = e_i {\uparrow} g_i$ for $i = 1, \ldots, k-1$; hence, such a walk is completely determined by the corresponding chain $(e, g_1 \ldots g_k)$: one lets $e_1 = e$ and $e_{i+1} = e \uparrow g_1 \ldots g_i$ for $i = 1, \ldots, k-1$. As a consequence, the *initial* and *terminal points* of a chain $\gamma = (e, w)$ are the edges

$$\gamma(0) = e \quad \text{and} \quad \gamma(1) = e \uparrow \operatorname{val} \gamma, \tag{1.33}$$

respectively, and the *inverse* of γ is the chain

$$\gamma^{-1} = (e \uparrow \operatorname{val} \gamma, w^{-1}), \tag{1.34}$$

where w^{-1} is the word obtained from w by replacing each letter g in w with g^{-1} and reordering the letters backwards. The *product* of two chains $\gamma_i = (e_i, w_i)$, $i = 1, 2$, is well defined whenever $e_2 = e_1 \uparrow \operatorname{val} \gamma_1$ and is given by the concatenation

$$\gamma_1 \cdot \gamma_2 := (e_1, w_1 w_2); \qquad \text{hence,} \quad \operatorname{val}(\gamma_1 \cdot \gamma_2) = \operatorname{val} \gamma_1 \operatorname{val} \gamma_2. \tag{1.35}$$

As a first consequence, we obtain the following theorem.

Theorem 1.36. *Given a marked skeleton $(\mathfrak{S}, e)$, the evaluation homomorphism* val *induces an isomorphism $\pi_1(\operatorname{Inf} \mathfrak{S}, e) = \operatorname{stab}_{\mathbb{G}} e$.*

Proof. Due to (1.33), a chain $\gamma = (e, w)$ is closed if and only if $\operatorname{val} \gamma \in \operatorname{stab}_{\mathbb{G}} e$. Hence, in view of (1.35), it suffices to show that the evaluation map val descends to the fundamental group and that the kernel of this descent is trivial.

An elementary contraction of a closed chain (e, w) corresponds to the cancellation of a subword of w of the form gg^{-1}, where $g = \mathbf{x}^{\pm 1}$ or $\mathbf{y}^{\pm 1}$; hence the image of w in $\mathbb{G}$ does not change. Conversely, since $\mathbf{x}$ and $\mathbf{y}$ form a free basis for $\mathbb{G}$, any cancelation of a word w in $\mathbb{G}$ is of the form as above, thus corresponding to an elementary contraction of the chain (e, w). Hence, the descent of val to the fundamental group is a monomorphism. □

Next statement generalizes the well known fact that a nontrivial normal subgroup of $\mathbb{G}$ is finitely generated if and only if it is of finite index, see, *e.g.*, [114].

Corollary 1.37. *A subgroup $H \in \mathbb{G}$ satisfying $(*)$ is finitely generated if and only if it is of finite index.*

Proof. Due to the Schreier index formula, see Theorem 1.9, if H is of finite index, it is finitely generated. For the converse, let $\mathfrak{S} = H\backslash\mathbb{G}$ and consider the geometric realization $|\operatorname{Inf}\mathfrak{S}|$. It is not difficult to see that $|\operatorname{Inf}\mathfrak{S}|$ is homotopy equivalent to the wedge of $|\mathfrak{S}|$ and a number of circles, one circle for each vertex v of $\mathfrak{S}$. (Locally, the contractible star of v is replaced with a circle.) Thus, for the group $H = \pi_1(\operatorname{Inf}\mathfrak{S})$ to be finitely generated, $\mathfrak{S}$ must have finitely many vertices and, hence, finitely many edges. The number of edges of $\mathfrak{S}$ equals $[\mathbb{G} : H]$. □

The inclusion $|\operatorname{Inf}\mathfrak{S}| \hookrightarrow \operatorname{Supp}^\circ\mathfrak{S}$, see Lemma 1.31, and retraction $\operatorname{Supp}^\circ\mathfrak{S} \to |\mathfrak{S}|$, see Proposition 1.25 (1), induce an epimorphism $\pi_1(|\operatorname{Inf}\mathfrak{S}|) \twoheadrightarrow \pi_1(|\mathfrak{S}|)$. Hence, taking into account Theorem 1.36, in the combinatorial treatment of the fundamental group of a skeleton $\mathfrak{S}$ it proves more convenient to use chains rather than walks. By definition, a chain starts and ends at an *edge* of $\mathfrak{S}$ rather than at a vertex; hence, it is an edge rather than a vertex that should be chosen for the reference point. Topologically, one can assume that the base point for $|\mathfrak{S}|$ is chosen inside the open 1-cell representing an edge e in $|\mathfrak{S}|$. Since this cell is contractible in $|\mathfrak{S}|$, all groups thus obtained are *canonically* identified by the topological translation homomorphisms along paths *within the cell.* With this said, we adopt the following convention.

Convention 1.38. When dealing with the fundamental group of a skeleton $\mathfrak{S}$ or its supporting surface $\operatorname{Supp}\mathfrak{S}$ or $\operatorname{Supp}^\circ\mathfrak{S}$, we choose for the reference point an *edge* rather than a vertex of $\mathfrak{S}$. Thus, we speak about the *fundamental group* $\pi_1(\mathfrak{S}, e)$ of a marked skeleton $(\mathfrak{S}, e)$.

Theorem 1.39. *Let $(\mathfrak{S}, e)$ be a marked regular skeleton of some type $\mathbf{t} \in \mathbb{N}^2$. Then the evaluation homomorphism* val *induces an isomorphism $\pi_1(\mathfrak{S}, e) \to \operatorname{stab}_{\mathbf{t}} e \subset \mathbb{G}_{\mathbf{t}}$.*

Proof. Consider the subspace S obtained by patching all regions of $\operatorname{Inf}\mathfrak{S}$ of the form $\operatorname{inf} v$, $v \in \operatorname{Vtx}\mathfrak{S}$. Geometrically, we inflate each vertex of $\mathfrak{S}$ to a disk in $\operatorname{Supp}\mathfrak{S}$, choosing all disks disjoint except at the centers of the edges of $\mathfrak{S}$. Due to Lemma 1.31, there are inclusions $|\mathfrak{S}| \subset S \subset \operatorname{Supp}^\circ\mathfrak{S}$, and the retraction given by Proposition 1.25 (1) restricts to a strict deformation retraction $S \to |\mathfrak{S}|$, inducing an isomorphism of the fundamental groups.

The group $\pi_1(S)$ can easily be computed using the classical Seifert–van Kampen theorem. Patching a region $R_v := \operatorname{inf} v$, where v is a $*$-vertex of $\mathfrak{S}$, $* = \bullet, \circ$, does not produce any new generators, while adding a relation $[\partial R_v] = 1$. Thus, the inclusion $|\operatorname{Inf}\mathfrak{S}| \hookrightarrow S$ and strict deformation retraction $S \to |\mathfrak{S}|$ induce an epimorphism $\varphi\colon \pi_1(\operatorname{Inf}\mathfrak{S}, e) \twoheadrightarrow \pi_1(\mathfrak{S}, e)$. Since $\operatorname{wd} R_v = \deg v = t_*$, the kernel $\operatorname{Ker}\varphi$ is contained in the subgroup normally generated (in $\operatorname{Stab}_{\mathbb{G}}\mathfrak{S}$) by the elements $\mathbf{g}_*^{t_*}$. On the other hand, the geometric realization of any chain γ of the form $(e, w\mathbf{g}_*^{t_*}w^{-1})$ is a lasso about a $*$-vertex of $\mathfrak{G}$, implying that $\operatorname{val}\gamma$, which can be any element conjugate (in $\mathbb{G}$) to $\mathbf{g}_*^{t_*}$, belongs to $\operatorname{Ker}\varphi$. Thus, $\operatorname{Ker}\varphi = \operatorname{Ker}[\operatorname{Stab}_{\mathbb{G}}\mathfrak{G} \twoheadrightarrow \operatorname{Stab}_{\mathbf{t}}\mathfrak{G}]$, and the statement follows. □

Corollary 1.40. *For each type* $\mathbf{t} \in \mathbb{N}^2$*, there is a unique, up to isomorphism, regular* $\mathbf{t}$*-tree* $\mathfrak{F}_{\mathbf{t}}$*; its stabilizer is the class of the trivial subgroup* $0 \subset \mathbb{G}_{\mathbf{t}}$. ◁

The tree $\mathfrak{F}_{\mathbf{t}}$ given by Corollary 1.40 will be called the (*generalized*) *Farey tree*. This tree is infinite unless at least one of the multiplicities $t_\bullet$, $t_\circ$ equals one. The classical Farey tree is $\mathfrak{F} := \mathfrak{F}_{(3,2)}$, see Lemma 2.12. The marked Farey tree $(\mathfrak{F}_{\mathbf{t}}, f) := (\mathbb{G}_{\mathbf{t}}, 1)$ is the *initial object* of the category of marked $\mathbf{t}$-skeletons. The *terminal object*, both for marked $\mathbf{t}$-skeletons and for all marked skeletons, is the 'trivial' skeleton $(\mathfrak{G}_0, e_0) := (\mathbb{G}\backslash\mathbb{G}, \mathbb{G}\backslash\mathbb{G})$; it has a single edge connecting two vertices: •—∘. Thus, for any marked $\mathbf{t}$-skeleton $(\mathfrak{G}, e)$, there is a unique pair of morphisms $(\mathfrak{F}_{\mathbf{t}}, f) \to (\mathfrak{G}, e) \to (\mathfrak{G}_0, e_0)$.

The orbifold structure

In order to extend Theorem 1.39 to irregular skeletons, let us fix a type $\mathbf{t} \in \mathbb{N}^2$ and confine ourselves to the category of $\mathbf{t}$-skeletons. In these settings, the spaces $|\mathfrak{G}|$ and $\operatorname{Supp}\mathfrak{G}$ are equipped with a natural orbifold structure, and it is the corresponding orbifold fundamental group π_1^{orb} that is isomorphic to the stabilizer $\operatorname{Stab}_{\mathbf{t}}\mathfrak{G}$. A good introduction to the theory of orbifolds, orbifold fundamental groups, and other related notions can be found, *e.g.*, in [1].

The *codegree* of a $*$-vertex v, $* = \bullet, \circ$, of a bipartite ribbon graph $\mathfrak{G}$ of a certain fixed type $\mathbf{t} \in \mathbb{N}^2$ is the ratio $\operatorname{codeg} v := t_*/\deg v$. If the type $\mathbf{t}$ is to be specified, we use the notation $\operatorname{codeg}_{\mathbf{t}} v$. A vertex v is called *regular*, or $\mathbf{t}$*-regular*, if $\operatorname{codeg} v = 1$; otherwise v is called *irregular*.

Definition 1.41. Let $\mathfrak{G}$ be a bipartite ribbon graph of a certain fixed type $\mathbf{t} \in \mathbb{N}^2$. Then the spaces $|\mathfrak{G}|$, $\operatorname{Supp}\mathfrak{G}$, $\operatorname{Supp}^\circ\mathfrak{G}$ *etc.* are regarded as orbifolds, with the orbifold structure obtained by declaring each vertex v of $\mathfrak{G}$ a ramification point with the isotropy group $\mathbb{Z}_r$, where $r = \operatorname{codeg} v$.

More precisely, in $\operatorname{Supp}\mathfrak{G}$, a $*$-vertex v of codegree r has a neighborhood U that is modeled on the open disk D^2 with the $\mathbb{Z}_r$-action given by $1\colon z \mapsto z\exp(2\pi i/r)$, so that U is identified with the quotient $D^2/\mathbb{Z}_r$ and the quotient projection $D^2 \to U$ is the universal covering of U. Similarly, in $|\mathfrak{G}|$, the same vertex v has a neighborhood $V := U \cap |\mathfrak{G}|$ modeled on the union W of the t_* radii $\{z = r\exp(2\pi ik/t_*)\} \subset D^2$, $r \in [0,1)$, $k = 1, \ldots, t_*$, with the restricted $\mathbb{Z}_r$-action, so that V is identified with the quotient $W/\mathbb{Z}_r$ and the quotient projection $W \to V$ is the universal covering. (Thus, strictly speaking, $|\mathfrak{G}|$ is an *orbispace* in the sense of Haefliger [85] rather than an orbifold. We will however use the commonly accepted orbifold terminology.) If the orbifold structure is to be emphasized (*e.g.*, as opposed to the plain underlying topological space), we use the notation $|\mathfrak{S}|_{\mathbf{t}}$, $\operatorname{Supp}_{\mathbf{t}} \mathfrak{S}$, *etc.*

The orbifold fundamental group of one of the spaces considered in Definition 1.41 is denoted by $\pi_1^{\mathbf{t}}(\,\cdot\,, e)$ (see Convention 1.38) or just $\pi_1^{\mathrm{orb}}(\,\cdot\,, e)$, if $\mathbf{t}$ is understood. (In most applications below, the type is $\mathbf{t} = (3,2)$.) We abbreviate $\pi_1^{\mathbf{t}}(\mathfrak{S}, e) := \pi_1^{\mathbf{t}}(|\mathfrak{S}|, e)$. The deformation retraction $\operatorname{Supp}^{\circ} \mathfrak{S} \to |\mathfrak{S}|$ given by Proposition 1.25 (2) is a homotopy equivalence in the orbifold category, as it restricts and lifts to a $\mathbb{Z}_r$-equivariant strict deformation retraction $D^2 \to W$. In particular, for any edge e of $\mathfrak{S}$, this deformation retraction induces an isomorphism $\pi_1^{\mathrm{orb}}(\operatorname{Supp}^{\circ} \mathfrak{S}, e) = \pi_1^{\mathrm{orb}}(\mathfrak{S}, e)$.

The proof of the following statement repeats the proof of Theorem 1.39; one only needs to replace the reference to the classical Seifert–van Kampen theorem with its orbifold version, see, *e.g.*, [84].

Theorem 1.42. *Given a type* $\mathbf{t} \in \mathbb{N}^2$ *and a marked* $\mathbf{t}$*-skeleton* $(\mathfrak{S}, e)$*, the evaluation homomorphism* val *induces an isomorphism* $\pi_1^{\mathbf{t}}(\mathfrak{S}, e) \to \operatorname{stab}_{\mathbf{t}} e \subset \mathbb{G}_{\mathbf{t}}$. ◁

For an alternative proof of Theorem 1.42, one can observe that, in the orbifold category, the map of the geometric realizations induced by a morphism of $\mathbf{t}$-skeletons is a covering. In particular, for any skeleton $\mathfrak{S} = H\backslash\mathbb{G}_{\mathbf{t}}$, the universal covering of $|\mathfrak{S}|_{\mathbf{t}}$ is the map $|\mathfrak{F}_{\mathbf{t}}|_{\mathbf{t}} \to |\mathfrak{S}|_{\mathbf{t}}$ corresponding to the inclusion $0 \subset H$. The deck translation group of this covering is H and, since $\pi_1^{\mathbf{t}}(\mathfrak{F}_{\mathbf{t}}) = \pi_1(\mathfrak{F}_{\mathbf{t}}) = 0$ (the Farey tree is regular and one can use Theorem 1.39), one has $\pi_1^{\mathbf{t}}(\mathfrak{S}) = H$.

In view of Theorem 1.42, the orbifold fundamental group $\pi_1^{\mathbf{t}}(\mathfrak{S}, e)$ can be defined combinatorially as the group of closed chains $\gamma := (e, w)$, $e \uparrow \operatorname{val}\gamma = e$, modulo the equivalence relation generated by elementary contractions (removing a consecutive pair g, g^{-1} from w) and the operation of removing from w a sequence of t_* consecutive identical letters $\mathbf{g}_*^{\pm 1}$, $* = \bullet, \circ$. Note that, in general, the elements of $\pi_1^{\mathbf{t}}(\mathfrak{S}, e)$ cannot be represented by walks in $\mathfrak{S}$, *cf.* Theorem 1.4. Roughly speaking, upon arrival at a vertex v, a walk passes *through* v and records the outgoing edge only, whereas a chain goes *around* v and keeps track of both the outgoing edge and the sheet of the covering containing this edge. Thus, a chain does represent an orbifold path.

Recall that *parabolic* are the elements of $\mathbb{G}_{\mathbf{t}}$ conjugate to a power of $\mathbf{xy}$. Torsion elements of $\mathbb{G}_{\mathbf{t}}$ are called *elliptic*, and all other elements are *hyperbolic*. Next three statements follow from the geometric representation of paths by chains, Lemma 1.23, and the well known description of the orbifold fundamental group of a surface.

Corollary 1.43. *Given a type* $\mathbf{t} \in \mathbb{N}^2$ *and a marked* $\mathbf{t}$*-skeleton* $(\mathfrak{S}, e)$*, the kernel of the inclusion homomorphism* $\operatorname{stab}_{\mathbf{t}} e = \pi_1^{\mathbf{t}}(\mathfrak{S}, e) \to \pi_1^{\mathbf{t}}(\operatorname{Supp} \mathfrak{S}, e)$ *is the subgroup of* $\operatorname{stab}_{\mathbf{t}} e$ *generated by its parabolic elements.* ◁

Corollary 1.44. *A subgroup* $H \subset \mathbb{G}_{\mathbf{t}}$, $\mathbf{t} \in \mathbb{N}^2$, *is generated by its parabolic and elliptic elements if and only if the minimal supporting surface* $\operatorname{Supp}(H\backslash\mathbb{G})$ *(with the orbifold structure disregarded) is homeomorphic to* D^2 *or* S^2. ◁

Corollary 1.45. *A subgroup $H \subset \mathbb{G}_{\mathbf{t}}$, $\mathbf{t} \in \mathbb{N}^2$, is generated by its parabolic elements if and only if the minimal supporting surface* $\operatorname{Supp}_{\mathbf{t}}(H\backslash\mathbb{G})$ *is homeomorphic to either*

1. *disk D^2 with the trivial orbifold structure, or*
2. *sphere S^2 with $r \leqslant 2$ ramification points so that, in the case $r = 2$, the isotropy groups of the two points are of coprime orders.*

In case (2) *with $r = 0$, any set of parabolic generators of H contains a representative of each but* (*any*) *one set of conjugacy classes of minimal parabolic elements of H. In all other cases, any set of parabolic generators must contain a representative of each conjugacy class.* ◁

1.2.4 First applications

In conclusion, we briefly discuss a few immediate applications of Theorems 1.4, 1.42, and 1.39 concerning properties of subgroups of $\mathbb{G}_{\mathbf{t}}$ and representation of some infinite skeletons by their finite subgraphs.

The structure of subgroups

Our first application concerns the structure of subgroups of $\mathbb{G}_{\mathbf{t}}$, regarded as abstract groups. Certainly, Theorem 1.46, as well as Corollaries 1.49, 1.54, and 1.55, can be proved in a purely algebraic way and must be well known to experts. (For example, Theorem 1.46 is a special case of the famous Kurosh theorem.) We mention these results here to make the exposition more self-contained and because they are simple consequences of a few observations used in the sequel.

Theorem 1.46. *Given a type $\mathbf{t} \in \mathbb{N}^2$, any subgroup $H \subset \mathbb{G}_{\mathbf{t}}$ is a free product*

$$H \cong \circledast \mathbb{Z} * \circledast \mathbb{Z}_{r_i},$$

where each finite factor $\mathbb{Z}_{r_i}$ corresponds to a vertex v of the skeleton $\mathfrak{S} := H\backslash\mathbb{G}_{\mathbf{t}}$ with $r_i := \operatorname{codeg} v > 1$. If the index $[\mathbb{G}_{\mathbf{t}} : H]$ is finite, then the number n_∞ of infinite factors $\mathbb{Z}$ is subject to the relation

$$n_\infty + \sum \left(1 - \frac{1}{r_i}\right) = 1 + \left(1 - \frac{1}{t_\bullet} - \frac{1}{t_\circ}\right) [\mathbb{G}_{\mathbf{t}} : H].$$

Proof. Given an integer $r \in \mathbb{N}$, define the *orbistick* I_r as the unit segment I with the right end 1 declared an orbifold point with the isotropy group $\mathbb{Z}_r$. In other words, I_r is the geometric realization $|\bullet\!\!-\!\!\!-\!\!\circ|_{(1,r)}$. We use orbisticks in the wedge construction, which requires a basepoint. We always assume that the basepoint in I_r is the left end 0, *i.e.*, the end with the trivial orbifold structure (respectively, the $\bullet$-vertex in the alternative description). One has $\pi_1^{\mathrm{orb}}(I_r, 0) = \mathbb{Z}_r$.

Lemma 1.47. *For a skeleton $\mathfrak{S}$ of type* $\mathbf{t}$, *the orbispace* $|\mathfrak{S}|_{\mathbf{t}}$ *is homotopy equivalent (in the orbifold category) to the wedge* $|\mathfrak{S}| \vee \bigvee I_{r_i}$, *with one orbistick* I_r *for each vertex* v *of* $\mathfrak{S}$ *with* $r = \operatorname{codeg} v > 1$. *(Here,* $|\mathfrak{S}|$ *is the geometric realization of* $\mathfrak{S}$ *with the trivial orbifold structure.)*

Proof. In the orbifold $\operatorname{Supp}^{\circ}_{\mathbf{t}} \mathfrak{S}$, consider a collection of disjoint neighborhoods U_v of all vertices v with $r = \operatorname{codeg} v > 1$. In each neighborhood U_v, perturb $|\mathfrak{S}|$ to a new subgraph $|\mathfrak{S}|'$ disjoint from v and connect v to its new copy v' by an orbistick I_r, disjoint from $|\mathfrak{S}|'$ except for the basepoint 0 identified with v'. Then, the strict deformation retraction $\operatorname{Supp}^{\circ}_{\mathbf{t}} \mathfrak{S} \to |\mathfrak{S}|_{\mathbf{t}}$, see Proposition 1.25 (2), can be modified, locally in each neighborhood U_v, to a strict deformation retraction $\operatorname{Supp}^{\circ}_{\mathbf{t}} \mathfrak{S} \to |\mathfrak{S}|' \cup \bigcup I_{r_i}$, which establishes a homotopy equivalence. □

Let $\mathfrak{S} := H\backslash\mathbb{G}_{\mathbf{t}}$ be the skeleton as in the statement. Due to Lemma 1.47 and the orbifold version of the Seifert–van Kampen theorem, see [84], one has isomorphisms $H \cong \pi_1^{\mathbf{t}}(\mathfrak{S}) = \pi_1(\mathfrak{S}) * \circledast\, \mathbb{Z}_{r_i}$, and the group $\pi_1(\mathfrak{S})$ is free of rank $n_\infty = 1 - \chi(|\mathfrak{S}|)$, see Corollary 1.8.

For the second statement, assume that $\mathfrak{S}$ is finite. Then $\chi(|\mathfrak{S}|) = \#_\bullet + \#_\circ - e$, where $\#_*$ is the number of $*$-vertices of $\mathfrak{S}$, $* = \bullet, \circ$, and $e = [\mathbb{G}_{\mathbf{t}} : H]$ is the number of edges. Since each edge has precisely one $\bullet$-end and one $\circ$-end, one has

$$e = \sum_{v \in \operatorname{Vtx}_* \mathfrak{S}} \deg v, \qquad \text{hence} \quad \frac{e}{t_*} = \sum_{v \in \operatorname{Vtx}_* \mathfrak{S}} \left(\frac{1}{\operatorname{codeg} v} - 1 \right) + \#_*. \tag{1.48}$$

In the latter expression, the summation can be restricted to the vertices $v \in \operatorname{Vtx}_* \mathfrak{S}$ with $\operatorname{codeg} v > 1$, as those of codegree 1 contribute zero. Substituting the resulting values of $\#_*$ to $n_\infty + \#_\bullet + \#_\circ = 1 + e$, one obtains the expression for n_∞. □

Corollary 1.49. *For a subgroup* $H \subset \mathbb{G}_{\mathbf{t}}$, $t \in \mathbb{N}^2$, *the following statements are equivalent: H is free, H is torsion free, and the skeleton* $H\backslash\mathbb{G}_{\mathbf{t}}$ *is* $\mathbf{t}$*-regular.* ◁

The statement of Theorem 1.46 takes an especially simple form if $\mathbf{t} = (p, q)$ with p and q distinct primes. In this case, the codegree of a $*$-vertex is either 1 or t_*, $* = \bullet, \circ$. Hence, denoting by $\#_*^1$ the number of monovalent $*$-vertices of the skeleton $\mathfrak{S} := H\backslash\mathbb{G}_{\mathbf{t}}$, one has

$$H \cong \underset{n_\infty}{\circledast}\, \mathbb{Z} * \underset{\#_\bullet^1}{\circledast}\, \mathbb{Z}_p * \underset{\#_\circ^1}{\circledast}\, \mathbb{Z}_q.$$

The vertex counts in (1.48) turn into

$$e = p\#_\bullet - (p-1)\#_\bullet^1 = q\#_\circ - (q-1)\#_\circ^1,$$

and the expression for n_∞ in Theorem 1.46 takes the form

$$pq(n_\infty - 1) + q(p-1)\#_\bullet^1 + p(q-1)\#_\circ^1 = (pq - p - q)[\mathbb{G}_{\mathbf{t}} : H].$$

It follows that, if H is free, see Corollary 1.49, its index must be divisible by pq.

Skeletons of finite type

Next application concerns a compact geometric representation of infinite skeletons corresponding to finitely generated subgroups.

Definition 1.50. Given a type $\mathbf{t} \in \mathbb{N}^2$, a $\mathbf{t}$-skeleton $\mathfrak{S}$ is said to be *of finite type*, or *almost contractible*, if the fundamental group $\pi_1^{\mathbf{t}}(\mathfrak{S})$ (equivalently, any subgroup in the class $\operatorname{Stab}_{\mathbf{t}} \mathfrak{S}$) is finitely generated.

Note that this property depends on $\mathbf{t}$: for example, the Farey tree $\mathfrak{F}_{\mathbf{t}}$ is almost contractible (in fact, contractible) as a $\mathbf{t}$-skeleton, but it is not as a $r\mathbf{t}$-skeleton for any integer $r > 1$. Any finite skeleton is almost contractible.

Let $\mathfrak{S}_* = (\mathcal{E}_*, \mathrm{nx}_*, \mathrm{op}_*)$, $* = \bullet, \circ$, be two skeletons, each with a distinguished monovalent $*$-vertex v_*. The *splice* of $\mathfrak{S}_\bullet$ and $\mathfrak{S}_\circ$ is the skeleton $\mathfrak{S}_\bullet \#_{v_\bullet = v_\circ} \mathfrak{S}_\circ = (\mathcal{E}, \mathrm{nx}, \mathrm{op})$ defined as follows:

- $\mathcal{E} = \mathcal{E}_\bullet \cup \mathcal{E}_\circ / e_\bullet \sim e_\circ$, where e_* is the (only) edge of $\mathfrak{S}_*$ incident to v_*;
- $\mathrm{nx}\, e = \mathrm{nx}_\circ\, e$ and $\mathrm{op}\, e = \mathrm{op}_\bullet\, e$ for the common edge $e = e_\bullet = e_\circ$;
- on all other edges, nx and op act as in the original skeletons.

Geometrically, we erase the distinguished vertices and splice the resulting 'hanging' edges to a single edge of $\mathfrak{S}_\bullet \# \mathfrak{S}_\circ$. The splice contains copies of $\mathfrak{S}_\bullet$ and $\mathfrak{S}_\circ$ as induced subgraphs.

A *Farey branch* of type $\mathbf{t} \in \mathbb{N}^2$ is a skeleton $\mathfrak{B}_*$, $* = \bullet, \circ$, whose stabilizer $\operatorname{Stab}_{\mathbf{t}} \mathfrak{B}_*$ contains the cyclic subgroup generated by $\mathbf{g}_*$. (Such a skeleton $\mathfrak{B}_*$ is unique up to isomorphism; it is a tree with all but one vertices regular; the only irregular vertex is monovalent and of type $*$.) A *contractible branch* in a $\mathbf{t}$-skeleton $\mathfrak{S}$ is a subgraph isomorphic, as a skeleton, to a Farey branch. (Since all but one vertices have maximal allowed valency, such a subgraph is necessarily induced.) The union of an increasing sequence of contractible branches is either a contractible branch or the Farey tree. Hence, as a standard application of Zorn's lemma, one concludes that either $\mathfrak{S} \cong \mathfrak{F}_{\mathbf{t}}$ or any contractible branch in $\mathfrak{S}$ is contained in a unique maximal one. If two contractible branches have a common regular vertex, then one of them is contained in the other. Thus, any two distinct maximal contractible branches are disjoint with the possible exception of a common monovalent (for both branches) vertex.

Theorem 1.51. *Let $\mathbf{t} \in \mathbb{N}^2$ be a type and $\mathfrak{S} \ncong \mathfrak{F}_{\mathbf{t}}$ an almost contractible $\mathbf{t}$-skeleton. Then there is a unique finite subskeleton $\mathfrak{S}^{\mathrm{c}} \subset \mathfrak{S}$ and a collection $v_1, \ldots, v_k$ of monovalent vertices of $\mathfrak{S}^{\mathrm{c}}$ such that*

- *$\mathfrak{S}$ is the splice $\mathfrak{S}^{\mathrm{c}} \#_{v_1 = b_1} \mathfrak{B}_1 \# \cdots \#_{v_k = b_k} \mathfrak{B}_k$, where each $\mathfrak{B}_i$ is a Farey branch and b_i is the monovalent vertex of $\mathfrak{B}_i$, $i = 1, \ldots, k$, and*
- *the image in $\mathfrak{S}$ of each $\mathfrak{B}_i$ is a maximal contractible branch.*

The subskeleton $\mathfrak{S}^{\mathrm{c}} \subset \mathfrak{S}$ is called the compact part *of $\mathfrak{S}$.*

Proof. It suffices to show that any skeleton $\mathfrak{S}$ of finite type can be represented as a splice of a finite subskeleton and a number of Farey branches. Then, extending each contractible branch to a maximal one, one obtains a unique representation as in the theorem.

Pick an edge e and a finite set of generators for the group $\pi_1^{\text{orb}}(\mathfrak{S}, e)$. Represent each generator by a chain. Each chain meets finitely many vertices. Let $\mathfrak{S}'$ be the induced subgraph spanned by the finite set V of the vertices met by at least one of the chosen chains, and let $\mathfrak{S}''$ be its complement, *i.e.*, the induced subgraph spanned by $\operatorname{Vtx}\mathfrak{S} \smallsetminus V$. Denote, further, by $U' \subset |\mathfrak{S}|$ the union of V and all open edges incident to at least one vertex in V, with the induced orbifold structure, and let $U'' \subset |\mathfrak{S}|$ be the similar subspace constructed from $\operatorname{Vtx}\mathfrak{S} \smallsetminus V$. By construction, the inclusion induces an epimorphism $\pi_1^{\text{orb}}(U', e) \twoheadrightarrow \pi_1^{\text{orb}}(\mathfrak{S}, e)$. Hence, each vertex in the complement $\operatorname{Vtx}\mathfrak{S} \smallsetminus V$ is regular for $\mathfrak{S}$, see Theorem 1.46 and especially Lemma 1.47. Furthermore, from the exact sequence

$$\pi_1^{\text{orb}}(U', e) \twoheadrightarrow \pi_1^{\text{orb}}(\mathfrak{S}, e) \to \pi_1^{\text{orb}}(|\mathfrak{S}|, U', e) \to \pi_0(U')$$

and the fact that U' is connected one concludes that $\pi_1^{\text{orb}}(|\mathfrak{S}|, U', e)$ is a one point set. It follows that each component U_i'' of U'' meets U' at a single open edge e_i and is contractible to this edge. Let $v_i' \in V$ and $v_i'' \in \operatorname{Vtx}\mathfrak{S} \smallsetminus V$ be the two ends of e_i. Then, the union $\bar{U}_i''$ of U_i and the end v_i'' of e_i contained in V is (the geometric realization of) a tree, and all its vertices except v_i'' are regular, *i.e.*, $\bar{U}_i''$ is a Farey branch. Denoting by $\bar{U}'$ the union of U' and all opposite ends v_i'', one obtains a desired decomposition $\mathfrak{S} = \bar{U}' \#_{v_1''=v_1'} \bar{U}_1'' \# \cdots \#_{v_k''=v_k'} \bar{U}_k''$, where k is the number of connected components of U''. □

Convention 1.52. In the drawings, we represent an almost contractible skeleton $\mathfrak{S}$ by its compact part $\mathfrak{S}^{\text{c}}$, marking each vertex v_i as in Theorem 1.51 with a $\vartriangle$. The actual type, $\bullet$ or $\circ$, of a $\vartriangle$-vertex v_i is opposite to the type of its immediate neighbor in $\mathfrak{S}$, and the type, $\mathfrak{B}_\circ$ or $\mathfrak{B}_\bullet$, of the Farey branch to be spliced at v_i is determined by the type of v_i.

Remark 1.53. The maximality requirement in Theorem 1.51 imposes one condition on $\mathfrak{S}^{\text{c}}$: this skeleton cannot have a regular vertex v with all but one neighbors marked with a $\vartriangle$. Indeed, otherwise the union of the contractible branches $\mathfrak{B}_i$ spliced at the $\vartriangle$-neighbors of v would extend to a larger contractible branch containing all $\mathfrak{B}_i$.

Next statement extends to $\mathbb{G}_{\mathbf{t}}$ a well-known property of free groups, *cf.* Corollary 1.37.

Corollary 1.54. *Any nontrivial finitely generated normal subgroup $H \subset \mathbb{G}_{\mathbf{t}}$, $\mathbf{t} \in \mathbb{N}^2$, is of finite index.*

Proof. Assume that $[\mathbb{G}_{\mathbf{t}} : H] = \infty$ and let $\mathfrak{S} = H\backslash\mathbb{G}_{\mathbf{t}}$. According to Theorem 1.51, this skeleton has at least one contractible branch $\mathfrak{B}$. Let e be an edge of $\mathfrak{B}$ not incident to its irregular vertex. Then both ends of e are regular, both in $\mathfrak{B}$ and in $\mathfrak{S}$, and e is not contained in any cycle in $\mathfrak{S}$. On the other hand, since H is normal, the group $\operatorname{Aut}\mathfrak{S}$ is transitive on the set of edges of $\mathfrak{S}$, see Corollary 1.20. Hence, all vertices of $\mathfrak{S}$ are regular and $\mathfrak{S}$ has no cycles, *i.e.*, $\mathfrak{S} = \mathfrak{F}_{\mathbf{t}}$ and $H = 0$. □

Corollary 1.55. *If $t_\bullet, t_\circ > 1$, then $\mathcal{Z}(\mathbb{G}_{\mathbf{t}}) = 0$. In fact, unless $t_\bullet = 1$, $t_\circ = 1$, or $t_\bullet = t_\circ = 2$, the group $\mathbb{G}_{\mathbf{t}}$ has no nontrivial normal abelian subgroups.*

Proof. Due to Theorem 1.46, any abelian subgroup $H \subset \mathbb{G}_{\mathbf{t}}$ is cyclic, $H \cong \mathbb{Z}$ or $H \cong \mathbb{Z}_r$. If H is also normal, then, due to Corollary 1.54, the index $[\mathbb{G}_{\mathbf{t}} : H]$ is finite and one can use the index formula in Theorem 1.46. If $H \cong \mathbb{Z}$, the formula implies $t_\bullet^{-1} + t_\circ^{-1} = 1$, hence $t_\bullet = t_\circ = 2$; if $H \cong \mathbb{Z}_r$, one obtains $t_\bullet^{-1} + t_\circ^{-1} > 1$, hence $t_\bullet = 1$ or $t_\circ = 1$.

The group $\mathbb{G}_{(2,2)}$ can be presented as $\langle \mathbf{x}, \mathbf{z} \mid \mathbf{x}^2 = 1,\ \mathbf{xzx} = \mathbf{z}^{-1} \rangle$, where $\mathbf{z} = \mathbf{xy}$. Thus, $\mathbb{G}_{(2,2)}$ is the infinite dihedral group $\mathbb{D}_\infty = \mathbb{Z} \rtimes \mathbb{Z}_2$, and it is clear that any normal abelian subgroup $H \subset \mathbb{G}_{(2,2)}$ is generated by a power $\mathbf{z}^a$ (corresponding to a 'cyclic' regular skeleton). Unless $a = 0$, this subgroup is not central. □

1.3 Pseudo-trees

In this section, we fix the type $\mathbf{t} = (3, 2)$ and denote $\Gamma = \mathbb{G}_{(3,2)}$, see Section 2.1.1 for more details concerning this remarkable group.

1.3.1 Admissible trees

The principal source of exponentially large examples of non-equivalent objects that appear further in the book are simple pseudo-trees, which are regular skeletons constructed from and enumerated by certain ribbon trees, which we call admissible.

Definition 1.56. An *admissible tree* is a finite ribbon tree $\mathfrak{T}$ with all vertices of valency three (*nodes*) or one (*leaves*). A *marking* of an admissible tree is a choice of one of its leaves v_1.

The bipartite subdivision $\bar{\mathfrak{T}}$ of an admissible tree $\mathfrak{T}$ is a finite bipartite ribbon tree of type $(3, 2)$ with all $\circ$-vertices regular. Conversely, any bipartite ribbon tree $\bar{\mathfrak{T}}$ with these properties is the bipartite subdivision of a unique admissible tree $\mathfrak{T}$, and the two categories are obviously equivalent. A marking of $\mathfrak{T}$ is equivalent to a marking of $\bar{\mathfrak{T}}$ with the additional requirement that the chosen edge e should be incident to a monovalent $\bullet$-vertex.

For each leaf v of $\mathfrak{T}$, there is a unique edge of $\bar{\mathfrak{T}}$ that is incident to v. In particular, it follows that a marked admissible tree has no nontrivial automorphisms.

An admissible tree $\mathfrak{T}$ has an even number $2k$ of vertices, $(k-1)$ being nodes and $(k+1)$, leaves. Given a marking v_1, the other leaves can be numbered consecutively in the clockwise direction, *cf.* Figure 1.5 (b), where the tree $\mathfrak{T}$ is shown in black and the indices are shown inside small grey circles. More precisely, let e_1 be the edge of the bipartite subdivision $\bar{\mathfrak{T}}$ that is incident to v_1. Since $\bar{\mathfrak{T}}$ is a tree, it has a single region, *i.e.*, any edge e of $\bar{\mathfrak{T}}$ can be written in the form $e = e_1 \uparrow (\mathbf{xy})^{-n}$, where $n \geqslant 0$ is an integer. Order the edges according to the increasing of the minimal exponent n with this property, and restrict this order to the edges incident to leaves, hence to the leaves themselves. The result $(v_1, \ldots, v_{k+1})$ is called the *canonical leaf order* of the marked admissible tree $(\mathfrak{T}, v_1)$. For a pair of *consecutive* leaves v_i, v_{i+1}, $i = 1, \ldots, k$, denote by e_i, e_{i+1} the incident edges of $\bar{\mathfrak{T}}$ and define the *circular distance* from v_i to v_{i+1} as the minimal positive integer m_i such that $e_{i+1} = e_i \uparrow (\mathbf{xy})^{1-m_i}$. The sequence $(m_1, \ldots, m_k)$ is called the *signature* of $(\mathfrak{T}, v_1)$. This sequence can be extended by the distance m_{k+1} from v_{k+1} to v_1; the extended sequence is called the *cyclic signature*. Since the width of the only region of $\bar{\mathfrak{T}}$ equals $4k-2$, one has $m_1 + \cdots + m_{k+1} = 5k - 1$.

More generally, for a pair of leaves v_i, v_j, $1 \leqslant i < j \leqslant k+1$, the *circular distance* from v_i to v_j is defined to be $m_{ij} := \sum_{s=i}^{j-1} m_s$. The meaning of these numbers is explained in Remark 1.62 below.

Lemma 1.57. *Two marked admissible trees are isomorphic if and only if they have equal signatures. Two unmarked admissible trees are isomorphic if and only if their cyclic signatures differ by a cyclic permutation.*

This lemma is proved in next subsection. We do *not* assert that any sequence $(m_1, \ldots, m_k)$ appears as the signature of an admissible tree; a realizability criterion is given by Theorem 1.73 below.

Contractions and cuts

We precede the proof of Lemma 1.57 with a few definitions and statements which are also used in the sequel. Let $\mathfrak{T}$ be an admissible tree and let (v', v'') be a pair of leaves of $\mathfrak{T}$ adjacent to the same node; we say that v' and v'' *share* this node. Then, the *elementary contraction* of $\mathfrak{T}$ (more precisely, of the pair (v', v'')) is the subtree $\mathfrak{T}' \subset \mathfrak{T}$ obtained from $\mathfrak{T}$ by removing v', v'' and the two incident edges, see Figure 1.4 (a). (The formal description of the bipartite subdivision $\bar{\mathfrak{T}}'$ as a Γ-set is left to the reader as an exercise.) Note that $\mathfrak{T}'$ is also an admissible tree. If $\mathfrak{T}$ is marked, we require in addition that v', v'' should be consecutive leaves, *i.e.*, we do not allow the pair (v_{k+1}, v_1). Then, the new tree $\mathfrak{T}'$ retains a natural marking: if the pair contracted is (v_1, v_2), we assign index 1 to their common node, becoming a leaf of $\mathfrak{T}'$; otherwise, v_1 remains the first leaf. If the tree is marked and the pair contracted is (v_i, v_{i+1}), the signature changes *via*

$$(\ldots, m_{i-1}, 3, m_{i+1}, \ldots) \mapsto (\ldots, m_{i-1} - 1, m_{i+1} - 1, \ldots). \tag{1.58}$$

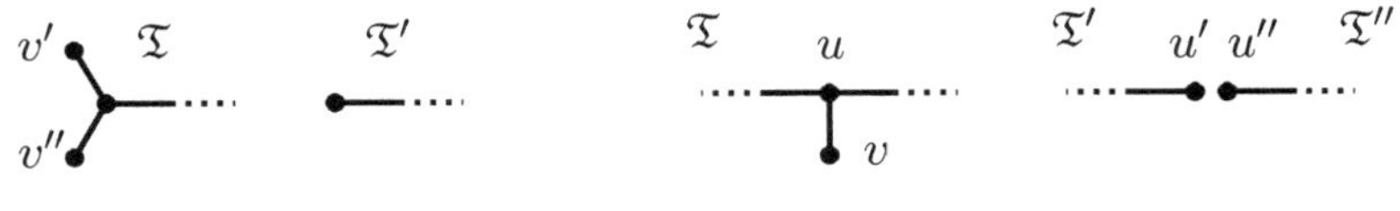

(a) Elementary contraction (b) Cut at a leaf

Figure 1.4. Operations on admissible trees.

(Note that two consecutive leaves share a node if and only if the circular distance between them equals three.)

If a subtree $\mathfrak{T}' \subset \mathfrak{T}$ can be obtained from $\mathfrak{T}$ by a sequence of elementary contractions, we say that $\mathfrak{T}$ *contracts to* $\mathfrak{T}'$. If $\mathfrak{T}'$ is isomorphic to the simplest admissible tree $\mathfrak{T}_0 :=$ •—• with two vertices, it can be identified with an edge e of $\mathfrak{T}$ and we say that $\mathfrak{T}$ *contracts to* e. If e is incident to a leaf v of the original tree $\mathfrak{T}$, we also say that $\mathfrak{T}$ *contracts towards* v. Note that, in general, the contraction (*i.e.*, the sequence of elementary contractions) is not uniquely determined by its result. A particular choice φ of a contraction to a subtree $\mathfrak{T}'$ will be indicated *via* a function like notation $\varphi\colon \mathfrak{T} \rightsquigarrow \mathfrak{T}'$ (respectively, $\varphi\colon \mathfrak{T} \rightsquigarrow e$ or $\varphi\colon \mathfrak{T} \rightsquigarrow v$).

Lemma 1.59. *Any admissible tree $\mathfrak{T}$ contracts to any of its edges. As a marked tree, $\mathfrak{T}$ contracts both towards its first leaf v_1 and towards its last leaf v_{k+1}. In particular, a marked admissible tree with $2k \geqslant 4$ vertices has a pair of consecutive leaves at circular distance three.*

Proof. For the first statement, it suffices to show that, if the tree $\mathfrak{T}$ has $2k \geqslant 4$ vertices, it has a pair of leaves sharing the same node and such that neither of them is incident to the chosen edge e. If $k = 2$, this assertion is obvious: the only admissible tree with four vertices has three leaves adjacent to a single node. If $k > 2$, no node is shared by more than two leaves and, according to the pigeonhole principle, there are at least two disjoint pairs of leaves, each sharing a node. At least one of these pairs is as desired.

For the second statement, if $k > 2$, we apply the same argument and observe that, if one of the two pairs found contains v_1 or v_{k+1}, then the other one is distinct from (v_{k+1}, v_1) and hence can be contracted. If $k = 2$, one can contract both (v_2, v_3) (towards v_1) and (v_1, v_2) (towards v_3). □

Corollary 1.60. *If $\mathfrak{T}$ is an admissible tree with $2k \geqslant 6$ vertices, its signature is primitive, i.e.,* $\text{g.c.d.}(m_1, \dots, m_k) = 1$.

Proof. According to Lemma 1.59, at least one circular distance m_i equals 3; hence, the greatest common divisor is either 3 or 1. Assume that it is 3. Then each circular distance m_i is either 3 or $\geqslant 6$. Furthermore, since $2k \geqslant 6$, three vertices cannot share the same node, *i.e.*, two consecutive distances m_i, m_{i+1} cannot be both equal to 3. Hence, all pairs of consecutive leaves at circular distance 3 can be contracted independently and, due to (1.58), in the signature of the new tree each member is at least 4, which contradicts to Lemma 1.59. □

Proof of Lemma 1.57. The first statement is proved by induction in the number $2k$ of vertices, and the second one is an immediate consequence.

The only tree with $2k = 2$ vertices is $\bullet\!\!-\!\!\!-\!\!\bullet$; its signature is (2). Let $2k \geqslant 4$ and consider a pair of trees $\mathfrak{T}_1$, $\mathfrak{T}_2$ with the same signature $(m_1, \ldots, m_k)$. According to Lemma 1.59, in this sequence there is an entry $m_i = 3$ and, contracting the corresponding pairs of leaves in both trees and using (1.58), one arrives at a pair of trees $\mathfrak{T}'_1$, $\mathfrak{T}'_2$ with $(2k-2)$ vertices each and the same signature. By the induction hypothesis, $\mathfrak{T}'_1$ and $\mathfrak{T}'_2$ are isomorphic; hence, so are their expansions $\mathfrak{T}_1$ and $\mathfrak{T}_2$. Note that, in both $\mathfrak{T}'_1$ and $\mathfrak{T}'_2$, it is the i-th leaf that is to be expanded, and the only isomorphism between the marked trees $\mathfrak{T}_1$ to $\mathfrak{T}_2$ necessarily takes these leaves into each other. □

Another useful operation is the cut of a tree at a leaf. Let $\mathfrak{T}$ be and admissible tree and v a leaf of $\mathfrak{T}$ adjacent to a node u. Then the *cut* of $\mathfrak{T}$ at v is defined as the pair $\mathfrak{T}'$, $\mathfrak{T}''$ of trees, necessarily admissible, obtained from $\mathfrak{T}$ by removing v, u, and the edge $[v, u]$ and attaching two separate copies u', u'' of u to the remaining 'hanging' edges incident to u in $\mathfrak{T}$, see Figure 1.4 (b). Formally, let $e \in \mathcal{E}$ be the edge of the bipartite subdivision $\bar{\mathfrak{T}}$ that is incident to v. Then the new bipartite graph $\bar{\mathfrak{T}}'$ has the set of edges $\mathcal{E}' = \mathcal{E} \smallsetminus \{e, e \uparrow \mathbf{y}\}$ and the restricted action of $\mathbf{x}$ and $\mathbf{y}$, except that $\mathbf{x}$ is redefined to act identically on the edges $e' := e \uparrow \mathbf{y}\mathbf{x}^{-1}$ and $e'' := e \uparrow \mathbf{y}\mathbf{x}$ (the two edges in the star of u that are not removed). The two components of this graph are $\mathfrak{T}'$ and $\mathfrak{T}''$. Note that these components are clearly distinguishable, as so are the edges e' and e'': with respect to the cyclic order at u in the original tree $\mathfrak{T}$, one has $e' = \operatorname{pred}(e \uparrow \mathbf{y})$ and $e'' = \operatorname{succ}(e \uparrow \mathbf{y})$. With Figure 1.4 (b) in mind, we will refer to $\mathfrak{T}'$ and $\mathfrak{T}''$ as the *left* and *right* components of the cut, respectively.

Pseudo-trees

Define a *simple loop* as the $(3,2)$-skeleton $\mathfrak{L}$ with three edges $\{e_1, e_2, e_3\}$, on which the group Γ acts *via* the permutations (in the cycle notation) $\mathbf{x} \mapsto (e_1\, e_2\, e_3)$ and $\mathbf{y} \mapsto (e_1\, e_2)(e_3)$, see Figure 1.5 (a). The stabilizer $\operatorname{stab} e_1$ is the index 3 subgroup of Γ generated by $\mathbf{x}\mathbf{y}$ and $\mathbf{x}^{-1}\mathbf{y}\mathbf{x}$. A simple loop has a unique monovalent vertex, which is $\circ$ (the $\mathbf{y}$-orbit (e_3)).

Definition 1.61. A *simple pseudo-tree* is a regular $(3,2)$-skeleton that is obtained from an admissible tree $\mathfrak{T}$ by splicing a copy of a simple loop $\mathfrak{L}$ at each monovalent $\bullet$-vertex of the bipartite subdivision $\bar{\mathfrak{T}}$. The skeleton obtained from an admissible tree $\mathfrak{T}$ in this way is denoted by $\operatorname{Sk} \mathfrak{T}$.

Geometrically, one draws $\mathfrak{T}$ on the plane and attaches a small loop at each leaf to make it a trivalent vertex, see Figure 1.5 (b). If the original tree $\mathfrak{T}$ has $2k$ vertices, then

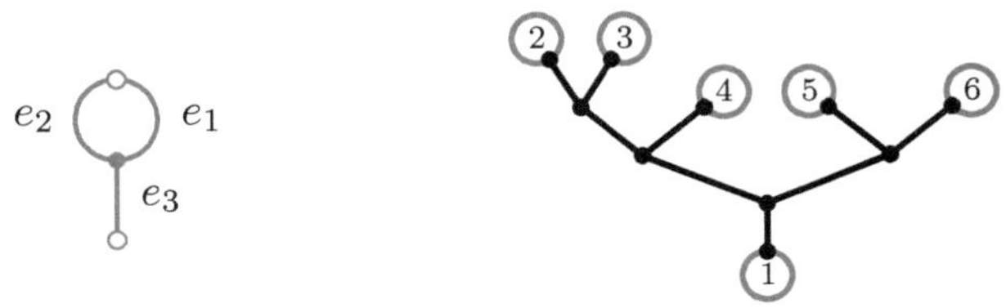

(a) A simple loop $\mathfrak{L}$ (b) Tree $\mathfrak{T}$ and skeleton $\operatorname{Sk}\mathfrak{T}$

Figure 1.5. Converting an admissible tree $\mathfrak{T}$ (black) to a simple pseudo-tree $\operatorname{Sk}\mathfrak{T}$ by splicing simple loops (shown in grey).

$\mathfrak{S} := \operatorname{Sk}\mathfrak{T}$ has $2k$ $\bullet$-vertices, $3k$ $\circ$-vertices, $6k$ edges, $(k+1)$ monogonal regions, and one *outer region* R of width $5k-1$. A marking $(\mathfrak{S}, e)$ of a simple pseudo-tree $\mathfrak{S}$ is called *proper* if e belongs to the outer region R of $\mathfrak{S}$. Note that all edges of the original bipartite subdivision $\bar{\mathfrak{T}}$ that are incident to the monovalent vertices of $\bar{\mathfrak{T}}$ do belong to R. Hence, a proper marking e of $\mathfrak{S}$ defines, in a canonical way, a marking of $\mathfrak{T}$: for example, one can take for v_1 the $\bullet$-vertex incident to the edge $e' = e{\uparrow}(\mathbf{xy})^{-n}$ with the minimal possible $n \geqslant 0$.

By construction, any isomorphism of admissible trees extends to an isomorphism of the corresponding simple pseudo-trees. Conversely, the monovalent vertices of $\mathfrak{T}$ can be characterized by the fact that such a vertex belongs to the boundary of a monogonal region of $\operatorname{Sk}\mathfrak{T}$; hence, any isomorphism of simple pseudo-trees restricts to an isomorphism of the original admissible trees. In particular, one has $\operatorname{Aut}\mathfrak{T} = \operatorname{Aut}\operatorname{Sk}\mathfrak{T}$.

Remark 1.62. The circular distance m_{ij} from a leaf v_i to a leaf v_j, $i < j$, introduced above is, in fact, the distance from v_i to v_j in the oriented boundary of the outer region of $\operatorname{Sk}\mathfrak{T}$. More precisely, denote by e_i the edge of $\bar{\mathfrak{T}}$ incident to v_i and regard it as an edge of $\operatorname{Sk}\mathfrak{T}$. Then m_{ij} is the minimal positive integer m such that $e_j = e_i{\uparrow}(\mathbf{xy})^{-m}$.

A simple pseudo-tree is a finite regular skeleton. For some applications, we need a more general construction, allowing, in addition to simple loops, other types of 'ends', such as Farey branches and/or monovalent vertices. Let $\mathfrak{T}$ be an admissible tree and let ℓ be a function defined on the leaves of $\mathfrak{T}$ and taking values in the four element set $\{0, \bullet, \circ, \vartriangle\}$; we call ℓ the *vertex function.* Define the *pseudo-tree* $\operatorname{Sk}(\mathfrak{T}, \ell)$ as the $(3,2)$-skeleton obtained from the bipartite subdivision $\bar{\mathfrak{T}}$ by the following operations on each monovalent $\bullet$-vertex v of $\bar{\mathfrak{T}}$ (*i.q.* leaf v of $\mathfrak{T}$):

- if $\ell(v) = 0$, splice a simple loop $\mathfrak{L}$ at v, as above;
- if $\ell(v) = \bullet$, leave v intact;
- if $\ell(v) = \circ$, remove v and the incident edge, leaving a monovalent $\circ$-vertex;
- if $\ell(v) = \vartriangle$, splice a Farey branch $\mathfrak{B}_\circ$ at v.

In the last case $\ell(v) = \vartriangle$, in the drawings we merely remove v and the incident edge and mark the resulting monovalent $\circ$-vertex u with a $\vartriangle$, see Convention 1.52. (Note that the maximal contractible branch containing v contains u as well.) As above, any iso-/automorphism of pairs $(\mathfrak{T}, \ell)$ extends to an iso-/automorphism of the pseudo-trees $\mathrm{Sk}(\mathfrak{T}, \ell)$. For the converse statement, to make $\mathfrak{T}$ uniquely recoverable from $\mathrm{Sk}(\mathfrak{T}, \ell)$, one should make an additional assumption that ℓ should not take the same value $\vartriangle$ on a pair of leaves sharing the same node, see Remark 1.53.

The next theorem gives a complete description of the monodromy group of a simple monodromy factorization in the modular group Γ (and related groups $\tilde{\Gamma}$ and $\mathbb{B}_3$), see Section 10.2 for precise definitions. It is used in Section 10.2.2.

Theorem 1.63. *A proper subgroup $H \subset \Gamma$ is generated by finitely many parabolic elements of width one* (Dehn twists) *if and only if the skeleton $\mathfrak{S} := H\backslash\Gamma$ is a pseudo-tree* $\mathrm{Sk}(\mathfrak{T}, \ell)$ *with the vertex function ℓ taking values in $\{0, \vartriangle\}$.*

Proof. Let $S = \mathrm{Supp}_{(3,2)}\,\mathfrak{S}$. In view of Corollary 1.45 and Lemma 1.23, it suffices to consider the following three cases.

Case 1: $S \cong D^2$ with the trivial orbifold structure. In this case, $\mathfrak{S}$ is a regular skeleton with all finite regions monogonal. If all regions are infinite, then $\mathfrak{S} \cong \mathfrak{F}$ and $H = 0$. Otherwise, shrinking each monogonal region to its only $\bullet$-vertex and each maximal contractible branch $\mathfrak{B}_\circ$ to its $\bullet$-vertex adjacent to its monovalent $\circ$-vertex, one obtains an admissible tree. Hence, $\mathfrak{S}$ is a pseudo-tree as stated. (Note that, unless $\mathfrak{S} \cong \bullet\!\!-\!\!\circ$, see the next case, each monogonal region of $\mathfrak{S}$ looks like that of a simple loop, see Figure 1.5 (a).)

Case 2: $S \cong S^2$ with $r = 1$ or 2 ramification points. If $\mathfrak{S}$ has two monovalent vertices adjacent to each other, then $\mathfrak{S} \cong \bullet\!\!-\!\!\circ$ and $H = \Gamma$. Otherwise, $\mathfrak{S}$ has an at least bigonal region (a region containing a monovalent vertex in its boundary) and H is not generated by Dehn twists.

Case 3: $S \cong S^2$ with the trivial orbifold structure. In this case, all but one regions of $\mathfrak{S}$ are monogons and, shrinking each monogon to its only vertex, one obtains an admissible tree. Hence, $\mathfrak{S}$ is a pseudo-tree (with $\ell \equiv 0$). □

Corollary 1.64 (of Case 3 in the proof). *Any regular skeleton of genus zero with all but one regions monogonal is a simple pseudo-tree.* ◁

1.3.2 The counts

Our immediate goal is to describe the automorphism groups of admissible trees and simple pseudo-trees and to count the number of such trees. We express the results in terms of the so-called Catalan numbers.

Digression: Catalan numbers (see [36, 96])

Recall that the n-th *Catalan number* $C(n)$, $n \geqslant 0$, is given by

$$C(n) := \frac{1}{n+1}\binom{2n}{n} = \frac{(2n)!}{(n+1)!\,n!}\,.$$

Using Stirling's approximation for $n!$, one can see that asymptotically these numbers grow as $C(n) \sim 4^n/n^{3/2}\sqrt{\pi} > a^n$ for any $a < 4$.

It is immediate that

$$(-4)^n\binom{1/2}{n} = -\frac{1}{2n-1}\binom{2n}{n} = -2C(n-1),$$

and, using the binomial series, one can see that $g(t) := (1-\sqrt{1-4t})/2t$ is the generating function for the Catalan sequence, *i.e.*, $g(t) = \sum_{n=0}^{\infty} C(n)t^n$. On the other hand, g satisfies the identity $g(t) = 1 + tg(t)^2$ and, comparing the two series, one obtains the recurrence relation

$$C(n+1) = \sum_{i=0}^{n} C(i)C(n-i). \tag{1.65}$$

Together with the initial condition $C(0) = 1$, this relation determines $C(n)$.

Catalan numbers, first discovered by Euler, play an important rôle in combinatorics: there is a great deal of counting problems to which this remarkable sequence provides a solution. For example, $C(n)$ counts

- the number of ways to arrange n pairs of correctly matched parentheses (same as the number of Dyck words of length $2n$);
- the number of ways to parenthesize completely a product of $(n+1)$ factors;
- the number of oriented rooted binary trees with n vertices (nodes or leaves);
- the number of rooted ribbon trees with $(n+1)$ vertices;
- the number of triangulations of a convex $(n+2)$-gon by its diagonals;
- the number of ways to connect $2n$ points in the circle by n disjoint chords.

Several dozens of combinatorial problems whose solution involves Catalan numbers can be found at `http://en.wikipedia.org/wiki/Catalan_number`. Our interest in these numbers is due to the fact that they count admissible trees.

Automorphisms and counts

Lemma 1.66. *The number of isomorphism classes of marked admissible trees with $2k$ vertices, $k \in \mathbb{N}$, is $C(k-1)$.*

Proof. We prove the statement by induction in k. If $k = 1$, it is obvious: there is a unique admissible tree with 2 vertices, $\mathfrak{T}_0 = \bullet\!\!-\!\!\!-\!\!\bullet$, and the two markings of this tree are isomorphic.

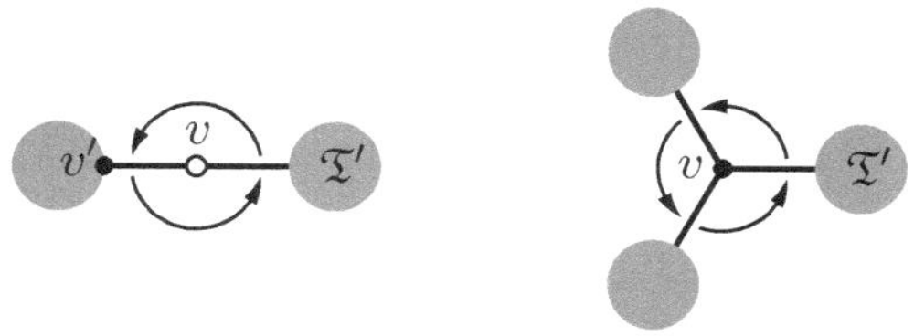

(a) ord $\varphi = 2$ (b) ord $\varphi = 3$

Figure 1.6. An automorphism φ of an admissible tree.

Assume that $k > 1$ and consider a marked admissible tree $(\mathfrak{T}, v_1)$ with $2k$ vertices. The leaf v_1 is adjacent to a node u and $\mathfrak{T}$ can be cut at v_1 into an *ordered* pair $(\mathfrak{T}', \mathfrak{T}'')$, see Figure 1.4 (b). The total number of vertices in $\mathfrak{T}'$ and $\mathfrak{T}''$ equals $2k$, and each of these trees can be canonically marked: one assigns index 1 to u' in $\mathfrak{T}'$ and to u'' in $\mathfrak{T}''$. Furthermore, the cut operation is invertible: given an ordered pair of marked trees with the total number of vertices $2k$, they can be joined together at their first leaves to produce a marked tree $\mathfrak{T}$. Thus, the statement follows from the recurrence relation (1.65) and the induction hypothesis. □

Alternatively, one can relate marked admissible trees with $2k$ vertices to the so-called oriented rooted binary trees with $(k-1)$ vertices and use the known count for the latter, see [52]. Note though that oriented rooted binary trees are usually counted along the same lines, cutting a tree into two and referring to (1.65).

Theorem 1.67. *For a simple pseudo-tree* $\mathfrak{S} = \operatorname{Sk}\mathfrak{T}$, *one has* $\operatorname{ord}(\operatorname{Aut}\mathfrak{S}) \leqslant 3$. *Let* $T_i(k)$ *be the number of isomorphism classes of simple pseudo-trees* $\mathfrak{S}$ *with* $2k$ $\bullet$*-vertices and* $\operatorname{ord}(\operatorname{Aut}\mathfrak{S}) = i$, $i = 1, 2, 3$. *Then, for* $i = 2, 3$, *one has*

$$T_i(k) = C(k'-1) \text{ if } k = ik' - 1, \qquad T_i(k) = 0 \text{ if } k \neq -1 \bmod i,$$

and the number $T_1(k)$ *is found from the relation*

$$T_1(k) + \frac{1}{2}T_2(k) + \frac{1}{3}T_3(k) = \frac{C(k-1)}{k+1},$$

where C *is the Catalan number. The total number* $T(k)$ *of isomorphism classes of simple pseudo-trees* $\mathfrak{S}$ *with* $2k$ $\bullet$*-vertices is* $T(k) = T_1(k) + T_2(k) + T_3(k)$.

Proof. As was explained right after Definition 1.56, instead of simple pseudo-trees one can consider corresponding admissible trees. Let $\mathfrak{T}$ be an admissible tree with $2k$ vertices, and let $\mathrm{id} \neq \varphi \in \operatorname{Aut}\mathfrak{T}$. By the Lefschetz fixed point theorem, the induced auto-homeomorphism of the compact contractible space $|\mathfrak{T}|$ must have a fixed point. This point must be a vertex v of the bipartite subdivision $\bar{\mathfrak{T}}$, as otherwise φ would fix an edge of $\bar{\mathfrak{T}}$ and thus be the identity. Hence, φ is the 'rotation' about v and $\operatorname{ord}\varphi = 2$ (if v is a $\circ$-vertex) or 3 (if v is a $\bullet$-vertex), see Figures 1.6 (a) and (b), respectively. It is also clear that $\mathfrak{T}$ has no other automorphisms, as the fixed vertex w of any such automorphism would be inside one of the shaded areas in the figure and the vertices of $\mathfrak{T}$ would be distributed unevenly with respect to w (in the sense that one of the components of the complement $|\mathfrak{T}| \smallsetminus w$ would contain more vertices than the others).

For the counts, observe that $\mathfrak{T}$ is the union of $\operatorname{ord}\varphi$ copies of an admissible subtree $\mathfrak{T}'$: it is the subtree in the shaded area shown in Figure 1.6 extended towards v, if $\operatorname{ord}\varphi = 3$, or towards the opposite $\bullet$-vertex v' adjacent to v if $\operatorname{ord}\varphi = 2$. This subtree $\mathfrak{T}'$ is naturally marked: one assigns index 1 to v or v', respectively, and the number $2k'$ of vertices of $\mathfrak{T}'$ is subject to the relation $k = (\operatorname{ord}\varphi)k' - 1$. Conversely, given a marked admissible tree $\mathfrak{T}'$ with $2k'$ vertices, an appropriate union of $m = 2$ or 3 copies of $\mathfrak{T}'$ produces a tree $\mathfrak{T}$ with an automorphism φ of order m. Together with Lemma 1.66, this fact proves the expressions for T_2 and T_3 in the statement.

The relation for T_1 is a simple orbit count: the number of marked admissible trees is $C(k-1)$, see Lemma 1.66, each tree $\mathfrak{T}$ has $(k+1)$ markings, and, $\operatorname{Aut}\mathfrak{T}$ acting freely on the leaves, these markings form $(k+1)/\operatorname{ord}(\operatorname{Aut}\mathfrak{T})$ isomorphism classes. □

Theorem 1.68. *The number $\tilde{T}(k)$ of isomorphism classes of properly marked simple pseudo-trees with $2k$ $\bullet$-vertices is*

$$\tilde{T}(k) = \frac{5k-1}{k+1} C(k-1),$$

where C is the Catalan number.

Proof. Arguing as in the proof of Theorem 1.67, since each simple pseudo-tree $\mathfrak{G}$ with $2k$ $\bullet$-vertices has $(5k-1)$ distinct proper markings and the group $\operatorname{Aut}\mathfrak{G}$ acts freely on these markings, one has

$$T_1(k) + \frac{1}{2}T_2(k) + \frac{1}{3}T_3(k) = \frac{\tilde{T}(k)}{5k-1}.$$

According to Theorem 1.67, the left hand side equals $C(k-1)/(k+1)$. □

Corollary 1.69. *For any integer $k \geqslant 1$ one has*

$$\tilde{T}(k) > T(k) \geqslant \frac{C(k-1)}{k+1}.$$

Hence, the sequences $T(k)$ and $\tilde{T}(k)$ grow faster than a^k for any positive $a < 4$. ◁

A few first values of $T(k)$ and $\tilde{T}(k)$ are shown in Table 1.1.

Table 1.1. First few values of $T(k)$ and $\tilde{T}(k)$.

k	1	2	3	4	5	6	7	8	...	11	...	16
$T(k)$	1	1	1	1	4	6	19	49	...	1424	...	570285
$\tilde{T}(k)$	2	3	7	19	56	174	561	1859	...	75582	...	45052515

Digression: Other counts

In general, a skeleton $\mathfrak{S}$ with e edges can be represented by a pair of permutations $\mathrm{nx}, \mathrm{op} \in \mathbb{S}(\mathcal{E}) = \mathbb{S}_e$ (the images of $\mathbf{x}$ and $\mathbf{y}$) satisfying the following conditions:

($*$) nx is a product of $\frac{1}{3}(e - \#^1_\bullet)$ cycles of length 3,
op is a product of $\frac{1}{2}(e - \#^1_\circ)$ transpositions, and
the subgroup generated by nx and op is transitive.

(Here, $\#^1_*$ is the number of monovalent $*$-vertices.) Two such pairs correspond to isomorphic skeletons if and only if they are conjugate in $\mathbb{S}_e$.

Let $\sigma := (\mathrm{nx} \circ \mathrm{op})^{-1}$ be the image of $\mathbf{xy}$ in $\mathbb{S}_e$. Then the cycles constituting σ (including those of length one) represent the regions of $\mathfrak{S}$, the length of a cycle being the width of the corresponding region. Hence, the conjugacy class of σ is determined by the multiset of the widths of the regions; it can be encoded by a partition λ of e. The genus g of $\mathfrak{S}$ is given by

$$2 - 2g = \frac{2}{3}\#^1_\bullet + \frac{1}{2}\#^1_\circ - \frac{1}{6}e + |\lambda|,$$

where $|\lambda|$ is the number of parts in λ, *i.q.* the number of regions of $\mathfrak{S}$. Thus, counting skeletons with given numbers $\#^1_\bullet$, $\#^1_\circ$, and e reduces to a purely combinatorial problem: one needs to count the conjugacy classes of pairs $(\mathrm{nx}, \mathrm{op})$ satisfying ($*$) above and, possibly, certain additional restrictions on the partition λ representing the conjugacy class of σ.

For example, the weighted count of all (not necessarily connected) ribbon graphs with a given partition λ is given by the Frobenius type formula [4] (see Remark 10.5):

$$\sum_{\mathfrak{S}} \frac{1}{\mathrm{Card}(\mathrm{Aut}\,\mathfrak{S})} = \frac{\mathrm{Card}[\![\mathrm{nx}]\!]\,\mathrm{Card}[\![\mathrm{op}]\!]\,\mathrm{Card}[\![\sigma]\!]}{(e!)^2} \sum_{\chi} \chi(\mathrm{nx})\chi(\mathrm{op})\chi(\sigma)\chi(1)^{-1},$$

where the first sum runs over all isomorphism classes of graphs and the second one, over all irreducible characters of $\mathbb{S}_e$. A recursive formula for the similar weighted number of regular ($\#^1_\bullet = \#^1_\circ = 0$) connected skeletons of a given genus g is obtained in [79]. For $g = 0$ the count is

$$\frac{4^e(3e-2)!!}{e(e+2)!(e-2)!!}.$$

1.3.3 The associated lattice

The associated lattice $\mathbf{Q}_{\mathfrak{T}}$ and characteristic functional $\xi_{\mathfrak{T}}$ introduced in this section are used in Chapter 9.

Given a marked admissible tree $(\mathfrak{T}, v_1)$ with $2k$ vertices, the *associated lattice* $\mathbf{Q}_{\mathfrak{T}}$ is the group freely generated by k vectors $\mathbf{q}_i$, $i = 1, \dots, k$, (informally corresponding to pairs (v_i, v_{i+1}) of consecutive leaves of $\mathfrak{T}$) with the products given by

$$\mathbf{q}_i^2 = m_i - 2, \quad \mathbf{q}_i \cdot \mathbf{q}_j = 1 \text{ if } |i-j| = 1, \quad \mathbf{q}_i \cdot \mathbf{q}_j = 0 \text{ if } |i-j| \geqslant 2, \qquad (1.70)$$

where $(m_1, \dots, m_k)$ is the signature of $\mathfrak{T}$. Define also the *characteristic functional*

$$\xi_{\mathfrak{T}}(x) := \sum_{i=1}^{k} m_i \mathbf{q}_i^* \in \mathbf{Q}_{\mathfrak{T}}^*.$$

Lemma 1.71. *Any contraction $\varphi\colon \mathfrak{T} \rightsquigarrow e$ to an edge e gives rise, in a canonical way, to an isomorphism $\varphi_*\colon \mathbf{Q}_{\mathfrak{T}} \cong \tilde{\mathbf{B}}_{k-1}^+$. More generally, any contraction $\psi\colon \mathfrak{T} \rightsquigarrow \mathfrak{T}'$ to a subtree $\mathfrak{T}'$ with $2k'$ vertices gives rise to an isomorphism $\psi_*\colon \mathbf{Q}_{\mathfrak{T}} \cong \mathbf{B}_{k-k'}^+ \oplus \mathbf{Q}_{\mathfrak{T}'}$.*

Proof. The former statement follows from the latter and the fact that $\mathbf{Q}_{\bullet\!-\!\bullet} = \tilde{\mathbf{B}}_0^+$. Hence, it suffices to consider one elementary contraction.

Consider the elementary contraction of a pair (v_i, v_{i+1}). The fact that this pair can be contracted means that $m_i = 3$, hence $\mathbf{q}_i^2 = 1$ and the sublattice $\mathbb{Z}\mathbf{q}_i \subset \mathbf{Q}_{\mathfrak{T}}$ splits as an orthogonal summand. The orthogonal complement $\mathbf{q}_i^\perp$ is generated by the vectors $\mathbf{q}_j'$, where $\mathbf{q}_j' = \mathbf{q}_j$ for $j < i-1$, $\mathbf{q}_{i-1}' = \mathbf{q}_{i-1} - \mathbf{q}_i$ (assuming $i > 1$), $\mathbf{q}_i' = \mathbf{q}_{i+1} - \mathbf{q}_i$ (assuming $i < k$), and $\mathbf{q}_j' = \mathbf{q}_{j-1}$ for $j > i+1$. Using (1.58), one finds that the Gram matrix of this basis equals that of the standard basis for $\mathbf{Q}_{\mathfrak{T}'}$ (where $\mathfrak{T}'$ is the result of the contraction), hence canonically $\mathbf{q}_i^\perp = \mathbf{Q}_{\mathfrak{T}'}$. □

According to Lemma 1.71, any contraction $\varphi\colon \mathfrak{T} \rightsquigarrow \mathfrak{T}'$ induces an isomorphism $\varphi_*\colon \mathbf{Q}_{\mathfrak{T}}^* \cong \mathbf{B}_{k-k'}^* \oplus \mathbf{Q}_{\mathfrak{T}'}^*$ of the dual groups. We say that a functional $\alpha \in \mathbf{Q}_{\mathfrak{T}}^*$ *contracts* to its image $\varphi_*(\alpha)$. The following statement is immediate, as $\mathbf{q}_i^*$ is only affected when the pair (v_i, v_{i+1}) is contracted.

Lemma 1.72. *For any contraction of a marked admissible tree $\mathfrak{T}$ towards its first leaf v_1 (last leaf v_{k+1}), the functional $\mathbf{q}_1^*$ (respectively, $\mathbf{q}_k^*$) contracts to $\mathbf{w}^*$.* ◁

Next theorem is not used in the sequel; we mention it here because it complements Lemma 1.57 and completes the characterization of marked admissible trees in terms of their signatures. For the statement, notice that the definition of the associated lattice $\mathbf{Q}_{\mathfrak{T}}$ uses the signature of $\mathfrak{T}$ only; hence, one can formally start with a sequence $\bar{m} = (m_1, \dots, m_k)$ of integers and use (1.70) to define a lattice $\mathbf{Q}_{\bar{m}}$.

Theorem 1.73. *A sequence $\bar{m} = (m_1, \dots, m_k)$ of integers is the signature of a marked admissible tree if and only if $\mathbf{Q}_{\bar{m}} \cong \tilde{\mathbf{B}}_{k-1}^+$.*

Proof. The 'only if' part is given by Lemma 1.71. The 'if' part is proved by induction. The case $k = 1$ is trivial: one must have $\bar{m} = (2)$. Thus, assume that $k \geqslant 2$ and that $\mathbf{Q}_{\bar{m}} \cong \tilde{\mathbf{B}}^+_{k-1}$.

First, observe that all $m_i \geqslant 3$, as otherwise $\mathbf{q}_i \in \mathbf{Q}_{\bar{m}}$ would be an element of square $\leqslant 0$ and not in the kernel $\ker \mathbf{Q}_{\bar{m}}$. We assert that at least one entry m_i equals 3. For proof, denote by d_i, $i = 1, \dots, k$, the determinant of the sublattice spanned by $\mathbf{q}_1, \dots, \mathbf{q}_i$. Then there is a recursive relation

$$d_i = (m_i - 2)d_{i-1} - d_{i-2}, \quad i = 1, \dots, k, \quad d_0 = 1, \quad d_i = 0 \text{ for } i < 0, \tag{1.74}$$

which can be rewritten in the form $(d_i - d_{i-1}) = (d_{i-1} - d_{i-2}) + (m_i - 4)d_{i-1}$. Assuming all $m_i \geqslant 4$ and proceeding by induction, one concludes that $d_k \geqslant \dots \geqslant d_i \geqslant \dots \geqslant d_0 = 1 > 0$. On the other hand, since $\mathbf{Q}_{\bar{m}}$ is degenerate, $d_k = 0$.

Consider an entry $m_i = 3$, so that $\mathbf{q}_i^2 = 1$. Then $\mathbf{Q}_{\bar{m}} = \mathbb{Z}\mathbf{q}_i \oplus \mathbf{q}_i^\perp$ and, as in the proof of Lemma 1.71, the sublattice $\mathbf{q}_i^\perp$ can be identified with $\mathbf{Q}_{\bar{m}'}$, where the new sequence $\bar{m}'$ is given by (1.58). According to Corollary A.8 (applied to the definite lattice $\mathbf{Q}_{\bar{m}}/\ker$), one has $\mathbf{Q}_{\bar{m}'} \cong \tilde{\mathbf{B}}^+_{k-2}$ and, by the induction assumption, $\bar{m}'$ is the signature of a certain marked admissible tree $\mathfrak{T}'$. Now, attach a pair of new leaves to the leaf v_i' of $\mathfrak{T}'$, converting v_i' to a node. The new tree $\mathfrak{T}$ has signature $\bar{m}$. □

Corollary 1.75 (of the proof). *Let $\mathfrak{T}$ be a marked admissible tree with $2k$ vertices. Then the kernel* $\ker \mathbf{Q}_{\mathfrak{T}}$ *is generated by* $\mathbf{k} := \sum_{i=1}^k c_i \mathbf{q}_i$*, where* $c_i = (-1)^{i-1} d_{i-1}$*, see* (1.74)*. One has* $c_1 = 1$*,* $c_k = (-1)^{k-1}$*, and* $\xi_{\mathfrak{T}}(\mathbf{k}) = 1 + (-1)^{k-1}$.

Proof. Due to (1.74), the sequence c_i is subject to the recursive relations

$$c_{i-1} + (m_i - 2)c_i + c_{i+1} = 0, \quad c_1 = 1, \quad c_i = 0 \text{ for } i \leqslant 0. \tag{1.76}$$

Hence, $\mathbf{k} \in \ker \mathbf{Q}_{\mathfrak{T}}$ and, since $\operatorname{rk} \ker \mathbf{Q}_{\mathfrak{T}} = 1$ (see Lemma 1.71) and $\mathbf{k}$ is a primitive vector (as $c_1 = 1$), this vector does generate the kernel. By Sylvester's criterion, one has $d_i > 0$ for $0 \leqslant i < k$; hence the sign of c_k is $(-1)^{k-1}$. On the other hand, one could have started from $c_k = \pm 1$ and solved (1.76) backwards to find another primitive vector generating $\ker \mathbf{Q}_{\mathfrak{T}}$; since such a vector is unique up to sign, it follows that $c_k = \pm 1$, hence $c_k = (-1)^{k-1}$. Finally, using (1.76) again, one has

$$\xi(\mathbf{k}) = \sum m_i c_i = \sum (2c_i - c_{i-1} - c_{i+1}) = c_1 + c_k. \qquad \square$$

As another consequence of Lemma 1.71, the isomorphism type of the lattice $\mathbf{Q}_{\mathfrak{T}}$ does not depend on the choice of the marking used for the construction.

Lemma 1.77. *Up to isomorphism, the sublattice* $\operatorname{Ker} \xi_{\mathfrak{T}} \subset \mathbf{Q}_{\mathfrak{T}}$ *does not depend on the choice of a marking of* $\mathfrak{T}$.

We postpone the proof of this lemma till Section 9.2.2, see page 286, where the lattice $\operatorname{Ker} \xi_{\mathfrak{T}}$ is given a simple geometric meaning.

Lemma 1.78. *Let $l = 2s - 1 \geqslant 3$ be an odd integer. Then there is a unique, up to automorphism of $\tilde{\mathbf{B}}_l^+$, primitive functional $\bar{\xi}_l \in \tilde{\mathbf{B}}_l^*$ with the following properties*:

1. $\ker \operatorname{Ker} \bar{\xi}_l \neq 0$,
2. $\det(\operatorname{Ker} \bar{\xi}_l / \ker) = 5l + 4$,
3. *the lattice* $\operatorname{Ker} \bar{\xi}_l / \ker$ *is even, and*
4. *the maximal root lattice contained in* $\operatorname{Ker} \bar{\xi}_l / \ker$ *is* $\mathbf{A}_{s-1}^+ \oplus \mathbf{A}_{s-2}^+$.

Up to reordering the generators, $\bar{\xi}_l = 3(\pm\mathbf{v}_1^* \pm \cdots \pm \mathbf{v}_s^*) \pm \mathbf{v}_{s+1}^* \pm \cdots \pm \mathbf{v}_l^*$.

Proof. Let $\bar{\xi}_l = \sum_i r_i \mathbf{v}_i^* + t\mathbf{w}_i^*$; due to (3) all $r_i \neq 0$, and we can assume $r_i > 0$.

Condition (1) is equivalent to the requirement that $t = 0$ and, factoring out the kernel, one can replace $\tilde{\mathbf{B}}_l^+$ with the unimodular lattice $\mathbf{B}_l^+$ and speak about the functional $x \mapsto x \cdot \mathbf{u}$, where $\mathbf{u} = \sum_i r_i \mathbf{v}_i \in \mathbf{B}_l^+$.

All even roots of $\mathbf{B}_l^+$ are of the form $\mathbf{v}_i \pm \mathbf{v}_j$, $i \neq j$, and such a root v belongs to $\mathbf{u}^\perp$ if and only if $v = \mathbf{v}_i - \mathbf{v}_j$ with $i \neq j$ and $r_i = r_j$. Hence, the maximal root lattice in $\mathbf{u}^\perp$ splits into irreducible summands $\mathbf{A}_p^+$, $p \geqslant 1$, with one summand for each $(p+1)$-fold value assumed by the sequence (r_i), $i = 1, \ldots, l$ (*cf.* Proposition 1.12). Thus, for $\bar{\xi}_l$ to satisfy (4), the sequence (r_i) must assume exactly two distinct values, an s-fold value $m > 0$ and an $(s-1)$-fold value $n > 0$. To satisfy (3), one also has $m = n \bmod 2$.

Finally, since $\mathbf{u}$ is primitive, one has $\det(\mathbf{u}^\perp) = \mathbf{u}^2$ and condition (2) means that $\sum_i r_i^2 = 5l + 4$. This results in the Diophantine equation $sm^2 + (s-1)n^2 = 10s - 1$, $m, n > 0$, $m = n \bmod 2$. Then immediately $m, n < 5$ (since $s \geqslant 2$) and, by trial and error, one finds that the only solution to this equation is $m = 3$, $n = 1$. It is easily seen that the functional $\bar{\xi}_l$ thus obtained does satisfy all conditions in the statement. □

Proposition 1.79. *Assume that k is even, and let $\varphi : \mathfrak{T} \rightsquigarrow e$ be a contraction of a marked admissible tree $\mathfrak{T}$ with $2k$ vertices. Then, up to reordering and changing signs of the generators, the characteristic functional $\xi_{\mathfrak{T}}$ contracts to the functional $\bar{\xi}_{k-1} \in \tilde{\mathbf{B}}_{k-1}^*$ given by Lemma* 1.78.

Proof. Due to Corollary 1.60 and Lemma 1.78, whenever the contraction of $\xi_{\mathfrak{T}}$ is as stated, this fact follows from the isomorphism type of the lattice $\operatorname{Ker} \xi_{\mathfrak{T}} / \ker$. Hence, for any given tree $\mathfrak{T}$, it suffices to prove the assertion for some choice of the marking and contraction; then for any other set of choices it would follow from Lemma 1.77. For the only tree with four vertices the statement of the proposition is immediate, and we proceed by induction, assuming that it holds for all contractions of all marked admissible trees with fewer that $2k$ vertices and proving it for *some* contraction of a given tree $\mathfrak{T}$ with $2k$ vertices and *some* particular marking.

Define a *loose end* as a leaf of $\mathfrak{T}$ sharing a node with an even number of other leaves. One has

$$\operatorname{Card}\{\text{loose ends of } \mathfrak{T}\} = (k-1) \bmod 2.$$

Since $k > 2$ is even, $\mathfrak{T}$ has at least one loose end v, which is the only leaf adjacent to its node u. Consider the cut of $\mathfrak{T}$ at v, see Figure 1.4 (b). We may assume that one of the subtrees obtained, say $\mathfrak{T}''$, contains no loose ends of the original tree $\mathfrak{T}$, as otherwise we could use that other loose end instead of v. Then, u'' is the only loose end of $\mathfrak{T}''$ and, due to the congruence above, the number of vertices of $\mathfrak{T}''$ is $4s'' = 0 \bmod 4$. By additivity, the number of vertices of $\mathfrak{T}'$ is $4s' := 4(s - s'') = 0 \bmod 4$.

Mark the two subtrees so that u'' is the last leaf of $\mathfrak{T}''$ and u' is the first leaf of $\mathfrak{T}''$. Then, mark $\mathfrak{T}$ by choosing for the first vertex that of $\mathfrak{T}''$. As a group, $\mathbf{Q}_{\mathfrak{T}}$ is canonically identified with $\mathbf{Q}_{\mathfrak{T}'} \oplus \mathbf{Q}_{\mathfrak{T}''}$. Note though that this is *not* an isomorphism of lattices: denoting by $\mathbf{q}'$ and $\mathbf{q}''$, respectively, the first generator of $\mathfrak{T}'$ and the last generator of $\mathfrak{T}''$, one has $\mathbf{q}' \cdot \mathbf{q}'' = 1$ in $\mathbf{Q}_{\mathfrak{T}}$, and the squares of $\mathbf{q}'$ and $\mathbf{q}''$ in $\mathbf{Q}_{\mathfrak{T}}$ differ by one from their squares in $\mathbf{Q}_{\mathfrak{T}'} \oplus \mathbf{Q}_{\mathfrak{T}''}$, as so do the corresponding vertex distances. For the same reason, the functional $\xi_{\mathfrak{T}}$ is identified with the sum $\xi_{\mathfrak{T}'} + (\mathbf{q}')^* + \xi_{\mathfrak{T}''} + (\mathbf{q}'')^*$.

Contract $\mathfrak{T}'$ and $\mathfrak{T}''$ to u' and u'' respectively. Then $\mathfrak{T}$ contracts to a four vertex subtree with a single node u, yielding an isomorphism $\mathbf{Q}_{\mathfrak{T}} \cong \mathbf{B}^+_{2s'-1} \oplus \mathbf{B}^+_{2s''-1} \oplus W$, where W is the lattice generated by two elements $\mathbf{w}'$, $\mathbf{w}''$ with $(\mathbf{w}')^2 = (\mathbf{w}'')^2 = 1$ and $\mathbf{w}' \cdot \mathbf{w}'' = 1$. Here, the products in W reflect the difference between $\mathbf{Q}_{\mathfrak{T}}$ and $\mathbf{Q}_{\mathfrak{T}'} \oplus \mathbf{Q}_{\mathfrak{T}''}$ explained in the previous paragraph. The generators of the $\mathbf{B}^+$ summands are those split off by the elementary contractions, see the proof of Lemma 1.71; in the obvious sense, these generators are shared by $\mathfrak{T}$ and $\mathfrak{T}'$, $\mathfrak{T}''$. By the induction hypothesis and Lemma 1.72, the characteristic functional $\xi_{\mathfrak{T}}$ contracts to $\bar{\xi}_{2s'-2} + (\mathbf{w}')^* + \bar{\xi}_{2s''-2} + (\mathbf{w}'')^*$. Now, one last contraction gives rise to an isomorphism $W \cong \tilde{\mathbf{B}}^+_1$ and takes $\xi_{\mathfrak{T}}$ to $\bar{\xi}_{2s'-2} + \bar{\xi}_{2s''-2} + \mathbf{v}_1^* = \bar{\xi}_{k-1}$, as stated. □

Lemma 1.80. *For an integer $l \geqslant 2$, there is a unique, up to automorphism of $\tilde{\mathbf{B}}^+_l$, primitive functional $\bar{\xi}'_l \in \tilde{\mathbf{B}}^*_l$ such that* $\operatorname{Ker} \bar{\xi}'_l \cong \mathbf{D}^+_l$. *One has $\bar{\xi}'_l = \sum_i r_i \mathbf{v}_i^* \pm 2\mathbf{w}^*$ for any collection of* odd *integers r_i, $i = 1, \dots, l$.*

Proof. Let $\bar{\xi}'_l = \sum_i r_i \mathbf{v}_i^* + t\mathbf{w}^*$. According to Proposition 1.12, one has $\operatorname{Ker} \bar{\xi}'_l \cong \mathbf{D}^+_l$ if and only if the projection $\tilde{\mathbf{B}}^+_l \to \tilde{\mathbf{B}}^+_l / \ker = \mathbf{B}^+_l$ maps $\operatorname{Ker} \bar{\xi}'_l$ onto the maximal even sublattice of $\mathbf{B}^+_l$, in particular, each even root $\mathbf{v}_i \pm \mathbf{v}_j$ of $\mathbf{B}^+_l$ belongs to the image. Hence, for each pair $i, j = 1, \dots, l$, one has $t \mid (r_i \pm r_j)$, and then $t \mid 2r_i$ and $t \mid 2r_j$. Thus, $\bar{\xi}'_l$ is primitive if and only if either $t = 1$ or $t = 2$ and all r_i are odd. In the former case, $\operatorname{Ker} \bar{\xi}'_l$ is not even. In the latter case, $\bar{\xi}'_l$ does satisfy the requirement and, moreover, all such functionals are isomorphic: the automorphism $\mathbf{v}_i \mapsto \mathbf{v}_i \pm \mathbf{w}$, $\mathbf{v}_j \mapsto \mathbf{v}_j$ for $j \neq i$ changes the coefficient r_i by ± 2. □

Proposition 1.81. *Assume that $k = 2s - 1$ is odd. Then, for any contraction of a marked admissible tree $\mathfrak{T}$ with $2k$ vertices, the characteristic functional $\xi_{\mathfrak{T}}$ contracts to a functional $\bar{\xi}'_{k-1} \in \tilde{\mathbf{B}}^*_{k-1}$ given by Lemma* 1.80. *If $\mathfrak{T}$ is contracted towards its last leaf v_{k+1}, then $\xi_{\mathfrak{T}}$ contracts to $\bar{\xi}_{k-1} + 2\mathbf{w}^*$, see Lemma* 1.78.

Proof. Convert the last leaf v_{k+1} to a node by attaching two extra leaves and contract the resulting tree $\mathfrak{T}'$ with $4s$ vertices towards its last leaf. Applying Proposition 1.79 and using Lemma 1.72 to compensate for the difference between $\mathfrak{T}$ and $\mathfrak{T}'$, one can conclude that $\xi_{\mathfrak{T}}$ contracts to $\bar{\xi}_{k-1}+2\mathbf{w}^*$. For any other contraction of $\mathfrak{T}$, the statement follows from the uniqueness given by Lemma 1.80. □

Remark 1.82. Simple examples show that, if k is odd, the precise expression for the image of $\xi_{\mathfrak{T}}$ in the basis of $\tilde{\mathbf{B}}^*_{k-1}$ given by the contraction *does* depend on the contraction used.

Chapter 2

The groups Γ and $\mathbb{B}_3$

In this chapter we introduce two related groups that play the central rôle in the rest of the book: the *modular group* $\Gamma := PSL(2, \mathbb{Z})$ and the *braid group* $\mathbb{B}_3$ on three strands.

2.1 The modular group $\Gamma := PSL(2, \mathbb{Z})$

The modular group $\Gamma := PSL(2, \mathbb{Z})$ is undoubtedly among the most remarkable objects discovered by the mathematicians, and one can trace its influence in virtually every area of mathematics. In this brief survey we merely touch the plethora of properties of Γ and its applications. A more comprehensive coverage of the classical results can be found in [103, 147, 150]. Note though that the subject is still rapidly developing and the reader interested in its modern state should consult recent papers.

2.1.1 The presentation of Γ

Denote by $\mathcal{H}$ the abelian group $\mathbb{Z}\mathbf{a} \oplus \mathbb{Z}\mathbf{b}$ with the skew-symmetric form $\bigwedge^2 \mathcal{H} \to \mathbb{Z}$, $\mathbf{a} \wedge \mathbf{b} \mapsto 1$. (In what follows, we usually identify $\mathcal{H}$ with the first homology group of a torus or punctured torus; under this identification, the above form on $\mathcal{H}$ is the intersection index.) Then the special linear group $\tilde{\Gamma} := SL(2, \mathbb{Z})$ can be identified with the group $\operatorname{Sp}(\mathcal{H})$ of symplectic isometries of $\mathcal{H}$. (Indeed, one can easily check that the condition that a matrix g should respect this particular bilinear form is equivalent to the requirement $\det g = 1$.) Introduce the isometries $\mathbb{X}, \mathbb{Y} \colon \mathcal{H} \to \mathcal{H}$ given by

$$\mathbb{X} := \begin{bmatrix} 1 & -1 \\ 1 & 0 \end{bmatrix}, \qquad \mathbb{Y} := \begin{bmatrix} 0 & 1 \\ -1 & 0 \end{bmatrix}. \tag{2.1}$$

One has $\mathbb{X}^3 = \mathbb{Y}^2 = -\operatorname{id}$. (As usual, we consider the *right* action, so that $\tilde{\Gamma}$ acts on *row* vectors by *right* matrix multiplication.) The *the modular group* Γ is defined to be $\Gamma := PSL(2, \mathbb{Z}) = \tilde{\Gamma}/\pm \operatorname{id}$.

We fix the notation $\mathcal{H}$, $\mathbf{a}$, $\mathbf{b}$ and $\mathbb{X}$, $\mathbb{Y}$ throughout the book. In most cases, we use the same symbol to represent an element of $\tilde{\Gamma}$ (a matrix) and its image in Γ.

It is well known that any auto-diffeomorphism of the torus T^2 is isotopic to a linear one. Hence, there is a canonical isomorphism $\operatorname{Map}^+(T^2) \cong \tilde{\Gamma}$, sending an auto-diffeomorphism φ to the induced homomorphism φ_* in the homology $\mathcal{H} = H_1(T^2)$.

Theorem 2.2. *The group $\tilde{\Gamma}$ is generated by $\mathbb{X}$ and $\mathbb{Y}$, the defining relations being $\mathbb{X}^3 = \mathbb{Y}^2$ and $\mathbb{Y}^4 = 1$. Hence, one has $\Gamma = \langle \mathbb{X}, \mathbb{Y} \,|\, \mathbb{X}^3 = \mathbb{Y}^2 = 1 \rangle$.*

Theorem 2.2 is very well known and is probably due to Gauss, although I could not find a precise reference in the literature. For the reader's convenience, we reproduce a proof here. We begin with a few observations that are of quite an interest on their own right.

It is often convenient to consider other pairs of generators, most commonly used being L, R given by

$$L := \mathbb{X}\mathbb{Y} = \begin{bmatrix} 1 & 1 \\ 0 & 1 \end{bmatrix}, \qquad R := (\mathbb{Y}\mathbb{X})^{-1} = \mathbb{X}^2\mathbb{Y} = \begin{bmatrix} 1 & 0 \\ 1 & 1 \end{bmatrix} \tag{2.3}$$

and S, T, where $S := \mathbb{Y}$ and $T := L$. (We try to use the commonly accepted notation here, but we do not fix it for the sequel.) Since one has

$$\mathbb{X} = RL^{-1}, \quad \mathbb{Y} = LR^{-1}L \qquad \text{and} \qquad \mathbb{X} = TS^{-1}, \quad \mathbb{Y} = S, \tag{2.4}$$

all three pairs do indeed generate the same (sub-)group of $\tilde{\Gamma}$.

Under the identification $\tilde{\Gamma} \cong \operatorname{Map}^+(T^2)$, the matrix R represents the so-called *Dehn twist* about a simple closed curve $a \subset T^2$ with $[a] = \mathbf{a} \in H_1(T^2)$. (In general, a simple closed curve a in an oriented surface X can be regarded as the core of a handle, and the Dehn twist t_a about a is the automorphism of X consisting in cutting the handle along a, twisting it through 2π, and gluing back; the induced automorphism of $H^1(X)$ is the symplectic reflection, or *transvection* $x \mapsto x - (x \circ [a])[a]$. In the mapping class group of the torus T^2, we have

$$t_{[p,q]} = \begin{bmatrix} 1 - pq & -q^2 \\ p^2 & 1 + pq \end{bmatrix}. \tag{2.5}$$

Any such Dehn twist is conjugate to R, *cf.* Proposition 2.6.) Similarly, L represents the negative Dehn twist about a simple closed curve b with $[b] = \mathbf{b}$. Up to sign, any upper (lower) triangle matrix in $\tilde{\Gamma}$ is a power of L (respectively, R): one has

$$L^n = \begin{bmatrix} 1 & n \\ 0 & 1 \end{bmatrix}, \qquad R^n = \begin{bmatrix} 1 & 0 \\ n & 1 \end{bmatrix}$$

for any $n \in \mathbb{Z}$.

Proposition 2.6. *The (sub-)group of $\tilde{\Gamma}$ generated by L, R acts transitively on the set of primitive vectors of $\mathcal{H}$.*

Proof. Let $[p,q] \in \mathcal{H}$ be a primitive vector, *i.e.*, such that g.c.d.$(p,q) = 1$. By the Euclidean algorithm, $[p,q]$ can be converted to $[\pm 1, 0]$ or $[0, \pm 1]$ by a sequence of moves $[a,b] \mapsto [a, b \pm a] = [a,b] \uparrow L^{\pm 1}$ or $[a,b] \mapsto [a \pm b, b] = [a,b] \uparrow R^{\pm 1}$, and the four vectors $[\pm 1, 0]$, $[0, \pm 1]$ are conjugate by powers of $\mathbb{Y}$. □

Corollary 2.7. *The group $\tilde{\Gamma}$ is generated by L, R (hence also by $\mathbb{X}$, $\mathbb{Y}$ or S, T).*

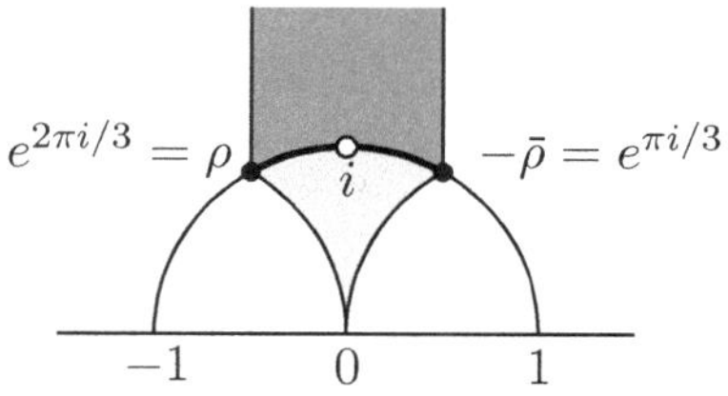

Figure 2.1. A fundamental domain D of Γ (dark grey) and its image $S \downarrow D$ (light grey).

Proof. Let $g \in \tilde{\Gamma}$. Then $[1,0] \uparrow g \in \mathcal{H}$ is a primitive vector and, by Proposition 2.6, there is an element h of the subgroup generated by L and R such that $[1,0] \uparrow gh = [1,0]$. Then gh is a power of R, *i.e.*, $g = \pm R^n h^{-1}$. □

According to Corollary 2.7, the map $\mathbf{x} \mapsto \mathbb{X}$, $\mathbf{y} \mapsto \mathbb{Y}$ establishes an epimorphism $\mathbb{G}_{(3,2)} \twoheadrightarrow \Gamma$, and we are aiming to prove that this map is an isomorphism.

The action on $\mathfrak{H}$

Let $\mathfrak{H} = \{z \in \mathbb{C} \mid \operatorname{Im} z > 0\}$ be the upper half plane. It is a straightforward verification that the assignment

$$g \downarrow z = \frac{az+b}{cz+d} \quad \text{for} \quad g = \begin{bmatrix} a & b \\ c & d \end{bmatrix} \tag{2.8}$$

defines a *left* Γ-action on $\mathfrak{H}$ by isometries (with respect to the standard hyperbolic metric on $\mathfrak{H}$). The generators S and T act *via* $S\colon z \mapsto -1/z$ and $T\colon z \mapsto z+1$. One can also easily check the relation

$$\operatorname{Im}(g \downarrow z) = \frac{\operatorname{Im} z}{|cz+d|^2} \tag{2.9}$$

for any $z \in \mathfrak{H}$. Denote by D the subset $\{z \in \mathfrak{H} \mid |z| \geqslant 1,\ |\operatorname{Re} z| \leqslant 1/2\}$, see Figure 2.1. (This drawing is a landmark that is supposed to appear in every book dealing with the modular group.) We assert that D is a fundamental domain of the above action. More precisely, the following statement holds.

Theorem 2.10 (see, *e.g.*, Serre [147]). *Let the domain D and the action $\Gamma \times \mathfrak{H} \to \mathfrak{H}$ be as above. Then:*

1. *for every $z \in \mathfrak{H}$, there exists $g \in \Gamma$ such that $g \downarrow z \in D$;*
2. *if $z, z' \in D$, $z \neq z'$, are conjugate by Γ, then either* $\operatorname{Re} z = \pm 1/2$ *and $z' = z \mp 1$ or $|z| = 1$ and $z' = -1/z$;*
3. *if $z \in D$, then* $\operatorname{stab}_\Gamma z = 0$ *except the following three cases:*

- $z = i$ and $\operatorname{stab}_\Gamma z$ is the order 2 subgroup generated by S, or
- $z = \rho := e^{2\pi i/3}$ and $\operatorname{stab}_\Gamma z$ is the order 3 subgroup generated by ST, or
- $z = -\bar{\rho} = e^{\pi i/3}$ and $\operatorname{stab}_\Gamma z$ is the order 3 subgroup generated by TS.

Proof. Given $z \in \mathfrak{H}$, the quantity $|cz + d|$, $c, d \in \mathbb{Z}$, attains its minimal value. Hence, due to (2.9), there exists an element $g \in \Gamma$ such that $\operatorname{Im}(g \downarrow z)$ is maximal. Next, there exists an integer n such that $|\operatorname{Re}(T^n g \downarrow z)| \leqslant 1/2$. Then the element $z' := T^n g \downarrow z$ belongs to D, as if $|z'| < 1$, one would have $\operatorname{Im}(-1/z') > \operatorname{Im} z'$, which would contradict to our choice of g. This proves Item 1.

For Items 2 and 3, consider a pair $z, g \downarrow z \in D$, where g is as in (2.8), and assume that $\operatorname{Im} z' \geqslant \operatorname{Im} z$. Then $|cz + d| \leqslant 1$, see (2.9), and hence $|c| < 2$, *i.e.*, $c = 0$ or 1. (Recall that in Γ matrices are defined up to sign.) If $c = 0$, then g is a power of L, *i.e.*, shift by an integer, and, assuming $g \neq \mathrm{id}$, we have the first choice: $\operatorname{Re} z = \pm 1/2$ and $z' = z \mp 1$. If $c = 1$, the inequality $|z + d| \leqslant 1$ implies that $d = 0$ unless $z = \rho$ (and then $d = 1$) or $z = -\bar{\rho}$ (and then $d = -1$). The last two cases contribute to the isotropy groups of ρ and $-\bar{\rho}$: for example, if $z = \rho$ and $c = d = 1$, then $a - b = 1$ and $z' = a - 1/(\rho + 1) = a + \rho$; hence $a = 0$ or 1. The case $d = 0$ implies $|z| \leqslant 1$, hence $|z| = 1$, and $b = -1$, *i.e.*, $z' = a - 1/z$ and $\operatorname{Re} z + \operatorname{Re} z' = a$. Thus, one has $a = 0$ unless $z = z' = \rho$ (and then $a = -1$) or $z = z' = -\bar{\rho}$ (and then $a = 1$). Finally, if $a = 0$, then $z' = z$ if and only if $z = i$. □

The Farey tree

In order to keep the notation consistent with the rest of the book, we compose the Γ-action on $\mathfrak{H}$ described in the previous paragraph with the outer automorphism

$$g \mapsto \bar{g} := CgC, \quad \text{where} \quad C := \begin{bmatrix} 0 & 1 \\ 1 & 0 \end{bmatrix} \in GL(2, \mathbb{Z}) \tag{2.11}$$

is an orientation reversing involution. One has $\bar{\mathbb{X}} = \mathbb{X}^{-1}$ and $\bar{\mathbb{Y}} = \mathbb{Y}^{-1}$.

Let l be the geodesic connecting $-\bar{\rho}$ and ρ, see the bold arc in Figure 2.1. Place a ∘-vertex at i and a pair of •-vertices at ρ and $-\bar{\rho}$, so that l splits into two edges, $[\rho, i]$ and $[-\bar{\rho}, i]$. Consider the union of all images of the resulting fragment under the action of Γ. In view of Theorem 2.10, this union is the geometric realization of a certain $(3, 2)$-regular skeleton $\mathfrak{F} = (\mathcal{E}, \mathrm{nx}, \mathrm{op})$, with the ribbon graph structure induced from the embedding of the geometric realization into $\mathfrak{H}$.

Lemma 2.12. *The skeleton* $\mathfrak{F}$ *is a tree*; *it is called the* Farey tree.

Proof. Each double edge (pair of edges incident to the same ∘-vertex) of $|\mathfrak{F}|$ is the common part of the boundary of a pair of fundamental domains of Γ; these domains are disjoint from all other edges of $|\mathfrak{F}|$ and divide $\mathfrak{H}$ into two connected components. (Up do an element of Γ, a double edge and corresponding pair of fundamental domains look like l, D, and $S \downarrow D$ shown in Figure 2.1.) It follows that an edge of $\mathfrak{F}$ cannot be part of a cycle. □

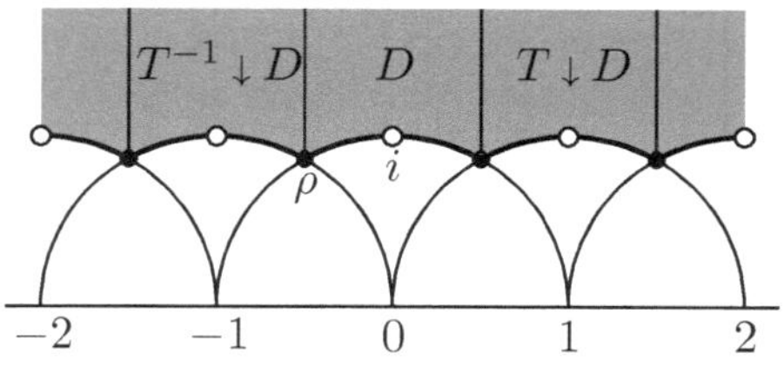

Figure 2.2. The region $|\operatorname{reg} e|$, $e = [\rho, i]$.

By construction and due to Theorem 2.10, the group Γ acts freely and transitively on the set $\mathcal{E}$ of edges of $\mathfrak{F}$. Label the edge represented by $[-\bar{\rho}, i]$ by $1 \in \Gamma$. Then any other edge has the form $\bar{g} \downarrow 1$ for a unique element $g \in \Gamma$, and we label this edge by g. Thus, we have $\mathcal{E} = \Gamma$ and Γ acts on $\mathcal{E}$ by the left multiplication. Then the ribbon graph structure on $\mathfrak{F}$ is given by the action of Γ by the right multiplication. Indeed, the element $\bar{\mathbb{X}} = \mathbb{X}^{-1}$ rotates $1 \in \mathcal{E}$ about $-\bar{\rho}$ in the counterclockwise direction, and $\bar{\mathbb{Y}} = \mathbb{Y}^{-1}$ reflects $1 \in \mathcal{E}$ about i. (It is the direction of the rotation why we had to combine the action with the conjugation by C.)

Lemma 2.13. *The minimal supporting surface* $\operatorname{Supp} \mathfrak{F}$ *is homeomorphic to* $\mathfrak{H}$.

Proof. The statement is almost obvious: each copy of $\mathfrak{H}$ used as a 2-cell in the construction of $\operatorname{Supp} \mathfrak{F}$ can be identified with a union of images of the fundamental domain D, each image appearing exactly once in the construction. For example, the geometric realization $|\operatorname{reg} e|$ for the edge e represented by the geodesic segment $[\rho, i]$ is the union $\bigcup_{i \in \mathbb{Z}} T^i \downarrow D$, see Figure 2.2. □

Proof of Theorem 2.2. Due to Corollary 2.7 and since $\tilde{\Gamma}$ is a central $\mathbb{Z}_2$-extension of Γ, it suffices to show that the epimorphism $\varphi\colon \mathbb{G}_{(3,2)} \twoheadrightarrow \Gamma$, $\mathbf{x} \mapsto \mathbb{X}$, $\mathbf{y} \mapsto \mathbb{Y}$, is an isomorphism. The kernel $\operatorname{Ker} \varphi$ lies in the stabilizer $\operatorname{stab}_{(3,2)} 1$ of the edge 1 of $\mathfrak{F}$. Since $\mathfrak{F}$ is a regular tree, see Lemma 2.12, one has $\operatorname{stab}_{(3,2)} 1 = 0$. □

The word problem and conjugacy problem

Since $\Gamma \cong \mathbb{Z}_3 * \mathbb{Z}_2$ is a free product of cyclic groups, both word problem and conjugacy problems in Γ are solvable, see, *e.g.*, [114].

Definition 2.14. A word w in the alphabet $\{\mathbb{X}^{\pm 1}, \mathbb{Y}^{\pm 1}\}$ is said to be *reduced* if any two consecutive letters in w are distinct, all exponents of $\mathbb{X}$ are ± 1, and all exponents of $\mathbb{Y}$ are 1. Such a word w is said to be *cyclically reduced* if any cyclic permutation of w is reduced.

Since $\mathbb{X}^3 = \mathbb{Y}^2 = 1$ in Γ, any element $g \in \Gamma$ can be represented by a reduced word and, up to conjugation, g can be represented by a cyclically reduced word. With the obvious exceptions of the one letter words $\mathbb{X}^{\pm 1}$ and $\mathbb{Y}$, a word is cyclically reduced if and only if it is reduced and its first and last letters are distinct.

Proposition 2.15. *Two elements of Γ are equal if and only if their reduced representatives coincide. Two elements are conjugate if and only if their cyclically reduced representatives differ by a cyclic permutation.* $\triangleleft$

An element $g \in \Gamma$ is called *elliptic*, *parabolic* (or *unipotent*), or *hyperbolic* if, respectively, $|\operatorname{trace} g| \leqslant 1$, $|\operatorname{trace} g| = 2$, or $|\operatorname{trace} g| > 2$. Since $\det g = 1$, these conditions are equivalent to the requirement that g should have no real eigenvalues, one real eigenvalue equal to 1, or two distinct (necessarily irrational) real eigenvalues, respectively. For yet another equivalent restatement, use (2.8) to extend the action of Γ to the closure $\mathfrak{H} \cup \bar{\mathbb{R}}$, where $\bar{\mathbb{R}} := \mathbb{R} \cup \{\infty\}$. Then an element $g \in \Gamma$ is elliptic, parabolic, or hyperbolic if it has one fixed point inside $\mathfrak{H}$, one rational fixed point in $\bar{\mathbb{R}}$, or two distinct (necessarily irrational) fixed points in $\mathbb{R}$, respectively.

In view of Proposition 2.6, any parabolic element $g \in \Gamma$ is conjugate to an element g' with the eigenvector $[1, 0]$, *i.e.*, to a power R^n, $n \in \mathbb{Z}$. The exponent n is called the *width* of g. Unless $g = 1$, its cyclically reduced representative is a sequence of n copies of $\mathbb{X}^{-1}\mathbb{Y}$ (for $n > 0$) or $|n|$ copies of $\mathbb{X}\mathbb{Y}$ (for $n < 0$). An element of negative width $n < 0$ is conjugate to the positive power L^{-n}. Parabolic elements of width 1 (respectively, width -1) are often referred to as *Dehn twists* (respectively, *negative Dehn twists*).

Any elliptic element g is conjugate to $\mathbb{X}^{\pm 1}$ or $\mathbb{Y}$, see Theorem 2.10; in particular, g has finite order. Conversely, any finite order element $g \neq 1$ must have an eigenvalue $\lambda \neq 1$ which is a root of unity, hence not real; such an element is elliptic.

It follows that an element $g \in \Gamma$ is hyperbolic if and only if its cyclically reduced representative contains both $\mathbb{X}$ and $\mathbb{X}^{-1}$. In other words, up to conjugation, such an element g can be expressed as a product of *positive* powers of L and R,

$$g = L^{a_1} R^{a_2} \ldots L^{a_{2n-1}} R^{a_{2n}}, \quad a_i > 0, \ \ i = 1, \ldots 2n, \tag{2.16}$$

with both L and R present. (Up to cyclic permutation, we can assume that the product starts with L and ends with R.) The sequence $[a_1\, a_2 \ldots a_{2n-1}\, a_{2n}]$ is called the *cutting period-cycle* of g, see [143]; it is well defined up to an even power $(1\,2 \ldots 2n{-}1\, 2n)^{2i}$ of cyclic permutation, and the corresponding equivalence class of cutting period-cycles determines the conjugacy class of g. There is a beautiful relation between the cutting period-cycle of a hyperbolic element and its fixed point $r \in \mathbb{R}$, see [143, 146]: if $r > 1$, then $[a_1\, a_2 \ldots a_{2n-1}\, a_{2n}]$ is a period (not necessarily minimal) of the continued fraction expansion of r. A geometric interpretation of the cutting period-cycle of g in terms of the position of the invariant geodesic of g with respect to the so-called *Farey tessellation* is also found in [143].

The word and conjugacy problems for $\tilde{\Gamma}$ are easily solved using the central extension

$$0 \longrightarrow \mathbb{Z}_2 \longrightarrow \tilde{\Gamma} \longrightarrow \Gamma \longrightarrow 0.$$

In view of Theorem 2.2, the abelianizations of Γ and $\tilde{\Gamma}$ are $\mathbb{Z}_6$ and $\mathbb{Z}_{12}$, respectively, and the *degree homomorphisms* dg mod 6: $\Gamma \twoheadrightarrow \mathbb{Z}_6$ and dg mod 12: $\tilde{\Gamma} \twoheadrightarrow \mathbb{Z}_{12}$ can be chosen to send $\mathbb{X}$ and $\mathbb{Y}$ to 2 and -3, respectively, so that $R \mapsto 1$ and $L \mapsto -1$. Hence, $\tilde{\Gamma}$ can be represented as a fibered product

$$\tilde{\Gamma} = \Gamma \times_{\mathbb{Z}_6} \mathbb{Z}_{12}, \tag{2.17}$$

i.e., the subgroup of all pairs $(g, n) \in \Gamma \times \mathbb{Z}_{12}$ such that $\operatorname{dg} g = n \bmod 6$. The degrees of opposite matrices $\pm g \in \tilde{\Gamma}$ differ by 6, and the following statement is straightforward.

Proposition 2.18. *Two elements $g_1, g_2 \in \tilde{\Gamma}$ are equal (respectively, conjugate) if and only if* $\operatorname{dg} g_1 = \operatorname{dg} g_1 \bmod 12$ *and the images of g_1, g_2 in Γ are equal (respectively, conjugate).* ◁

2.1.2 Subgroups

In view of the isomorphism $\mathbb{G}_{(3,2)} \to \Gamma$, $\mathbf{x} \mapsto \mathbb{X}$, $\mathbf{y} \mapsto \mathbb{Y}$, see Theorem 2.2, one can apply to Γ the machinery developed in Section 1.2.1 and study the subgroups of Γ in terms of their skeletons. The *skeleton* $H\backslash\Gamma$ of a subgroup $H \subset \Gamma$ will be denoted by Sk H; it is a skeleton of type $(3, 2)$. Since this case is our main application, we restate some results of Section 1.2.1 here. From now on, the term *regular* is reserved for $(3, 2)$-regular skeletons.

1. Under the map $H \mapsto \operatorname{Sk} H$, finite skeletons correspond to (the conjugacy classes of) finite index subgroups, and regular skeletons correspond to free subgroups.
2. The monovalent •-vertices of Sk H are in a one-to-one correspondence with the H-conjugacy classes of order three subgroups of H.
3. The monovalent ∘-vertices of Sk H are in a one-to-one correspondence with the H-conjugacy classes of order two elements of H.
4. The finite regions of Sk H are in a one-to-one correspondence with the H-conjugacy classes of minimal parabolic elements of H, the width of a region being equal to the width of the corresponding minimal parabolic element.
5. There is an isomorphism $\operatorname{Aut}(\operatorname{Sk} H) = \mathcal{N}_\Gamma(H)$. In particular, H is normal if and only if the group $\operatorname{Aut}(\operatorname{Sk} H)$ is transitive on the edges of Sk H.

The *genus* of a finite index subgroup $H \subset \Gamma$ is defined as the genus of its skeleton Sk H. Below we show (see Lemma 2.22) that this definition of genus is equivalent to the conventional one in terms of modular curves.

Remark 2.19. The idea to study the subgroups of Γ *via* their skeletons resembles the results of [22]. However, the two constructions differ: in [22], finite index subgroups of the congruence subgroup $\Gamma(2)$ are encoded using bipartite ribbon graphs with vertices of arbitrary valency. Our approach is closer to that of [25], where the modular

j-invariant on a modular curve is described in terms of a certain special triangulation of the curve. The correspondence between subgroups of Γ of finite index and finite graphs was first studied in [102].

The *skeleton* $\operatorname{Sk} H$ of a subgroup $H \subset \tilde{\Gamma}$ is defined as the skeleton of the projection of H to Γ, and the *genus* of H is the genus of $\operatorname{Sk} H$. In more details, subgroups of $\tilde{\Gamma}$ are treated in Section 2.2.3 below, see Remark 2.61.

Proposition 2.20. *Given a subgroup $H \subset \Gamma$, the map* $\operatorname{reg}(Hg) \mapsto [0,1]\uparrow g^{-1}$, $g \in \Gamma$, *establishes a one-to-one correspondence between the set of regions of* $\operatorname{Sk} H$ *and the set of H-orbits of pairs of opposite primitive vectors $\pm v \in \mathcal{H}$.*

Proof. Due to Proposition 2.6, the map in question is surjective. The stabilizer of $[0,1]$ is the cyclic subgroup generated by $\mathbb{X}\mathbb{Y}$; hence, two elements $g_1, g_2 \in \Gamma$ are in the same coset $\Gamma/\langle\mathbb{X}\mathbb{Y}\rangle$ if and only if $[0,1]\uparrow g_1^{-1} = [0,1]\uparrow g_2^{-1}$. It follows that the map is well defined and injective. □

Remark 2.21. In Proposition 2.20, instead of $(g, v) \mapsto v\uparrow g^{-1}$, one can use any other reasonable *left* Γ-action; with appropriate modifications, the statement still holds. For example, for the action $\Gamma \times \mathcal{H} \to \mathcal{H}$ by the left matrix multiplication on column vectors, the bijection is established by the map $\operatorname{reg}(Hg) \mapsto g\downarrow[1,0]^{\mathrm{t}}$, since in this case it is $[1,0]^{\mathrm{t}}$ whose stabilizer is the cyclic group generated by $\mathbb{X}\mathbb{Y}$. Yet another version of this statement is discussed in Remark 2.23 below.

Modular curves

Consider the left action $\Gamma \times \mathfrak{H} \to \mathfrak{H}$, $(g, z) \mapsto \bar{g}\downarrow z$, see (2.11). It is discrete and all its isotropy groups are finite; hence, the quotient $\Gamma\backslash\mathfrak{H}$ is a complex analytic curve. In view of Theorem 2.10, the quotient is homeomorphic to the open disk D^2, and it can be shown that, in fact, $\Gamma\backslash\mathfrak{H} \cong \mathbb{C}^1$. A detailed proof is found in [150]: the compactification $\mathbb{P}^1$ of $\Gamma\backslash\mathfrak{H}$ is constructed as the quotient $\Gamma\backslash\mathfrak{H}^*$, where $\mathfrak{H}^*$ is the union $\mathfrak{H} \cup \mathbb{Q} \cup \{\infty\}$ with an appropriate topology and analytic structure, *not* the ones induced from $\mathbb{C}^1$. (For example, a 'typical' neighborhood of ∞ is the shaded region in Figure 2.2, and a base of neighborhoods of ∞ is obtained from the one shown by vertical shifts. Any finite rational point has the form $r = \bar{g}\downarrow\infty$ for some $g \in \Gamma$, and its neighborhoods are the images under g of those of ∞; they look like open disks tangent to $\mathbb{R}$ at r.)

The quotient projection $\tilde{\jmath}\colon \mathfrak{H} \to \mathbb{C}^1$ is called the *modular j-invariant*. Strictly speaking, $\tilde{\jmath}$ is only well defined up to linear transformation $z \mapsto az + b$ in the target. We use Kodaira's normalization, with respect to which $\tilde{\jmath}(\rho) = 0$ and $\tilde{\jmath}(i) = 1$.

If $H \subset \Gamma$ is a subgroup, the quotient $H\backslash\mathfrak{H}$ is also a complex analytic curve. If the index $[\Gamma : H]$ is finite, the quotient admits a natural compactification, which can be obtained as $H\backslash\mathfrak{H}^*$. Compact Riemann surfaces obtained in this way are called *modular curves*. The genus of $H\backslash\mathfrak{H}^*$ is called the *genus* of H. This definition of genus is equivalent to the definition in terms of Sk H given above due to the following statement, the proof of which repeats almost literally that of Lemma 2.13.

Lemma 2.22. *For a subgroup* $H \subset \Gamma$, *the punctured minimal supporting surface* $\mathrm{Supp}^\circ(\mathrm{Sk}\, H)$ *is homeomorphic to* $H\backslash\mathfrak{H}$. *If* H *is a subgroup of finite index, there is a homeomorphism* $\mathrm{Supp}(\mathrm{Sk}\, H) \cong H\backslash\mathfrak{H}^*$. ◁

Remark 2.23. As yet another version of Proposition 2.20 (see Remark 2.21), one can identify the regions of Sk H with the H-orbits of the Γ-action on the union $\mathbb{Q} \cup \{\infty\}$. In fact, this latter action essentially coincides with the left matrix multiplication: if $g{\downarrow}[p,q]^{\mathrm{t}} = [p',q']^{\mathrm{t}}$ (for $[p,q] \in \mathcal{H}$), then $g{\downarrow}(p/q) = (p'/q')$ and hence $\bar{g}{\downarrow}(q/p) = q'/p'$ (for $p/q, q/p \in \mathbb{Q} \cup \{\infty\}$). The corresponding version of Proposition 2.20 has a geometric meaning: each cofinite orbit projects to a single point in $\mathrm{Supp}(\mathrm{Sk}\, H)$, and the image can be regarded as the center of the corresponding region. A cofinite orbit is formed by the fixed points of an H-conjugacy class of parabolic elements of H; these fixed points are called the *cusps* of H, and the widths of the corresponding regions are traditionally referred to as the *cusp widths* or *cusp amplitudes* of H.

Congruence subgroups

Given an integer $N \in \mathbb{N}$, the *principal congruence subgroup* of level N is the subgroup $\tilde{\Gamma}(N) := \{g \in \tilde{\Gamma} \mid g = \mathrm{id} \bmod N\}$. Principal congruence subgroups are normal and one has $\tilde{\Gamma}/\tilde{\Gamma}(N) = SL(2, \mathbb{Z}_N)$ (an easy exercise similar to Proposition 2.6). Hence

$$[\tilde{\Gamma} : \tilde{\Gamma}(N)] = N^3 \prod_{p|N} \left(1 - \frac{1}{p^2}\right),$$

the product running over all prime divisors p of N.

The image of $\tilde{\Gamma}(N)$ in Γ is denoted by $\Gamma(N)$. (A similar naming convention applies to all other subgroups that bear a proper name.) Since $\tilde{\Gamma}(N) \ni -\mathrm{id}$ if and only if $N = 2$, one has $[\Gamma : \Gamma(2)] = 6$ and $[\Gamma : \Gamma(N)] = \frac{1}{2}[\tilde{\Gamma} : \tilde{\Gamma}(N)]$ for $N \geqslant 3$. Since $\Gamma(N)$ is normal, its skeleton is regular and all its region have the same with, obviously equal to N, see Corollary 1.22. Thus, the numbers of vertices and regions of the skeleton are easily found and one arrives at the following expression for the genus:

$$\text{genus of } \Gamma(N) = 1 + \frac{N-6}{12N}[\Gamma : \Gamma(N)].$$

It follows that $\Gamma(N)$ is of genus zero if and only if $N \leqslant 5$. The skeletons of $\Gamma(N)$, $N = 3, 4, 5$, are the three Platonic solids with simplicial vertices, *i.e.*, the tetrahedron, hexahedron, and dodecahedron, respectively.

A subgroup H of Γ or $\tilde{\Gamma}$ is called a *congruence subgroup* if H contains, respectively, $\Gamma(N)$ or $\tilde{\Gamma}(N)$ for some $N \in \mathbb{N}$. If this is the case, the minimal integer N

with this property is called the *level* of H. Any congruence subgroup is of finite index, but the converse is not true: unlike $SL(n, \mathbb{Z})$ with $n \geqslant 3$ or many other similar groups, Γ does *not* have the congruence subgroup property. In fact, according to the next statement, *most* finite index subgroups of Γ are not congruence subgroups.

Theorem 2.24 (Thompson [157], Cox and Parry [38]). *For each integer $g \geqslant 0$, there are finitely many congruence subgroups of Γ of genus g.* ▷

The complete list of congruence subgroups of genus $g \leqslant 24$ is found in [40] or at

`http://www.math.tu-berlin.de/~pauli/congruence/congruence.html.`

There are 133 conjugacy classes of congruence subgroups of Γ of genus zero.

Remark 2.25. It is clear that all cusp widths of a congruence subgroup $H \subset \Gamma$ of level N must divide N. According to [104], the maximal cusp width of H *equals* N, and the minimal one is the greatest common divisor of all cusp widths. This observation gives a rather strong necessary condition for H to be a congruence subgroup. The condition is far from sufficient, see, *e.g.*, Remark 6.18 below.

2.2 The braid group $\mathbb{B}_3$

In this elementary introduction to braid groups, we mainly follow E. Artin's founding paper [14]. For further details and a comprehensive coverage of the current state of the subject, we refer to [92] and [23].

2.2.1 Artin's braid groups $\mathbb{B}_n$

Given an integer $n \in \mathbb{N}$, a *geometric braid* on n strands is an injective piecewise linear map $\beta \colon I \times \{1, \dots, n\} \to \mathbb{R}^3$ with the following properties:

- the x-coordinate of $\beta(x, k)$ equals x for all $x \in I$ and $k \in \{1, \dots, n\}$;
- for each $k \in \{1, \dots, n\}$, the y-coordinate of the image $\beta(0, k)$ equals k and the y-coordinate of the image $\beta(1, k)$ is in $\{1, \dots, n\}$.

An example of a braid on four strands is shown in Figure 2.3. Two geometric braids are said to be *equivalent* if they are isotopic in the class of geometric braids. An equivalence class of geometric braids is called a *braid*. The *product* $\beta_1 \cdot \beta_2$ of two geometric braids β_1, β_2 and the *inverse* β^{-1} of a geometric braid β are defined similar to the corresponding operations for paths:

$$\beta_1 \cdot \beta_2(x, k) = \begin{cases} \beta_1(2x, k) & \text{if } x \leqslant 1/2 \\ \beta_2(2x - 1, k) & \text{if } x \geqslant 1/2 \end{cases}, \qquad \beta^{-1}(x, k) = \beta(1 - x, k).$$

In other words, β^{-1} is the mirror image of β and $\beta_1 \cdot \beta_2$ is obtained by drawing the two braids next to each other and rescaling the picture back so that it fits over the

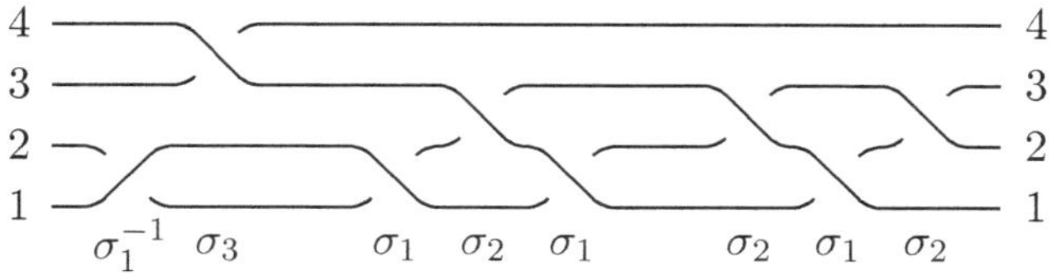

Figure 2.3. A geometric braid.

segment $x \in [0, 1]$. With respect to these operations, the set $\mathbb{B}_n$ of equivalence classes of geometric braids turns into a group, called the *braid group* on n strands.

It is quite clear geometrically that the braid group $\mathbb{B}_n$ is generated by the so-called *Artin generators* σ_i, $i = 1, \ldots, n-1$, see Figure 2.3: the braid σ_i twists the i-th and $(i+1)$-st strands through an angle of π in the counterclockwise direction while leaving the other strands intact. In this basis, the defining relations for $\mathbb{B}_n$ are

$$[\sigma_i, \sigma_j] = 1 \text{ if } |i - j| > 1, \qquad \sigma_i\sigma_{i+1}\sigma_i = \sigma_{i+1}\sigma_i\sigma_{i+1}. \tag{2.26}$$

The fact that the Artin generators are indeed subject to these relations is also clear geometrically, see Figure 2.3 again; the fact that relations (2.26) define the braid group $\mathbb{B}_n$ is proved in [26]. (The proof uses the algebraic representation of $\mathbb{B}_n$ as a subgroup of $\operatorname{Aut} \mathbb{F}_n$, see below.) A simple explanation for $\mathbb{B}_3$ is given at the beginning of Section 2.2.3.

It follows from (2.26) that the abelianization of $\mathbb{B}_n$ is $\mathbb{Z}$, the abelianization (degree) epimorphism $\operatorname{dg}\colon \mathbb{B}_n \twoheadrightarrow \mathbb{Z}$ sending each Artin generator σ_i to $1 \in \mathbb{Z}$. The image $\operatorname{dg} \beta \in \mathbb{Z}$ of a braid β is called the *degree* of β.

The center $\mathcal{Z}(\mathbb{B}_n)$ is the infinite cyclic group generated by Δ^2, where Δ is the *Garside element* $(\sigma_1 \ldots \sigma_{n-1})(\sigma_1 \ldots \sigma_{n-2}) \ldots (\sigma_1)$. Accidentally, Δ depends on the choice of an Artin basis, whereas its square $\Delta^2 = (\sigma_1 \ldots \sigma_{n-1})^n$ does not.

Sending each Artin generator $\sigma_i \in \mathbb{B}_n$ to the transposition $(i, i+1) \in \mathbb{S}_n$, one obtains the so-called *permutation representation* $\mathbb{B}_n \twoheadrightarrow \mathbb{S}_n$. (Clearly, relations (2.26) do hold in the symmetric group as well.) Its kernel is called the *pure braid group*. The geometric meaning of this representation is explained in the subsequent sections.

Braid group as a mapping class group

The braid group $\mathbb{B}_n$ can be identified with the mapping class group $\operatorname{Map}_n^+(\bar{D}^2, \partial\bar{D}^2)$, where the latter notation stands for the group of isotopy classes of orientation preserving auto-diffeomorphisms of $\bar{D}^2$ preserving n punctures in the interior of $\bar{D}^2$ as a set and the boundary $\partial\bar{D}^2$ pointwise. Geometrically, consider a geometric braid β and denote by $B \subset \mathbb{R}^3$ its image. Identify the disk $\bar{D}^2$ with a large disk D in the yz-plane encompassing the n punctures $(0, k, 0)$, $k \in \{1, \ldots, n\}$; one can assume that the projection of B to the yz-plane is in the interior of D. Then, the projection to the x-axis restricts to a locally trivial fibration $p: (I \times D, B) \to I$. Since the segment I is contractible, p is trivial and, choosing a trivialization, *i.e.*, a fiberwise diffeomorphism $(I \times D, B) \cong (I \times D, I \times \{1, \ldots, n\})$, and restricting it to the fibers over 0 and 1, which are canonically identified with $(D, \{1, \ldots, n\})$ one obtains an element of the mapping class group. Certainly, one should take care that the trivialization chosen is identical on the boundary $I \times \partial D$ and that such a trivialization is unique up to homotopy. All these statements are rather standard applications of the obstruction theory; crucial is the fact that the mapping class group $\operatorname{Map}_0^+(\bar{D}^2, \partial\bar{D}^2)$ is trivial. More details and a rigorous proof can be found in [14].

The *permutation representation* $\mathbb{B}_n \twoheadrightarrow \mathbb{S}_n$ is obtained by disregarding the action of a braid on the disk and keeping track of the punctures only.

The inclusion $\mathbb{B}_n \hookrightarrow \operatorname{Aut} \mathbb{F}_n$

The isomorphism $\mathbb{B}_n = \operatorname{Map}_n^+(\bar{D}^2, \partial\bar{D}^2)$ described in the previous paragraph gives rise to a homomorphism $\mathbb{B}_n \to \operatorname{Aut} \pi_1(\bar{D}_n^\circ, b)$, where $\bar{D}_n^\circ$ is the closed disk $\bar{D}^2$ punctured at n points and $b \in \partial\bar{D}^2$ is a fixed basepoint. Recall that $\pi_1(\bar{D}_n^\circ, b)$ is a free group on n generators. For a particular isomorphism $\pi_1(\bar{D}_n^\circ, b) \cong \mathbb{F}_n$ one usually employs a so-called *geometric basis* $\{\alpha_1, \ldots, \alpha_n\}$, which is formed by the classes of n lassoes, one about each puncture, disjoint except at the common point b and indexed so that the product $\alpha_1 \ldots \alpha_n$ is the class $[\partial\bar{D}^2]$ of the boundary, *cf.* Figure 2.4, left. (By a *lasso* about a point p in an oriented surface S we mean any loop of the form $\zeta[\partial U]\zeta^{-1}$, where ∂U is the boundary of a regular neighborhood U of p, with its boundary orientation, and ζ is a simple path in $S \smallsetminus U$ connecting the basepoint b and a point in ∂U.) In the particular basis shown in Figure 2.4, the action of $\mathbb{B}_n$ on $\mathbb{F}_n$ is as follows:

$$\sigma_i\colon\ \alpha_i \mapsto \alpha_i \alpha_{i+1} \alpha_i^{-1}, \quad \alpha_{i+1} \mapsto \alpha_i, \quad \alpha_j \mapsto \alpha_j \ \text{if } j \neq i, i+1, \tag{2.27}$$

see Figure 2.4, right. As usual, we assume that both $\mathbb{B}_n$ and $\operatorname{Aut} \mathbb{F}_n$ act on $\mathbb{F}_n$ *from the right*. A simple implementation of this action in `GAP` [76] is shown in Listing 2.5, the function "`Braid`". (The meaning of the function "`BraidRelations`" is explained in Section 5.1.2 below; it returns a list of relations of the form (5.29).)

One can see that a braid takes each generator to a conjugate of a generator. Hence, (2.27) induces a certain action on the set of n conjugacy classes of geometric generators. This induced action factors through the *permutation representation* $\mathbb{B}_n \twoheadrightarrow \mathbb{S}_n$.

The homomorphism $\mathbb{B}_n \to \operatorname{Aut} \mathbb{F}_n$ is injective. (In fact, what is proved in [26] is that relations (2.26) define a subgroup of $\operatorname{Aut} \mathbb{F}_n$.) In particular, this fact gives one an effective solution of the word problem for the braid group $\mathbb{B}_n$, see, *e.g.*, "`IsBraidID`" in Listing 2.5. The image of the inclusion $\mathbb{B}_n \hookrightarrow \operatorname{Aut} \mathbb{F}_n$ is given by the following remarkable theorem, which essentially states that an automorphism is a braid if and only if it takes a fixed geometric basis to another geometric basis.

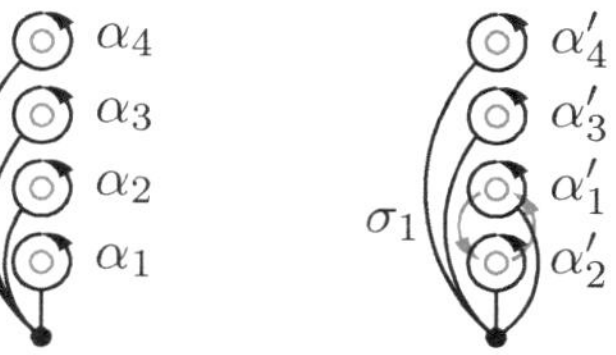

Figure 2.4. A geometric basis (left) and its image under σ_1 (right).

"Braid(b, G, x)" evaluates a braid "b" on element(s) "x" of group "G" (which is assumed to be a "FreeGroup")

```
Braid := function(b, G, x)  # There must be a way to recover "G" from "x"?
 local i, res, g;
 if IsList(x) then return List(x, i -> Braid(b, G, i)); fi;
 g := GeneratorsOfGroup(G); res := x;
 for i in LetterRepAssocWord(b) do
  if i > 0 then  # this is σ_i
   res := MappedWord(res, [g[i],g[1+i]],[g[i]*g[1+i]*g[i]^-1,g[i]]);
  else           # this is σ_{-i}^{-1}
   res := MappedWord(res, [g[-i],g[1-i]],[g[1-i],g[1-i]^-1*g[-i]*g[1-i]]);
  fi;
 od;
 return res;
end;
```

"BraidRelations(b, G[, list])" returns the list of $b(x)x^{-1}$ for all "x" in "list" (which must all be elements of group "G"). By default, "list = GeneratorsOfGroup(G)"

```
BraidRelations := function(arg)
 if Length(arg) = 2 then arg[3] := GeneratorsOfGroup(arg[2]); fi;
 return List(arg[3], x -> Braid(arg[1], arg[2], x)*x^-1);
end;
```

"IsBraidID(b, G)" decides if a braid "b" acts trivially on a group "G"

```
IsBraidID := function(b, G)
 local x;
 for x in GeneratorsOfGroup(G) do
  if x <> Braid(b, G, x) then return false; fi;
 od;
 return true;
end;
```

Listing 2.5. Braid manipulation in GAP ("braid.txt").

The $\mathbb{B}_n$ action on $\mathbb{F}_n$ has invariant element $\rho := \alpha_1 \dots \alpha_n$, and the central element Δ^2 acts *via* $\alpha \uparrow \Delta^2 = \rho\alpha\rho^{-1}$.

Theorem 2.28 (Artin [14]). *An automorphism φ of the free group $\mathbb{F}_n = \langle \alpha_1, \dots, \alpha_n \rangle$ is a braid (acting via (2.27)) if and only if each image $\alpha_i' := \varphi(\alpha_i)$ is a conjugate of one of the generators and one has $\alpha_1' \cdots \alpha_n' = \alpha_1 \cdots \alpha_n$.*

Proof (see [14]). The necessity is obvious: a geometric basis is taken by a braid to another geometric basis. For the sufficiency, consider the reduced representations $\alpha_i' = \gamma_i^{-1}\alpha_{j(i)}\gamma_i$, let r_i be the length of γ_i (as a word in the generators), and call the sum $\sum_i r_i$ the *length* of φ. If the length is zero, the product relation holds if and only if $\alpha_i' = \alpha_i$, *i.e.*, if $\varphi = \mathrm{id}$. Assume the assertion proved for all automorphisms of length less than that of φ and prove it for φ. In the relation

$$\gamma_1^{-1}\alpha_{j(1)}\gamma_1 \cdots \gamma_n^{-1}\alpha_{j(n)}\gamma_n = \alpha_1 \cdots \alpha_n,$$

the left hand side is longer than the right hand side; hence, for the relation to hold, cancellation must occur. If the cancellation does not affect any of the 'middle' terms $\alpha_{j(i)}$, the first word γ_1^{-1} cannot be canceled (whereas it must, as otherwise the left hand side would still be longer than the right hand side) and one must have $\gamma_1 = 1$; then, by induction, also $\gamma_2 = \cdots = \gamma_n = 1$, *i.e.*, φ is of length zero.

Thus, if φ is of positive length, at a certain place $\cdots \alpha_{j(i)}\gamma_i\gamma_{i+1}^{-1}\alpha_{j(i+1)} \cdots$ in the product $\alpha_1' \cdots \alpha_n'$, the cancellation must affect one or both middle terms $\alpha_{j(i)}, \alpha_{j(i+1)}$. Since both terms have positive exponents, they cannot be hit simultaneously. Then, assuming that $\alpha_{j(i)}$ (respectively, $\alpha_{j(i+1)}$) is canceled first, one can easily see that the reduced form of the pair $\alpha_i'\alpha_{i+1}'\alpha_i'^{-1}$, α_i (respectively, α_{i+1}', $\alpha_{i+1}'^{-1}\alpha_i'\alpha_{i+1}'$) is shorter than that of the original pair α_i', α_{i+1}'. By the induction hypothesis and (2.27), φ is the composition of two braids, hence a braid. □

Remark 2.29. It is worth mentioning that a slightly stronger statement has been proved: one does not need to assume *a priori* that φ is an automorphism, *i.e.*, that $\{\alpha_1', \dots, \alpha_n'\}$ is a basis for $\mathbb{F}_n$. As long as each α_i' is a conjugate of a generator and the product relation $\alpha_1' \cdots \alpha_n' = \alpha_1 \cdots \alpha_n$ holds, the sequence $\alpha_1', \dots, \alpha_n'$ is the image of $\alpha_1, \dots, \alpha_n$ under a certain braid; in particular, it is a free basis for $\mathbb{F}_n$.

Another slight generalization of Artin's theorem is Theorem 10.7 in Chapter 10.

2.2.2 The Burau representation

Identify the free group $\mathbb{F}_n$ with the fundamental group $\pi_1(\bar{D}_n^\circ, b)$ of a punctured disk and let $\{\alpha_1, \dots, \alpha_n\}$ be a geometric basis for $\mathbb{F}_n$. Define the *degree homomorphism* $\deg\colon \mathbb{F}_n \to \mathbb{Z}$ as the homomorphism sending each geometric generator to $1 \in \mathbb{Z}$. This homomorphism is independent of the choice of a geometric basis.

Definition 2.30. The *universal Alexander module* A_n is the abelianization of the kernel $\operatorname{Ker}[\deg\colon \mathbb{F}_n \to \mathbb{Z}]$.

The universal Alexander module is a module over the ring $\Lambda := \mathbb{Z}[t, t^{-1}]$ of integral Laurent polynomials; the module structure is given by $t\colon \mathrm{A}_n \to \mathrm{A}_n$, $[h] \mapsto [\alpha h \alpha^{-1}]$, where $h \in \operatorname{Ker}\deg$ and $\alpha \in \mathbb{F}_n$ is any element of degree 1. Using the Reidemeister–Schreier method, one can see that A_n is freely generated over Λ by the $(n-1)$ vectors $\mathbf{e}_i := [\alpha_{i+1}\alpha_i^{-1}]$, $i = 1, \ldots, n-1$. As an abelian group, A_n can be identified with the homology $H_1(\tilde{D}_n^\circ)$ of the infinite cyclic covering $\tilde{D}_n^\circ \to \bar{D}_n^\circ$ corresponding to deg; then the multiplication by t is the homomorphism induced by the deck translation.

The braid group action on $\mathbb{F}_n$ commutes with deg (and the trivial action on $\mathbb{Z}$) and hence induces a certain $\mathbb{B}_n$-action on A_n, which commutes with t.

Definition 2.31 (Burau [30]). The *(reduced) Burau representation* of the braid group $\mathbb{B}_n$ is the homomorphism $\mathbb{B}_n \to GL(\mathrm{A}_n) \cong GL(n-1, \Lambda)$ corresponding to the induced $\mathbb{B}_n$-action on A_n.

If the bases $\{\sigma_i\}$, $\{\alpha_j\}$ are chosen so that the action is given by (2.27), then

$$\sigma_i\colon\ \mathbf{e}_{i-1} \mapsto \mathbf{e}_{i-1} + t\mathbf{e}_i, \quad \mathbf{e}_i \mapsto -t\mathbf{e}_i, \quad \mathbf{e}_{i+1} \mapsto \mathbf{e}_i + \mathbf{e}_{i+1}, \tag{2.32}$$

where we let $\mathbf{e}_j = 0$ for $j \leqslant 0$ or $j \geqslant n$. (It is understood that $\sigma_i\colon \mathbf{e}_j \mapsto \mathbf{e}_j$ whenever $|i-j| > 1$.) Representing this action by matrices, one has $\det\sigma_i = -t$, $i = 1, \ldots, n-1$; hence $\det\beta = (-t)^{\operatorname{dg}\beta}$ for any braid $\beta \in \mathbb{B}_n$.

The Burau representation is known to be faithful for $n \leqslant 3$, see Section 2.2.3 below, and unfaithful for $n \geqslant 5$, see [21, 112]. The case $n = 4$ remains open.

Historically, the Burau representation (more precisely, its specializations at roots of unity $t \in \mathbb{C}$) is the first linear representation of the braid group. Our particular interest in this representation is due to its relation to the Alexander invariants of algebraic curves (as well as knots and links), see Section 6.2 below. At present, a great deal of other representations of $\mathbb{B}_n$ is known, see [23, 92] for references; in particular, it is known that all braid groups are linear.

The Burau group

We will also consider the extended group $\mathbb{B}_n \cdot \operatorname{Inn}\mathbb{F}_n \subset \operatorname{Aut}\mathbb{F}_n$, where $\operatorname{Inn}\mathbb{F}_n \cong \mathbb{F}_n$ is the group of inner automorphisms of $\mathbb{F}_n$, $g\colon \alpha \mapsto g^{-1}\alpha g$ for $g, \alpha \in \mathbb{F}_n$. (Here, the dot stands for the product of two subgroups of $\operatorname{Aut}\mathbb{F}_n$. Note that, since $\operatorname{Inn}\mathbb{F}_n$ is normal, the product is indeed a subgroup.)

Lemma 2.33. *The intersection $\mathbb{B}_n \cap \operatorname{Inn}\mathbb{F}_n$ is the center $\mathcal{Z}(\mathbb{B}_n)$, i.e., the cyclic group generated by the element $\Delta^2 = (\sigma_1 \cdots \sigma_{n-1})^n \in \mathbb{B}_n$, which conjugates each generator by $(\alpha_1 \cdots \alpha_n)^{-1}$. In particular, one has $\mathbb{B}_n \cdot \operatorname{Inn}\mathbb{F}_n / \operatorname{Inn}\mathbb{F}_n = \mathbb{B}_n/\Delta^2$.*

Proof. The statement follows from Theorem 2.28: any braid stabilizes the product $\rho := \alpha_1 \cdots \alpha_n$, and the centralizer of this product in $\mathbb{F}_n$ is the cyclic subgroup generated by ρ. □

Due to Lemma 2.33, one can extend the degree homomorphism dg to

$$\mathrm{dg}\colon \mathbb{B}_n \cdot \operatorname{Inn}\mathbb{F}_n \to \mathbb{Z}, \quad \beta \cdot \alpha \mapsto \mathrm{dg}\,\beta - (n-1)\deg\alpha, \quad \beta \in \mathbb{B}_n,\ \alpha \in \mathbb{F}_n.$$

Sending $\alpha \in \operatorname{Inn}\mathbb{F}_n$ to $t^{-\deg g}$, one can also extend the Burau representation to a homomorphism $\mathbb{B}_n \cdot \operatorname{Inn}\mathbb{F}_n \to GL(n-1, \Lambda)$. The image of this homomorphism will be called the *Burau group* and denoted by Bu_n. If n is odd, one still has $\det\beta = (-1)^{\mathrm{dg}\,\beta}$ for any $\beta \in \mathrm{Bu}_n$.

Definition 2.34. Given a ring homomorphism $\Lambda \to \mathcal{R}$, $t \mapsto r$, the induced homomorphisms $\mathbb{B}_n \to GL(n-1, \mathcal{R})$ and $\mathrm{Bu}_n \to GL(n-1, \mathcal{R})$ are called the *specializations* at $t = r \in R$. The images of an element $\beta \in \mathrm{Bu}_n$ or a subgroup $G \subset \mathrm{Bu}_n$ are denoted by $\beta(r)$ and $G(r)$, respectively.

The specialization of the Burau representation at $t = 1 \in \mathbb{Z}$ factors through the permutation representation $\mathbb{B}_n \twoheadrightarrow \mathbb{S}_n$. The resulting representation of the symmetric group $\mathbb{S}_n$ is the orthogonal complement of the invariant vector $\mathbf{v}_1 + \cdots + \mathbf{v}_n$ in the lattice $\mathbf{V}_n$ on which $\mathbb{S}_n$ acts *via* the permutations of the n basis vectors.

Squier's sesquilinear form

Since $\tilde{D}_n^\circ$ is an oriented surface, the identification $\mathrm{A}_n = H_1(\tilde{D}_n^\circ)$ gives rise to the *intersection index pairing* $\bigwedge_{\mathbb{Z}}^2 \mathrm{A}_n \to \mathbb{Z}$, $a \wedge b \mapsto a \circ b$, which is a skew-symmetric $\mathbb{Z}$-bilinear form with the property $ta \circ tb = a \circ b$ and $(a \uparrow \beta) \circ (b \uparrow \beta) = a \circ b$ for any pair $a, b \in \mathrm{A}_n$ and any braid $\beta \in \mathbb{B}_n$. (The latter property is due to the fact that both t and β are induced by continuous maps.)

Definition 2.35 (Squier [156]). *Squier's form* is the pairing $\mathrm{A}_n \otimes_{\mathbb{Z}} \mathrm{A}_n \to \Lambda$ given by $a \otimes b \mapsto (a, b) := \sum_{i \in \mathbb{Z}} t^{-i}(t^i a \circ b)$.

Note that, since homology classes are represented by compact cycles, all but finitely many terms in Definition 2.35 vanish and the sum is essentially finite. Note also that Definition 2.35 differs slightly from the definition actually given in [156], where the coefficient ring Λ is extended to $\mathbb{Z}[s, s^{-1}]$, $s^2 = t$, and a Hermitian form is constructed over this extended ring.

A simple computation shows that, in the basis $\{\mathbf{e}_i\}$, $i = 1, \ldots, n-1$, Squier's form is given by the following expressions:

$$(\mathbf{e}_i, \mathbf{e}_i) = t^{-1} - t, \quad (\mathbf{e}_i, \mathbf{e}_{i+1}) = t - 1, \quad (\mathbf{e}_i, \mathbf{e}_j) = 0 \text{ if } |i - j| > 1.$$

Next two statements follow immediately from the definitions and the fact that the action on A_n of any braid β is induced by an auto-diffeomorphism of $\bar{D}_n^\circ$ and hence commutes with both t and $\circ$. For Proposition 2.36, define the *conjugation* $\Lambda \to \Lambda$ as the involutive automorphism $t \mapsto t^{-1}$; as above, we use the notation $\lambda \mapsto \lambda(t^{-1})$.

Proposition 2.36. *Squier's form is skew-Hermitian, i.e., for all $a, b \in \mathrm{A}_n$ and $\lambda \in \Lambda$ one has $(\lambda a, b) = \lambda(a, b) = (a, \lambda(t^{-1})b)$ and $(b, a) = -(a, b)(t^{-1})$.* $\triangleleft$

Proposition 2.37. *Each element $\beta \in \mathrm{Bu}_n$ of the Burau group is orthogonal with respect to Squier's form, i.e., one has $(a \uparrow \beta, b \uparrow \beta) = (a, b)$ for $a, b \in \mathrm{A}_n$.* $\triangleleft$

2.2.3 The group $\mathbb{B}_3$

The braid group $\mathbb{B}_3$ on three strands is closely related to the modular group Γ. Due to (2.26), one has

$$\mathbb{B}_3 = \langle \sigma_1, \sigma_2 \mid \sigma_1\sigma_2\sigma_1 = \sigma_2\sigma_1\sigma_2 \rangle = \langle u, v \mid u^3 = v^2 \rangle, \tag{2.38}$$

where $u := \sigma_1\sigma_2$ and $v := \sigma_1\sigma_2\sigma_1$; hence, due to Theorem 2.2, the assignment $u \mapsto \mathbb{X}$, $v \mapsto \mathbb{Y}^{-1}$ defines epimorphisms $\mathbb{B}_3 \twoheadrightarrow \tilde{\Gamma} \twoheadrightarrow \Gamma$. Artin generators map to the Dehn twists

$$\sigma_1 \mapsto (\mathbb{Y}\mathbb{X})^{-1} = R, \qquad \sigma_2 \mapsto (\mathbb{X}\mathbb{Y})^{-1} = L^{-1}. \tag{2.39}$$

It is easy to see that $\mathbb{B}_3 \twoheadrightarrow \Gamma$ is the universal central extension of Γ.

The Burau representation $\mathbb{B}_3 \to GL(\mathrm{A}_2) = GL(2, \Lambda)$, see (2.32), results in the matrices

$$\sigma_1 = \begin{bmatrix} -t & 0 \\ 1 & 1 \end{bmatrix}, \qquad \sigma_2 = \begin{bmatrix} 1 & t \\ 0 & -t \end{bmatrix} \tag{2.40}$$

(acting by the *right* multiplication), which we identify with σ_1, σ_2 themselves. For future references, we compute a few products

$$\sigma_1\sigma_2 = \begin{bmatrix} -t & -t^2 \\ 1 & 0 \end{bmatrix}, \quad \sigma_1\sigma_2\sigma_1 = \begin{bmatrix} 0 & -t^2 \\ -t & 0 \end{bmatrix}, \quad (\sigma_1\sigma_2)^3 = \begin{bmatrix} t^3 & 0 \\ 0 & t^3 \end{bmatrix} \tag{2.41}$$

and powers

$$\sigma_1^n = \begin{bmatrix} (-t)^n & 0 \\ \tilde{\varphi}_n(-t) & 1 \end{bmatrix}, \qquad \sigma_2^n = \begin{bmatrix} 1 & t\tilde{\varphi}_n(-t) \\ 0 & (-t)^n \end{bmatrix}, \tag{2.42}$$

where $\tilde{\varphi}_n(t) := (t^n - 1)/(t - 1)$, $n \in \mathbb{Z}$.

Comparing (2.41) to (2.1), one concludes that the homomorphism $\mathbb{B}_3 \twoheadrightarrow \tilde{\Gamma}$ above can be regarded as the specialization of the Burau representation at $t = -1$. Using this observation, one can extend the homomorphism to the Burau group $\mathrm{Bu}_3 \twoheadrightarrow \tilde{\Gamma}$. The further projection $\mathrm{Bu}_3 \twoheadrightarrow \Gamma$ will be denoted by pr_Γ. (We will also use the self-explanatory notation $\beta \bmod \Delta^2$ for elements and H/Δ^2 for subgroups, even if H does not contain Δ^2.) Besides, since $(\sigma_1\sigma_2)^3 \in GL(2, \Lambda)$ is obviously of infinite order, the Burau representation is exact (and that is why we identify $\mathbb{B}_3$ with its image) and relations (2.38) define $\mathbb{B}_3$. Furthermore, in view of Corollary 1.55, the center $\mathcal{Z}(\mathbb{B}_3)$ is indeed generated by $\Delta^2 = (\sigma_1\sigma_2)^3$.

Using the central extension $\mathbb{B}_3 \twoheadrightarrow \Gamma$, one obtains the following simple solution to the conjugacy problem in $\mathbb{B}_3$.

Proposition 2.43. *Two braids $\beta_1, \beta_2 \in \mathbb{B}_3$ are conjugate if and only if* $\deg \beta_1 = \deg \beta_2$ *and the images* $\beta_1 \bmod \Delta^2$, $\beta_2 \bmod \Delta^2$ *are conjugate in Γ, see Proposition* 2.15. $\triangleleft$

To my knowledge, the conjugacy problem for the other braid groups $\mathbb{B}_n$, $n \geqslant 4$, is still an open question.

Garside normal form (see [133])

Denote by $\mathbb{B}_n^+ \subset \mathbb{B}_n$ the submonoid generated by σ_i, $i = 1, \ldots, n-1$; the elements of $\mathbb{B}_n^+$ are called *positive braids*. A positive braid β is said to be *left* (*right*) *divisible* by a positive braid α if $\beta = \alpha\alpha'$ (respectively, $\beta = \alpha'\alpha$) for some positive braid α'.

Theorem 2.44 (Garside [77]). *The defining relations for $\mathbb{B}_n^+$ are $\sigma_i\sigma_j = \sigma_j\sigma_i$ for $|i-j| > 1$ and $\sigma_i\sigma_{i+1}\sigma_i = \sigma_{i+1}\sigma_i\sigma_{i+1}$, cf.* (2.26). $\triangleright$

From now on, we confine ourselves to the group $\mathbb{B}_3$ and monoid $\mathbb{B}_3^+$.

The conjugation by the Garside element $\Delta = \sigma_1\sigma_2\sigma_1 = \sigma_2\sigma_1\sigma_2 \in \mathbb{B}_3$ acts on the basis *via* $\sigma_1 \leftrightarrow \sigma_2$; in particular, it takes $\mathbb{B}_3^+$ to $\mathbb{B}_3^+$. It follows that the left and right divisibility by Δ are equivalent; furthermore, an element $\beta \in \mathbb{B}_3^+$ is divisible by Δ if and only if $\beta = \alpha'\Delta\alpha''$ for some $\alpha', \alpha'' \in \mathbb{B}_3^+$.

Corollary 2.45. *An element $\beta \in \mathbb{B}_3^+$ not divisible by Δ is represented by a* unique *positive word in $\{\sigma_1, \sigma_2\}$.* $\triangleleft$

Definition 2.46. A *Garside decomposition* of a braid $\beta \in \mathbb{B}_3$ is a representation of β in the form $\beta = \beta^+\Delta^m$, where $\beta_+ \in \mathbb{B}_3^+$ and $m \in \mathbb{Z}$. If β_+ is not divisible by Δ, the Garside decomposition is called the *Garside normal form* of β.

Theorem 2.47 (Orevkov [133]). *Any braid $\beta \in \mathbb{B}_3$ has a unique Garside normal form $\beta_+\Delta^m$, with $m \in \mathbb{Z}$ and β_+ represented by a unique positive word in $\{\sigma_1, \sigma_2\}$.*

Proof. Represent β by a word in $\{\sigma_1^{\pm 1}, \sigma_2^{\pm 1}\}$, replace each occurrence of σ_1^{-1} or σ_2^{-1} with $\sigma_2\sigma_1\Delta^{-1}$ or, respectively, $\sigma_1\sigma_2\Delta^{-1}$, and push all factors Δ^{-1} to the right to obtain a Garside decomposition $\beta = \beta'_+\Delta^{m'}$. Then, replace each occurrence of $\sigma_1\sigma_2\sigma_1$ or $\sigma_2\sigma_1\sigma_2$ in β'_+ with Δ and push this factor to the right; we obtain a Garside normal form. If $\beta'_+\Delta^{m'} = \beta''_+\Delta^{m''}$ are two Garside normal forms of β, then $m' = m''$ (since neither β'_+ nor β''_+ is divisible by Δ) and hence $\beta'_+ = \beta''_+$ in $\mathbb{B}_3^+$; due to Corollary 2.45, this element is represented by a unique positive word. $\square$

Subgroups *via* skeletons

The epimorphism $\operatorname{pr}_\Gamma\colon \mathbb{B}_3 \twoheadrightarrow \Gamma$ commutes with dg: $\operatorname{dg}(\beta \bmod \Delta^2) = \operatorname{dg}\beta \bmod 6$ for any $\beta \in \mathbb{B}_3$. The homomorphism $\operatorname{pr}_\Gamma\colon \operatorname{Bu}_3 \twoheadrightarrow \Gamma$ commutes with dg mod 2 only, as the element $t\operatorname{id} \in \operatorname{Bu}_3$ maps to $1 \in \Gamma$. It follows that the groups $\mathbb{B}_3$ and Bu_3 can be represented as fibered products, *cf.* (2.17),

$$\mathbb{B}_3 = \Gamma \times_{\mathbb{Z}_6} \mathbb{Z}, \qquad \operatorname{Bu}_3 = \Gamma \times_{\mathbb{Z}_2} \mathbb{Z}. \tag{2.48}$$

The kernel $\operatorname{Ker}\operatorname{pr}_\Gamma$ is the central cyclic subgroup generated by $t\operatorname{id}$.

Definition 2.49. The *skeleton* $\operatorname{Sk} H$ of a subgroup $H \subset \operatorname{Bu}_3$ is the skeleton of the projection $(H/\Delta^2) \subset \Gamma$. The *genus* of H is the genus of $\operatorname{Sk} H$, whenever defined.

Skeletons of conjugate subgroups are isomorphic. To recover a conjugacy class of subgroups from its skeleton we need a few extra data.

Definition 2.50. The *depth* $\operatorname{dp} G$ of a subgroup $H \subset \operatorname{Bu}_3$ is the degree of the positive generator of the intersection $H \cap \operatorname{Ker}\operatorname{pr}_\Gamma$, or zero if this intersection is trivial. One has $\operatorname{dp} H = 0 \bmod 2$ and $\operatorname{dp} H = 0 \bmod 6$ if $H \subset \mathbb{B}_3$.

Consider a subgroup $H \subset \operatorname{Bu}_3$, let $2d = \operatorname{dp} H$, and let H_d be the image of H under the projection $\operatorname{pr}_d := \operatorname{pr}_\Gamma \times (\operatorname{dg} \bmod 2d)\colon \operatorname{Bu}_3 \to \Gamma \times \mathbb{Z}_{2d}$. (We let $\mathbb{Z}_0 = \mathbb{Z}$.) Then $H = \operatorname{pr}_d^{-1} H_d$ and H_d projects isomorphically onto H/Δ^2; in other words, H_d is the graph of a certain homomorphism $\varphi\colon H/\Delta^2 \to \mathbb{Z}_{2d}$. This construction is summarized by the following definition and proposition.

Definition 2.51. The homomorphism $\varphi\colon H/\Delta^2 \to \mathbb{Z}_{2d}$, $2d = \operatorname{dp} H$, as above is called the *slope* of a subgroup $H \subset \operatorname{Bu}_3$.

Proposition 2.52. *The map* $H \mapsto (H/\Delta^2, \text{slope})$ *establishes a bijection between the set of subgroups* $H \subset \operatorname{Bu}_3$ *and the set of pairs* $(\bar{H}, \varphi)$*, where* $\bar{H} \subset \Gamma$ *is a subgroup and* φ *is a homomorphism* $\bar{H} \to \mathbb{Z}_{2d}$ *with the property* $\varphi = \operatorname{dg} \bmod 2$. *One has* $H \subset \mathbb{B}_3$ *if and only if* $d = 0 \bmod 3$ *and* $\varphi = \operatorname{dg} \bmod 6$. $\lhd$

Each subgroup $\bar{H} \subset \Gamma$ admits three 'canonical' slopes, *viz.* the restrictions to $\bar{H}$ of the homomorphisms $\pm\operatorname{dg}\colon \Gamma \to \mathbb{Z}_6$ and $\operatorname{dg} \bmod 2\colon \Gamma \to \mathbb{Z}_2$. We denote the corresponding subgroups of Bu_3 by $(\bar{H})^{\pm}$ and $(\bar{H})^{\mathrm{bu}}$, respectively. The subgroups $(\bar{H})^{\mathrm{bu}} = \operatorname{pr}_\Gamma^{-1} \bar{H}$ and $(\bar{H})^{+} = (\bar{H})^{\mathrm{bu}} \cap \mathbb{B}_3$ are merely the full preimages of $\bar{H}$ under the projections $\operatorname{pr}_\Gamma\colon \operatorname{Bu}_3 \twoheadrightarrow \Gamma$ and $\operatorname{pr}_\Gamma\colon \mathbb{B}_3 \twoheadrightarrow \Gamma$, respectively.

If $H \subset \operatorname{Bu}_3$ is a subgroup of genus zero, its slope can be described more geometrically in terms of the so-called *type specification*. (The meaning of this term is explained in Section 5.3.1.) In this case, the projection $\bar{H} := H/\Delta^2$ is generated by its parabolic and elliptic elements, see Corollary 1.44. On the other hand, since the range of the slope $\varphi\colon \bar{H} \to \mathbb{Z}_{2d}$ is an abelian group, the values of φ on conjugate elements are equal. Thus, φ can be regarded as a function defined on the regions of $\operatorname{Sk} H$ (the conjugacy classes of minimal parabolic elements, Lemma 1.23) and its monovalent vertices (the conjugacy classes of elliptic elements).

Definition 2.53. The *type specification* of a subgroup $H \subset \mathrm{Bu}_3$ is the $\mathbb{Z}_{\mathrm{dp}\, H}$-valued function tp on the set of all finite regions and monovalent vertices of the skeleton $\operatorname{Sk} H$ defined as follows: each region or monovalent vertex is sent to the degree of (any) lift to H of the corresponding generator (respectively, parabolic or elliptic) of H/Δ^2 or, equivalently, to the value of the slope φ of H on the corresponding generator.

Proposition 2.54. *Let $H \subset \mathrm{Bu}_3$ be a subgroup of genus zero, and let $d = 6$ if $H \subset \mathbb{B}_3$ and $d = 2$ otherwise. Then one has:*

1. $\mathrm{dp}\, H = 0 \bmod d$;
2. $\mathrm{tp}(R) = \mathrm{wd}\, R \bmod d$ *for any region R of* $\operatorname{Sk} H$;
3. $\mathrm{tp}(\bullet) = 2 \bmod d$ *and* $3\,\mathrm{tp}(\bullet) = 0 \bmod \mathrm{dp}\, H$ *for any $\bullet$-vertex*;
4. $\mathrm{tp}(\circ) = 3 \bmod d$ *and* $2\,\mathrm{tp}(\circ) = 0 \bmod \mathrm{dp}\, H$ *for any $\circ$-vertex*;
5. *the sum of all values of* tp *equals* $0 \bmod \mathrm{dp}\, H$.

Any pair $(\mathrm{dp}, \mathrm{tp})$ *satisfying these restrictions defines a unique slope, hence a unique conjugacy class of subgroups $H \subset \mathrm{Bu}_3$. Such a pair results in a class of subgroups $H \subset \mathbb{B}_3$ if and only if it satisfies the restrictions in Items* 1–4 *with $d = 6$.*

Proof. Since $\operatorname{Supp}(\operatorname{Sk} H) \cong S^2$, the group $H/\Delta^2 = \pi_1^{\mathrm{orb}}(\operatorname{Sk} H)$ has a presentation

$$H/\Delta^2 = \left\langle \beta_R, \gamma_v \mid (\gamma_v^{\bullet})^3 = (\gamma_v^{\circ})^2 = 1,\ \prod \beta_R \prod \gamma_v = 1 \right\rangle,$$

where the indices R and v run, respectively, over all regions and monovalent vertices of $\operatorname{Sk} H$ and the superscript in the relations indicates the type of a vertex. The product in the last relation is in a certain order depending on the particular choice of a basis; in fact, $\{\beta_R, \gamma_v\}$ is merely a geometric basis for the fundamental group of a punctured sphere. Thus, Item 5 and the congruences $3\,\mathrm{tp}(\bullet) = 0$, $2\,\mathrm{tp}(\circ) = 0$ follow from the relations in the above presentation. All other congruences follows from the properties of slopes, see Proposition 2.52.

A type specification assigns a value to each generator of H/Δ^2 and thus defines a homomorphism $H/\Delta^2 \to \mathbb{Z}_{\mathrm{dp}\, H}$. Hence, the converse statement on the existence of a subgroup follows from Proposition 2.52. □

Given an integer m, a type specification is said to be *trivial modulo m* if it satisfies the congruences in Items 1–4 in Proposition 2.54 with $d = m$. Thus, Proposition 2.54 states that any type specification is trivial modulo 2 and that a subgroup $H \subset \mathrm{Bu}_3$ is in $\mathbb{B}_3$ if and only if its type specification is trivial modulo 6.

Convention 2.55. In the drawings, we indicate the type specification (inside a region or next to a vertex) only when it is not trivial modulo 0.

Remark 2.56. If H is not a subgroup of genus zero, its type specification satisfies all condition in Proposition 2.54 (except Item 5 if $\operatorname{Sk} H$ is infinite). In this case, the slope (and hence the subgroup itself) can no longer be uniquely recovered from the type specification, as the projection H/Δ^2 unavoidably has hyperbolic generators.

Subgroups of Bu_3 *vs.* $\mathbb{B}_3$

Consider a subgroup $H \subset \mathrm{Bu}_3$ and assume that $H \not\subset \mathbb{B}_3$. Then the intersection $H \cap \mathbb{B}_3$ is a normal subgroup of H of index three (as so is the subgroup $\mathbb{B}_3 \subset \mathrm{Bu}_3$) and the depth $\operatorname{dp}(H \cap \mathbb{B}_3)$ equals either $\operatorname{dp} H$ or $3 \operatorname{dp} H$.

Lemma 2.57. *Let $H \subset \mathrm{Bu}_3$ be a subgroup, $H \not\subset \mathbb{B}_3$, and let $H' = H \cap \mathbb{B}_3$.*

1. *If $\operatorname{dp} H' = 3 \operatorname{dp} H$, then $\operatorname{dp} H \neq 0 \bmod 3$, one has $\operatorname{Sk} H = \operatorname{Sk} H'$, and the slope/type specification of H is the reduction of that of H' modulo $\operatorname{dp} H$.*
2. *If $\operatorname{dp} H' = \operatorname{dp} H$, then one has $\operatorname{Sk} H = (\operatorname{Sk} H')/T$ for a certain order three automorphism $T \in \operatorname{Aut}(\operatorname{Sk} H')$.*

Proof. If $\operatorname{dp} H' = 3 \operatorname{dp} H$, then H contains an element of the form $t^s \operatorname{id}$, $s \neq 0 \bmod 3$, and the projections of H and H' to Γ coincide. Otherwise, H contains an element $t^s g$ for some $s \neq 0 \bmod 3$ and $g \in \mathbb{B}_3 \smallsetminus H'$, $g^3 \in H'$. In this case, the projection $g \bmod \Delta^2$ normalizes H'/Δ^2 and thus defines an automorphism $T \in \operatorname{Aut}(\operatorname{Sk} H')$, see Corollary 1.21. On the other hand, $g \bmod \Delta^2$ generates H/Δ^2 over H'/Δ^2; hence one has $\operatorname{Sk} H = (\operatorname{Sk} H')/T$. □

Extended skeletons

The map $\mathbf{x} \mapsto u$, $\mathbf{y} \mapsto v^{-1}$, see (2.38), represents $\mathbb{B}_3$ as a quotient of $\mathbb{G}$. Hence, any subgroup $H \subset \mathbb{B}_3$ defines its own skeleton $H \backslash \mathbb{B}_3$, which typically *differs* from $\operatorname{Sk} H$. Occasionally, we will make use of this skeleton, see, *e.g.*, the automated computation of the fundamental group in Listing C.3.

Definition 2.58. Given a subgroup $H \subset \mathbb{B}_3$, the skeleton $H \backslash \mathbb{B}_3$ is called the *extended skeleton* of H and is denoted by $\operatorname{Sk}^{\sim} H$.

The principal advantage of $\operatorname{Sk}^{\sim} H$ is the fact that one does not need to keep track of the type specification. The following statement is similar to Lemma 2.57; the automorphism T is induced by the multiplication by the central element $t^3 \operatorname{id}$ of the group $\mathbb{B}_3$.

Lemma 2.59. *For a subgroup $H \subset \mathbb{B}_3$, there is an automorphism $T \in \operatorname{Aut}(\operatorname{Sk}^{\sim} H)$ of order $(\operatorname{dp} H)/6$ and such that $\operatorname{Sk} H = (\operatorname{Sk}^{\sim} H)/T$.* $\triangleleft$

Warning 2.60. It is worth emphasizing that the genus of $H \subset \mathbb{B}_3$ is defined as the genus of $\operatorname{Sk} H$, not that of $\operatorname{Sk}^{\sim} H$; typically, the latter is much larger.

Remark 2.61. A subgroup $H \subset \tilde{\Gamma}$ can be identified with a subgroup of $\mathbb{B}_3$ of depth 6 (if $-\operatorname{id} \in H$) or 12 (if $-\operatorname{id} \notin H$). Hence, everything said in this section applies to such subgroups literally: one can speak about their slopes, type specifications, extended skeletons, *etc*.

Chapter 3

Trigonal curves and elliptic surfaces

This chapter contains most algebro-geometric prerequisites needed in the sequel. For algebraic geometry in general, we refer to the standard textbooks [81, 86] and to the excellent monograph [88], which, in my opinion, contains almost everything that a topologist may need to know about algebraic geometry. All varieties are analytic (not necessarily projective) over $\mathbb{C}$, and we freely use topological methods.

3.1 Trigonal curves

Traditionally, a *trigonal curve* is a compact algebraic curve equipped with a (linear) pencil of degree three. Any such curve C can be embedded into a geometrically ruled rational surface $\Sigma \to \mathbb{P}^1$, so that the degree three pencil is cut on C by the ruling of Σ. With some applications in mind, we consider a slight generalization of this notion and deal with curves equipped with an arbitrary, possibly irrational, pencil of degree three. On the other hand, instead of abstract trigonal curves, we consider only curves embedded to geometrically ruled surfaces $\Sigma \to B$.

3.1.1 Basic definitions and properties

Throughout this chapter, we consider a curve B and a geometrically ruled surface $p\colon \Sigma \to B$ with a distinguished section E, called the *exceptional section*. (We identify a section and its image, regarded as a divisor.) In most cases we assume that B is compact and that $d := -E^2 \geqslant 0$, see Remark 3.7. Under these assumptions, if $d > 0$, the exceptional section is unique. In fact, any other section of Σ is homologous to $[E + kF]$, $k \geqslant d$, where F is a fiber, and hence has self-intersection at least d.

Fibers of Σ are those of the projection p, which is usually not mentioned explicitly. Fibers are often referred to by the self-explanatory colloquialism 'vertical'; thus, for example, an expression like 'the curve has a vertical tangent' means that the curve is tangent to a fiber.

To avoid excessive notation in the statements, we usually identify fibers of Σ and their projections to B. When it is important to refer to the fiber over $b \in B$ as a curve, we use the notation F_b. The *punctured fiber* is the complement $F_b^\circ := F_b \smallsetminus E$. Since F_b° is a complex affine line, one can speak about angles, length ratios, circles, convexity, *etc.* For a subset $S \subset \Sigma \smallsetminus E$, we denote by $\operatorname{conv}_b S$ the convex hull in F_b° of the intersection $S \cap F_b^\circ$ and define the *convex hull* $\operatorname{conv} S$ to be the union of $\operatorname{conv}_b S$ over all points $b \in B$.

Definition 3.1. A *trigonal curve* is a reduced curve $C \subset \Sigma$ not containing E or a fiber of Σ as a component and such that the restriction $p\colon C \to B$ is a map of degree three. A trigonal curve is said to be *proper*, or *genuine*, if it is disjoint from the exceptional section E.

A *singular fiber* of a trigonal curve $C \subset \Sigma$ is a fiber F of Σ intersecting $C + E$ geometrically at fewer than four points. Thus, F is singular if either C passes through $E \cap F$, or C is tangent to F, or C has a singular point in F, the three cases being not exclusive. A singular fiber F is called *proper* if C does not pass through $E \cap F$; thus, C is proper if and only if so are all its singular fibers. A more detailed description of the topological types of singular fibers is given in Section 3.1.2 below.

Singular fibers of C form a discrete subset in the base B. With C fixed, we denote by $B^\sharp$ the set of all nonsingular fibers; it is a punctured curve.

Definition 3.2. An *elementary deformation* of a trigonal curve $C \subset \Sigma \to B$ is a deformation, in the sense of Kodaira–Spencer, of the quintuple $(p\colon \Sigma \to B, E, C)$.

We emphasize that neither Σ nor B are assumed fixed during the deformation: the analytic structure may change. As usual, (*equisingular*) *deformation equivalence* of trigonal curves is the minimal equivalence relation generated by the (equisingular) elementary deformations and fiberwise isomorphisms.

Remark 3.3. Since in general normal forms of singular fibers have continuous moduli, the term 'equisingular' in the definition is not quite appropriate, as it would be too restrictive to demand that the singular fibers should remain diffeomorphic during the deformation. One way to overcome this difficulty is to switch to the topological or *PL*-category when comparing singular fibers. Another, somewhat safer, approach that we adopt in this book is to require that the singular fibers do not split or merge. Thus, we call an elementary deformation *equisingular* if, at each time t, each singular fiber F of C, regarded as a point of the topological surface B, has a neighborhood in which F is the only singular fiber at all times close to t.

An elementary deformation over the disk D^2 is called a *degeneration* or *perturbation* if its restriction to $D^2 \smallsetminus \{0\}$ is equisingular. (We say that C_t, $t \neq 0$, *degenerates* to C_0 and C_0 *perturbs* to C_t.) A degeneration is *nontrivial* if it is not equisingular.

Nagata transformations and induced curves

Let $p\colon \Sigma \to B$ be a ruled surface. A positive (negative) *Nagata transformation* (also called *elementary transformation*) is the fiberwise birational transformation $\Sigma \dashrightarrow \Sigma'$ consisting in blowing up a point P on (respectively, not on) the exceptional section and blowing down the proper transform of the fiber through P. The result of a Nagata transformation is a ruled surface Σ' over the same base B and with an exceptional section E' (the proper transform of E) of self-intersection $E^2 \pm 1$. An *m-fold Nagata transformation*, $m \geqslant 1$, is a sequence of m Nagata transformations over the same point of the base and *of the same sign*.

Two trigonal curves C_1, C_2 over the same base B are said to be *Nagata equivalent* (*m-Nagata equivalent*) if they are related by a sequence of Nagata transformations (respectively, m-fold Nagata transformations).

Since all points of the intersection $C \cap E$ can be resolved, any trigonal curve C is Nagata equivalent to a proper trigonal curve C' over the same base; any such proper curve is called a *proper model* of C.

Another useful construction is that of induction. Let $\varphi\colon B' \to B$ be a nonconstant morphism between two compact curves and let $C \subset \Sigma \to B$ be a trigonal curve over B. Then $\Sigma' := \varphi^*\Sigma = \Sigma \times_B B'$ is a ruled surface over B', the pull-back $E' := \varphi^* E \subset \Sigma'$ is a section of Σ', and $C' := \varphi^* C \subset \Sigma'$ is a trigonal curve. This curve is said to be *induced* from C by φ.

Weierstraß equation

Let $C \subset \Sigma \to B$ be a proper trigonal curve. Sending each point $b \in B$ to the barycenter of the intersection $C \cap F_b^\circ$ (with each intersection point weighted according to its multiplicity) defines another section $Z \subset \Sigma$, which we call the *zero section*. It is disjoint from E. Hence, the 2-bundle whose projectivization is Σ splits and, after a renormalization (multiplying by an appropriate line bundle), one has $\Sigma = \mathbb{P}(1 \oplus \mathcal{Y})$. We choose the normalization so that Z is the projectivization of $\mathcal{Y}$; then $\deg \mathcal{Y} = -E^2$. Note that $\mathcal{Y}$ is the normal bundle of Z and the conormal bundle of E.

The curve C can be given by a *Weierstraß equation*; in appropriate affine charts it has the form

$$x^3 + g_2 x + g_3 = 0, \tag{3.4}$$

where g_2 and g_3 are certain sections of $\mathcal{Y}^2$ and $\mathcal{Y}^3$, respectively, and x is an affine coordinate such that $Z = \{x = 0\}$ and $E = \{x = \infty\}$. (In other words, we identify the complement $\Sigma \smallsetminus B$ with the total space of $\mathcal{Y}$ and take for x a local trivialization of this bundle. Certainly, we do not assert that x can be chosen globally over b. However, g_2 and g_3 *are* global sections.) The bundle $\mathcal{Y}$ is determined by the curve uniquely, and the sections g_2, g_3 are unique up to the transformation

$$(g_2, g_3) \mapsto (s^2 g_2, s^3 g_3), \quad s \in H^0(B; \mathcal{O}_B^*). \tag{3.5}$$

The singular fibers of the proper trigonal curve given by (3.4) are the zeroes of its *discriminant section*

$$\Delta := -4g_2^3 - 27g_3^2 \in H^0(B; \mathcal{O}_B(\mathcal{Y}^6)). \tag{3.6}$$

The curve is reduced if and only if Δ is not identically zero.

A positive Nagata transformation over a point $b \in B$ replaces $\mathcal{Y}$ with $\mathcal{Y} \otimes \mathcal{O}_B(b)$ and transforms g_2 and g_3 to $s^2 g_2$ and $s^3 g_3$, respectively, where $s \in H^0(B; \mathcal{O}_B(b))$ is any section vanishing at b.

Remark 3.7. The existence of a proper trigonal curve $C \subset \Sigma$ implies that at least one bundle $\mathcal{Y}^2$ or $\mathcal{Y}^3$ has a nontrivial section and hence $d = \deg \mathcal{Y} \geqslant 0$. For this reason we always assume that $E^2 \leqslant 0$ in the compact case.

Digression: the j-invariant

We recall briefly a few well known properties of the so-called j-invariant, which we regard as an invariant of an unordered quadruple of points in the projective line $\mathbb{P}^1$.

Recall that the *cross-ratio* (also called *double ratio* or *anharmonic ratio*) of an ordered quadruple of pairwise distinct points z_1, z_2, z_3, z_4 in the extended complex plane $\bar{\mathbb{C}} := \mathbb{C} \cup \{\infty\}$ is defined to be

$$(z_1, z_2; z_3, z_4) := \frac{(z_1 - z_3)(z_2 - z_4)}{(z_2 - z_3)(z_1 - z_4)} \in \mathbb{C}.$$

This quantity is invariant under *Möbius transformations*

$$z \mapsto \frac{az + b}{cz + d}, \qquad a, b, c, d \in \mathbb{C}, \quad ad - bc \neq 0,$$

i.e., under all analytic automorphisms of the projective line $\mathbb{P}^1 \cong \bar{\mathbb{C}}$. Bringing the four points to more or less special position, one has

$$(z_1, z_2; z_3, \infty) = (z_1 - z_3)/(z_2 - z_3), \qquad (\lambda, 1; 0, \infty) = \lambda.$$

The cross ratio does depend on the order of the points. Interchanging the points and bringing them back to the form $\lambda', 1, 0, \infty$, one subjects λ to the transformations $\lambda \mapsto 1 - \lambda$, $\lambda \mapsto 1/\lambda$ and iterations thereof, the full orbit being

$$\lambda, \quad 1 - \lambda, \quad \frac{1}{\lambda}, \quad \frac{1}{1 - \lambda}, \quad 1 - \frac{1}{\lambda}, \quad \frac{\lambda}{\lambda - 1}. \tag{3.8}$$

A direct substitution shows that the expression

$$j := \frac{4\,(\lambda^2 - \lambda + 1)^3}{27\,\lambda^2(\lambda - 1)^2} \tag{3.9}$$

is invariant under (3.8) and, since (3.9) is an equation of degree 6 in λ for any given value of j, its all six solutions are given by (3.8). It follows that j determines the unordered quadruple $z_1, z_2, z_3, z_4 \in \mathbb{P}^1$ uniquely up to Möbius transformation. This expression is called the *j-invariant* of unordered quadruple $z_1, z_2, z_3, z_4 \in \mathbb{P}^1$. Once again, in general, the parameter λ in (3.9) is the cross-ratio of the quadruple.

We will also speak about the j-invariant of a triple $z_1, z_2, z_3 \in \mathbb{C}^1$, meaning that of the quadruple $z_1, z_2, z_3, \infty \in \mathbb{P}^1 := \mathbb{C}^1 \cup \{\infty\}$.

Remark 3.10. In algebraic geometry, the value given by (3.9) is usually multiplied by $(12)^3$ to make it well-behaved over fields of characteristic 2 and 3. Since we are working over $\mathbb{C}$, we use Kodaira's normalization, which makes the 'special values' of j (see Proposition 3.12 below) equal to 0 and 1.

A straightforward but tedious computation using Vieta's formula shows that, if $z_1, z_2, z_3 \in \mathbb{C}^1$ are the three roots of an equation of the form $z^3 + g_2 z + g_3 = 0$, the j-invariant is given by

$$j = -\frac{4g_2^3}{\Delta}, \quad \text{where } \Delta = -4g_2^3 - 27g_3^2 \tag{3.11}$$

is the discriminant of the equation.

It is clear from (3.9) that $j \to \infty$ whenever $\lambda \to 0$, 1, or ∞, *i.e.*, when two of the four points coincide. The two other special values of j are 0 and 1.

Proposition 3.12. *A triple $z_1, z_2, z_3 \in \mathbb{C}^1$ has a nontrivial symmetry (i.e., a nontrivial permutation of z_1, z_2, z_3 extends to a linear transformation $z \mapsto az + b$) if and only if its j-invariant equals* 0 *(and then there is a symmetry of order three) or* 1 *(and then there is a symmetry of order two).*

Proof. The triple has a symmetry if and only if at least two of the six members of the orbit (3.8) coincide or, equivalently, if (3.9) has a multiple solution. Hence, one can either equate various pairs in (3.8) or compute the discriminant of (3.9); after clearing the denominators, the discriminant is $2^{12}3^{21}j^4(j-1)^3$. The kind of the symmetry is better seen from (3.11): one has $j = 0$ or 1 if and only if, respectively, $g_2 = 0$ (and the roots are $ae^{2\pi ik/3}$, $k = 0, 1, 2$, $a \in \mathbb{C}$) or $g_3 = 0$ (and the roots are $0, \pm a$). □

Recall that an *elliptic curve* is a smooth analytic curve E of genus 1. Fixing a point $0 \in E$ turns E into abelian group, and the quotient $E/\pm 1$ is a rational curve $\mathbb{P}^1$. By the Riemann–Hurwitz formula, the quotient projection $E \to \mathbb{P}^1$ has four branch points. The j-invariant $j(E)$ of this quadruple of points is called the *j-invariant* of E.

The automorphism group of pair $(E, 0)$ is a cyclic group of order 6, 4, or 2 if $j(E) = 0$, 1, or neither, respectively.

Any elliptic curve E can be represented as the quotient $\mathbb{C}/L$, where L is the lattice $\mathbb{Z}\cdot 1 + \mathbb{Z}\cdot\tau$, $\tau \in \mathfrak{H}$; in fact, τ is well defined up to the action of Γ, see (2.8), and hence can be chosen within the fundamental domain D, see Figure 2.1. Then the j-invariant of E equals $\tilde{\jmath}(\tau)$, where $\tilde{\jmath}$ is the modular j-invariant.

Real values of the j-invariant

For the next statement, observe that the Möbius transformations preserving the point $\infty \in \mathbb{P}^1$ are affine, $z \mapsto az + b$, $a \neq 0$, and in the affine complex line $\mathbb{C}^1$ one can speak about well defined angles and length ratios.

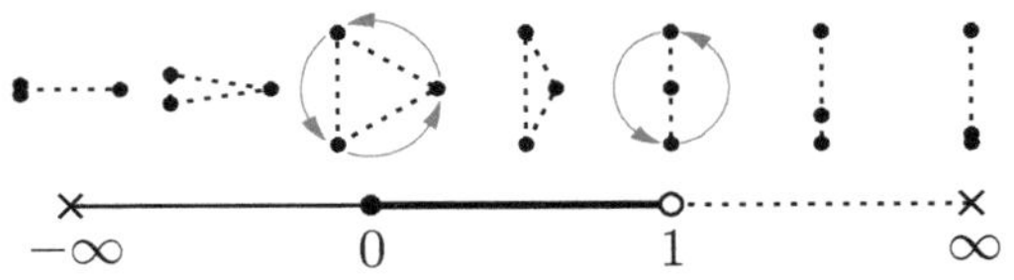

Figure 3.1. Real values of the j-invariant.

Proposition 3.13. *Assume that the j-invariant j of a triple $z_1, z_2, z_3 \in \mathbb{C}^1$ is real. Then one has one of the following two cases*:

1. $j \geqslant 1$ *and the points z_1, z_2, z_3 are collinear. In this case, the length ratio of the two segments formed by the points (shorter to longer) decreases from $1/2$ to 0 with j increasing from 1 to ∞.*
2. $j \leqslant 1$ *and the points z_1, z_2, z_3 form an isosceles triangle. In this case, the angle θ at the vertex decreases from 2π to 0 with j decreasing from 1 to $-\infty$.*

Conversely, the j-invariant of any triple as above is real.

Remark 3.14. The 'evolution' of a triple $z_1, z_2, z_3 \in \mathbb{C}^1$ with real j-invariant is shown in Figure 3.1, the grey arrows representing the symmetries. (For the reader's convenience, the points and segments in the real line $\mathbb{P}^1_{\mathbb{R}}$ are marked according to the convention used in Chapter 4.) Over $j = -\infty = \infty$, we have equivalent configurations, with two points coinciding and the third one distinct. Proposition 3.13 essentially states that j is real if and only if the triple z_1, z_2, z_3 is real in the sense that, after an affine transformation, the set $\{z_1, z_2, z_3\}$ becomes invariant under the complex conjugation $z \mapsto \bar{z}$. The j-invariant of an elliptic curve E is real if and only if E admits a real structure, see Theorems 3.77 and 3.79.

Proof of Proposition 3.13. Certainly, if λ in (3.9) is real, so is j. In this case the three points $\lambda, 1, 0$ are collinear and we are in Case 1. Then the whole orbit (3.8) is real and, choosing an appropriate representative, one can assume that $\lambda \in [1/2, 1)$. Now, using freshman calculus, one can see that the restriction of (3.9) to this interval is an increasing function taking values in $[1, \infty)$. (We will not insult the reader by computing the derivative.)

Assume that λ is not real. Since all coefficients in (3.9) are real, its solutions split into complex conjugate pairs. Equating $\bar{\lambda}$ to one of the other five members of (3.8), one arrives at the following equations.

- $\bar{\lambda} = 1 - \lambda$ implies $\operatorname{Re} \lambda = 1/2$.
- $\bar{\lambda} = \lambda^{-1}$ implies $|\lambda| = 1$, and then $\operatorname{Re}(1 - \lambda)^{-1} = 1/2$.
- The equations $\bar{\lambda} = (1 - \lambda)^{-1}$ and $\bar{\lambda} = 1 - \lambda^{-1}$ result in $\lambda = 1 + |\lambda|^2$ and $\bar{\lambda} = 1 + |\lambda|^2$, respectively; hence, λ is real in these two cases.
- Finally, $\bar{\lambda} = \lambda/(\lambda - 1)$ implies $2 \operatorname{Re} \lambda = |\lambda|^2$, in which case $\operatorname{Re} \lambda^{-1} = 1/2$.

Thus, up to permutation of the points, one can assume that $\operatorname{Re}\lambda = 1/2$, *i.e.*, λ is the vertex of an isosceles triangle over $[0,1]$, Case 2. Substituting $\lambda = 1/2 + bi$ to (3.9), one obtains

$$j = -\frac{(4b^2-3)^3}{27\,(4b^2+1)^2}.$$

Hence, j is indeed real and, using freshman calculus once again, one finds that its restriction to $b \in [0,\infty)$ is a decreasing function taking values in $(-\infty, 1]$. □

Corollary 3.15. *If the j-invariant of a triple $z_1, z_2, z_3 \in \mathbb{C}^1$ is* not *real, the points z_1, z_2, z_3 form a triangle with all three edges distinct. Hence, they can be ordered* canonically, *e.g., according to the decreasing of the opposite edge of the triangle.* ◁

Proposition 3.16. *The numbering of a triple $z_1, z_2, z_3 \in \mathbb{C}^1$ with non-real j-invariant j suggested in Corollary* 3.15 *is clockwise* (*with respect to the complex orientation of the line $\mathbb{C}^1$*) *if* $\operatorname{Im} j > 0$ *and counterclockwise if* $\operatorname{Im} j < 0$.

Proof. Since j is continuous, it suffices to consider one example, *e.g.*, one can see that $\operatorname{Im} j(\lambda) > 0$ for $\lambda = bi$, $0 < b < 1$. □

The j-invariant of a trigonal curve

The (*functional*) *j-invariant* j_C of a trigonal curve $C \subset \Sigma \to B$ is the analytic continuation of the function $B^\sharp \to \mathbb{C}$ sending each nonsingular fiber $b \in B^\sharp$ to the j-invariant of the triple of intersection points $C \cap F_b^\circ \subset F_b^\circ \cong \mathbb{C}^1$. Thus, j_C is a meromorphic function $B \to \mathbb{P}^1 = \mathbb{C} \cup \{\infty\}$. By definition, j_C is invariant under Nagata transformations. If C is proper and is given by its Weierstraß equation (3.4), then

$$j_C = -\frac{4g_2^3}{\Delta}, \quad \text{where } \Delta = -4g_2^3 - 27g_3^2, \tag{3.17}$$

see (3.11). The curve C is called *isotrivial* if $j_C = \text{const}$. For a sufficiently generic curve (if g_2 and g_3 have no common zeroes), j_C is a map of degree $6d$, $d = -E^2$.

If a curve C' over B' is induced from C over B by a nonconstant base change $\varphi\colon B' \to B$, one obviously has $j_{C'} = j_C \circ \varphi$.

Let Σ_1 be the projective plane $\mathbb{P}^2$ blown up at one point. Introduce affine coordinates $(\bar t, \bar x)$ in $\mathbb{P}^2$ and Σ_1 so that the point blown up is $(0,\infty)$. The projection $(\bar t, \bar x) \mapsto \bar t$ represents Σ_1 as a ruled surface over the Riemann sphere $\mathbb{P}^1 = \mathbb{C} \cup \{\infty\}$. Define the *universal cubic* as the trigonal curve $\bar C \subset \Sigma_1$ given by the Weierstraß equation

$$\bar x^3 - 3\bar t(\bar t - 1)\bar x + 2\bar t(\bar t - 1)^2 = 0. \tag{3.18}$$

It is straightforward that the j-invariant $j_{\bar C}$ is the identity map $\bar t \mapsto \bar t$.

Theorem 3.19. *Up to Nagata equivalence, any non-isotrivial trigonal curve C is induced by its j-invariant j_C from the universal cubic $\bar C$.*

Proof. Up to Nagata equivalence, we can assume the curve C proper, hence given by a Weierstraß equation (3.4). Since $j_C \neq$ const, neither g_2 nor g_3 is identically trivial. Let $s := -18g_2g_3/\Delta \in H^0(B;\mathcal{Y}^{-1})$. Then, up to the extra factor s^3, which can obviously be ignored, the equation (3.4) of C is obtained from (3.18) by the substitution $\bar{t} = j_C(t)$, $\bar{x} = sx$, where t is a local parameter in B. (Here, the change of the x-coordinate corresponds partially to a sequence of Nagata transformations, which is usually unavoidable.) □

Theorem 3.20. *Given a compact curve B and a nonconstant meromorphic function $j\colon B \to \mathbb{P}^1$, there is a unique, up to Nagata equivalence, ruled surface $\Sigma \to B$ and proper trigonal curve $C \subset \Sigma$ such that j is the j-invariant of C.*

Proof. Let $D_\bullet$ and $D_\circ$ be the divisors of zeroes of j and $j-1$, respectively. Since j and $j-1$ have common poles, one has $D_\bullet \sim D_\circ$. Let, further, D be any effective divisor such that $D'_\bullet := D_\bullet + D$ is divisible by 3 and $D'_\circ := D_\circ + D$ is divisible by 2, *i.e.*, $D'_\bullet = 3G_\bullet$, $D'_\circ = 2G_\circ$. (Since g.c.d.$(3,2) = 1$, such a divisor D always exists and is unique modulo 6, see Remark 3.21 below.) Still $3G_\bullet \sim 2G_\circ$ and hence $G_\bullet \sim 2(G_\circ - G_\bullet)$ and $G_\circ \sim 3(G_\circ - G_\bullet)$. Let $\mathcal{Y} = \mathcal{O}_B(G_\circ - G_\bullet)$ and define Σ as $\mathbb{P}(1 \oplus \mathcal{Y})$, with the exceptional section E equal to the projectivization of 1.

Let $\tilde{g}_2 \in H^0(B;\mathcal{O}_B(\mathcal{Y}^2))$ and $\tilde{g}_3 \in H^0(B;\mathcal{O}_B(\mathcal{Y}^3))$ be nontrivial sections with the divisors of zeroes $G_\bullet$ and $G_\circ$, respectively; they are defined up to constant factors. Pick $r, s \in \mathbb{C}^*$ and let $g_2 = r\tilde{g}_2$ and $g_3 = s\tilde{g}_3$. The sections $4g_2^3 j^{-1}$ and $27g_3^2(j-1)^{-1}$ of $\mathcal{Y}^6$ are regular and have the same divisor of zeroes, which is the common divisor of poles of j and $j-1$. It follows that r and s can be chosen so that the sum of the two sections is identically zero; this choice is unique up to (3.5). Then j is given by (3.17) and hence is the j-invariant of the trigonal curve (3.4).

Using (3.17), one can easily see that any proper trigonal curve with the prescribed j-invariant is given by the above construction. The only ambiguity, the choice of the auxiliary divisor D, results in Nagata transformations. □

Remark 3.21. Any 'extra' divisor D as in the proof of Theorem 3.20 has the form $D = D_{\min} + 6D_{\text{ex}}$, where D_{ex} is an effective divisor and $D_{\min} := 6\{\frac{1}{3}D_\bullet\} + 6\{\frac{1}{2}D_\circ\}$. (Here, for a rational divisor $Q = \sum q_b b$, $b \in B$, $q_b \in \mathbb{Q}$, we denote by $\{Q\}$ its fractional part $\sum\{q_b\}b$.) It follows that any meromorphic function $j\colon B \to \mathbb{P}^1$ gives rise to a unique proper trigonal curve C that is *minimal*, in the sense that any other proper curve with the same j-invariant is obtained from C by a sequence of positive Nagata transformations. This minimal curve can be characterized by the property that it has type $\tilde{\mathbf{A}}$ singular fibers only, see Section 3.1.2.

Due to (3.17), the pair (B, j_C) changes analytically under *equisingular* deformations of C. (In general, the degree of j_C may jump under deformations: this happens when roots of g_2 and g_3 collide.) Conversely, the construction described in the proof of Theorem 3.20 is obviously continuous, and one has the following statement.

Corollary 3.22. *Any deformation, in the sense of Kodaira–Spencer, of pairs (B, j), where B is a compact curve and $j: B \to \mathbb{P}^1$ is a nonconstant meromorphic function on B, results in a deformation of the minimal trigonal curves $C \subset \Sigma \to B$ given by Theorem* 3.20 (*see also Remark* 3.21). $\triangleleft$

Remark 3.23. More generally, a family of not necessarily minimal proper trigonal curves can be obtained from a deformation of triples (B, j, D_{ex}), where D_{ex} is an effective divisor on B. The equisingularity property of this family is discussed in the next section, see page 74.

3.1.2 Singular fibers

The topological classification of singular fibers of trigonal curves (actually, of elliptic surfaces) is due to Kodaira [95]. In the case of curves, the task is somewhat easier as one has the Weierstraß canonical form (3.4). (Since an improper singular fiber is an iterated negative Nagata transform of a proper one, we consider proper singular fibers only. A partial description of improper fibers is found in Section 5.2.2.)

The topological type of a proper singular fiber F is designated by $\tilde{\mathbf{T}}$, where $\mathbf{T}$ stands for the type of the singular point of C in F, in the notation of [6]. When just $\mathbf{T}$ is not enough, a number of *'s, indicating the 'extra tangency' of the fiber and the curve, is used. For many statements, it is convenient to extend Arnol′d's **J**–**E** notation for non-simple triple points to the simple **A**–**D**–**E** singularities; we explain this extension convention in the list below.

For the $\mathbf{J}_{2,*}$ series, see (3.26) below, we will also use the one index notation $\mathbf{J}_\mu := \mathbf{J}_{2,\mu-10}$ and $\tilde{\mathbf{J}}_\mu := \tilde{\mathbf{J}}_{2,\mu-10}$, $\mu \geqslant 10$, referring to the Milnor number.

A singular fiber F is called *simple* if the singularity of C in F is simple, see [6, 66]. A simple singular fiber F of a proper trigonal curve is related to a singular fiber $\tilde{F}$ of the covering elliptic surface, see Section 3.2.1 and Table 3.2 on page 82, and for such a fiber we indicate parenthetically Kodaira's classical notation for the type of $\tilde{F}$. In this case, the $\tilde{\mathbf{A}}$–$\tilde{\mathbf{D}}$–$\tilde{\mathbf{E}}$ notation refers as well to the adjacency graph of the irreducible components of $\tilde{F}$, see page 81, which is an affine Dynkin diagram, see Section 1.1.3. It is this fact that explains the tilde in the notation.

Remark 3.24. It should be noted that, strictly speaking, singular fibers *per se* are never simple from the point of view of the singularity theory, *i.e.*, they are not null modal: the versal deformation of each topological type splits into a countable union of connected analytic families of positive dimension.

For some statements related to simple singular fibers, the classical **A**–**D**–**E** notation is more suitable, and sometimes the most natural 'grouping' is that by Kodaira's series I, I*, *etc*. For this reason we will use all three notation systems. With this said, the topological types of proper singular fibers of trigonal curves are as follows:

- $\tilde{\mathbf{A}}_0 = \tilde{\mathbf{J}}_{0,0}$ (Kodaira's I_0): a nonsingular fiber;
- $\tilde{\mathbf{A}}_0^* = \tilde{\mathbf{J}}_{0,1}$ (Kodaira's I_1): a simple vertical tangent;
- $\tilde{\mathbf{A}}_0^{**} = \tilde{\mathbf{E}}_0$ (Kodaira's II): a vertical inflection tangent;
- $\tilde{\mathbf{A}}_1^* = \tilde{\mathbf{E}}_1$ (Kodaira's III): a node of C with one of the branches vertical;
- $\tilde{\mathbf{A}}_2^* = \tilde{\mathbf{E}}_2$ (Kodaira's IV): a cusp of C with vertical tangent;
- $\tilde{\mathbf{A}}_p = \tilde{\mathbf{J}}_{0,p+1}$ (Kodaira's I_{p+1}), $p \geqslant 1$,
  $\tilde{\mathbf{D}}_q = \tilde{\mathbf{J}}_{1,q-4}$ (Kodaira's I_{q-4}^*), $q \geqslant 4$,
  $\tilde{\mathbf{E}}_6$ (Kodaira's IV^*), $\tilde{\mathbf{E}}_7$ (Kodaira's III^*), $\tilde{\mathbf{E}}_8$ (Kodaira's II^*): a singular point of C of the same type with minimal possible local intersection index with the fiber;
- $\tilde{\mathbf{J}}_{r,p}$, $r \geqslant 2$, $p \geqslant 0$,
  $\tilde{\mathbf{E}}_{6r+\epsilon}$, $r \geqslant 2$, $\epsilon = 0, 1, 2$: a non-simple singular point of C of the same type.

Singular fibers of type $\tilde{\mathbf{E}}$ (including $\tilde{\mathbf{E}}_0$, $\tilde{\mathbf{E}}_1$, and $\tilde{\mathbf{E}}_2$) are called *exceptional*. Singular fibers of types $\tilde{\mathbf{A}}_0^{**} = \tilde{\mathbf{E}}_0$, $\tilde{\mathbf{A}}_1^* = \tilde{\mathbf{E}}_1$, and $\tilde{\mathbf{A}}_2^* = \tilde{\mathbf{E}}_2$ are called *unstable*, whereas all others are called *stable*. A proper trigonal curve is called *stable* if all its singular fibers are stable.

Warning 3.25. Our definition of 'stable' differs completely from the one commonly used in algebraic geometry, according to which stable turn out the fibers of types $\tilde{\mathbf{A}}_0^*$ and $\tilde{\mathbf{A}}_p$, $p \geqslant 1$, only, *i.e.*, those of Kodaira type I. We mean stability with respect to equisingular, but not necessarily fiberwise, deformations. The three offending types may split, $\tilde{\mathbf{A}}_0^{**} \mapsto 2\tilde{\mathbf{A}}_0^*$, $\tilde{\mathbf{A}}_1^* \mapsto \tilde{\mathbf{A}}_1 + \tilde{\mathbf{A}}_0^*$, and $\tilde{\mathbf{A}}_2^* \mapsto \tilde{\mathbf{A}}_2 + \tilde{\mathbf{A}}_0^*$, without affecting the topology of the curve.

Introducing a local parameter t in B so that $b = \{t = 0\}$ (and ignoring higher order terms and extra parameters, both continuous and discrete, see Remark 3.24), the local normal forms of singular fibers can be written as follows:

$$\tilde{\mathbf{J}}_{r,p} : x^3 + x^2 t^r + t^{3r+p} = 0, \quad r, p \geqslant 0 \tag{3.26}$$

(for $r = 0$ the curve has a double point and an extra branch transversal to the fiber),

$$\tilde{\mathbf{E}}_{6r} : x^3 + t^{3r+1} = 0, \quad \tilde{\mathbf{E}}_{6r+1} : x^3 + xt^{2r+1} = 0, \quad \tilde{\mathbf{E}}_{6r+2} : x^3 + t^{3r+2} = 0. \tag{3.27}$$

(For the $\tilde{\mathbf{J}}$ type fibers, the local normal form given above is *not* the Weierstraß normal form (3.4), as $\{x = 0\}$ is not the zero section Z. It can be converted to the Weierstraß normal form by the substitution $x \mapsto \bar{x} + \frac{1}{3}$.)

A positive Nagata transformation over a point $b \in B$ transforms the topological type of the fiber F_b over b as follows:

$$\tilde{\mathbf{J}}_{r,p} \mapsto \tilde{\mathbf{J}}_{r+1,p}, \; r, p \geqslant 0, \qquad \tilde{\mathbf{E}}_{6r+\epsilon} \mapsto \tilde{\mathbf{E}}_{6r+6+\epsilon}, \; r \geqslant 0, \; \epsilon = 0, 1, 2. \tag{3.28}$$

Thus, $\mathbf{J}_{r,p}$ and $\mathbf{E}_{6r+\epsilon}$, $r \geqslant 2$, are triple singular points that eventually blow up to $\mathbf{D}_{p+4}$ and $\mathbf{E}_{6+\epsilon}$, respectively. This fact may give the reader an idea about the nature of these

Table 3.1. Proper singular fibers.

Type of F_b	$j_C(b)$	$\operatorname{ind}_b j_C$	$\Delta \deg j_C$	$\operatorname{mult} F_b$
$\tilde{\mathbf{J}}_{r,0}$	?	?	$-6r$	$6r$
$\tilde{\mathbf{J}}_{r,p}$ $(p > 0)$	∞	p	$-6r$	$6r + p$
$\tilde{\mathbf{E}}_{6r}$	0	1 mod 3	$-6r - 2$	$6r + 2$
$\tilde{\mathbf{E}}_{6r+1}$	1	1 mod 2	$-6r - 3$	$6r + 3$
$\tilde{\mathbf{E}}_{6r+2}$	0	2 mod 3	$-6r - 4$	$6r + 4$

singularities. It follows that by a sequence of negative Nagata transformations any proper trigonal curve can be converted to one with $\tilde{\mathbf{A}}$ type singular fibers only. The minimal curve with a given j-invariant, see Theorem 3.20 and Remark 3.21, has this property.

Some other properties of singular fibers, especially the relation between their types and the j-invariant, are listed in Table 3.1. Below are a few comments.

The value $j_C(b)$ and ramification index $\operatorname{ind}_b j_C$ The types $\tilde{\mathbf{J}}_{r,0}$, $r \geqslant 1$, are the Nagata transforms of a nonsingular fiber $\tilde{\mathbf{A}}_0$; they are not detected by the j-invariant except that each such fiber decreases the degree $\deg j_C$, see below. At all other singular fibers, the j-invariant takes one of the three special values 0, 1, or ∞ and, up to Nagata transformation, the type of F_b is determined by the value $j_C(b)$ and the ramification index $\operatorname{ind}_b j_C$ at b. Exceptional fibers are over the finite values $j = 0$ or 1, and $\tilde{\mathbf{J}}$ type fibers are over $j = \infty$ (and conversely, any fiber F_b with $j_C(b) = \infty$ is singular). If $j_C(b) = 0$ and $\operatorname{ind}_b j_C = 0 \bmod 3$ or $j_C(b) = 1$ and $\operatorname{ind}_b j_C = 0 \bmod 2$, the fiber F_b is nonsingular (unless it is of type $\tilde{\mathbf{J}}_{r,0}$, $r \geqslant 1$), but the triple of points $C \cap F_b^\circ$ has a nontrivial symmetry.

If B is compact and $j_C \neq \text{const}$, then j_C must take value ∞. Hence, any non-isotrivial trigonal curve over a compact base has at least one singular fiber.

The defect $\Delta \deg j_C$ Generically, if g_2 and g_3 in (3.4) have no common zeroes, the degree $\deg j_C$ equals $6d$, where $d = -E^2$. Each exceptional singular fiber b of C is a common zero of g_2 and g_3, and the degree of j_C drops by a certain amount, shown in the table. Additionally, each positive Nagata transformation increases d by one while leaving j_C unchanged.

The multiplicity $\operatorname{mult} F_b$ By a small deformation (usually changing the analytic structure of B), any proper trigonal curve can be perturbed to one with the simplest, *i.e.*, type $\tilde{\mathbf{A}}_0^*$ singular fibers only. (In the framework of elliptic surfaces, this statement is due to Moishezon [121], see Theorem 3.55. The 'cut-and-paste' argument found, *e.g.*, in [75] extends literally to trigonal curves.) The number of such simplest fibers emerging from F_b is shown in the table. This number can be computed as the degree of the local braid monodromy about the fiber, see Chapter 5 for details. Each positive Nagata transformation over $b \in B$ increases $\operatorname{mult} F_b$ by six, and for a simple singular fiber F_b the multiplicity equals as well the Euler characteristic of the covering singular elliptic fiber $\tilde{F}_b$. The sum $\sum_F \operatorname{mult} F$ over all singular fibers F equals $6d$, which is the number of type $\tilde{\mathbf{A}}_0^*$ singular fibers (the degree $\deg j_C$) of a generic curve C.

Remark 3.29. Analyzing Table 3.1, one can see that the j-invariant j_C of a generic trigonal curve C has ramification index 3, 2, or 1 at each point $b \in B$ with $j_C(b) =$ 0, 1, or ∞, respectively. After a further perturbation, one can assume that all other critical values of j_C are simple. A function $j\colon B \to \mathbb{P}^1$ with these properties is said to have *generic branching behavior.* This generic branching behavior is, in fact, highly non-generic: a truly generic function $j\colon B \to \mathbb{P}^1$ would result in a trigonal curve with a large number of type $\tilde{\mathbf{A}}_0^{**}$ and $\tilde{\mathbf{A}}_1^*$ singular fibers.

Equisingular deformations

Using Table 3.1, it is easy to understand when the deformation of proper trigonal curves $C \subset \Sigma \to B$ arising from a deformation of pairs (B, j) or, more generally, triples (B, j, D_{ex}), see Corollary 3.22 and Remark 3.23, is equisingular. Given an effective divisor $D = \sum r_b b$ on B, $b \in B$, $r_b \geqslant 0$, and an integer $p > 0$, denote $D \bmod p := \sum (r_b \bmod p) b$, where $\bmod p$ is the residue in the range $0, \dots, p-1$. As in the proof of Theorem 3.20, consider the divisors $D_\bullet$ and $D_\circ$ of zeroes of the meromorphic functions j and $j-1$, respectively, and let $D_\times$ be their common divisor of poles. Then we have the following statement.

Proposition 3.30. *The deformation of proper trigonal curves $C \subset \Sigma \to B$ resulting from a deformation of triples (B, j, D_{ex}), see Corollary 3.22 and Remark 3.23, is equisingular if and only if so is the corresponding deformation of pairs $(B, \bar{D})$, where $\bar{D} := (D_\bullet \bmod 3) + (D_\circ \bmod 2) + D_\times + D_{\mathrm{ex}}$.* ◁

Paraphrasing, the condition states that the points constituting $\bar{D}$ should keep their multiplicities, neither splitting nor merging. For $D_\bullet$ and $D_\circ$, it is only the multiplicities modulo 3 and 2, respectively, that matter.

Induced curves

Assume that a trigonal curve $C' \subset \Sigma' \to B'$ is induced from a curve $C \subset \Sigma \to B$ by a non-constant base change $\varphi\colon B' \to B$. Pick a point $b' \in B'$, let $b = \varphi(b')$, and let m be the ramification index of φ at b'. Then the topological type of the fiber $F_{b'}$ of C' is recovered from that of F_b by the substitution $t \mapsto t^m$ in the local normal forms (3.26) and (3.27). For example, if F_b is of type $\tilde{\mathbf{J}}_{r,p}$, then $F_{b'}$ is of type $\tilde{\mathbf{J}}_{mr,mp}$. The ultimate statement for the $\tilde{\mathbf{E}}$ type fibers is less regular and we will not try to write it down.

As a consequence, we have the following simple characterization of Kodaira's type I (equivalently, non-exceptional type $\tilde{\mathbf{A}}$ or type $\tilde{\mathbf{J}}_{0,p}$, $p \geqslant 0$).

Lemma 3.31. *For a proper trigonal curve C, the following statements are equivalent:*

- *any curve C' induced from C by a non-constant base change is simple;*
- *all singular fibers of C are of Kodaira's type* I.

If this is the case, all singular fibers of any curve C' induced from C by a non-constant base change are also of Kodaira's type I. ◁

Isotrivial curves

Isotrivial curves occupy a separate niche in the world of trigonal curves: almost none of the general statements applies to the isotrivial case. On the other hand, such curves have relatively simple structure and are easy to classify, using Weierstraß equation (3.4). There are three cases to be considered.

Case 1: $g_2 \equiv 0$ and $j_C \equiv 0$ In this case $g_3 \not\equiv 0$ (as we assume C reduced) and the curve is given by an equation of the form $x^3 + g_3 = 0$. Hence, the curve is uniquely determined by an effective divisor G_3 (the divisor of zeroes of g_3) and a line bundle $\mathcal{Y}$ such that $\mathcal{Y}^3 \cong \mathcal{O}_B(G_3)$. After a sequence of negative Nagata transformations one can assume that the multiplicity of each point in G_3 does not exceed 2. Each point of G_3 of multiplicity $3r$, $3r+1$, or $3r+2$, $r \geqslant 0$, gives rise to a singular fiber of C of type $\tilde{\mathbf{J}}_{r,0}$, $\tilde{\mathbf{E}}_{6r}$, or $\tilde{\mathbf{E}}_{6r+2}$, respectively; these are all singular fibers of C.

Case 2: $g_3 \equiv 0$ and $j_C \equiv 1$ As in the previous case, one has $g_2 \not\equiv 0$, the curve is given by an equation of the form $x^3 + g_2 x = 0$ (thus containing the zero section Z as a component) and is uniquely determined by an effective divisor G_2 (the divisor of zeroes of g_2) and a line bundle $\mathcal{Y}$ with the property $\mathcal{Y}^2 \cong \mathcal{O}_B(G_2)$. After a sequence of negative Nagata transformations one can assume that all points of G_2 are simple. Each point of G_2 of multiplicity $2r$ or $2r+1$, $r \geqslant 0$, gives rise to a singular fiber of C of type $\tilde{\mathbf{J}}_{r,0}$ or $\tilde{\mathbf{E}}_{6r+1}$, respectively.

Remark 3.32. In Cases 1 and 2, the line bundle $\mathcal{Y}$ does *not* need to have a regular section; for example, in Case 1, if $G_3 = 0$, the bundle $\mathcal{Y}$ may represent a 3-torsion element of the Picard group Pic B, resulting in a curve C without singular fibers but with nontrivial monodromy of the triple covering $p: C \to B$. In spite of the absence of singular fibers, this curve C is irreducible. Similarly, in Case 2 with $G_2 = 0$, one can take for $\mathcal{Y}$ a 2-torsion element of Pic B and obtain a curve C without singular fibers and with two rather than three irreducible components.

Case 3: $g_3^2/g_2^3 = \mathrm{const} \neq 0$ In this case, the divisors G_2 and G_3 of zeroes of g_2 and g_3, respectively are related *via* $3G_2 = 2G_3$. Hence, the curve is determined by a certain effective divisor G on B, so that one has $G_2 = 2G$, $G_3 = 3G$, and $\mathcal{Y} = \mathcal{O}_B(G)$. By a sequence of negative Nagata transformations the curve can be reduced to the union of three disjoint sections of the trivial ruled surface $\Sigma = B \times \mathbb{P}^1$; we call such curves *trivial*. Each point of G of multiplicity r gives rise to a type $\tilde{\mathbf{J}}_{r,0}$ singular fiber.

3.1.3 Special geometric structures

In this section, we introduce several special geometric properties or structures that a trigonal curve may have and that are discussed in more details in subsequent chapters. Most of the time, we consider the case of the rational base only, *i.e.*, we assume that $B \cong \mathbb{P}^1$. In this case, the ruled surface Σ is one of the so-called *Hirzebruch surfaces* Σ_d; up to isomorphism, it is determined by the parameter $d := -E^2 \geqslant 0$. We recall that the classes E and F of the exceptional section and a fiber, respectively, generate both the group of classes of divisors of Σ_d and its semigroup of classes of effective divisors. One has $E^2 = -d$, $F^2 = 0$, $E \circ F = 1$, and any proper trigonal curve belongs to the linear system $|3E + 3dF|$.

Torus structures

A proper trigonal curve $C \subset \Sigma_{2n}$ is said to be of *torus type* if there are sections f_i of $\mathcal{O}(E + inF)$, $i = 0, 2, 3$, such that C is the zero set of the section $f := f_2^3 + f_0 f_3^2$. Informally, in affine coordinates (x, t) such that $E = \{x = \infty\}$ the equation of C has the form

$$[x + a_2(t)]^3 + [xa_1(t) + a_3(t)]^2 = 0 \tag{3.33}$$

for some polynomials $a_i(t)$ of degree $\deg a_i \leqslant in$, $i = 1, 2, 3$. Each representation of the equation of C in this form (up to obvious equivalence) is called a *torus structure*. A torus structure is uniquely determined by the section $\{f_2 = 0\}$, which is necessarily irreducible. Each point of intersection $\{f_2 = 0\} \cap \{f_3 = 0\}$ is singular for C; such points are called *inner* (with respect to the given torus structure), whereas the other singular points of C are called *outer*.

Proposition 3.34. *Torus structures are preserved by rational base changes and even Nagata equivalence.*

Proof. A rational base change is given by $t = u(t')/v(t')$, $x = x'/v^{2n}(t')$, and a positive 2-fold Nagata transformation (over $t = 0$) is given by $t = t'$, $x = x'/t'^2$. Substituting to (3.33) and clearing the denominators, we obtain again an equation in the form (3.33). A negative 2-fold Nagata transformation is given by $t = t'$, $x = x't'^2$. For the transform to be proper, the singularity of C at the origin must be adjacent to $\mathbf{J}_{10}$, *i.e.*, all terms of the form $t^\alpha x^\beta$, $2\alpha + \beta < 6$, must vanish. Evaluating at $t = 0$, we have $a_1(0) = a_2(0) = a_3(0) = 0$; then, step by step, we conclude that $a_i(t) = t^i a_i'(t)$ for some polynomials a_i', $i = 1, 2, 3$. Substituting and canceling t'^6, we arrive at (3.33). □

Lemma 3.35. *A torus structure on a proper trigonal curve $C \subset \Sigma_{2n}$ gives rise to a triple covering $X \to \Sigma_{2n}$ ramified at $C + E$ and with the full monodromy group $\mathbb{S}_3$.*

Proof. In the affine coordinates $(x, t; y)$ in $\mathbb{C}^2 \times \mathbb{C}$, the covering surface X is given by $y^3+3(x+a_2)y+2(a_1x+a_3) = 0$. More formally, X is the normalization of the triple section of $\mathcal{O}(E+nF)$ given by $y^3+3f_0f_2y+2f_0^2f_3 = 0$. Restricting to a generic fiber $t = \text{const}$, one obtains a covering $C_t \to \mathbb{P}^1$, where $C_t \subset \Sigma_1$ is a proper trigonal curve with the set of singular fibers $\tilde{\mathbf{A}}_1^* + 3\tilde{\mathbf{A}}_0^*$ (essentially, a nodal plane cubic projected from a generic point in the tangent to one of the branches at the node). All such curves are deformation equivalent, their skeletons are as shown in Figure 4.3 (e) (with two $\times$-vertices inside the bigonal region), and their monodromy groups $MG(C)/\Delta^2$ are easily found to be Γ, see Section 5.2.1, which projects onto $\mathbb{S}_3$. □

According to Proposition 3.34, when studying torus structures we can confine ourselves to a simple curve C. Fix a torus structure and consider $Q_2 := \{f_2 = 0\}$ and $Q_3 := \{f_3 = 0\}$; the former curve is irreducible, the latter may contain fibers of Σ_{2n} as components. Local analysis shows that an inner singular point P of C is of type

- $\mathbf{A}_{3p-1}$, if Q_3 is nonsingular at P and $(Q_2 \circ Q_3)_P = p$; then $(Q_2 \circ C)_P = 2p$, or
- $\mathbf{E}_6$, if Q_3 has a node at P and $(Q_2 \circ Q_3)_P = 2$; then $(Q_2 \circ C)_P = 4$.

Since $Q_2 \circ C = 6n$, a torus structure is determined by its inner singularities (as otherwise, for two distinct sections Q_2', Q_2'', one would have $Q_2' \circ Q_2'' = 3n$).

Corollary 3.36. *A torus structure on a proper trigonal curve $C \subset \Sigma_{2n}$ is uniquely determined by the monodromy of the covering given by Lemma* 3.35.

Proof. One can easily see that a simple inner singular point P of C is characterized by the property that the composition $\pi_1(U \smallsetminus C) \to \pi_1(\Sigma_{2n} \smallsetminus (C \cup E)) \twoheadrightarrow \mathbb{S}_3$ should be onto, where U is a Milnor ball about P. □

Z-splitting sections

We recall briefly a few definitions from [149], which we adjust to the case of splitting sections of trigonal curves. Let $C \subset \Sigma_{2n}$ be a proper trigonal curve, and let $X \to \Sigma_{2n}$ be the minimal resolution of singularities of the double covering of Σ_{2n} ramified at $C + E$. Identify C and E with their proper transforms in X and let $\tilde{F}_\infty$ be the pull-back of a generic fiber F_∞ of Σ_{2n}. Let, further, $\mathbf{S} \subset H_2(X)$ be the lattice spanned by the classes of the exceptional divisors contracted by the projection $X \to \Sigma_{2n}$. Consider the sublattice $\mathbf{S}' := \mathbf{S} \oplus \mathbb{Z}[E] \oplus \mathbb{Z}[\tilde{F}_\infty]$ and let $\tilde{\mathbf{S}}' := (\mathbf{S}' \otimes \mathbb{Q}) \cap H_2(X)$ be its primitive hull. (If C is simple, the lattice extension $\mathbf{S}' \subset H_2(X)$ is essentially equivalent to the homological type of C introduced in Section 6.1.2.)

Definition 3.37. An analytic section S of Σ_{2n} is said to be *splitting* for C if its proper transform in X splits into two components S' and S'' (possibly equal) interchanged by the deck translation $\tau\colon X \to X$. A splitting section is *pre-Z-splitting* if the classes $[S']$, $[S'']$ belong to $\tilde{\mathbf{S}}'$; in this case, the common order of $[S']$, $[S'']$ in $\tilde{\mathbf{S}}'/\mathbf{S}'$ is called the *class order* of S. Finally, a pre-Z-splitting section is *Z-splitting* if $[S'] \neq [S'']$.

A section is splitting if and only if its local intersection index with C at each intersection point is even. Intuitively, such a section is pre-Z-splitting if it follows all equisingular deformations of C.

Similar to Proposition 3.34, one can easily prove that splitting and (pre-)Z-splitting sections remain invariant under rational base changes and even Nagata transformations. In particular, when studying such sections, we can confine ourselves to simple trigonal curves. In this case, X is the Jacobian elliptic surface ramified at $C + E$, see Theorem 3.51 below, and the deck translation is the fiberwise multiplication by (-1); hence, the splitting sections are in a natural one-to-one correspondence with the pairs of opposite elements of the Mordell–Weil group $MW(X)$. Furthermore, in view of Theorem 3.60 below, pre-Z-splitting sections correspond to the torsion elements of $MW(X)$; more precisely, we have the following statement (where the term *linear component* means a component of C that is a section of Σ_{2n}).

Theorem 3.38. *Let $C \subset \Sigma_{2n}$ be a simple proper trigonal curve. The pre-Z-splitting sections for C of class order* 1 *and* 2 *are, respectively, E and the linear components of C. The pre-Z-splitting sections of class order $m \geqslant 3$ are Z-splitting; such sections are in a one-to-one correspondence with the pairs of opposite order m elements of the Mordell–Weil group of the covering elliptic surface ramified at $C + E$.* $\triangleleft$

Relation to plane curves

An ample source of trigonal curves is given by the following construction. Consider a plane curve $D \subset \mathbb{P}^2$ with a distinguished singular point P of multiplicity $(\deg D - 3)$ and assume that D has no linear components through P. The plane $\mathbb{P}^2$ blown up at P is a Hirzebruch surface Σ_1 (the exceptional divisor being the exceptional section), and the proper transform of D is a certain trigonal curve $\tilde{C} \subset \Sigma_1$; this curve is called the *trigonal model* of D, and its minimal proper model $C \subset \Sigma_d$ is called the *proper model* of D. The latter is equipped with a number of distinguished fibers $F_{\mathrm{I}}, F_{\mathrm{II}}, \ldots$, corresponding to the improper fibers of $\tilde{C}$, and the original plane curve D is recovered from $C, F_{\mathrm{I}}, \ldots$ and some additional information about the blow-up centers.

Clearly, the blow-up induces a diffeomorphism $\mathbb{P}^2 \smallsetminus D \cong \Sigma_1 \smallsetminus (\tilde{C} \cup E)$. This relation between plane and trigonal curves is exploited in Chapters 7 and 8.

Maximal trigonal curves

In this section we do *not* assume that the base B is rational.

Definition 3.39. A non-isotrivial trigonal curve $C \subset \Sigma \to B$ is called *maximal* if

- C has no singular fibers of type $\tilde{\mathbf{J}}_{r,0}$, $r \geqslant 1$,
- the j-invariant j_C has no critical values other than 0, 1, and ∞,
- each point in the pull-back $j_C^{-1}(0)$ has ramification index $\leqslant 3$, and
- each point in the pull-back $j_C^{-1}(1)$ has ramification index $\leqslant 2$.

A meromorphic function $j_C\colon B \to \mathbb{P}^1$ satisfying the last three conditions is said to have *extremal branching behavior*.

Proposition 3.40. *If $C \subset \Sigma \to B$ is a maximal trigonal curve, then B, Σ, and C are defined over an algebraic number field.*

Proof. Since the function $j_C\colon B \to \mathbb{P}^1$ has three critical values only, by Belyĭ's theorem [18] both B and j_C are defined over an algebraic number field. Analyzing the construction of the curve in the proof of Theorem 3.20, one immediately concludes that C is defined over the splitting field of j_C. Note that, due to the first condition in Definition 3.39, one has $D_{\text{ex}} = 0$, see Remark 3.21. □

Since any curve equisingular deformation equivalent to a maximal one is obviously also maximal, we have the following corollary.

Corollary 3.41. *If C_t is an equisingular elementary deformation of a maximal trigonal curve C_0, then all curves C_t are isomorphic to C_0.* ◁

Further properties of maximal curves are discussed in Chapter 4, see page 123.

3.2 Elliptic surfaces

Elliptic surfaces are generously covered in the literature, standard references being [95, 75, 16], and we just give a brief introduction, mainly to fix the terminology and notation. Very few statements are proved here; most are merely explained (when a simple topological explanation is available), and some are just mentioned.

3.2.1 The local theory

Let X be a smooth analytic surface and B a smooth curve. A proper morphism $p\colon X \to B$ is called a *fibration of genus g* if a generic fiber of p is a connected smooth curve of genus g. Our primary concern is the case $g = 1$, *i.e.*, that of the so-called *elliptic fibration*, but a few preliminary results extend to any genus $g \geqslant 1$. An orientation preserving diffeo-/homeomorphism $\varphi\colon X_1 \to X_2$ between two fibrations $X_i \to B_i$, $i = 1, 2$, is called *bi-oriented* if it takes fibers to fibers and preserves their orientation. (We extend this notion to smooth fibrations as well, *cf.* Section 3.3.3.) If φ is analytic, it is called an *isomorphism*.

In the local setting, we assume that B is the disk D^2 and that p is a submersion everywhere except possibly $0 \in D^2$. If p is *not* a submersion, we call $F_0 := p^{-1}(0)$ a *singular fiber*. Let $E_1, \ldots, E_n$ be the irreducible components of F_0. Since F_0 is a strict deformation retract of X, one has $H_2(X) = \bigoplus \mathbb{Z}[E_i]$; we regard this group as a lattice with respect to the intersection index form. Let F be a generic fiber. Then $[F] = \sum n_i[E_i]$ for some integers $n_i \geqslant 0$ and, since obviously $F^2 = 0$ and F_0 is connected, all coefficients in this expansion are positive, $n_i > 0$. First classical result states that $H_2(X)$ is negative semidefinite.

Proposition 3.42. *Under the assumptions above, the lattice $H_2(X)$ is negative semidefinite and its kernel is generated by a rational multiple of the fiber $[F] = \sum n_i[E_i]$.*

Proof. Obviously, $[F] \in \ker H_2(X)$, as F is disjoint from all E_i.

Let $G = [g_{ij}]$ be the Gram matrix of the intersection form, regarded as a self-adjoint operator on the vector space $V := H_2(X) \otimes \mathbb{R} = \bigoplus \mathbb{R}[E_i]$ with the standard Euclidean metric, which we denote by $(\,\cdot\,,\,\cdot\,)$. Let $\lambda_1 \geqslant \lambda_2 \geqslant \cdots$ be the eigenvalues of G. Then for any $u \in V$ one has $u^2 \leqslant \lambda_1 \|u\|$, the equality holding if and only if u belongs to the eigenspace V_1 of G corresponding to the maximal eigenvalue λ_1. In view of this characterization of V_1 and since $g_{ij} \geqslant 0$ for $i \neq j$, whenever a vector $u = \sum u_i[E_i]$ belongs to V_1, the vector $u' := \sum |u_i|[E_i]$ also belongs to V_1. Then $(u', [F]) > 0$ and, on the other hand,

$$\lambda_1(u', [F]) = (Gu', [F]) = (u', G[F]) = 0.$$

Hence $\lambda_1 = 0$, $V_1 = \ker(H_2(X) \otimes \mathbb{R})$, and the form is negative semidefinite.

Assume that $\dim V_1 \geqslant 2$ and let $v \in V_1$ be an eigenvector orthogonal to $[F]$. Then $v_+ := v + v'$ is also in V_1, one has $v_+ = \sum v_i[E_i]$ with all $v_i \geqslant 0$, and the components E_i appearing in this expansion form a proper subset of $\{E_1, \dots, E_n\}$ (as otherwise $(v, [F]) \neq 0$). Since F_0 is connected, there is a component E_j not appearing in the expansion of v_+ and intersecting one of those that do appear. Then $v_+ \circ [E_j] \neq 0$, which is a contradiction. □

According to Proposition 3.42, the class $[F]$ has the form mr, where r is a primitive generator of the kernel $\ker H_2(X)$ and $m \in \mathbb{N}$. If $m > 1$, the fiber F_0 is said to be *multiple*.

A fibration $p\colon X \to B$ is *relatively minimal* if its fibers do not contain (-1)-curves. If p is not relatively minimal, one can contract consecutively all (-1)-curves contained in its fibers and obtain the so-called *relatively minimal model* of p. If $g \geqslant 1$, this model is indeed unique, and here is a brief explanation of this fact. Assume that, at some moment, we have a choice and the fiber F_0 contains two (-1)-curves E_1, E_2 that intersect, so that contracting one of them would prevent us from contracting the other. To keep the intersection form negative semidefinite one must have $E_1 \circ E_2 = 1$; but then $(E_1 + E_2)^2 = 0$ and hence $E_1 + E_2$ is the whole fiber, see Proposition 3.42. On the other hand, $p_a(E_1 + E_2) = 0$ and hence p is a fibration of genus $g = 0$.

Lemma 3.43. *If $p\colon X \to D^2$ is a relatively minimal fibration of genus $g \geqslant 1$, then $K_X \circ E_i \geqslant 0$ for each irreducible component E_i of F_0.*

Proof. Assume that $K_X \circ E_i < 0$. By Proposition 3.42 and the adjunction formula, $2p_a(E_i) - 2 = K_X \circ E_i + E_i^2 < 0$; hence $p_a(E_i) = 0$ and E_i is a smooth rational curve. If $E_i^2 = 0$, then E_i is the whole fiber F_0 and p is a fibration of genus zero. Otherwise, one has $E_i^2 = -1$ and E_i is a (-1)-curve contained in a fiber. □

Lemma 3.44. *Let $p\colon X \to D^2$ be a relatively minimal* elliptic *fibration and F_0 its singular fiber. Then for each irreducible component E_i of F_0 one has $K_X \circ E_i = 0$. Hence, either F_0 is irreducible and $p_a(F_0) = 1$, or each component E_i of F_0 is a smooth rational (-2)-curve.*

Proof. Let F be a nonsingular fiber of p. Since $p_a(F) = 1$ and $F^2 = 0$, we have $K_X \circ F = 0$ and then $\sum n_i(K_X \circ E_i) = 0$ (see the beginning of this section). As $n_i > 0$ and $K_X \circ E_i \geqslant 0$ for all i, see Lemma 3.43, this implies $K_X \circ E_i = 0$. If $E_i^2 = 0$ for some i, then $E_i = F_0$, see Proposition 3.42, and $p_a(F_0) = p_a(E_i) = 1$. Otherwise $E_i^2 < 0$, and then $p_a(E_i) = 0$ and $E_i^2 = -2$. □

Classification of singular fibers

We are now ready to complete Kodaira's classification of singular fibers of elliptic fibrations. As above, let $p\colon X \to D^2$ be a *relatively minimal* elliptic fibration, let $F_0 = p^{-1}(0)$, and let $E_1, \dots, E_n$ be the irreducible components of F_0. Denote by $\mathfrak{G}$ the *adjacency graph* of F_0: the vertices of $\mathfrak{G}$ are the components $E_1, \dots, E_n$, and two distinct vertices E_i, E_j, $i \neq j$, are connected by $E_i \circ E_j$ edges. In view of Lemma 3.44 one has $H_2(X) = \mathbf{L}_{\mathfrak{G}}$, where $\mathbf{L}_{\mathfrak{G}}$ is the lattice introduced in Section 1.1.3.

If F_0 is irreducible, then $p_1(F_0) = 1$, see Lemma 3.44, and F_0 is either a smooth elliptic curve (type $\tilde{\mathbf{A}}_0 = \mathrm{I}_0$; we ignore the multiplicity of F_0), or a rational curve with a single node (type $\tilde{\mathbf{A}}_0^* = \mathrm{I}_1$), or a rational curve with a single cusp (type $\tilde{\mathbf{A}}_0^{**} = \mathrm{II}$); any other singularity would contribute more than one to the genus formula.

If F_0 is reducible, Propositions 3.42 and 1.10 imply that $\mathfrak{G}$ must be one of the simply laced affine Dynkin diagrams, see Section 1.1.3. If F_0 is a divisor with normal intersections, *i.e.*, the components of F_0 meet each other transversally at pairwise distinct points, the type of F_0 bears the same name as the corresponding Dynkin diagram: one has types $\tilde{\mathbf{A}}_p = \mathrm{I}_{p+1}$, $p \geqslant 1$, $\tilde{\mathbf{D}}_q = \mathrm{I}_{q-4}^*$, $q \geqslant 4$, $\tilde{\mathbf{E}}_6 = \mathrm{IV}^*$, $\tilde{\mathbf{E}}_7 = \mathrm{III}^*$, and $\tilde{\mathbf{E}}_8 = \mathrm{II}^*$. In most cases, the normal intersection property follows from the shape of $\mathfrak{G}$. The two exceptions are

- type $\tilde{\mathbf{A}}_1^* = \mathrm{III}$: the two components of F_0 are tangent to each other, and
- type $\tilde{\mathbf{A}}_2^* = \mathrm{IV}$: the three components of F_0 meet at a single point.

An explicit construction showing the existence of all types is found in [95] or [75]. A very simple approach is explained in Remark 3.50.

Singular fibers of type I_p, $p \geqslant 0$, are not simply connected and hence can be multiple. A fiber I_p of multiplicity $m > 1$ is denoted by ${}_m\mathrm{I}_p$. All other singular fibers are simply connected (as their components are smooth rational curves and their adjacency graphs are trees) and their multiplicities are necessarily equal to one.

Table 3.2. Monodromy of singular elliptic fibers.

Type of F_0	Monodromy	Type of F_0	Monodromy
I_p $(\tilde{\mathbf{A}}_0^*, \tilde{\mathbf{A}}_{p-1})$	$(\mathbb{YX})^{-p} = \begin{bmatrix} 1 & 0 \\ p & 1 \end{bmatrix}$	I_p^* $(\tilde{\mathbf{D}}_{p+4})$	$-(\mathbb{YX})^{-p} = -\begin{bmatrix} 1 & 0 \\ p & 1 \end{bmatrix}$
II $(\tilde{\mathbf{A}}_0^{**})$	$\mathbb{X} = \begin{bmatrix} 1 & -1 \\ 1 & 0 \end{bmatrix}$	II^* $(\tilde{\mathbf{E}}_8)$	$\mathbb{X}^5 = \begin{bmatrix} 0 & 1 \\ -1 & 1 \end{bmatrix}$
III $(\tilde{\mathbf{A}}_1^*)$	$\mathbb{Y}^3 = \begin{bmatrix} 0 & -1 \\ 1 & 0 \end{bmatrix}$	III^* $(\tilde{\mathbf{E}}_7)$	$\mathbb{Y} = \begin{bmatrix} 0 & 1 \\ -1 & 0 \end{bmatrix}$
IV $(\tilde{\mathbf{A}}_2^*)$	$\mathbb{X}^2 = \begin{bmatrix} 0 & -1 \\ 1 & -1 \end{bmatrix}$	IV^* $(\tilde{\mathbf{E}}_6)$	$\mathbb{X}^4 = \begin{bmatrix} -1 & 1 \\ -1 & 0 \end{bmatrix}$

A component E_i of a singular fiber F_0 is called *simple* (*multiple*) if one has $n_i = 1$ (respectively, $n_i > 1$) in the expansion $[F] = \sum_i n_i[E_i]$, see Proposition 3.42. One has $[F] = m\sum_i n_i'[E_i]$, where m is the multiplicity of F_0 and n_i' are the coefficients shown in Figure 1.1. It follows that E_i is simple if and only if both the multiplicity m of F_0 and the corresponding coefficient n_i' in Figure 1.1 are equal to one.

The topological types of singular fibers can also be described in terms of their *local monodromy*, *i.e.*, monodromy of the locally trivial fibration $p\colon X \smallsetminus F_0 \to D^2 \smallsetminus 0$ in the 1-homology $H_1(F)$ of a nonsingular fiber. (See Section D.1.2, page 361 for more details concerning the concept of monodromy.) Since the group $\pi_1(D^2 \smallsetminus 0) = \mathbb{Z}$ is generated by the class $[\partial D^2]$ of the boundary, it suffices to consider the image $\mathfrak{m}$ of this class. Identifying $H_1(F)$ with $\mathcal{H}$, we obtain a well defined conjugacy class $[\![\mathfrak{m}]\!]$ in $\tilde{\Gamma}$. The computation is explained in Chapter 5 and the result is shown in Table 3.2. (For the experts we remind that our groups act from the right and the matrices shown in the table are the transposed of what is usually found in the literature.) The exceptional types II–IV* realize all six conjugacy classes of elliptic elements of $\tilde{\Gamma}$, whereas the monodromy about a type I or I* fiber is parabolic (plus or minus a nonnegative power of a positive Dehn twist). The local monodromy is never hyperbolic.

We mention two consequences of Table 3.2. First, for a relatively minimal singular fiber F_0 one has

$$\chi(F_0) = \operatorname{dg}\mathfrak{m} \bmod 12. \tag{3.45}$$

Second, the inclusion $F \hookrightarrow X$ and strict deformation retraction $X \to F_0$ induce an isomorphism

$$H^1(F_0) = \operatorname{Ker}[(\mathfrak{m}^* - 1) : H^1(F) \to H^1(F)]. \tag{3.46}$$

Both are obtained by a direct computation using the table.

3.2.2 Compact elliptic surfaces

From now on, we assume that $p\colon X \to B$ is a *relatively minimal elliptic fibration without multiple fibers and with compact base* B. We denote by $B^\sharp \subset B$ the set of regular values of p and let $X^\sharp = p^{-1}(B^\sharp)$. Using Lemma 3.44, one can easily show that, for X compact, the relative minimality condition is equivalent to $K_X^2 = 0$.

A compact surface X of positive Kodaira dimension admits at most one elliptic fibration. Any compact surface of Kodaira dimension 1 is elliptic.

An *elementary deformation* of elliptic fibrations consists of a smooth 3-fold $\mathcal{X}$, a smooth surface $\mathcal{B}$, a proper holomorphic map $p\colon \mathcal{X} \to \mathcal{B}$, and a deformation (in the sense of Kodaira–Spencer) $\pi\colon \mathcal{B} \to D^2$ such that $p \circ \pi$ is a submersion and each restriction $p_t\colon \mathcal{X}_t \to \mathcal{B}_t$ of p to the slices $\mathcal{B}_t := \pi^{-1}(t)$ and $\mathcal{X}_t := p^{-1}(\mathcal{B}_t)$, $t \in D^2$, is an elliptic fibration. We will speak about *deformation equivalence* and *equisingular deformation equivalence* of elliptic fibrations, both being the equivalence relations generated by elementary deformations and isomorphism. In the latter case, the deformations are required to preserve the topology of the singular fibers, *cf.* Remark 3.3.

Any deformation of a surface of Kodaira dimension 1 is uniquely a deformation of elliptic fibrations. By Kodaira's results on the stability of exceptional curves, any elementary deformation of a relatively minimal elliptic fibration is relatively minimal.

The j-invariant and homological invariant

The *(functional)* j*-invariant* of an elliptic fibration $p\colon X \to B$ is the analytic continuation $j_X\colon B \to \mathbb{P}^1$ of the function $B^\sharp \to \mathbb{C}$ sending each nonsingular fiber F, which is an elliptic curve, to its j-invariant. The fibration p is called *isotrivial* if $j_X = \text{const}$.

The *homological invariant* of X is the sheaf $R^1p_*\mathbb{Z}_X$ on B. This invariant may appear in several other disguises. First, one can restrict $R^1p_*\mathbb{Z}_X$ to $B^\sharp$, obtaining a contravariant local system $\mathcal{M}_X^*$. No information is lost as, according to (3.46), one has $R^1p_*\mathbb{Z}_X = i_*\mathcal{M}_X^*$, where $i\colon B^\sharp \hookrightarrow B$ is the inclusion. The fiber of $\mathcal{M}_X^*$ is $H^1(\text{torus})$, *i.e.*, $\mathcal{M}_X^*$ is a $\tilde{\Gamma}$-bundle. Switching to homology, one obtains a covariant local system $\mathcal{M}_X$; after fixing an isomorphism $H_1(F_b) = \mathcal{H}$ for a generic fiber F_b, this local system $\mathcal{M}_X$ can be represented by its *monodromy* $\mathfrak{h}_X\colon \pi_1(B^\sharp, b) \to \tilde{\Gamma}$, which is also referred to as the homological invariant of X.

Due to Theorem 2.10, the modular j-invariant $\tilde{\jmath}\colon \mathfrak{H} \to \mathbb{P}^1$ endows $\mathbb{P}^1 \smallsetminus \{0, 1, \infty\}$ with a principal Γ-bundle $\mathcal{P}$. A $\tilde{\Gamma}$-bundle $\mathcal{M}^*$ is said to *belong* to a nonconstant holomorphic map $j\colon B \to \mathbb{P}^1$ if the principal Γ-bundle associated with $\mathcal{M}^*$ equals $j^*\mathcal{P}$ (after restricting to a common Zariski open set). The homological invariant of an elliptic fibration belongs to its j-invariant. In more geometric terms this relation is explained in Chapter 5, see Proposition 5.69 and Remark 5.70. As a consequence, the value and the ramification index of the j-invariant of X at a singular fiber F depend on the type of F only (except type I_0^*) and are those given by Table 3.1.

Since $\tilde{\Gamma} \twoheadrightarrow \Gamma$ is a central $\mathbb{Z}_2$-extension, the local systems $\mathcal{M}$ over $B^\sharp$ belonging to a given nonconstant function $j_X\colon B \to \mathbb{P}^1$ form a principal homogeneous space over $H^1(B^\sharp; \mathbb{Z}_2)$. If the types of all singular fibers are fixed, this set reduces to a principal homogeneous space over $H^1(B; \mathbb{Z}_2)$. Furthermore, since the sum of the classes realized by small loops about the punctures of $B^\sharp$ equals $0 \in H_1(B^\sharp; \mathbb{Z}_2)$, the type specification must satisfy the condition $\sum_F \operatorname{dg} \mathfrak{m}_F = 0 \bmod 12$, where $\mathfrak{m}_F$ is the local monodromy about F, see Table 3.2, and the summation runs over all singular fibers F of X. On the other hand, by the (generalized) Riemann–Hurwitz formula, $\chi(X) = \sum_F \chi(F)$. Using (3.45), we arrive at the following well known statement.

Theorem 3.47. *If $p\colon X \to B$ is a relatively minimal elliptic fibration without multiple fibers and over a compact base, then $\chi(X) = 0 \bmod 12$. If p has at least one singular fiber, then $\chi(X) > 0$.* ◁

Jacobian fibrations

A *Jacobian elliptic fibration* is an elliptic fibration $p\colon J \to B$ with a distinguished section $S \subset J$. Since S is a section, it intersects each fiber transversally at a simple component; in particular, p has no multiple fibers. Morphisms, deformations, *etc.* in the category of Jacobian fibrations are required to respect the sections. The automorphism group of a Jacobian elliptic fibration is a cyclic group of order 2, 4, or 6, the latter two cases occurring only if J is isotrivial (and $j_J \equiv 1$ or 0, respectively). In all cases, the order two element is the multiplication by (-1) in each nonsingular fiber F, which, in the presence of the distinguished point $F \cap S$, turns into an abelian variety. Any Jacobian elliptic surface is projective. (According to Kodaira, an elliptic surface is projective if and only if it admits a multi-section, *i.e.*, an irreducible curve S not contained in a fiber, so that the projection $S \to B$ is finite-to-one.)

To each non-isotrivial elliptic fibration $p\colon X \to B$ without multiple fibers one can assign, in a functorial way, a Jacobian fibration $J(X) \to B$ (*basic elliptic surface* in Kodaira's terminology; the isotrivial case needs extra work and we systematically ignore it here). To construct $J(X)$, one replaces each nonsingular fiber F of X with its Jacobian $\operatorname{Pic}^0 F$. Using the so-called *universal elliptic curve*, one can show that this procedure is analytic with respect to the base. Then, by a case by case analysis using the classification of singular fibers and their local normal forms, one shows that each singular fiber can be patched in a unique way. In fact, the types of the singular fibers of $J(X)$ are the same as those of the corresponding fibers of X.

An elliptic curve E is isomorphic to its Jacobian $\operatorname{Pic}^0 E$; hence $j_X = j_{J(X)}$. The groups $H^1(E)$ and $H^1(\operatorname{Pic}^0 E)$ are *canonically* isomorphic; hence also $\mathfrak{h}_X = \mathfrak{h}_{J(X)}$.

Let $J \to B$ be a Jacobian elliptic fibration with the distinguished section S. For each singular fiber F, let F' be the union of the components of F that are disjoint from S. If nonempty, the adjacency graph of the components of F' is the elliptic Dynkin diagram of the same name as the affine Dynkin diagram representing F (see Figure 1.1: to obtain F', one removes one of the vertices labeled with 1). By the Grauert criterion [80], the divisor F' contracts to a (singular) point, which is a rational double point, see [66]. Applying this contraction procedure to each reducible fiber, one obtains a singular surface J^{w}, called the *Weierstraß model* of J. It is still fibered over B. The automorphism $-\operatorname{id}$ of J (the multiplication by (-1) in the nonsingular fibers) descends to J^{w}, and the quotient $J^{\mathrm{w}}/\pm\operatorname{id}$ is a ruled surface $\Sigma \to B$ with a distinguished exceptional section E, the image of S. The quotient projection $J^{\mathrm{w}} \to \Sigma$ is a double covering ramified at E and a certain proper trigonal curve $C \subset \Sigma$. Hence, the line bundle $\mathcal{Y}$ defining Σ (*i.q.* the conormal bundle of $E \cong B$) equals $\mathcal{L}^2$, where $\mathcal{L}$ is the conormal bundle of S. It can be shown that $\mathcal{L}^{-1} \cong R^1 p_* \mathcal{O}_J$. With a slight abuse of the language, we refer to C as the *ramification curve* of J.

Since J^{w} has at worst simple singularities, so does C. In terms of the Weierstraß equation (3.4), this condition is equivalent to the requirement

$$\min\{3\operatorname{ord}_b(g_2), 2\operatorname{ord}_b(g_3)\} < 12 \quad \text{at each point } b \in B, \tag{3.48}$$

where $\operatorname{ord}_b(s)$ stands for the order of vanishing of a section s at point b.

Comparing various definitions of the j-invariant, one concludes that $j_C = j_J$. One also has $\chi(J) = 12\deg\mathcal{L} = -12S^2$, *cf.* the explanation preceding Theorem 3.47 and the discussion of mult F on page 73.

Conversely, given a ruled surface $\Sigma = \mathbb{P}(1 \oplus \mathcal{Y}) \to B$ with exceptional section $E = \mathbb{P}(1)$ and a trigonal curve $C \subset \Sigma$ with at worst simple singularities, a choice of a square root $\mathcal{L}$ of $\mathcal{Y}$ defines a unique double covering $J^{\mathrm{w}} \to \Sigma$ ramified at E and C. This double covering can be embedded to the 3-fold $\mathbb{P}(1 \oplus \mathcal{Y} \oplus \mathcal{L}^3)$, where, in appropriate affine coordinates, it is given by the equation

$$y^2 = x^3 + g_2 x + g_3, \tag{3.49}$$

cf. (3.4). (Here y is a section of $\mathcal{L}^3$. Each term of this equation is naturally a section of $\mathcal{L}^6 = \mathcal{Y}^3$, where the four sections should sum up to zero.) The double covering J^{w} is the Weierstraß model of a Jacobian elliptic fibration $J \to B$; as a surface, J is the minimal resolution of singularities of J^{w}.

Remark 3.50. The Weierstraß model provides a simple way to construct all singular elliptic fibers listed in Table 3.2. In fact, the singular fibers of a Jacobian elliptic surface J ramified at E and a proper trigonal curve C are precisely (the resolved double coverings of) the singular fibers of C, the type of a fiber F of J being the same as Kodaira's type of the corresponding fiber of C. This fact is proved by considering the simple singular fibers one by one and resolving the singularities.

Clearly, a deformation of Jacobian fibrations $J \to B$ gives rise to a deformation of their Weierstraß models and, hence, of triples $(B, \mathcal{L}, C)$ as above. Conversely, an *equisingular* deformation of triples $(B, \mathcal{L}, C)$ lifts to an equisingular deformation of the corresponding Jacobian fibrations. Summarizing, we obtain the following statement.

Theorem 3.51. *There is a canonical one-to-one correspondence between the set of isomorphism classes (equisingular deformation classes) of Jacobian elliptic fibrations* $J \to B$ *and the set of isomorphism classes (respectively, equisingular deformation classes) of triples* $(B, \mathcal{L}, C)$, *where* B *is a compact curve,* $\mathcal{L} \in \operatorname{Pic} B$ *is a line bundle, and* $C \subset \mathbb{P}(1 \oplus \mathcal{L}^2)$ *is a proper trigonal curve with at worst simple singularities.* $\lhd$

If a curve $C \subset \Sigma$ is fixed, the construction just described establishes a map from the set of square roots $\mathcal{L}$ of $\mathcal{Y}$ to the set of homological invariants $\mathfrak{h}$ that belong to j_C and extend the given type specification. Since both sets are principal homogeneous spaces over $H^1(B; \mathbb{Z}_2)$ (or both empty), this map is a bijection and Theorem 3.51 takes the following classical form.

Theorem 3.52 (Kodaira [95]). *For each triple* $(B, j, \mathfrak{h})$, *where* B *is a compact curve,* $j\colon B \to \mathbb{P}^1$ *is a non-constant meromorphic function, and* $\mathfrak{h}$ *is a homological invariant that belongs to* j, *there is a Jacobian elliptic fibration* $J \to B$ *such that* $j_J = j$ *and* $\mathfrak{h}_J = \mathfrak{h}$. *This fibration is unique up to isomorphism.* $\lhd$

Remark 3.53. In Theorem 3.51, instead of a trigonal curve C one can speak about a pair of sections $g_2 \in H^0(B; \mathcal{O}_B(\mathcal{L}^4))$ and $g_3 \in H^0(B; \mathcal{O}_B(\mathcal{L}^6))$, defined up to rescaling (3.5) and subject to (3.48) and the condition $\Delta \not\equiv 0$, where Δ is the discriminant section, see (3.6).

Remark 3.54. Assume that $\Sigma = \mathbb{P}(1 \oplus \mathcal{Y})$. A square root $\mathcal{L}$ of $\mathcal{Y}$ exists if and only if $\deg \mathcal{Y}$ is even, and in this case all square roots form a principal homogeneous space over $H^1(B; \mathbb{Z}_2)$, *cf.* the discussion prior to Theorem 3.52. Thus, given an elementary deformation $C_t \subset \Sigma_t = \mathbb{P}(1 \oplus \mathcal{Y}_t)$ of trigonal curves and a square root $\mathcal{L}_0 \in \operatorname{Pic} B_0$ of $\mathcal{Y}_0$ over $t = 0$, it lifts to a unique elementary deformation of triples $(B_t, \mathcal{L}_t, C_t)$.

If an elementary deformation of triples $(B, \mathcal{L}, C)$ is not equisingular (but singular fibers of C remain simple), the singularities of the coverings J^{w} can be resolved simultaneously after an appropriate base change, see [28, 29]. Hence, Theorem 3.51 extends to not necessarily equisingular deformation classes as well.

We conclude with a few statements concerning the deformation classification (not equisingular) of Jacobian elliptic fibrations. Later, these results are extended to arbitrary elliptic fibrations without multiple fibers using the connectedness of the Tate–Shafarevich group, see Corollary 3.64.

Theorem 3.55 (Moishezon [121]). *By an arbitrarily small deformation in the class of Jacobian fibrations, any non-isotrivial Jacobian elliptic fibration can be perturbed to a Jacobian fibration whose* j*-invariant has generic branching behavior.* $\rhd$ [121, 75]

Theorem 3.56 (Seiler [144, 145]). *Given a pair of integers* $g \geqslant 0$ *and* $n > 0$, *any two relatively minimal Jacobian elliptic fibrations* $J_i \to B_i$, $i = 1, 2$, *over bases* B_i *of genus* g *and with* $\chi(J_i) = 12n$ *are deformation equivalent.* $\rhd$ [144, 145, 75]

Modulo Theorem 3.55, Seiler's Theorem 3.56 reduces to showing that the space of pairs (B, j), where B is a compact curve of genus g and $j\colon B \to \mathbb{P}^1$ is a function with generic branching behavior, is irreducible. (As usual, the isotrivial case needs special treatment.) By the Riemann existence theorem, the latter reduces to showing that, given a closed topological surface B, any two ramified coverings $B \to \mathbb{P}^1$ of the same degree and having generic branching behavior are isotopic in the class of such coverings. In its spirit, this statement is close to Hurwitz Theorem 10.10, and it can be proved similarly, by a computation in the symmetric group, see, *e.g.*, [75]; the algebraic counterpart of this theorem is stated in Section 10.1.3, see Theorem 10.15.

Theorem 3.57. *If $p\colon J \to B$ is a Jacobian elliptic fibration with at least one singular fiber, then $p_*\colon \pi_1(J) \to \pi_1(B)$ is an isomorphism.* ▷ [121, 75]

In [75], this theorem is proved using Theorem 3.56, by considering a particular surface suitable for the computation of the induced homomorphism. Theorem 3.57 extends to arbitrary elliptic fibrations without multiple fibers, see Corollary 3.64. In the most general situation, one has an isomorphism $p_*\colon \pi_1(J) \to \pi_1^{\mathrm{orb}}(B)$, where the orbifold structure on B takes into account the multiple fibers.

Corollary 3.58. *Let $J \to B$ be a relatively minimal Jacobian elliptic fibration over a compact base B of genus $g \geqslant 0$ and with $\chi(J) = 12n > 0$. Then one has*

$$b_1(J) = b_3(J) = 2g, \quad \sigma_+(J) = 2n + 2g - 1, \quad \sigma_-(J) = 10n + 2g - 1,$$

where $\sigma_\pm$ are the inertia indices of the intersection lattice $H_2(J)$. Furthermore, one has $w_2(J) = n[\tilde{F}] \bmod 2$, where $\tilde{F}$ is a fiber.

Proof. Consider the Weierstraß model $\pi\colon J^{\mathrm{w}} \to \Sigma$. After a small deformation, we can assume that the branch locus $S := C + E$ is nonsingular and, hence, $J^{\mathrm{w}} = J$. Then $E^2 = -2n$ and $[S] = [4E + 6nF]$ in $H_2(\Sigma)$.

The Betti numbers $b_1 = b_3$ are given by Theorem 3.57 and Poincaré duality. Since $b_0 = b_4 = 1$, we have $\sigma_+ + \sigma_- = b_2(J) = 12n + 4g - 2$. The signature $\sigma(J) = \sigma_+ - \sigma_-$ can be found using Hirzebruch's signature theorem [87] applied to the deck translation of the ramified covering π: it states that $\sigma(J) + S_J^2 = 2\sigma(\Sigma)$, where S_J^2 is the self-intersection of S *in* J. Since $\sigma(\Sigma) = 0$ and $S_J^2 = \frac{1}{2}S^2 = -4E^2 = 8n$, the expressions for $\sigma_\pm$ follow.

Since the group $H^2(\Sigma)$ is torsion free, the induced map $\pi^*\colon H^2(\Sigma) \to H^2(J)$ is a monomorphism and the assertion on w_2 follows from the topological version of the projection formula: $w_2(J) = \pi^* w_2(\Sigma) + [S] = \pi^*(w_2(\Sigma) + 3nF) \bmod 2$. □

The Mordell–Weil group and Tate–Shafarevich group

Fix a Jacobian elliptic fibration $p\colon J \to B$ with a distinguished section S and let J^{ab} be the surface obtained from J by removing all singular points of all fibers, including multiple components. Choosing $S \cap F$ for zero converts each fiber F of the restricted fibration $J^{\mathrm{ab}} \to B$ into an abelian group, not necessarily connected.

Definition 3.59. The *Mordell–Weil group* $MW(J)$ of a Jacobian elliptic fibration J is the group of holomorphic sections of the fibration $J^{\mathrm{ab}} \to B$. The (*analytic*) *Tate–Shafarevich group* $\text{Ш}(J)$ is the set of isomorphism classes of pairs (X, φ), where $X \to B$ is an elliptic fibration and φ is an isomorphism $J(X) \to J$. (The group structure on $\text{Ш}(J)$ is described below.)

Denote by $\mathcal{J}$ the sheaf of germs of holomorphic sections of the restricted fibration $J^{\mathrm{ab}} \to B$. Then $MW(J) = H^0(B; \mathcal{J})$ (immediately from the definition) and the set $\text{Ш}(J)$ can be identified with $H^1(B; \mathcal{J})$ (and it is this identification that defines the group structure on Ш). In the analytic setting, the latter statement is almost a tautology: one covers B by appropriate Stein manifolds and represents cohomology classes by Čech cocycles.

We outline the proofs of Theorems 3.60 and 3.62 below. Both theorems assume that the induced homomorphism $p_*\colon H_1(J) \to H_1(B)$ is an isomorphism. In view of Theorem 3.57, this is always the case if J has at least one singular fiber.

Theorem 3.60 (Shioda [152]). *If $p_*\colon H_1(J) \to H_1(B)$ is an isomorphism, the map $S' \mapsto [S'] \in NS(J) \subset H_2(J)$ establishes an isomorphism $MW(J) = NS(J)/\mathbf{S}'$, where $\mathbf{S}' \subset NS(J)$ is the sublattice generated by the class $[S]$ of the distinguished section and the classes $[E_i]$ of all irreducible components of all fibers of J.*

Corollary 3.61. *If the fibration $p\colon J \to B$ has at least one singular fiber, the group $MW(J)$ is discrete and one has* $\operatorname{Tors} MW(J) = \operatorname{Tors}(H_2(J)/\mathbf{S}')$. ◁

Theorem 3.62 (see, *e.g.*, [75]). *If $p_*\colon H_1(J) \to H_1(B)$ is an isomorphism, there is a canonical short exact sequence $H^2(J) \to H^2(J; \mathcal{O}_J) \to \text{Ш}(J) \to 0$, the first arrow being induced by the map $\mathbb{Z}_J \to \mathcal{O}_J$, $n \mapsto 2\pi i n$, cf. the exponential exact sequence.*

Corollary 3.63. *If the fibration $p\colon J \to B$ has at least one singular fiber, the group $\text{Ш}(J)$ is connected. Hence, any elliptic fibration $X \to B$ with $\chi(X) > 0$ and without multiple fibers is deformation equivalent to its Jacobian $J(X)$.* ◁

Corollary 3.64. *Theorems* 3.56 *and* 3.57 *and Corollary* 3.58 *extend to arbitrary* (*not necessarily Jacobian*) *elliptic fibrations without multiple fibers.* ◁

Proof of Theorem 3.60 (the idea). Recall that the set of isomorphism classes of line bundles (invertible sheaves) on a smooth analytic variety X is canonically identified with $H^1(X; \mathcal{O}_X^*)$. Hence, the fiberwise Abel–Jacobi map and fiberwise degree of a line bundle give rise to a short exact sequence

$$0 \longrightarrow \mathcal{J} \longrightarrow (R^1 p_* \mathcal{O}_J^*)/\mathcal{S} \xrightarrow{\deg} \mathbb{Z}_B \longrightarrow 0. \tag{3.65}$$

$$
\begin{array}{ccccccccc}
 & & 0 & & 0 & & & & \\
 & & \downarrow & & \downarrow & & & & \\
H^1(B) & \longrightarrow & H^1(B;\mathcal{O}_B) & \longrightarrow & H^1(B;\mathcal{O}_B^*) & \longrightarrow & \mathbb{Z} & \longrightarrow & 0 \\
{\scriptstyle p^*}\downarrow{\scriptstyle \cong} & & {\scriptstyle p^*}\downarrow{\scriptstyle \cong} & & \downarrow & & \downarrow{\scriptstyle 1\mapsto[F]} & & \\
H^1(J) & \longrightarrow & H^1(J;\mathcal{O}_J) & \longrightarrow & H^1(J;\mathcal{O}_J^*) & \longrightarrow & \mathit{NS}(J) & \longrightarrow & 0 \\
 & & & & \downarrow & & & & \\
 & & & & H^0(B;R^1p_*\mathcal{O}_J^*) & & & & \\
 & & & & {\scriptstyle d_2}\downarrow & & & & \\
 & & & & 0 & & & &
\end{array}
$$

Diagram 3.2. Proof of Theorem 3.60.

Here, $\mathcal{S}$ is the sheaf associated to the presheaf sending an open set $U \subset B$ to the set of the isomorphism classes of invertible sheaves of the form $\mathcal{O}_{p^{-1}(U)}(\sum_i r_i E_i)$, where E_i are components of the singular fibers of J and $r_i \in \mathbb{Z}$. It is naturally a subsheaf of $R^1p_*\mathcal{O}_J^*$. Furthermore, this sequence is split, the splitting sending a constant section n over $U \subset B$ to the class of $\mathcal{O}_{p^{-1}(U)}(nS)$.

Note that $\mathcal{S}$ is a skyscraper concentrated at the reducible fibers of J.

Observe also that

$$R^0p_*\mathcal{O}_J = \mathcal{O}_B, \quad R^0p_*\mathcal{O}_J^* = \mathcal{O}_B^*, \quad R^ip_*\mathcal{O}_J^* = 0 \text{ if } i \geqslant 2 \tag{3.66}$$

since the fibers of p are compact and connected and $H^i(D;\mathcal{O}_D^*) = 0$ for any curve D and any $i \geqslant 2$ (an exercise using the exponential exact sequence).

Consider commutative Diagram 3.2, where we use the identifications (3.66). The horizontal exact sequences are obtained from the exponential exact sequence on B and J; in the second row we also use the definition of $\mathit{NS}(J)$. The long column is the five term exact sequence resulting from the Leray spectral sequence for $\mathcal{O}_J$; the differential d_2 vanishes as so does its target $H^2(B;\mathcal{O}_B^*)$. The first arrow marked with p^* is an isomorphism by the hypotheses, and the second one is a restriction of the induced isomorphism $p^*\colon H^1(B;\mathbb{C}) \to H^1(J;\mathbb{C})$ of Hodge structures.

The diagram gives an isomorphism $H^0(B;R^1p_*\mathcal{O}_J^*) = \mathit{NS}(J)/\mathbb{Z}[F]$, where F is a generic fiber. On the other hand, from the split exact sequence (3.65) we have $H^0(B;\mathcal{J}) = H^0(B;R^1p_*\mathcal{O}_J^*)/(H^0(B;\mathcal{S}) \oplus \mathbb{Z}[S])$, and it remains to notice that $H^0(B;\mathcal{S})$ is generated by the components $[E_i]$ of reducible fibers. □

Proof of Theorem 3.62 (the idea). We use the description of $\mathcal{J}$ given by the split exact sequence (3.65). Since $\mathcal{S}$ is a skyscraper, it can be ignored when computing the homology H^i with $i \geqslant 1$: one has $H^i(B; (R^1 p_* \mathcal{O}_J^*)/\mathcal{S}) = H^i(B; R^1 p_* \mathcal{O}_J^*)$.

In the Leray spectral sequence for $\mathcal{O}_J^*$, the terms $E_2^{p,q}$ with either $q \geqslant 2$ or $q = 0$ and $p \geqslant 2$ vanish, see (3.66) and explanation thereafter, and we have isomorphisms $H^i(B; R^1 p_* \mathcal{O}_J^*) = H^{i+1}(J; \mathcal{O}_J^*)$ for all $i \geqslant 1$. The fiberwise degree homomorphism deg in (3.65) is a quotient of the boundary homomorphism $R^1 p_* \mathcal{O}_J^* \to R^2 p_* \mathbb{Z}_J$ in the exact sequence of higher direct images associated to the exponential exact sequence. Hence, from the Leray spectral sequence we have a commutative diagram

$$\begin{array}{ccccccccc} H^2(J) & \longrightarrow & H^2(J; \mathcal{O}_J) & \longrightarrow & H^2(J; \mathcal{O}_J^*) & \longrightarrow & H^3(J) & & \\ & & & & \downarrow \cong & & \downarrow \cong & & \\ 0 & \longrightarrow & H^1(B; \mathcal{J}) & \longrightarrow & H^1(B; R^1 p_* \mathcal{O}_J^*) & \longrightarrow & H^1(B) & \longrightarrow & 0 \end{array}$$

with exact rows, which are associated with the exponential exact sequence for $\mathcal{O}_J^*$ and with (3.65), respectively. Finally, using the multiplicative properties of the Leray sequence, it can be shown that the last vertical arrow is Poincaré dual to $p_* \colon H_1(J) \to H_1(B)$, which is an isomorphism by the hypotheses, and the statement of the theorem follows from the identification $\text{Ш}(J) = H^1(B; \mathcal{J})$. □

Extremal elliptic surfaces

A non-isotrivial compact Jacobian elliptic surface J is *extremal* if it has maximal Picard rank $\operatorname{rk} NS(J) = h^{1,1}(J)$ and minimal Mordell–Weil rank $\operatorname{rk} MW(J) = 0$. The following theorem is a restatement of the criterion found in [126].

Theorem 3.67 (Nori [126]). *A Jacobian elliptic surface J is extremal if and only if its ramification curve C is a maximal trigonal curve without unstable singular fibers.*

Proof. Due to Theorem 3.60, $\operatorname{rk} MW(J) = 0$ if and only if $NS(J)$ is spanned over $\mathbb{Q}$ by the section and the components of the singular fibers of J. Since these classes span a sublattice of rank $\mu(C) + 2$ and $h^{1,1}(J) = \sigma_-(J) + 1$, the statement easily follows from comparing the counts given by Corollaries 3.58 and 4.37. □

3.3 Real structures

In this section, we give a brief overview of topology of real algebraic varieties and specialize some of the general results to real trigonal curves and real elliptic surfaces. Unfortunately, there still seems to be no comprehensive textbook on this fascinating subject. For more details and further references, we suggest the reader to consult monographs [59, 153] and recent survey [62].

3.3.1 Real varieties

Recall that a *real structure* on a complex analytic variety X is an anti-holomorphic involution, *i.e.*, a map $c\colon X \to X$ with $c^2 = \mathrm{id}$, differentiable at smooth points of X and such that dc is $\mathbb{C}$-anti-linear at each smooth point. A *real variety* is a pair (X, c), where X is a complex variety and c is a real structure on X. (For the obvious reason, we avoid the term 'real analytic variety'.) If c is understood, we omit it in the notation.

The *real part* $X_{\mathbb{R}}$ of a real variety X is the fixed point set $\operatorname{Fix} c$ of the real structure. If nonempty, $X_{\mathbb{R}}$ is a submanifold of X (nonsingular if X is nonsingular) of pure real dimension $n = \dim_{\mathbb{C}} X$. The motivation for these definitions is the coordinatewise complex conjugation $z \mapsto \bar{z}$ on $\mathbb{C}^n$ or $\mathbb{P}^n$: an affine (projective) variety X is real in the conventional sense (defined by real polynomials) if and only if it is invariant under the conjugation, and real points of X are those fixed by the conjugation. In the algebraic setup, c can be regarded as the generator of the Galois group $\operatorname{Gal}(\mathbb{C}/\mathbb{R})$, and one speaks about varieties defined over $\mathbb{C}$ and their real forms.

Morphisms, deformations, *etc.* of real varieties are called *real*, or *equivariant*, if they preserve the real structures. From the topological point of view, interesting is the study of equivariant deformation classes of real structures within a given complex deformation family $\mathcal{X}$ (which is usually assumed equisingular). The ultimate end is the classification, an intermediate goal being the understanding of the restrictions on the topology of $X_{\mathbb{R}}$ or, more generally, of the pair (X, c) imposed by the topology and/or analytic structure of X. A complex family $\mathcal{X}$ is called *quasi-simple* if, within this family, the equivariant deformation type of a real structure c is determined by the topology of (X, c). Most families for which the complete classification is known (*e.g.*, abstract curves, surfaces of Kodaira dimension $\leqslant 0$, and certain elliptic surfaces) are quasi-simple; an example of a family (of surfaces of general type) that is not quasi-simple is found in [100]. In general, it is not even known if the number of deformation classes is necessarily finite.

As a common practice, one usually studies equivariant deformations of *nonsingular* real varieties, as opposed to equisingular deformations of *singular* complex varieties (within the same deformation family). The two problems are of approximately equal level of difficulty and, to much surprise, often use similar techniques.

One of the very first and most important restrictions to the topology of the real part of a real variety is the *Thom–Smith inequality*

$$b_*(X_{\mathbb{R}}; \mathbb{Z}_2) \leqslant b_*(X; \mathbb{Z}_2), \tag{3.68}$$

where b_* stands for the total Betti number. Usually (3.68) is derived from the so-called Smith exact sequence; the easiest, but requiring most prerequisites proof is a reference to Kalinin's spectral sequence $H_*(X; \mathbb{Z}_2) \Rightarrow H_*(X_{\mathbb{R}}; \mathbb{Z}_2)$, see [91, 59]. The difference $b_*(X; \mathbb{Z}_2) - b_*(X_{\mathbb{R}}; \mathbb{Z}_2)$ is always even; if this difference equals $2d$, the variety is called an *$(M-d)$-variety*. (Here M stands for 'maximal' and d is a number.) Usually it is the *M-varieties* that exhibit most interesting topological properties.

As a byproduct of any of the above proofs, one can deduce that, if (X, c) is an M-variety, the induced homomorphism $c_*\colon H_*(X;\mathbb{Z}_2) \to H_*(X;\mathbb{Z}_2)$ is the identity. The converse is not true. On the other hand, if X is compact, $\dim_{\mathbb{C}} X = n$, the image of $[X_{\mathbb{R}}]$ in $H_n(X;\mathbb{Z}_2)$ is the characteristic class of the *twisted intersection form* $a \otimes b \mapsto a \circ c_* b$. (It is a routine check that this expression does define a symmetric bilinear form, and the assertion can easily be proved geometrically: if $a = [A]$ is realized by a submanifold, then $\operatorname{Card}(A \cap c(A)) = \operatorname{Card}(A \cap [X_{\mathbb{R}}]) \bmod 2$. Both the observation and its proof are due to V. Arnol′d.) As a consequence, for an M-variety X one has $[X_{\mathbb{R}}] = u_n(X) \in H_n(X;\mathbb{Z}_2)$, where u_n is the n-th Wu class.

Definition 3.69. A compact real variety (X, c) of dimension $\dim_{\mathbb{C}} X = n$ is said to be of *type* I if $[X_{\mathbb{R}}] = u_n(X) \in H_n(X;\mathbb{Z}_2)$, where u_n is the n-th Wu class.

For curves, the condition $[X_{\mathbb{R}}] = u_n(X)$ is equivalent to $[X_{\mathbb{R}}] = 0 \in H_1(X;\mathbb{Z}_2)$, *cf.* the definition of separating curves on page 93. For surfaces, the condition turns into $[X_{\mathbb{R}}] = w_2(X) = K_X \bmod 2 \in H_2(X;\mathbb{Z}_2)$.

A *real sheaf* on a real variety (X, c_X) is a sheaf $\mathcal{A}$ equipped with a morphism $c\colon \mathcal{A} \to c_X^*\mathcal{A}$ involutive in the sense that $c \circ c_X^* c = \mathrm{id}$, where $c_X^* c\colon c_X^*\mathcal{A} \to \mathcal{A}$ is the pull-back of c. Informally, c can be regarded as a sheaf morphism over the morphism c_X of the bases; this one is involutive in the usual sense. Denote by $\pi\colon X \to X/c_X$ the quotient projection and consider the induced sheaf morphism $c' := \pi_* c\colon \pi_*\mathcal{A} \to \pi_*\mathcal{A}$. Then c is involutive if and only if $c'^2 = \mathrm{id}$.

For any abelian group G, the constant sheaf G_X has a canonical real structure, which is the identity $G_X = c_X^* G_X$.

Given a real sheaf $(\mathcal{A}, c)$, the sequence

$$\pi_*\mathcal{A}^*: \quad 0 \longrightarrow \pi_*\mathcal{A} \xrightarrow{1-c'} \pi_*\mathcal{A} \xrightarrow{1+c'} \pi_*\mathcal{A} \xrightarrow{1-c'} \cdots \tag{3.70}$$

is a complex of sheaves over X/c_X. (We assign degree 0 to the leftmost copy of $\pi_*\mathcal{A}$.) To avoid excessive notation, we denote the hypercohomology $\mathbf{H}^*(X/c_X; \pi_*\mathcal{A}^*)$ by $\mathbf{H}^*(\mathcal{A}, c)$, or just $\mathbf{H}^*(\mathcal{A})$ if c is understood. Since the projection π is finite-to-one, one has $H^*(X;\mathcal{A}) = H^*(X/c_X; \pi_*\mathcal{A})$ and the hypercohomology spectral sequence turns into

$$H^p(\mathbb{Z}_2; H^q(X;\mathcal{A})) \Longrightarrow \mathbf{H}^{p+q}(\mathcal{A}, c). \tag{3.71}$$

Similarly, one defines a *real line bundle* over a real variety (X, c_X) as a complex line bundle $\mathcal{L}$ equipped with an involutive anti-linear bundle morphism c over c_X. Restricting $\mathcal{L}$ to the real part $\mathbb{X}_{\mathbb{R}}$, one obtains a topological 1-bundle with an anti-linear involution c. The invariant subspaces of c form a real vector bundle $\mathcal{L}_{\mathbb{R}}$ of dimension one; this bundle is called the *real part* of $\mathcal{L}$.

The next theorem is a 'typical' restatement of a known complex result in the real setting, *cf.* also Theorem 3.84 *vs.* Theorem 3.62; we outline its proof as an example.

Theorem 3.72. *The set of isomorphism classes of real line bundles over X is in a canonical one-to-one correspondence with the group $\mathbf{H}^1(\mathcal{O}_X^*)$.*

Proof. Let c be the natural real structure on $\mathcal{O}_X^*$, regarded as a morphism over c_X. Cover X by a c_X-invariant system $\{U_i\}$ of Stein manifolds and trivialize $\mathcal{L}$ over each U_1. Let (C^*, d) be the Čech complex for $\mathcal{O}_X^*$. As in the complex setting, the bundle structure of $\mathcal{L}$ is defined by a 1-cocycle $\delta_1 = \{s_{ij}\}$. The real structure on $\mathcal{L}$ can be defined by a 0-cochain $\delta_0 = \{s_i\}$, so that a section $u \mapsto z(u)$ over U_i is taken to the section $c_X(u) \mapsto s_i(u)\bar{z}(u)$ over $c_X(U_i)$. It is a routine check that this map is well defined (the pieces agree over the intersections $U_i \cap U_j$) if and only if $d\delta_0 + (1 - c)\delta_1 = 0$, and it is an involution if and only if $(1 + c)\delta_0 = 0$. (To be consistent with (3.70) and avoid complicated expressions, we use the additive notation for $\mathcal{O}_X^*$.) In other words, the pair (δ_1, δ_0) must be a 1-cocycle in the double complex associated with (3.70) and thus determines a cohomology class in $\mathbf{H}^1(\mathcal{O}_X^*)$. Any other trivialization of $\mathcal{L}$ differs from the chosen one by a 0-cochain δ, resulting in adding $d\delta$ to δ_1 and $(1 - c)\delta$ to δ_0, *i.e.*, changing (δ_1, δ_0) by a coboundary. □

Corollary 3.73. *If B is a compact real curve with $B_{\mathbb{R}} \neq \varnothing$, then there is a canonical exact sequence*

$$0 \longrightarrow \operatorname{Pic}^0_{\mathbb{R}} B \longrightarrow \mathbf{H}^1(\mathcal{O}_B^*) \xrightarrow{\ \alpha\ } H^1(B_{\mathbb{R}}; \mathbb{Z}_2) \oplus H^2(B) \xrightarrow{\ \beta\ } \mathbb{Z}_2,$$

where $\operatorname{Pic}^0_{\mathbb{R}} B$ is the component of zero of the real part $\operatorname{Pic}_{\mathbb{R}} B$ (with respect to the induced real structure), α is the homomorphism $\mathcal{L} \mapsto w_1(\mathcal{L}_{\mathbb{R}}) \oplus \deg \mathcal{L}$, and β is the homomorphism $(w, d) \mapsto w[B_{\mathbb{R}}] + d \bmod 2$.

Proof. The exponential exact sequence and the fact that $\mathbf{H}^2(\mathcal{O}_B) = 0$, see (3.71), give us an exact sequence

$$\longrightarrow \mathbf{H}^1(\mathbb{Z}_B^-) \xrightarrow{\ \gamma\ } \mathbf{H}^1(\mathcal{O}_B) \longrightarrow \mathbf{H}^1(\mathcal{O}_B^*) \longrightarrow \mathbf{H}^2(\mathbb{Z}_B^-) \longrightarrow 0,$$

where $\mathbb{Z}_B^-$ is the constant sheaf with the real structure multiplied by (-1). One has $\operatorname{Coker}\gamma = \operatorname{Pic}^0_{\mathbb{R}} B$ and $\mathbf{H}^2(\mathbb{Z}_B^-) = H^1(\mathbb{Z}_2; H^1(B; \mathbb{Z}_B^-)) \oplus H^2(B)$ due to (3.71). If $B_{\mathbb{R}} \neq \varnothing$, then $H^1(\mathbb{Z}_2; H^1(B; \mathbb{Z}_B^-)) = \operatorname{Ker} \beta'$, where $\beta' \colon H^1(B_{\mathbb{R}}; \mathbb{Z}_2) \to \mathbb{Z}_2$ is the map $w \mapsto w[B_{\mathbb{R}}]$; this fact can be proved using integral Kalinin's spectral sequence, see [59] for details. □

Abstract curves

Let X be an abstract connected compact curve of genus g with a real structure c. The real part $X_{\mathbb{R}}$ consists of a certain number l of topological circles; according to (3.68), one has $l \leqslant g + 1$. (In the context of curves, the latter statement is known as the *Harnack–Klein inequality*; it can easily be proved by cutting X along $X_{\mathbb{R}}$, patching the cuts with disks, and counting the Euler characteristic.) The complement $X \setminus X_{\mathbb{R}}$

either is connected or consists of two components, interchanged by c. In the former case, the curve is said to be of *type* II, or *non-separating*, in the latter case it is of *type* I, or *separating*. Since in dimension one an anti-holomorphic map reverses the complex orientation, X is separating if and only if the orbit space X/c is orientable. The proof of the Harnack–Klein inequality indicated above shows as well that

- if $l = g + 1$, then X is of type I, and
- if X is of type I, then $l > 0$ and $l = g + 1 \bmod 2$.

The real part of a separating curve X bounds any of the two components of $X \smallsetminus X_{\mathbb{R}}$, inheriting from it a pair of opposite orientations, called the *complex orientations*. A *directed curve* is a separating curve with one of the complex orientations chosen; an orientation preserving diffeo-/homeomorphism of such curves is *directed* if it also preserves the chosen orientations of the real parts.

Next statement, known as Weichold's theorem, is merely the classification of compact 2-manifolds with boundary applied to the quotient X/c.

Theorem 3.74 (Weichold). *Any pair* (l, type) *satisfying the restrictions above and the inequality* $l \leqslant g+1$ *is realized by an orientation reversing involution* $c\colon X \to X$, *unique up to conjugation by an element of* $\text{Map}^+(X)$. ◁

In the analytic category, the equivariant deformation classification of abstract real curves is covered by the following theorem. For proof, one shows that the real curves with a fixed topological type of a real structure form the real part of an appropriate anti-linear real structure on the Teichmüller space (a period space of complex curves of a fixed genus). The connectedness of this real part follows directly from (3.68), as the Teichmüller space is known to be contractible.

Theorem 3.75 (Natanzon [123]). *Any topological type* (l, type) *of orientation reversing involutions* $c\colon X \to X$ *can be realized by a real structure. The real curves* (X, c) *with a given topological type of* c *admit a period space which is a ball of real dimension* $3g - 3$ *if* $g > 1$ *or* 1 *if* $g = 1$. *In particular, this space is connected.* ▷ [123]

In the case of rational and elliptic curves, one can easily obtain a complete analytic classification of real structures. The next three statements are so classical that I find it difficult to give a precise attribution.

Theorem 3.76. *Up to automorphism, there are two real structures on the projective line* $\mathbb{P}^1 \cong \mathbb{C} \cup \{\infty\}$; *they are* $c_0\colon z \mapsto \bar{z}$ *with* $\text{Fix}\, c_1 = \mathbb{P}^1_{\mathbb{R}}$ *and* $c_1\colon z \mapsto -1/\bar{z}$ *with* $\text{Fix}\, c_1 = \varnothing$.

Proof. Since any two anti-automorphisms differ by an automorphism, any real structure is the composition of the conjugation $z \mapsto \bar{z}$ and a Möbius transformation, *i.e.*, has the form

$$c\colon z \mapsto \frac{a\bar{z} + b}{c\bar{z} + d}, \quad a, b, c, d \in \mathbb{C}, \quad ac - bd \neq 0.$$

If $\operatorname{Fix} c \neq \varnothing$, we can choose the coordinates so that $0, 1, \infty \in \operatorname{Fix} c$; then $c\colon z \to \bar{z}$. If $\operatorname{Fix} c = \varnothing$, choose the coordinates so that $c(0) = \infty$. Then $c\colon z \mapsto a/\bar{z}$ and, since c is an involution, $a \in \mathbb{R}$. After a further rescaling this turns into $z \mapsto \pm 1/\bar{z}$, and the first choice $z \mapsto 1/\bar{z}$ has fixed points, hence is conjugate to c_0. □

A *Jacobian elliptic curve* is an elliptic curve J with a distinguished point $0 \in J$; such a curve is an abelian group and, in particular, has an automorphism $-\mathrm{id}\colon J \to J$. Automorphisms and real structures of such a curve are required to preserve 0. If c is a real structure on J, then $-c := c \circ (-\mathrm{id})$ is also a real structure.

Theorem 3.77. *A Jacobian elliptic curve J admits a real structure if and only if the j-invariant $j := j(J)$ is real. One has:*

- *if $j \neq 0, 1$, then J has two non-isomorphic real structures $\pm c$, which are both M-structures if $j > 1$ and both $(M-1)$-structures if $j < 1$;*
- *if $j = 0$, then J has six $(M-1)$-structures $\pm c_i$, $i = 1, 2, 3$; the adjoint action of $\operatorname{Aut} J$ on the set of real structures has two orbits: c_1, c_2, c_3 and $-c_1, -c_2, -c_3$;*
- *if $j = 1$, then J has two M-structures $\pm c_1$ and two $(M-1)$-structures $\pm c_2$; the structures $\pm c_1$ are conjugate (by an order four automorphism), and so are $\pm c_2$.*

Proof. Let $c\colon J \to J$ be a real structure. It commutes with $-\mathrm{id}$ (*e.g.*, because $c \circ (-\mathrm{id}) \circ c$ is an order two automorphism of J, which is unique) and hence descends to the quotient $E/\pm\mathrm{id} \cong \mathbb{P}^1 = \mathbb{C} \cup \{\infty\}$. Let $Z := \{z_1, z_2, z_3, \infty\} \subset \mathbb{P}^1$ be the branch locus, so that ∞ is the image of $0 \in J$. This quadruple of points must be real with respect to a certain real structure c' on $\mathbb{P}^1$ (the descent of c) preserving ∞; hence its j-invariant is real, see Remark 3.14. If $j \neq 0$ or 1, the quadruple Z has no automorphisms and such a real structure c' is unique; it lifts to two real structures $\pm c$ on J. In the unstable cases $j = 0$ and $j = 1$, there are three (respectively, two) real structures c', which lift to six (respectively, four) real structures on J. The case by case analysis of the action of the automorphism group and of the fixed point sets is left to the reader; Figure 3.1 is a clue. □

Remark 3.78. M- and $(M-1)$- real structures on J can be told apart by the induced action on the homology $H_1(J)$. Under an appropriate identification $H_1(J) = \mathcal{H}$, an M-structure c induces the automorphism $(\mathbf{a}, \mathbf{b}) \mapsto (\mathbf{a}, -\mathbf{b})$, and an $(M-1)$-structure induces $(\mathbf{a}, \mathbf{b}) \mapsto (\mathbf{b}, \mathbf{a})$. Both are anti-symplectic and both have (± 1)-eigenspaces of dimension one each. However, in the case of an M-structure, primitive eigenvectors generate $\mathcal{H}$, whereas in the case of an $(M-1)$-structure they only generate an index two subgroup.

Now, let E be an elliptic curve without a distinguished point, and let $J = \operatorname{Pic}^0 E$ be its Jacobian. Any real structure c on E induces a real structure c_J on J. If $\operatorname{Fix} c \neq \varnothing$, one can fix a *real* point $0 \in E$ and identify c with c_J; since $\operatorname{Aut} E = J \rtimes \operatorname{Aut} J$, such real structures are counted by Theorem 3.77 up to translations.

Theorem 3.79. *An elliptic curve E admits a real structure if and only if $j := j(E)$ is real. In addition to Theorem 3.77, counting real structures up to translation, in the case $j \geqslant 1$ there are two (up to translations) real structures $\pm c$ with empty real parts; they induce M-structures on the jacobian J or E.*

Proof. Represent J as the quotient $\mathbb{C}/L$, where $L \subset \mathbb{C}$ is a lattice. The real structure c_J lifts to $\mathbb{C}$, and one can assume that the lift is given by $z \mapsto \bar{z}$ and $L = \mathbb{Z} \cdot 1 + \mathbb{Z} \cdot bi$, $b \in \mathbb{R}^+$, if $j \geqslant 1$ (the M-structures) or $L = \mathbb{Z} \cdot \tau + \mathbb{Z} \cdot \bar{\tau}$, $\operatorname{Im} \tau > 0$, if $j \leqslant 1$ (the $(M-1)$-structures). If $j = 0$, there also is an exceptional lattice $L = \mathbb{Z} \cdot 1 + \mathbb{Z} \cdot \rho$, $\rho^3 = 1$, also resulting in an $(M-1)$-structure. Then c is given by $z \mapsto \bar{z} + a$ for some fixed $a \in \mathbb{C}$ with $2 \operatorname{Re} a \in L$. It is easy to check that, up to translations, the only choice resulting in $\operatorname{Fix} c = \varnothing$ is $j \geqslant 1$ and $a = 1/2$. □

Under the canonical identification $H_1(E) = H_1(J)$, the automorphisms induced by c and c_J coincide. Hence, homologically a real structure with the empty real part is indistinguishable from an M-structure.

Remark 3.80. If (E, c) is a real elliptic curve or, more generally, a topological torus with orientation reversing involution, an isomorphism $H_1(E) \cong \mathcal{H}$ can be chosen canonically up to sign. Let $e_\pm \in H_1(E)$ be primitive (± 1)-eigenvectors of c_*, and assume that $e_+ \circ e_- > 0$; this assumption determines $e_\pm$ up to *simultaneous* change of sign. Then the isomorphism in question is obtained by choosing the basis $\mathbf{a} = e_+$, $\mathbf{b} = e_-$ in the case of an M- or $(M-2)$-structure or $\mathbf{a} = \frac{1}{2}(e_+ - e_-)$, $\mathbf{b} = \frac{1}{2}(e_+ + e_-)$ in the case of an $(M-1)$-structure.

3.3.2 Real trigonal curves and real elliptic surfaces (see [60])

Consider a real geometrically ruled surface $p\colon \Sigma \to B$ with a real exceptional section E. In other words, we fix two real structures $c_\Sigma\colon \Sigma \to \Sigma$ and $c_B\colon B \to B$, commuting with p and such that E is c_Σ-invariant. Denote by $p_\mathbb{R}$ the restriction $\Sigma_\mathbb{R} \to B_\mathbb{R}$. Due to the existence of a section, all fibers of $p_\mathbb{R}$ are isomorphic to $\mathbb{P}^1_\mathbb{R}$, see Theorem 3.76, *i.e.*, are topological circles. Hence, $p_\mathbb{R}$ establishes a bijection between the set of connected components Σ_i of $\Sigma_\mathbb{R}$ and that of connected components B_i of $B_\mathbb{R}$, and each component Σ_i is either a torus or a Klein bottle.

Assume that $\Sigma = \mathbb{P}(1 \oplus \mathcal{Y})$ for some real line bundle $\mathcal{Y}$ over B, and denote by $\mathcal{Y}_i$ the restriction of the real part $\mathcal{Y}_\mathbb{R}$ to B_i. Then a component Σ_i is orientable if and only if $\mathcal{Y}_i$ is topologically trivial, *i.e.*, $w_1(\mathcal{Y}_i) = 0$. Note that $\sum_i w_1(\mathcal{Y}_i) = \deg \mathcal{Y} \bmod 2$. (For example, if $\mathcal{Y} = \mathcal{O}_B(D)$ for some *real* divisor D, then $w_1(\mathcal{Y}_i)$ is the class Poincaré dual to $[D \cap B_i]$.) In particular, $\Sigma_\mathbb{R}$ is nonorientable whenever $\deg \mathcal{Y}$ is odd.

Real trigonal curves

Let $p\colon \Sigma \to B$ be a real ruled surface as above, and consider a proper real, *i.e.*, c_Σ-invariant, trigonal curve $C \subset \Sigma$. Then the line bundle $\mathcal{Y}$ in $\Sigma = \mathbb{P}(1 \oplus \mathcal{Y})$ is also real,

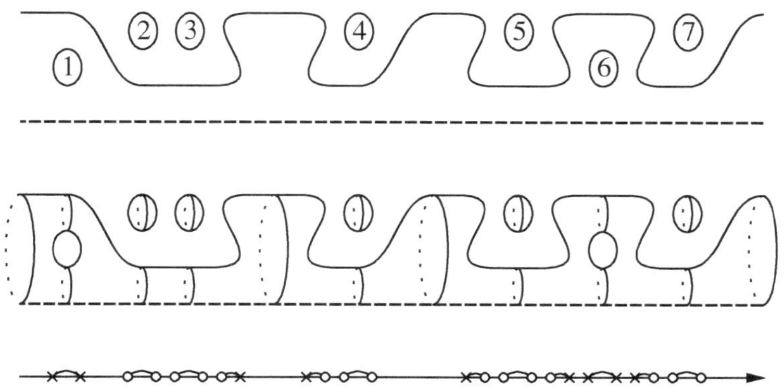

Figure 3.3. A non-hyperbolic trigonal curve (top), a covering Jacobian surface (middle), and its uncoated necklace diagram (bottom); the horizontal dotted lines represent the distinguished sections.

and in the Weierstraß equation (3.4) of C the sections g_2, g_3 can be chosen real; they are well defined up to rescaling (3.5) with real section s.

The j-invariant of a real trigonal curve C is obviously real (with respect to the standard real structure $z \mapsto \bar{z}$ in the target $\mathbb{C} \cup \{\infty\}$). Conversely, analyzing the proof of Theorem 3.20, one arrives at the following statement.

Theorem 3.81. *The trigonal curve $C \subset \Sigma$ corresponding to a real triple (B, j, D_{ex}) (see Theorem* 3.20 *and Remark* 3.21*) can be chosen real, uniquely up to rescaling* (3.5) *with real section s. Equivariant deformations of trigonal curves result in equivariant deformations of the corresponding triples (B, j, D_{ex}) and vice versa.* ◁

In the real setting, we usually assume that *all singular fibers of C are of type* $\tilde{\mathbf{A}}_0^*$; in particular, C is nonsingular. A real trigonal curve with this property is called *almost generic*. Denote by $q_{\mathbb{R}}\colon C_{\mathbb{R}} \to B_{\mathbb{R}}$ the restriction of p. The real part $C_{\mathbb{R}}$ splits into *groups of components* $C_i := q_{\mathbb{R}}^{-1}(B_i)$, and each restriction $q_i\colon C_i \to B_i$ of $q_{\mathbb{R}}$ is onto. A component B_i and the corresponding group C_i are called *hyperbolic* if q_i is generically three-to-one; otherwise, B_i and C_i are called *non-hyperbolic*. The curve C is called *hyperbolic* if $C_{\mathbb{R}} \neq \varnothing$ and *all* groups C_i are hyperbolic.

The brief description of the topology of the real part $C_{\mathbb{R}} \subset \Sigma_{\mathbb{R}}$ given below is a simple consequence of the fact that $C_{\mathbb{R}}$ intersects each fiber of the projection $p_{\mathbb{R}}$ in at most three points, which are away from $E_{\mathbb{R}}$.

Over a hyperbolic component B_i, the group C_i of an almost generic curve consists of a 'central' component, mapped onto B_i homeomorphically, and two (if Σ_i is orientable) or one (if Σ_i is nonorientable) additional components; the restriction of q_i to the additional components is a double covering, trivial in the former case and nontrivial in the latter case.

The group C_i of an almost generic curve over a non-hyperbolic component B_i looks like shown in Figure 3.3, top. More precisely, C_i has one 'long' component that is mapped onto B_i and a few contractible components, commonly called *ovals.* In addition, the long component may have a few *zigzags* (the Z-shaped fragments in Figure 3.3), which are also preserved by equisingular equivariant deformations. For trigonal curves, it is convenient to define ovals and zigzags as portions of the base component B_i.

Definition 3.82. *Ovals* and *zigzags* of a non-hyperbolic group C_i of an almost generic real trigonal curve C are the connected components of the set

$$\{b \in B_i \mid \operatorname{Card} q_i^{-1}(b) \geqslant 2\};$$

ovals are those components whose pull-back is disconnected.

The set of all ovals and zigzags of a non-hyperbolic group C_i inherits from the circle B_i a natural pair of opposite cyclic orders. Fix one of the orders and consider a continuous section $\sigma\colon B_i \to \Sigma_i$ disjoint from $E_{\mathbb{R}}$ and taking each oval inside the corresponding contractible component of C_i; such a section exists and is unique up to homotopy. A set $o_1 \prec o_2 \prec \cdots \prec o_k$ of consecutive ovals (possibly separated by zigzags) is said to form a *chain* if, between any two neighbors o_j, o_{j+1}, $j = 1, \ldots, k-1$, the section σ intersects the long component an even number of times. For example, in Figure 3.3, the maximal chains are $[1]$, $[2,3,4,5]$, $[6]$ and $[7]$ (assuming that Σ_i is orientable; otherwise, ovals 7 and 1 form a single maximal chain).

Real Jacobian surfaces

A Jacobian elliptic fibration $p\colon J \to B$ is called *real* if both the projection p and the distinguished section S are real (with respect to a pair of fixed real structures $c_J\colon J \to J$ and $c_B\colon B \to B$). Under these assumptions, the conormal bundle $\mathcal{L}$ of S, the automorphism $-\operatorname{id}\colon J \to J$, and the Weierstraß model J^{w} are also real, and hence J^{w} is a double covering of a real ruled surface $\Sigma_{\mathbb{R}} := J^{\mathrm{w}}/\pm\operatorname{id}$ ramified over the exceptional section E (the image of S) and a real proper trigonal curve $C \subset \Sigma$.

Conversely, if $C \subset \Sigma = \mathbb{P}(1 \oplus \mathcal{Y}) \to B$ is a real trigonal curve with at worst simple singularities and $\mathcal{L}$ is a *real* square root of $\mathcal{Y}$, the minimal resolution of singularities of the surface J^{w} given by (3.49) inherits two *opposite* real structures $\pm c_J$, which differ by $-\operatorname{id}$ (the deck translation of the covering). The real part $\Sigma_{\mathbb{R}}$ splits into two *halves* $\Sigma_{\mathbb{R}}^{\pm}$, disjoint except for the common boundary $C_{\mathbb{R}}$; they are the projections of the real parts $J_{\mathbb{R}}^{\pm}$ corresponding to the two real structures. A choice of one of the two real structures is equivalent to a choice of one of the two halves. (Note that, in spite of the commonly used notation $\pm$, neither of the real structures is distinguished.)

Assume that $B_{\mathbb{R}} \neq \varnothing$. Then, due to Theorem 3.72 and Corollary 3.73, a real square root $\mathcal{L}$ of $\mathcal{Y}$ exists if and only if $w_1(\mathcal{Y}_{\mathbb{R}}) = 0$, *i.e.*, if $\Sigma_{\mathbb{R}}$ is orientable. If this is the case, a similar computation shows that all real roots of $\mathcal{Y}$ form a principal homogeneous space over $\mathbf{H}^1(B; \mathbb{Z}_2)$. We do not use this statement, and we refer to [60] for further details. Note though that, due to Corollary 3.73 again, the real roots $\mathcal{L}$ are partially distinguished by the class $w_1(\mathcal{L}_{\mathbb{R}}) \in H^1(B_{\mathbb{R}}; \mathbb{Z}_2)$, which can take any value subject to the restriction $w_1(\mathcal{L}_{\mathbb{R}})[B_{\mathbb{R}}] = \frac{1}{2} \deg \mathcal{Y} \bmod 2$.

Fix a root $\mathcal{L}$ of $\mathcal{Y}$ and one of the two real structures on the corresponding surface J. Assume that J is *almost generic*, *i.e.*, that all its singular fibers are of type $\tilde{\mathbf{A}}_0^*$. Then the real part $J_{\mathbb{R}}$ is a double of the corresponding half of $\Sigma_{\mathbb{R}}$, see Figure 3.3, middle. The real part splits into groups $J_i := p_{\mathbb{R}}^{-1}(B_i)$, consisting of whole components. Each group J_i has a distinguished *principal* component, which contains the points of S over B_i. It is orientable if and only if the restriction $\mathcal{L}_i$ of $\mathcal{L}_{\mathbb{R}}$ to B_i is topologically trivial, *i.e.*, if $w_1(\mathcal{L}_i) = 0$. If B_i is non-hyperbolic, all other components of J_i are spheres. If B_i is hyperbolic, there is exactly one extra component in J_i, which is homeomorphic to the principal one; both are either tori or Klein bottles.

Lemma 3.83. *An almost generic real Jacobian surface J is an $(M-d)$-variety if and only if so is the branch locus of its Weierstraß model.*

Proof. Using Theorem 3.57, Poincaré duality, and the Riemann–Hurwitz formula $\chi(J) = 2\chi(\Sigma) - \chi(C \cup E) = 4\chi(B) - \chi(C \cup E)$, one obtains $b_*(J; \mathbb{Z}_2) = b_*(C; \mathbb{Z}_2) + b_*(B; \mathbb{Z}_2)$. On the other hand, from the topological description of $J_{\mathbb{R}}$ one has $b_*(J_{\mathbb{R}}; \mathbb{Z}_2) = b_*(C_{\mathbb{R}}; \mathbb{Z}_2) + b_*(B_{\mathbb{R}}; \mathbb{Z}_2)$: both equal $4b_0(B_{\mathbb{R}}) + 2l + 4h$, where l is the number of ovals of the real part $C_{\mathbb{R}}$ and h is the number of its hyperbolic components. □

The Jacobian $J(X) \to B$ of a real elliptic fibration $X \to B$ is naturally real. Hence, given a real Jacobian fibration $J \to B$, one can speak about the set of the isomorphism classes of pairs (X, φ), where $X \to B$ is a real elliptic fibration and $\varphi\colon J(X) \to J$ is a real isomorphism. We denote this set by $\mathbb{R}Ш(J)$ and call it the *real Tate–Shafarevich group* of J. This group is computed in [60], and we cite a few results here. Observe that the sheaf $\mathcal{J}$ of germs of holomorphic sections of J^{ab}, see page 88, has a natural real structure.

Theorem 3.84. *There is a natural isomorphism $\mathbb{R}Ш(J) = \mathbf{H}^1(\mathcal{J})$.* ▷ [60]

The proof of Theorem 3.84 is quite similar to that of Theorem 3.72, *cf.* also the isomorphism $Ш(J) = H^1(B; \mathcal{J})$ right after Definition 3.59.

From the point of view of the deformation classification, of primary interest is the 'discrete part' $\mathbb{R}Ш^{\mathrm{top}} := \mathbb{R}Ш/\mathbb{R}Ш^0$, where $\mathbb{R}Ш^0$ is the component of zero. A rather straightforward computation, which involves more homological algebra and topology of involutions than we intend to use in this book, yields Theorem 3.85 below. For the

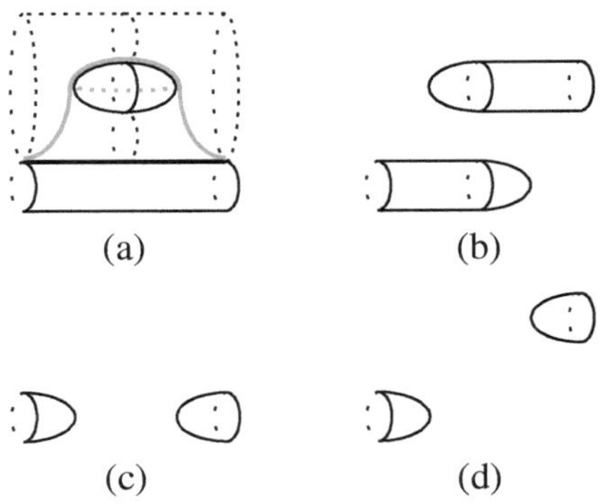

Figure 3.4. Modifications of the real part of an elliptic fibration.

statement, denote by $J_{\mathbb{R}}^{\mathrm{ext}} \subset J_{\mathbb{R}}$ the union of all connected components of $J_{\mathbb{R}}$ other than the principal components of the groups J_i, and let $B_{\mathbb{R}}^{\mathrm{ext}} := p(J_{\mathbb{R}}^{\mathrm{ext}})$.

Theorem 3.85. *Let $J \to B$ be a compact non-isotrivial real Jacobian elliptic fibration with irreducible fibers and nonempty real part. Then there is a natural exact sequence*

$$0 \longrightarrow H^1(B_{\mathbb{R}}^{\mathrm{ext}}, \partial B_{\mathbb{R}}^{\mathrm{ext}}; \mathbb{Z}_2) \longrightarrow \mathbb{R}\text{Ш}^{\mathrm{top}}(J) \longrightarrow H^0(B_{\mathbb{R}}^{\mathrm{ext}}; \mathbb{Z}_2) \longrightarrow 0$$

and an isomorphism $H^0(B_{\mathbb{R}}^{\mathrm{ext}}; \mathbb{Z}_2) = \bigoplus_i \mathbb{R}\text{Ш}^{\mathrm{top}}(F_i)$, where F_i are the fibers over some points $b_i \in B_{\mathbb{R}}$, one point inside each connected component of $B_{\mathbb{R}}^{\mathrm{ext}}$. ▷ [60]

Other elliptic fibrations

If the fibration J is almost generic, Theorem 3.85 has a very transparent and quite expectable geometric meaning, providing a clear description of the topology of the real part $X_{\mathbb{R}}$ of an arbitrary, not necessarily Jacobian, almost generic real elliptic fibration X with the fixed real Jacobian fibration $J(X) = J$. Up to equivariant deformation, such fibrations are obviously classified by the group $\mathbb{R}\text{Ш}^{\mathrm{top}}(J)$.

Each extra component $U \subset J_{\mathbb{R}}$ contributes two $\mathbb{Z}_2$ summands to $\mathbb{R}\text{Ш}^{\mathrm{top}}(J)$: one to $H^1 := H^1(B_{\mathbb{R}}^{\mathrm{ext}}, \partial B_{\mathbb{R}}^{\mathrm{ext}}; \mathbb{Z}_2)$ and one to $H^0 := H^0(B_{\mathbb{R}}^{\mathrm{ext}}; \mathbb{Z}_2)$. The nonzero element of H^1 represents a *real modification of the fibration*, given by a real 1-cocycle, *cf.* the proof of Theorem 3.72. If U is a sphere, Figure 3.4(a), the resulting real part $X_{\mathbb{R}}$ is shown in Figure 3.4(b), the cocycle being the partial section shown by a grey dotted line in Figure 3.4(a). If U is a torus or a Klein bottle, the two components of J_i over the corresponding hyperbolic component B_i are intertwined into one. In both cases, since the restriction of H^1 to each real fiber of J is trivial, the real structures in the real fibers remain unchanged.

The nonzero element of H^0 restricts nontrivially to each M-fiber of J, resulting in a new real structure with empty real part, *cf.* Theorem 3.79. This new real structure is obtained from c_J *via* the shift by a section real with respect to the opposite real structure $-c_J$ on J (shown in grey in Figure 3.4(a); the real part $J_{\mathbb{R}}^-$ is represented by dotted lines). If U is a torus or a Klein bottle, the result has empty real part over the corresponding hyperbolic component B_i. If U is a sphere, the result is shown in Figure 3.4(c). The remaining Figure 3.4(d), formally homeomorphic to (c), indicates the fact that one can also change *both* the fibration and the real structure.

Here are two immediate consequences of this description.

Corollary 3.86. *A compact non-isotrivial real elliptic fibration* $p: X \to B$ *without multiple fibers and with all fibers irreducible is equivariantly deformation equivalent to its Jacobian fibration* $J(X) \to B$ *if and only if the restriction* $p_{\mathbb{R}}: X_{\mathbb{R}} \to B_{\mathbb{R}}$ *admits a continuous section.*

Proof. None of the fibrations shown in Figure 3.4(b), (c), or (d) has a section. Similarly, over a hyperbolic component B_i, a section exists if and only if the corresponding group X_i is a pair of tori or Klein bottles. □

Corollary 3.87. *If a real elliptic fibration* $p: X \to B$ *as in Corollary* 3.86 *is an* M*- or* $(M-1)$*-variety, it is equivariantly deformation equivalent to its Jacobian.*

Proof. Each elementary modification shown in Figure 3.4, from (a) to (b), (c), or (d), either leaves the total Betti number unchanged or reduces it by 4, depending on whether a new component is created or not. The very first modification of $J(X)$ does not create a new component and hence reduces the total Betti number. Similarly, a nontrivial modification over a hyperbolic component reduces the total Betti number by 4 or 8. □

3.3.3 Lefschetz fibrations

We conclude this chapter with a brief discussion of the so-called Lefschetz fibrations, which are a topological counterpart of analytic fibrations dealt with in Section 3.2. The topological setting is usually restricted to the simplest type $\tilde{\mathbf{A}}_0^*$ singular fibers.

Definition 3.88. Let X be a compact connected oriented smooth 4-manifold and B a compact connected smooth oriented surface. A *Lefschetz fibration* is a surjective smooth map $p: X \to B$ with the following properties:

- $p(\partial X) = \partial B$ and the restriction $p: \partial X \to \partial B$ is a submersion;
- p has but finitely many critical points, which are all in the interior of X, and all critical values are pairwise distinct;
- about each critical point x, there are charts $(U, x) \cong (\mathbb{C}^2, 0)$ and $(V, b) \cong (\mathbb{C}^1, 0)$, $b = p(x)$, in which p is given by $(z_1, z_2) \mapsto z_1^2 + z_2^2$.

The restriction of a Lefschetz fibration to the set $B^\sharp$ of regular values of p is locally trivial, all fibers being closed connected oriented surfaces; the genus of a generic fiber is called the *genus* of p. Lefschetz fibrations of genus one are called *elliptic*.

An *isomorphism* between two Lefschetz fibrations is a bi-oriented diffeomorphism. The *monodromy* of a Lefschetz fibration is the monodromy of its restriction to $B^\sharp$. As it follows from the local normal form in the definition, the local monodromy about a singular fiber is the positive Dehn twist about a certain simple closed curve, well defined up to isotopy; this curve is called the *vanishing cycle*. (In singularity theory this local monodromy is known as the *Picard–Lefschetz transformation*.) The singular fiber itself is obtained from a close nonsingular one by contracting the vanishing cycle to a point to form a single node. A singular fiber is irreducible (remains connected after resolving the node) if and only if its vanishing cycle is not null-homologous. If the vanishing cycle bounds a disk, the singular fiber contains a sphere, which necessarily has self-intersection (-1), *i.e.*, is a topological analogue of a (-1)-curve. As in the analytic case, such a sphere can be blown down. The fibration is called *relatively minimal* if its singular fibers do not contain (-1)-spheres.

From now on, we confine ourselves to relatively minimal elliptic Lefschetz fibrations over the sphere $B \cong S^2$. After choosing a base point $b \in B^\sharp$ and fixing an isomorphism $H_1(F_b) = \mathcal{H}$, the monodromy of such a fibration is a homomorphism $\pi_1(B^\sharp, b) \to \tilde{\Gamma}$, and it is more or less clear (see [121] for a complete proof) that, up to isomorphism, the fibration is determined by its monodromy. By the Riemann–Hurwitz formula, $\chi(X) = r$, where r is the number of singular fibers.

Theorem 3.89 (Moishezon, Livné [121]). *Up to isomorphism, a relatively minimal elliptic Lefschetz fibration $X \to S^2$ is determined by the Euler characteristic $\chi(X)$, which is subject to the restrictions $\chi(X) \geqslant 0$ and $\chi(X) = 0 \bmod 12$.* ▷

The restrictions on $\chi(X)$ are proved in literally the same way as in the analytic case, see Theorem 3.47. The uniqueness is a purely algebraic statement about monodromy factorizations of central elements of the braid group $\mathbb{B}_3$, see Theorem 10.18.

Remark 3.90. As an immediate consequence of Theorem 3.89, any elliptic Lefschetz fibration over S^2 admits an analytic structure, as for any integer $n \geqslant 0$ there does exist an elliptic surface $E(n)$ with $12n$ type $\tilde{\mathbf{A}}_0^*$ singular fibers. If $n \geqslant 1$, then $E(n)$ is the so-called *fiber sum* of n copies of $E(1)$, which is the plane $\mathbb{P}^2$ blown up at the nine basepoints of a generic pencil of cubics.

Real elliptic Lefschetz fibrations (see [141, 142])

Many known restrictions on the topology of a real algebraic variety can be proved by purely topological means, using very little information about the real structure (see, *e.g.*, Theorem 3.74). Define a *real structure* on a smooth oriented $2n$-manifold X as an involutive autodiffeomorphism $c\colon X \to X$ satisfying the following conditions:

- c is orientation preserving (reversing) if n is even (respectively, odd);
- the *real part* $X_{\mathbb{R}} := \operatorname{Fix} c$ is either empty or of pure dimension n.

Next, define a *real Lefschetz fibration* as a Lefschetz fibration $p\colon X \to B$ equipped with a pair of real structures $c_X\colon X \to X$ and $c_B\colon B \to B$ commuting with p.

Isomorphisms of real Lefschetz fibrations are required to commute with the real structures. If the bases are directed curves, we can also speak about *directed* diffeo-/homeo-/isomorphisms. As a special case, this terminology applies to ruled surfaces.

Real elliptic Lefschetz fibrations over the sphere S^2 are studied in details in [141, 142], and we cite a few results here. To simplify the principal statements, we need a few assumptions and extra structures.

Definition 3.91. A *directed fibration* is a relatively minimal real elliptic Lefschetz fibration $p\colon X \to B \cong S^2$ over a directed real sphere S^2 (so that $B_{\mathbb{R}} \cong S^1$, *cf.* Theorems 3.74 and 3.76) equipped with an equivariant section $s\colon B \to X$. A directed fibration is *totally real* if all its singular fibers are real. An *isomorphism* of directed fibrations is a directed isomorphism commuting with s.

Consider a directed fibration $p\colon X \to B \cong S^2$. Due to the existence of a section, the real part $F_{\mathbb{R}}$ of each nonsingular real fiber F is nonempty, thus consisting of one or two circles, and the number of circles determines the topological type of the restricted real structure $c_F\colon F \to F$, see Theorem 3.74. The restriction $p_{\mathbb{R}}\colon X_{\mathbb{R}} \to B_{\mathbb{R}}$ can be regarded as an S^1-valued Morse function, and one can assign an index 0, 1, or 2 to each singular fiber, *i.q.* critical point of $p_{\mathbb{R}}$. The type of the real structure c_F alternates at each singular fiber. Assume that p has at least one real singular fiber and define its *uncoated necklace diagram* as the following decoration of the oriented circle $B_{\mathbb{R}}$:

- each singular fiber of index 0 or 2 is marked with a $\circ$, and each singular fiber of index 1 is marked with a $\times$;
- each segment connecting two consecutive singular fibers over which the real structure in the nonsingular fibers is an M-structure is doubled.

For example, the uncoated necklace diagram of the surface shown in Figure 3.3 is depicted in the same figure at the bottom.

As was explained above, the types of the segments (single *vs.* double) alternate at each singular fiber. Replace each double segment with a single *stone* as shown in Table 3.3 and call the result the *necklace diagram* of p. (Note that stones of type $>$ and $<$ depend on the chosen orientation of $B_{\mathbb{R}}$, which is shown by arrows in the table.) More precisely, we give the following definitions.

Table 3.3. Necklace stones.

Segment	Stone	Dual	Inverse	Monodromy
	$\bigcirc$	$\square$	$\bigcirc$	$\mathbb{Y}\mathbb{X}^2\mathbb{Y}\mathbb{X}^2\mathbb{Y}$
	$\square$	$\bigcirc$	$\square$	$\mathbb{X}^2\mathbb{Y}\mathbb{X}^2$
	$>$	$<$	$<$	$\mathbb{X}\mathbb{Y}$
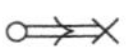	$<$	$>$	$>$	$\mathbb{Y}\mathbb{X}$

Definition 3.92. A *broken necklace diagram* is a nonempty word in the *stone* alphabet $\{\bigcirc, \square, >, <\}$. Assign to each stone its *dual*, *inverse*, and *monodromy* (an element of Γ) as shown in Table 3.3. For a broken necklace diagram $\mathcal{N}$, define its

- *monodromy* $\mathfrak{m}(\mathcal{N}) \in \Gamma$, obtained by replacing each stone with its monodromy and evaluating the resulting word in Γ,
- *dual diagram* $\mathcal{N}^*$, obtained by replacing each stone with its dual, and
- *inverse diagram* $\mathcal{N}^{-1}$, obtained by replacing each stone with its inverse and reversing the order of the stones.

An (*oriented*) *necklace diagram* is a class of broken necklace diagrams that differ by a cyclic permutation. The inverse, dual, and monodromy (as a conjugacy class in Γ) of a necklace diagram are defined in the obvious way. A *non-oriented necklace diagram* is obtained by the further identification of $\mathcal{N}$ and $\mathcal{N}^{-1}$.

For example, the necklace diagram of the surface shown in Figure 3.3 is

$$-\square-\bigcirc-\bigcirc-<-->-\bigcirc->-\bigcirc-<-\square->-\bigcirc-.$$

(In [141], necklace diagrams are drawn in the oriented circle $B_{\mathbb{R}}$, and we respect this convention by drawing a 'broken' necklace; certainly, the approach of [141] is equivalent to considering cyclic words.)

The sphere B splits into two disks with common boundary $B_{\mathbb{R}}$; let B^+ be the one whose orientation induces the distinguished orientation of $B_{\mathbb{R}}$. Consider the double segment I represented by a stone S, pick a pair of regular real values $b_{\pm}$ of p, right before I and right after I (with respect to the distinguished orientation of $B_{\mathbb{R}}$, see Figure 3.5), and let $F_{\pm}$ be the corresponding nonsingular fibers. Consider a path γ_S in $B^+ \cap B^{\sharp}$ from b_- to b_+, path homotopic to the segment $[b_-, b_+] \subset B_{\mathbb{R}}$ connecting b_- to b_+, and let $\mathfrak{m}_S \colon H_1(F_-) \to H_1(F_+)$ be the monodromy along γ_S. Since $F_{\pm}$ are tori with real structures, their homology groups are identified with $\mathcal{H}$ *canonically* up to sign, see Remark 3.80; hence, $\mathfrak{m}_S$ can be regarded as an element of Γ. It is these elements, depending on the type of the stone, that are listed in Table 3.3. (The computation using the local normal forms is found in [142].) The product $\gamma := \prod \gamma_S$ (in the order of appearance in the diagram) over all stones S is a loop freely homotopic to the boundary $\partial U \cap B^+$, where U is a closed regular neighborhood of $B_{\mathbb{R}}$ in B disjoint

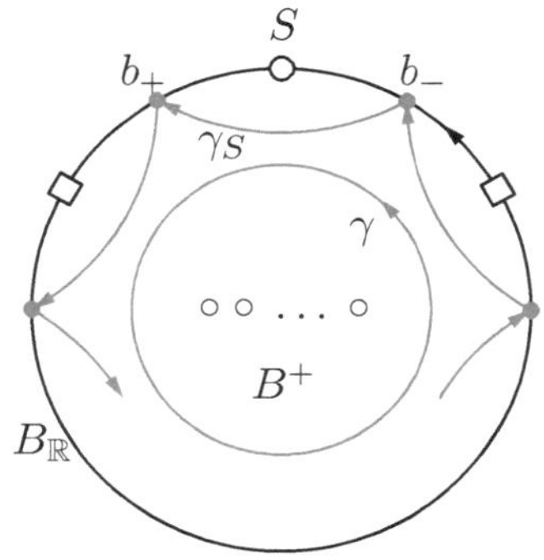

Figure 3.5. Monodromies of stones.

from the non-real critical values (if present; these critical values are represented by white dots in Figure 3.5). Hence, the total monodromy of the necklace diagram in Definition 3.92 is, up to conjugation, the Γ-valued reduction of the monodromy of p along γ.

Theorem 3.93 (Salepci [141]). *For each $n \in \mathbb{N}$, the map sending a fibration p to its necklace diagram establishes a one-to-one correspondence between the isomorphism classes of totally real directed fibrations with $12n$ singular fibers and the oriented necklace diagrams of length $6n$ whose monodromy equals $1 \in \Gamma$.*

Proof (the idea; see [141] for details). Local analysis shows that a necklace diagram defines a unique, up to isomorphism, germ $p_{\mathbb{R}}$ of directed fibrations over (a regular neighborhood U of) the circle $B_{\mathbb{R}}$, and it remains to extend this germ to a fibration $p^+ : X^+ \to B^+$; the extension to the other disk $B^- := c_B(B^+)$ is then $c_B^* p^+$. Since we need a totally real fibration, the extension should be locally trivial over the disk $B' := B^+ \smallsetminus \operatorname{Int} U$; such an extension exists (and then is unique up to isomorphism identical over $\partial B'$) if and only if the monodromy $\mathfrak{m}$ of $p_{\mathbb{R}}$ along $\partial B'$ is $1 \in \tilde{\Gamma}$. The image $\operatorname{pr}_\Gamma \mathfrak{m}$ is the monodromy of the diagram, see Figure 3.5 and the discussion prior to the statement; hence, the latter must be trivial. For the converse, one can show that $\operatorname{dg} \mathfrak{m} = r \bmod 12$, where r is the number of real singular fibers; hence, if $\operatorname{pr}_\Gamma \mathfrak{m} = 1$ and the diagram has $6n$ stones, then also $\mathfrak{m} = 1$ and an extension exists. □

The topology of the projection $X_{\mathbb{R}} \to B_{\mathbb{R}}$ is easily recovered from the necklace diagram, *cf.* Figure 3.3; the ○ type stones correspond to the spherical components of $X_{\mathbb{R}}$. One can also compute simple topological invariants of $X_{\mathbb{R}}$; for example,

$$\chi(X_{\mathbb{R}}) = 2(\#_{\bigcirc} - \#_{\square}), \qquad b_*(X_{\mathbb{R}}; \mathbb{Z}_2) = 2(\#_{\bigcirc} + \#_{\square}) + 4, \tag{3.94}$$

where $\#_*$ is the number of stones of type $* \in \{\bigcirc, \square, >, <\}$. A directed fibration is called *maximal* if c_X is an M-involution, *i.e.*, if $b_*(X_{\mathbb{R}}; \mathbb{Z}_2) = b_*(X; \mathbb{Z}_2)$.

Corollary 3.95 (of (3.68) and (3.94)). *The necklace diagram representing a totally real directed fibration $X \to B$ has at least two arrow type stones, and the fibration is maximal if and only if its diagram has exactly two such stones.* $\triangleleft$

For another observation, note that $\mathfrak{m}(\mathcal{N}^*) = \mathbb{Y}\mathfrak{m}(\mathcal{N})\mathbb{Y}$ for any broken necklace diagram $\mathcal{N}$. Hence, $\mathfrak{m} = \mathrm{id}$ if and only if $\mathfrak{m}^* = \mathrm{id}$ and, if one of the two diagrams represents a totally real directed fibration, then so does the other. From the point of view of the map $p: X_{\mathbb{R}} \to B_{\mathbb{R}}$, the passage from $\mathcal{N}$ to $\mathcal{N}^*$ is similar to the passage to the opposite real structure on a real Jacobian elliptic fibration, even though neither $-\mathrm{id}$ nor the opposite real structure are defined for real Lefschetz fibrations.

We discuss briefly the effect on Theorem 3.93 of removing the extra structures and/or assumptions imposed in Definition 3.91.

The distinguished orientation of $B_{\mathbb{R}}$ Undirected fibrations are classified by their non-oriented necklace diagrams. Note that $\mathfrak{m}(\mathcal{N}^{-1}) = 1$ if and only if $\mathfrak{m}(\mathcal{N}) = 1$; hence, if one of the two diagrams represents a totally real elliptic Lefschetz fibration, then so does the other.

The existence of a real section Even though the Jacobian fibration $J(X)$ is not defined in the topological category, it can be shown that any real elliptic Lefschetz fibration over S^2 can be obtained from one with a real section by the modifications discussed on page 100, see Figure 3.4, and all four modifications differ topologically (see [142, 141]). It follows that, in order to classify such fibrations, one merely needs to *refine* the ○ type stones into four subtypes, corresponding to Figure 3.4(a)–(d). This refinement does not affect the monodromy of the stone (as M- and $(M-2)$-structures on a torus do not differ homologically). Thus, each necklace diagram with trivial monodromy gives rise to $4^{\#\bigcirc}$ refined diagrams. Of course, if the original diagram has symmetries, some of its refinements may turn out to be equivalent.

The total reality condition The existence part of Theorem 3.93 can be modified to cover directed fibrations with not necessarily real singular fibers.

Theorem 3.96. *A necklace diagram $\mathcal{N}$ with r stones represents a directed fibration with $2r$ real and w pairs of c_B-conjugate singular fibers if and only if the monodromy of $\mathcal{N}$ can be represented as a product of w Dehn twists.*

Proof. One repeats the proof of the existence part of Theorem 3.93. The w Dehn twists in the statement are the monodromies about the singular fibers inside the disk B^+, see Figure 3.5 and the proof of Theorem 3.93. $\square$

Theorem 3.96 reduces the problem of existence of directed fibrations to an algebraic question to which the answer is known: the cases $w = 0$ or 1 are obvious, see Proposition 2.15, the case $w = 2$ is discussed in Section 10.2.2, see Theorem 10.27, and in the general case the existence of a factorization is given by Theorem 10.17. In view of (2.4), any element of Γ is a product of *some number* of Dehn twists; hence any necklace diagram appears as the diagram of some directed fibration.

Table 3.4. Totally real fibrations with 12 singular fibers (see [142, 141]).

	$\chi = -8$	$\chi = 0$	$\chi = 8$
$b_* = 12$	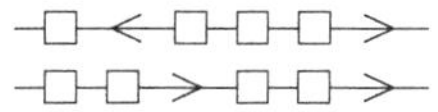		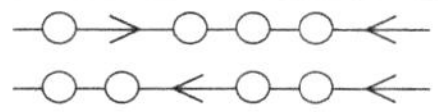
$b_* = 10$	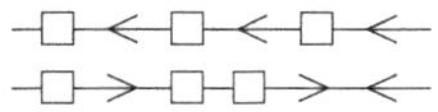		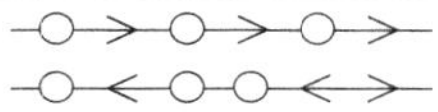
$b_* = 8$	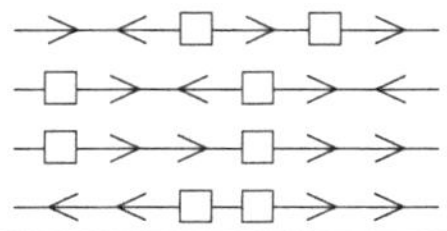	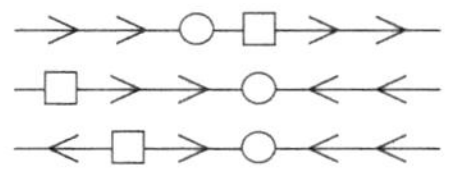	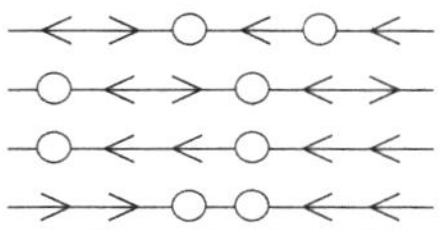
$b_* = 6$	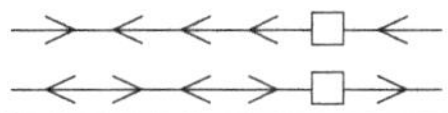		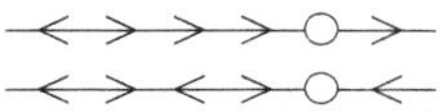
$b_* = 4$		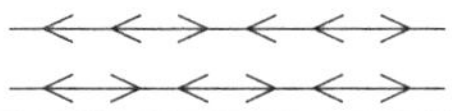	

The total reality assumption is crucial for the uniqueness part of Theorem 3.93: one of the principal results of this book states that the number of isomorphism classes of directed fibrations with w pairs of conjugate singular fibers and the same necklace diagram (depending on w) may grow exponentially with w, see Example 10.87 in Chapter 10 (where we also refine Theorem 3.96 and reduce the classification to a purely algebraic, although difficult problem, see Theorem 10.86).

The classification for $n = 1$ and 2 (see [142, 141])

A complete classification of totally real relatively minimal elliptic Lefschetz fibrations with a section and with $12n = 12$ or 24 singular fibers is also found in [142, 141].

For $n = 1$ (the topological analogue of rational elliptic surfaces), there are 25 (undirected) isomorphism classes; their necklace diagrams are listed in Table 3.4, where $\chi := \chi(X_{\mathbb{R}})$ and $b_* := b_*(X_{\mathbb{R}}; \mathbb{Z}_2)$.

For $n = 2$ (the analogue of elliptic $K3$-surfaces), the number of classes is 8421.

Further analysis of these results reveals a sharp contrast with Theorem 3.89 and its consequences discussed in Remark 3.90.

For $n = 1$, eight of the 25 isomorphism classes do not admit an analytic structure with respect to which they are real Jacobian elliptic fibration. This statement is proved using the techniques of dessins, and we discuss it in more details in Chapter 4, see Corollary 4.77. In general, it appears that 'most' elliptic Lefschetz fibrations are not algebraic, *cf.* Corollary 4.78.

For $n = 2$, there are fibrations that do not split into a fiber sum of two fibrations with 12 singular fibers each. The *fiber sum* operation corresponds to the *connected sum* of the corresponding *uncoated* necklace diagrams: one cuts the two circles at a pair of segments of the same type (both simple or both double) and glues them into a single circle respecting the orientations. If the segments chosen are both simple, this procedure can be regarded as the concatenation of the coated necklace diagrams, possibly after a cyclic permutation; this operation is called a *mild sum* in [141]. If the segments are both double, the operation destroys a pair of stones and recreates another pair, possibly of different types; it is called a *harsh sum*. The complete list of the modifications, depending on the pair of stones involved and the orientations, is found in [141]; this list is easily obtained from Table 3.3. The following example, rebuilding two arrow type stones into a pair ○, □, is used in the sequel:

$$(-\mathcal{N}_1 \rightarrowtail -) \sqcup (-\leftarrowtail \mathcal{N}_2 -) \longmapsto (-\mathcal{N}_1 -\square- \mathcal{N}_2 -\bigcirc-). \tag{3.97}$$

Here, $\mathcal{N}_1$ and $\mathcal{N}_2$ are arbitrary fragments and all orientations are left to right.

One indecomposable directed fibration with 24 singular fibers has necklace diagram

$$-\square-\square-\square-\bigcirc-\leftarrowtail-\bigcirc-\rightarrowtail-\bigcirc-\rightarrowtail-\bigcirc-\square-\rightarrowtail-.$$

Its indecomposability is proved by considering the possible splittings of the diagram and comparing the results to Table 3.4.

Chapter 4

Dessins

Historically, *dessins d'enfants* seem to have first been introduced by Felix Klein [93]. A century later, they were rediscovered (and named) by A. Grothendieck [83] as an intuitive means of studying rational maps with three critical points and, ultimately, the absolute Galois group $\mathrm{Gal}(\bar{\mathbb{Q}}/\mathbb{Q})$. Finally, S. Orevkov [131] suggested a modified overdecorated version applicable to equivariant rational maps with many critical values *of which only three matter*, which is precisely the case for the j-invariant of a trigonal curve or elliptic surface. For the purpose of this book, we reserve the term 'dessin' for the latter construction suggested by Orevkov; in the case of maximally singular curves/surfaces, when just the bold part of the dessin (which *is* the dessin d'enfants in the original sense of Grothendieck) suffices to adequately describe the situation, we call it the *skeleton*.

4.1 Dessins (see [60])

In the exposition below, we mainly follow [60] but with a certain shift of terminology, as we need to cover both real (usually generic) and complex (usually singular) trigonal curves. Thus, our dessins are much more general than those in [60] and the terms 'generic', 'reduced', *etc.* have a completely different meaning.

4.1.1 Trichotomic graphs

Let S be a compact connected surface, not necessarily orientable or closed. (Unless specified otherwise, in the topological part of this section we are working in the *PL*-category.) We use the term *real* for points, segments, *etc.* situated in the boundary ∂S; in the drawings, the real part ∂S is represented by wide grey lines. For an embedded graph $\mathfrak{D} \subset S$, we denote by $\operatorname{Cut}\mathfrak{D}$ the *cut* of S along $\mathfrak{D}$. Formally, the cut can be defined as follows: triangulate the pair $(S, \mathfrak{D})$ and let $\operatorname{Cut}\mathfrak{D}$ be the disjoint union of the triangles identified along the edges that are common in S *but do not belong to* $\mathfrak{D}$. The connected components of $\operatorname{Cut}\mathfrak{D}$ are called the *regions* of $\mathfrak{D}$.

Definition 4.1. A *trichotomic graph* on a compact surface S is an embedded finite directed graph $\mathfrak{D} \subset S$ decorated with the following additional structures (referred to as *colorings* of the edges and vertices of $\mathfrak{D}$, respectively):

- each edge of $\mathfrak{D}$ is of one of the three kinds: *solid*, *bold*, or *dotted*;
- each vertex of $\mathfrak{D}$ is of one of the four kinds: $\bullet$, $\circ$, $\times$, or *monochrome* (the vertices of the first three kinds being called *essential*),

and satisfying the following conditions:

1. the boundary ∂S is a union of edges and vertices of $\mathfrak{D}$;
2. each essential vertex of $\mathfrak{D}$ is incident to at least two edges, and each monochrome vertex of $\mathfrak{D}$ is incident to at least three edges;
3. the orientations of the edges of $\mathfrak{D}$ form an orientation of the boundary $\partial \operatorname{Cut} \mathfrak{D}$; this orientation extends to an orientation of $\operatorname{Cut} \mathfrak{D}$;
4. all edges incident to a monochrome vertex are of the same kind;
5. $\times$-vertices are incident to incoming dotted edges and outgoing solid edges;
6. $\bullet$-vertices are incident to incoming solid edges and outgoing bold edges;
7. $\circ$-vertices are incident to incoming bold edges and outgoing dotted edges.

In Items 5–7, the lists are complete, *i.e.*, vertices cannot be incident to edges of other kinds or with a different orientation.

As in the case of bipartite graph, we will use $*$- to refer to an *essential* vertex of an unspecified type. We avoid stating Item 2 in terms of the valency because we are planning to redefine this notion for trichotomic graphs, see Convention 4.7.

The boundary of each region R of a trichotomic graph $\mathfrak{D}$ contains a certain number $n = 3k \geqslant 0$ of *essential* vertices (and possibly some monochrome vertices), which alternate according to the pattern $\bullet, \circ, \times, \bullet, \ldots$. We call such a region R an *n-gon*, or *n-gonal* region.

According to Item 4 of Definition 4.1, the monochrome vertices of $\mathfrak{D}$ can further be subdivided into solid, bold, and dotted, according to their incident edges. The sets of solid, bold, and dotted monochrome vertices of $\mathfrak{D}$ will be denoted by $\operatorname{Vtx}_{\text{solid}} \mathfrak{D}$, $\operatorname{Vtx}_{\text{bold}} \mathfrak{D}$, and $\operatorname{Vtx}_{\text{dotted}} \mathfrak{D}$, respectively. The *monochrome part* of $\mathfrak{D}$ of a given kind (solid, bold, or dotted) is the union of (open) edges and monochrome vertices of the corresponding kind. Thus, essential vertices *never* belong to a monochrome part. The monochrome parts are denoted by $\mathfrak{D}_{\text{solid}}$, $\mathfrak{D}_{\text{bold}}$, and $\mathfrak{D}_{\text{dotted}}$.

A path in a trichotomic graph $\mathfrak{D}$ is called *monochrome* if it is contained in a monochrome part of $\mathfrak{D}$. Define a binary relation $\prec$ on the monochrome vertices of $\mathfrak{D}$ as follows: $u \prec v$ if and only if there is a directed monochrome path from u to v. (Clearly, only monochrome vertices of the same kind can be compatible.) The graph $\mathfrak{D}$ is called *admissible* if $\prec$ is a partial order. Since $\prec$ is obviously transitive, $\mathfrak{D}$ is admissible if and only if it does not have directed monochrome cycles.

Definition 4.2. A trichotomic graph $\mathfrak{D}$ is called a *dessin* if

1. $\mathfrak{D}$ is admissible, *i.e.*, it has no directed monochrome cycles, and
2. each triangular region of $\mathfrak{D}$ is homeomorphic to a disk.

Remark 4.3. The orientation of a trichotomic graph $\mathfrak{D}$ is almost superfluous. Indeed, $\mathfrak{D}$ may have at most two orientations satisfying Item 3 of Definition 4.1, and if $\mathfrak{D}$ has at least one essential vertex, its orientation is uniquely determined by Items 5–7. Each connected component of an *admissible* graph $\mathfrak{D}$ does have an essential vertex (hence it has essential vertices of all three kinds), as otherwise the boundary $\partial \operatorname{Cut} \mathfrak{D}$ would contain directed monochrome cycles.

Convention 4.4. In fact, all three decorations of a dessin $\mathfrak{D}$ (orientation and the two colorings) can be recovered from any of the colorings. For clarity, we retain both colorings in the figures, but we do omit the orientation.

Complex *vs.* real dessins

If $\mathfrak{D} \subset S$ is a trichotomic graph in an orientable surface S, a choice of an orientation of S defines a *chessboard coloring* of $\operatorname{Cut} \mathfrak{D}$: a region R is said to be *positive* or *negative* if the orientation of R induced from that of S coincides with (respectively, is opposite to) its orientation defined by $\mathfrak{D}$ according to Item 3 in Definition 4.1. Conversely, a chessboard coloring of $\operatorname{Cut} \mathfrak{D}$ defines an orientation of S.

Since we need to cover both complex and real trigonal curves, we will consider dessins or, more generally, trichotomic graphs in two mutually exclusive versions. In the *complex* setting, we assume the underlying surface S closed and *oriented*, so that the cut $\operatorname{Cut} \mathfrak{D}$ has a canonical chessboard coloring. In the *real* case, we *exclude* closed orientable surfaces, *i.e.*, we assume that either $\partial S \neq \varnothing$, or S is nonorientable, or both. The reason for this discrimination is that real dessins are supposed to represent complex ones that are invariant under an orientation reversing involution $c\colon S \to S$. More precisely, we have the following two constructions.

Let $\mathfrak{D} \subset S$ be a real trichotomic graph. Denote by $\tilde{S}$ the *oriented double* of S, *i.e.*, the orientation double covering of S with the two preimages of each real point $s \in \partial S$ identified. Note that $\tilde{S}$ is canonically oriented and closed, and it is connected *unless* S is closed and oriented. Denote by $p\colon \tilde{S} \to S$ the projection and by $c\colon \tilde{S} \to \tilde{S}$ the deck translation; c is an orientation reversing involution, *i.e.*, a real structure in the topological sense. Let $\tilde{\mathfrak{D}} = p^{-1}(\mathfrak{D}) \subset \tilde{S}$, and equip this graph with the decorations induced by p. Clearly, c preserves all three decorations, including the orientation.

With an abuse of the language, we will also call the graph $\tilde{\mathfrak{D}} \subset \tilde{S}$ the *oriented double* of $\mathfrak{D} \subset S$, referring to the involution $c\colon \tilde{S} \to \tilde{S}$ as the *real structure*.

Conversely, let $\tilde{\mathfrak{D}} \subset \tilde{S}$ be a complex trichotomic graph, and let $c\colon \tilde{S} \to \tilde{S}$ be a real structure (in the topological sense, see page 102) such that $c(\tilde{\mathfrak{D}}) = \tilde{\mathfrak{D}}$ and c preserves all three decorations of $\tilde{\mathfrak{D}}$; we say that $\tilde{\mathfrak{D}}$ is *c-invariant*. Consider the quotient $S := \tilde{S}/c$ and let $\mathfrak{D} \subset S$ be the image of $\tilde{\mathfrak{D}}$. We assert that $\mathfrak{D}$ is a real trichotomic graph.

Proposition 4.5. *Given a real trichotomic graph $\mathfrak{D} \subset S$, its oriented double $\tilde{\mathfrak{D}} \subset \tilde{S}$ is a deck translation invariant trichotomic graph. Conversely, given a c-invariant* *trichotomic graph* $\tilde{\mathfrak{D}} \subset \tilde{S}$ *(where* $c\colon \tilde{S} \to \tilde{S}$ *is a fixed real structure), its image* $\mathfrak{D}$ *in the quotient* $S := \tilde{S}/c$ *is a real trichotomic graph. In these settings,* $\tilde{\mathfrak{D}}$ *is a dessin if and only if so is* $\mathfrak{D}$.

Proof. The direct statement is obvious, and for the converse, we need the following *separation property* of invariant trichotomic graphs.

Lemma 4.6. *Each open region* R *of a c-invariant trichotomic graph* $\tilde{\mathfrak{D}} \subset \tilde{S}$ *is disjoint from its image* $c(R)$.

Proof. Since c preserves the orientation of $\tilde{\mathfrak{D}}$ but reverses the orientation of $\tilde{S}$, the regions R and $c(R)$ have opposite signs with respect to the canonical chessboard coloring of Cut $\tilde{\mathfrak{D}}$. □

Let $p\colon \tilde{S} \to S$ be the quotient projection. Due to Lemma 4.6, the restriction of p to each region R of $\tilde{\mathfrak{D}}$ is a one-to-one map onto the image $p(R)$; hence, as the restriction $p\colon \tilde{\mathfrak{D}} \to \mathfrak{D}$ is orientation preserving, Item 3 in Definition 4.1 for $\mathfrak{D}$ follows from that for $\tilde{\mathfrak{D}}$. Lemma 4.6 implies also that $\tilde{\mathfrak{D}}$ contains the fixed point set Fix c; this yields Items 1 and 2 for $\mathfrak{D}$. All other requirements are straightforward.

Since p preserves the decorations of $\tilde{\mathfrak{D}}$ and $\mathfrak{D}$, the admissibility of one of the graphs implies the admissibility of the other. Furthermore, each region of $\tilde{\mathfrak{D}}$ is mapped homeomorphically onto a region of $\mathfrak{D}$; hence, Item 2 in Definition 4.2 for one of the graphs also implies this condition for the other. □

Index of a vertex

According to Item 3 in Definition 4.1, the orientations of the edges at each vertex alternate. In particular, each vertex is incident to an even number of edges, half of them being incoming and half outgoing. In the sequel, we will often consider, at the same time, a dessin $\mathfrak{D}$ and the skeleton $\mathfrak{S}$ formed by the bold edges and •- and ○-vertices of $\mathfrak{D}$; clearly, the valency of each vertex in $\mathfrak{S}$ is half of its valency in $\mathfrak{D}$. To avoid confusion, we accept the following valency convention for trichotomic graphs.

Convention 4.7. The *valency*, or *index* ind v of a vertex v of a complex trichotomic graph $\mathfrak{D} \subset S$ is half the number of edges of $\mathfrak{D}$ incident to v (equivalently, the number of incoming edges at v or the number of outgoing edges at v).

To avoid ambiguity, we will mainly use the term index rather than valency; however, it is often convenient to speak about mono-, bi-, *etc.* valent vertices. As yet another justification for this convention and the term, we will see that the index of a vertex v equals the ramification index at v of a certain ramified covering associated with $\mathfrak{D}$, see Remark 4.9.

For a complex trichotomic graph $\mathfrak{D}$, we denote by $\#_\bullet(\mathfrak{D})$, $\#_\circ(\mathfrak{D})$, and $\#_\times(\mathfrak{D})$ the numbers of its •-, ○-, and ×-vertices, respectively.

When working with a *real* trichotomic graph $\mathfrak{D} \subset S$, we always keep in mind that $\mathfrak{D}$ is merely a compact way to represent a complex graph $\tilde{\mathfrak{D}} \subset \tilde{S}$ invariant with respect to a certain real structure. Thus, we define the indices (valencies) and vertex counts for $\mathfrak{D}$ as those for $\tilde{\mathfrak{D}}$.

Definition 4.8. The *full valency*, or *full index* $\operatorname{ind} v$ of a vertex v of a real trichotomic graph $\mathfrak{D} \subset S$ is the index $\operatorname{ind} \tilde{v}$ of any preimage of v in the oriented double $\tilde{\mathfrak{D}} \subset \tilde{S}$. The counts $\#_\bullet(\mathfrak{D})$, $\#_\circ(\mathfrak{D})$, and $\#_\times(\mathfrak{D})$ for $\mathfrak{D}$ are defined as those for $\tilde{\mathfrak{D}}$. In other words, $\#_*(\mathfrak{D})$ is the weighted number of $*$-vertices, with each inner vertex counted twice.

With further applications in mind, define a *singular vertex* of a trichotomic graph as a $\bullet$-vertex of full index $\neq 0 \bmod 3$, a $\circ$-vertex of full index $\neq 0 \bmod 2$, or a $\times$-vertex. The set of singular vertices of $\mathfrak{D}$ is denoted by $\operatorname{Sing} \mathfrak{D}$.

The dessin of a ramified covering

Let S be an oriented closed connected surface and $c\colon S \to S$ a (topological) real structure. We are interested in orientation preserving ramified coverings $j\colon S \to \mathbb{P}^1$, where $\mathbb{P}^1$ is the standard Riemann sphere $\mathbb{C} \cup \{\infty\}$ with the standard real structure $z \mapsto \bar{z}$; such a covering is called real, or c-equivariant, if $j \circ c = {}^{-} \circ j$.

A ramified covering j as above gives rise to a trichotomic graph $\mathfrak{D} := \mathfrak{D}(j)$, called the *dessin* of j. As a set, $\mathfrak{D}$ is the preimage $j^{-1}(\mathbb{P}^1_{\mathbb{R}})$. The $\bullet$-, $\circ$-, and $\times$-vertices of $\mathfrak{D}$ are the pull-backs of 0, 1, and ∞, respectively, and the monochrome vertices are the ramification points of j whose image is in $\mathbb{P}^1_{\mathbb{R}} \smallsetminus \{0, 1, \infty\}$. An edge of $\mathfrak{D}$ is declared solid, bold, or dotted if its image lies in the segment $[\infty, 0]$, $[0, 1]$, or $[1, \infty]$, respectively. (The corresponding decoration of the real line $\mathbb{P}^1_{\mathbb{R}}$ as the dessin of the identity map is shown in Figure 3.1.) Finally, the orientation of $\mathfrak{D}$ is that induced from the positive orientation of $\mathbb{P}^1_{\mathbb{R}}$, *i.e.*, the order of $\mathbb{R}$.

Remark 4.9. According to the construction, the index $\operatorname{ind} v$ of each vertex v of $\mathfrak{D}(j)$ is equal to the ramification index of j at v. Note that j may also have ramification points with non-real critical values; such points are located inside the regions of $\mathfrak{D}(j)$ and are not detected by $\mathfrak{D}(j)$, except that the number of such points inside a region R can be recovered from $\chi(R)$ using the Riemann–Hurwitz formula.

Lemma 4.10. *The trichotomic graph $\mathfrak{D} := \mathfrak{D}(j)$ constructed above is a dessin. If, in addition, j is real with respect to a real structure $c\colon S \to S$, then $\mathfrak{D}$ is c-invariant.*

Proof. Each region R of $\operatorname{Cut} \mathfrak{D}$ is mapped by j onto one of the disks $\{\operatorname{Im} z \geqslant 0\}$ or $\{\operatorname{Im} z \leqslant 0\}$; declaring R positive (negative) in the former (respectively, latter) case defines a chessboard coloring compatible with the orientation of $\mathfrak{D}$, proving Item 3 in Definition 4.1. All other axioms of trichotomic graphs hold automatically. If R is a triangle, the restriction $j\colon \partial R \to \mathbb{P}^1_{\mathbb{R}}$ is one-to-one; hence, the restriction of j to R is a homeomorphism onto one of the above disks, which proves Item 2 in Definition 4.2. Finally, since $j\colon \mathfrak{D} \to \mathbb{P}^1_{\mathbb{R}}$ is orientation preserving, the relation $\prec$ is a subset of the partial order induced by the linear order on the intervals $(1, \infty)$, $(\infty, 0)$, $(0, 1)$; hence, $\mathfrak{D}$ is admissible, see Item 1 in Definition 4.2.

The last statement is a tautology. □

For a ramified covering $j\colon S \to \mathbb{P}^1$ equivariant with respect to a real structure $c\colon S \to S$, the quotient $\mathfrak{D}_c(j) := \mathfrak{D}(j)/c \subset S/c$ is called the *real dessin* of j. In view of Lemma 4.10 and Proposition 4.5, $\mathfrak{D}_c(j)$ is indeed a dessin.

Theorem 4.11. *Let S be an oriented connected closed surface (and let $c\colon S \to S$ be a real structure). A (c-invariant) trichotomic graph $\mathfrak{D} \subset S$ is a dessin if and only if it has the form $\mathfrak{D}(j)$ for some (c-equivariant) orientation preserving ramified covering $j\colon S \to \mathbb{P}^1$. Furthermore, the map j is determined by $\mathfrak{D}$ uniquely up to homotopy in the class of (c-equivariant) ramified coverings with a fixed dessin.*

Proof. We will prove the theorem in the equivariant setting. In the 'plain' case, the proof is literally the same, with the real structure c ignored.

The 'if' part is given by Lemma 4.10. For the 'only if' part, we construct a map j and, at each step, check that the construction is unique up to homotopy.

Any map $j\colon S \to \mathbb{P}^1$ as in the statement has an orientation preserving restriction $j'\colon \mathfrak{D} \to \mathbb{P}^1_{\mathbb{R}}$ and hence an orientation preserving descent $\bar{\jmath}\colon \bar{\mathfrak{D}} \to \mathbb{P}^1_{\mathbb{R}}$, $\bar{\mathfrak{D}} := \mathfrak{D}/c$, *cf.* Proposition 4.5. The images of the essential vertices of $\bar{\mathfrak{D}}$ are given by the definition, and the extension to, say, the dotted part of $\bar{\mathfrak{D}}$ is determined, uniquely up to regular homotopy, by a monotonous map $\bar{\jmath}\colon (\mathrm{Vtx}_{\mathrm{dotted}}\,\bar{\mathfrak{D}}, \prec) \to ((1, \infty), <)$. The set of such maps is defined by the linear inequalities $\bar{\jmath}(u) < \bar{\jmath}(v)$ whenever $u \prec v$; hence, it is a convex subset of a Cartesian power of the interval $(1, \infty)$ and thus is connected. For the existence, extend $\prec$ to a linear order on $\mathrm{Vtx}_{\mathrm{dotted}}\,\bar{\mathfrak{D}}$ and, according to this order, map dotted vertices to consecutive integers.

The c-equivariant restriction $j'\colon \mathfrak{D} \to \mathbb{P}^1_{\mathbb{R}}$ is the composition of the map $\bar{\jmath}$ just constructed and the quotient projection $\mathfrak{D} \to \bar{\mathfrak{D}}$. If no real structure is present, the construction is applied directly to the original dessin $\mathfrak{D}$.

Consider the chessboard coloring of the cut $\mathrm{Cut}\,\mathfrak{D}$. For each positive region R, the restriction $j'\colon \partial R \to \mathbb{P}^1_{\mathbb{R}}$ is an orientation preserving covering; hence, it extends to an orientation preserving ramified covering $j\colon R \to \{\operatorname{Im} z \geqslant 0\}$, and this extension is unique up to homotopy in the class of ramified coverings identical on ∂R, see Corollary 10.13 and Item 2 in Definition 4.2. The extension to the negative regions is given by $\bar{\ } \circ j \circ c\colon c(R) \to \{\operatorname{Im} z \leqslant 0\}$, and Lemma 4.6 assures that this extension is well defined and is c-equivariant. (If no real structure is present, the negative regions of $\mathrm{Cut}\,\mathfrak{D}$ are treated in the same way as the positive ones: they are mapped onto the disk $\{\operatorname{Im} z \leqslant 0\}$ using Corollary 10.13 directly.)

Each inner point of an edge of $\mathfrak{D}$ is regular for any map j constructed as above (as the two components of Cut $\mathfrak{D}$ incident to the edge have opposite signs). It follows that j has isolated critical points only and hence is a ramified covering. □

Theorem 4.11 and the Riemann existence theorem result in two corollaries, which we state separately in the complex and real settings.

Corollary 4.12. *Given a complex dessin $\mathfrak{D} \subset S$, there are a complex structure on S and a holomorphic map $j\colon S \to \mathbb{P}^1$ such that $\mathfrak{D}(j) = \mathfrak{D}$. Both the complex structure and the map j are unique up to deformation.* ◁

Corollary 4.13. *Given a real dessin $\mathfrak{D} \subset S$, there are a complex structure on the oriented double $\tilde{S}$ and a holomorphic map $j\colon \tilde{S} \to \mathbb{P}^1$ such that the deck translation c is a real structure, j is c-real, and $\mathfrak{D}_c(j) = \mathfrak{D}$. Both the complex structure and the map j are unique up to equivariant deformation.* ◁

4.1.2 Deformations

Fix an oriented connected closed surface S. For the real framework, fix also a real structure $c\colon S \to S$ and let $\bar{S} = S/c$. We are interested in (c-equivariant) orientation preserving ramified coverings $j\colon S \to \mathbb{P}^1$, where $\mathbb{P}^1 = \mathbb{C} \cup \{\infty\}$ is equipped with the standard real structure $z \mapsto \bar{z}$. A *deformation* of coverings is a homotopy $S \times I \to \mathbb{P}^1$ in the class of (equivariant) ramified coverings. A deformation is called *simple* if it preserves the multiplicities of the pull-backs of 0, 1, and ∞ and the multiplicities of the ramification points with real critical values. Any deformation is locally simple with the exception of finitely many values of the parameter $t \in I$. (Recall that we are working in the *PL*-category.)

The next statement is a consequence of Theorem 4.11 and related definitions.

Proposition 4.14. *Two (c-equivariant) ramified coverings $j_0, j_1\colon S \to \mathbb{P}^1$ can be connected by a simple deformation if and only if their dessins $\mathfrak{D}(j_0), \mathfrak{D}(j_1) \subset S$ (or, respectively, real dessins $\mathfrak{D}_c(j_0), \mathfrak{D}_c(j_1) \subset \bar{S}$) are isotopic.* ◁

A deformation of ramified coverings is said to be *equisingular* if it preserves the multiplicities of the pull-backs of 0, 1, and ∞ modulo 3, 2, and 0, respectively, *cf.* Proposition 3.30. More precisely, define a *divisor* on the topological surface S as a finite integral linear combination of points of S, *i.e.*, a formal expression of the form

$$D = \textstyle\sum_{s \in S} m_s s, \quad m_s \in \mathbb{Z}, \quad \operatorname{Card}\{s \in S \mid m_s \neq 0\} < \infty.$$

The *support* of D is the set $\{s \in S \mid m_s \neq 0\} \subset S$; the residue $D \bmod p$, $p \geqslant 0$, is the divisor $\sum_{s \in S}(m_s \bmod p)s$, where $\bmod p$ is the residue in the range $0, \ldots, p-1$. For a ramified covering $\varphi\colon S \to S'$, the *divisorial pull-back* of a point $s' \in S'$ is the divisor $\varphi^*(s') := \sum_{s \in \varphi^{-1}(s')} r_s s$, where r_s is the ramification index of φ at s.

Definition 4.15. A deformation $j_t\colon S \to \mathbb{P}^1$ of ramified coverings is *equisingular* if the union of the supports of the divisors $(j_t^*(0) \bmod 3) + (j_t^*(1) \bmod 2) + j_t^*(\infty)$, $t \in I$, regarded as a subset of $S \times I$, is an isotopy.

Deformations of trichotomic graphs

Let $\mathfrak{D}_0 \subset S$ be a trichotomic graph, complex or real. Pick disjoint regular neighborhoods U_v of the vertices v of $\mathfrak{D}_0$ (assuming that $U_v \cap \partial S = \varnothing$ unless $v \in \partial S$) and replace each intersection $\mathfrak{D}_0 \cap U_v$ with another decorated graph, so that the result $\mathfrak{D}_1$ is again a trichotomic graph. If each intersection $\mathfrak{D}_1 \cap U_v$ contains essential vertices of at most one kind, the graph $\mathfrak{D}_1$ is called a *perturbation* of $\mathfrak{D}_0$, and $\mathfrak{D}_0$ is called a *degeneration* of $\mathfrak{D}_1$. A perturbation of a dessin is a dessin if and only if it is admissible, *i.e.*, if none of the intersections $\mathfrak{D}_1 \cap U_v$ contains a directed monochrome cycle. (The condition in Item 2 in Definition 4.2 holds automatically.) There seems to be no simple local criterion for a degeneration of a dessin to be a dessin: both conditions in Definition 4.2 need to be checked.

Remark 4.16. Speaking about perturbations of a real dessin $\mathfrak{D}_0 \subset S$ is equivalent to speaking about invariant perturbations of its oriented double $\tilde{\mathfrak{D}}_0 \subset \tilde{S}$: one should assume that the union of the neighborhoods chosen is preserved by the canonical real structure c, each neighborhood $\tilde{U}_{\tilde{v}}$ is disjoint from $\operatorname{Fix} c$ unless $\tilde{v}$ is fixed by c, and the new graph $\tilde{\mathfrak{D}}_1$ is also c-invariant; then $\mathfrak{D}_1 = \tilde{\mathfrak{D}}_1/c$.

If $\mathfrak{D}_1$ is an admissible perturbation of $\mathfrak{D}_0$, each intersection $\mathfrak{D}_1 \cap U_v$ is connected, *cf.* Remark 4.3. Since the intersection $\mathfrak{D}_1 \cap \partial U_v$ is fixed, it follows that $\mathfrak{D}_1 \cap U_v$ either is monochrome, if v is monochrome, or consists of monochrome vertices, essential vertices of the same kind as v, and edges of the two kinds incident to v.

A perturbation/degeneration is called *equisingular* if each intersection $\mathfrak{D}_1 \cap U_v$ contains at most one singular vertex.

Any deformation $j_t\colon S \to \mathbb{P}^1$, $t \in I$, of ramified coverings with simple restriction to $S \times (0,1]$ gives rise to a perturbation of the dessin $\mathfrak{D}_0 := \mathfrak{D}(j_0)$. (The restriction that each intersection $\mathfrak{D}_1 \cap U_v$ should contain essential vertices of at most one kind is due to the fact that essential vertices have predefined distinct images in $\mathbb{P}^1$.) The next theorem states the converse.

Theorem 4.17. *Given a dessin $\mathfrak{D}_0 \subset S$ and its admissible perturbation $\mathfrak{D}_1$, there is a deformation $j_t\colon S \to \mathbb{P}^1$, $t \in I$, with the following properties*:

- *one has $\mathfrak{D}_0 = \mathfrak{D}(j_0)$ and $\mathfrak{D}_1 = \mathfrak{D}(j_1)$;*
- *the restrictions of all maps j_t to $S \smallsetminus \bigcup_v U_v$ coincide;*
- *the restriction of the deformation to $S \times (0,1]$ is simple.*

The deformation as above is equisingular if and only if so is the perturbation $\mathfrak{D}_1$. If both $\mathfrak{D}_0$ and $\mathfrak{D}_1$ are c-invariant, the deformation can be chosen equivariant.

Proof. As usual, we prove the statement in the equivariant settings only. Let j_0 be any ramified covering given by Theorem 4.11. We can assume that the restriction of j_0 to each neighborhood U_v has no ramification points other than v itself. Then it suffices to construct a desired homotopy, fixed on the boundary, on each U_v.

Assume that v is a •-vertex, so that $j_0(v) = 0$. (In the other cases the proof is literally the same after reordering the colors and a coordinate change in $\mathbb{P}^1$.) First, assume that v is real and let $d = \operatorname{ind} v$. Regard U_v as a hemisphere in a real sphere $\bar{U}_v \cong S^2$ and extend both $\mathfrak{D}_0 \cap U_v$ and $\mathfrak{D}_1 \cap U_v$ to invariant trichotomic graphs $\bar{\mathfrak{D}}_0, \bar{\mathfrak{D}}_1 \subset \bar{U}_v$ by adding a real d-valent ×-vertex $\bar{v}$ (opposite to v), d bivalent ∘-vertices, and appropriate edges. Both graphs are dessins, and Corollary 4.13 gives rise to two real analytic maps $f_0, f_1 \colon \bar{U}_v \cong \mathbb{P}^1 \to \mathbb{P}^1$ corresponding to $\bar{\mathfrak{D}}_0$, $\bar{\mathfrak{D}}_1$, respectively. Clearly, $f_0(z) = z^d$ and $f_1(z)$ is a real polynomial of degree d, so that the family $f_t(z) = t^d f_1(z/t)$ is a desired homotopy. More precisely, we can assume that all critical points of f_1 other than $\bar{v}$ are mapped to the disk $\{|z| \leqslant 1/2\}$ (otherwise, replace f_1 with some $\varepsilon^d f_1(\,\cdot\,/\varepsilon)$); then $f_t^{-1}\{|z| \leqslant 1/2\}$, $t \in I$, is a disk bundle over I, and it can be identified with $U_v \times I$ so that the restriction of the homotopy to the boundary $\partial U_v \times I$ is constant.

If v is not real, we apply the same construction (with c ignored) and extend the result to the neighborhood $c(U_v)$ of $c(v)$ by symmetry. If no real structure is present, we just treat each neighborhood separately. The statement concerning equisingular deformations and perturbations follows from the definitions. □

Corollary 4.18. *Let S be a compact curve and $j \colon S \to \mathbb{P}^1$ a nonconstant holomorphic map. Assume that there is a chain $\mathfrak{D}(j) = \mathfrak{D}_0, \mathfrak{D}_1, \dots, \mathfrak{D}_n$ of dessins on S such that each $\mathfrak{D}_i$, $i = 1, \dots, n$, is a perturbation of, a degeneration of, or isotopic to $\mathfrak{D}_{i-1}$. Then there is a piecewise analytic deformation $j_t \colon S_t \to \mathbb{P}^1$, $t \in I$, of $j = j_0$ such that $\mathfrak{D}(j_1) = \mathfrak{D}_n$.*

Corollary 4.19. *Let (S, c) be a compact real curve and $j \colon (S, c) \to (\mathbb{P}^1, {}^-)$ a nonconstant real map. Assume that there is a chain $\mathfrak{D}_c(j) = \mathfrak{D}_0, \mathfrak{D}_1, \dots, \mathfrak{D}_n$ of real dessins on S/c such that each $\mathfrak{D}_i$, $i = 1, \dots, n$, is a perturbation of, a degeneration of, or isotopic to $\mathfrak{D}_{i-1}$. Then there is a piecewise analytic equivariant deformation $j_t \colon (S_t, c) \to (\mathbb{P}^1, {}^-)$, $t \in I$, of $j = j_0$ such that $\mathfrak{D}_c(j_1) = \mathfrak{D}_n$.*

We emphasize that, in both corollaries, the analytic structure on S may change during the deformation.

Remark 4.20. In Corollaries 4.18 and 4.19, each analytic piece of the deformations can be chosen as a closed real subinterval of an (equivariant) deformation in the sense of Kodaira–Spencer over an open complex disk.

Remark 4.21. Due to Theorem 4.17, the deformation j_t given by Corollaries 4.18 and 4.19 is equisingular in the sense of Definition 4.15 if and only if all perturbations and degenerations of the dessins involved are equisingular.

Proof of Corollaries 4.18 *and* 4.19. Using Proposition 4.14 and Theorem 4.17 (see also Remark 4.16), there is a topological (equisingular) deformation $S \times I \to \mathbb{P}^1 \times I$ as in the statement. According to the construction (in terms of polynomials), the branch locus in $\mathbb{P}^1 \times I$ is piecewise analytic and, for each piece, the (equivariant) ramified covering extends to a small open neighborhood in the complexification of the piece. Hence, the Grauert–Riemann theorem applies to produce a complex structure. □

4.2 Trigonal curves *via* dessins

We keep following [60], extending most statements to singular curves, both real and complex. The principal idea is the fact that a trigonal curve is (almost) determined by its j-invariant, whereas the latter, up to deformation of the complex structure, is adequately described by its dessin.

4.2.1 The correspondence theorems

Let $C \subset \Sigma \to B$ be a non-isotrivial trigonal curve. We define the (complex) *dessin* $\operatorname{Dssn} C$ of C as the dessin $\mathfrak{D}(j_C) \subset B$ of its j-invariant. If C is real with respect to a certain real structure $c\colon B \to B$, the *real dessin* $\operatorname{Dssn}_c C$ of C is the real dessin $\mathfrak{D}_c(j_C) \subset S := B/c$. When speaking about dessins, we regard B as a topological (or, more precisely, *PL*-) surface, disregarding its analytic structure.

Throughout this section, we fix the following convention. The underlying surface of a complex dessin $\mathfrak{D}$ is denoted by B. For a real dessin $\mathfrak{D} \subset S$, we denote by B and $c\colon B \to B$ the oriented double of S and its deck translation involution, respectively.

Any (real) dessin $\mathfrak{D}$ is of the form $\operatorname{Dssn} C$ (respectively, $\operatorname{Dssn}_c C$) for some (real) trigonal curve $C \subset \Sigma \to B$ with type $\tilde{\mathbf{A}}$ singular fibers only; such a curve is unique up to (equivariant) deformation in the class of curves with a fixed (real) dessin. (In view of Proposition 3.30, such deformations are necessarily equisingular.) In the complex case, this assertion follows from Corollary 4.12 and Theorem 3.20, see also Corollary 3.22; in the real case, it follows from Corollary 4.13 and Theorem 3.81. Furthermore, due to Corollaries 4.18 and 4.19, any sequence of isotopies, perturbations, or degenerations of (real) dessins gives rise to a piecewise analytic (equivariant) deformation of (real) trigonal curves; this deformation is equisingular if and only if all perturbations and degenerations used are equisingular.

Definition 4.22. A dessin $\mathfrak{D}$ is said to be *reduced* if it has the following properties:

- the full index of each •-vertex of $\mathfrak{D}$ is at most three;
- the full index of each ∘-vertex of $\mathfrak{D}$ is at most two;
- all monochrome vertices of $\mathfrak{D}$ are real and of full index two.

A reduced dessin is called *generic* if all its •- and ∘-vertices are nonsingular and all its ×-vertices are of full index one. A reduced complex dessin is called *maximal* if all its regions are triangular.

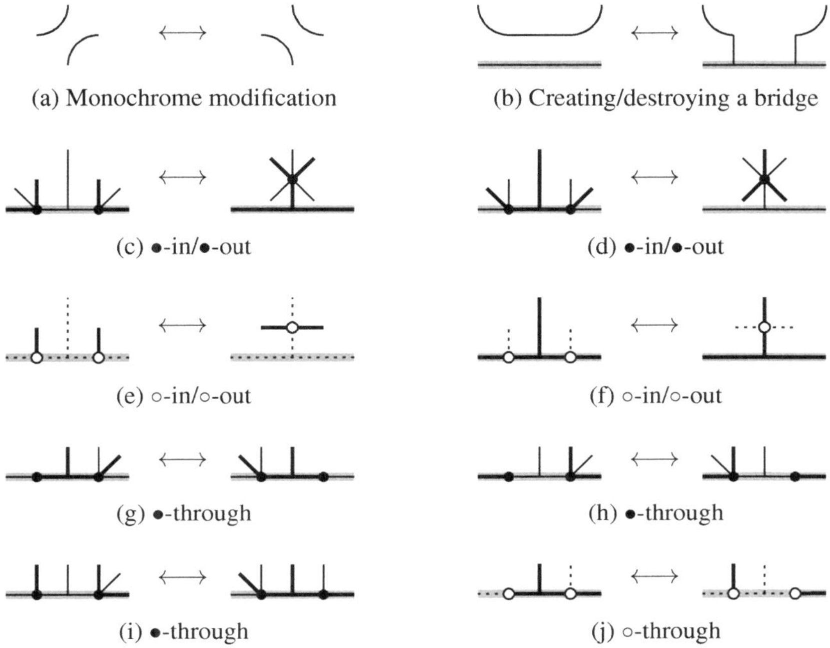

(a) Monochrome modification (b) Creating/destroying a bridge (c) •-in/•-out (d) •-in/•-out (e) ∘-in/∘-out (f) ∘-in/∘-out (g) •-through (h) •-through (i) •-through (j) ∘-through

Figure 4.1. Elementary moves of dessins.

Any dessin admits an equisingular perturbation to a reduced one. In the complex case, this statement is obvious. In the real case, perturbations of a real •-vertex of full index four to six and those of a real ∘-vertex of full index three to four are shown in Figure 4.1 (c)–(j). (In each case, up to isotopy, there are two perturbations, which are both shown in the figures.) Vertices of larger full index are reduced to these cases ($0 < \operatorname{ind} v \leqslant 6$ for •-vertices or $0 < \operatorname{ind} v \leqslant 4$ for ∘-vertices) by introducing an appropriate number of inner vertices of index three or two, respectively.

Definition 4.23. Two reduced dessins are *equivalent* if, after a homeomorphism of the underlying surfaces (orientation preserving in the complex case) they are connected by a sequence of isotopies and *elementary moves* shown in Figure 4.1.

Analyzing 'events of codimension one', one can see that, after a homeomorphism, two reduced dessins can be connected by a sequence of isotopies and/or equisingular perturbations and degenerations if and only if they are equivalent in the sense of the above definition. In the complex case, the only move required is the monochrome modification, Figure 4.1 (a). For *generic* real dessins, the set of moves is restricted to Figure 4.1 (a)–(f). The remaining four moves, Figure 4.1 (g)–(j), deal with singular real curves; they do not appear in [60].

Warning 4.24. A monochrome modification, see Figure 4.1 (a), or creating a bridge, see Figure 4.1 (b), are valid moves only if the result is a dessin. One needs to check both restrictions in Definition 4.2.

Combining the observations made so far, we obtain the following three correspondence theorems, which we state separately for the complex, real, and real nonsingular cases. Recall that, when speaking about deformation equivalence of trigonal curves over a base B, we assume that the analytic structure on B may change; thus, in the statements below B should be regarded as an oriented topological surface (in the real case, equipped with an orientation reversing involution $c\colon B \to B$).

Theorem 4.25. *There is a one-to-one correspondence between the set of equisingular deformation classes of non-isotrivial proper trigonal curves $C \subset \Sigma \to B$ with $\tilde{\mathbf{A}}$ type singular fibers only and the set of equivalence classes of reduced dessins $\mathfrak{D} \subset B$.* $\lhd$

Theorem 4.26. *There is a one-to-one correspondence between the set of equivariant equisingular deformation classes of non-isotrivial proper real trigonal curves $C \subset \Sigma \to (B, c)$ with $\tilde{\mathbf{A}}$ type singular fibers only and the set of equivalence classes of reduced real dessins $\mathfrak{D} \subset B/c$.* $\lhd$

Theorem 4.27. *There is a one-to-one correspondence between the set of equivariant equisingular deformation classes of almost generic real trigonal curves $C \subset \Sigma \to (B, c)$ and the set of equivalence classes of generic real dessins $\mathfrak{D} \subset B/c$.* $\lhd$

4.2.2 Complex curves

From now on, in order to obtain more detailed statements, we treat complex and real curves separately. In the complex case, we introduce the *type specification* that would let us describe curves with arbitrary, more general than just type $\tilde{\mathbf{A}}$, singular fibers, discuss various counts related to dessins, and explain how maximal curves can be described in terms of their skeletons, which, unlike dessins, should be considered up to homeomorphism only, with no extra moves involved.

More general singular fibers

As explained in Remarks 3.21 and 3.23, in order to describe a proper non-isotrivial trigonal curve $C \subset \Sigma \to B$ with more general than type $\tilde{\mathbf{A}}$ only singular fibers, in addition to the j-invariant j_C, one needs to specify an effective divisor D_{ex} on B: if $D_{\mathrm{ex}} = \sum_{b \in B} m_b b$, an m_b-fold positive Nagata transformation is to be performed over each point $b \in B$ with $m_b > 0$.

We are interested in equisingular deformations of such curves. Observe that the divisors D_*, $* = \bullet, \circ, \times$, considered in Proposition 3.30 are the 'divisors of $*$-vertices' of the dessin of C, *i.e.*, $D_* = \sum_v (\operatorname{ind} v) v$, the summation running over all $*$-vertices of $\mathfrak{D} := \operatorname{Dssn} C$, and the sum $(D_\bullet \bmod 3) + (D_\circ \bmod 2) + D_\times$ is the 'divisor of singular vertices' of $\mathfrak{D}$. It follows that important are the multiplicities of the points in D_{ex} and the intersection of D_{ex} with the set of *singular* vertices of $\mathfrak{D}$, and we encode this information in terms of the so-called type specification on $\mathfrak{D}$. To be consistent with Definition 2.53 and Proposition 2.54, we give the following definition.

Definition 4.28. Let $\mathfrak{D}$ be a dessin. The *trivial type specification* on $\mathfrak{D}$ is the function $\operatorname{tp}_0\colon \operatorname{Sing}\mathfrak{D} \to \mathbb{N}$ sending each singular $\bullet$-, $\circ$-, or $\times$-vertex v to $2 \operatorname{ind} v \bmod 6$, 3, or $\operatorname{ind} v$, respectively. (Note that the values at $\bullet$-vertices are in fact 2 or 4.) The *type specification* of a proper singular curve C defined by a j-invariant j_C and extra divisor $D_{\text{ex}} = \sum_b m_b b$ is the function tp_C sending each singular vertex v to $\operatorname{tp}_0(v) + 6m_v$.

It follows that any valid type specification tp satisfies the restrictions

$$\operatorname{tp}(v) \geqslant \operatorname{tp}_0(v), \quad \operatorname{tp}(v) = \operatorname{tp}_0(v) \bmod 6 \tag{4.29}$$

and, conversely, any function tp satisfying these restrictions can appear as the type specification of a proper trigonal curve. At first sight Definition 4.28 looks awkward, but in fact it assigns to each singular vertex the degree of the braid monodromy about the corresponding singular fiber of C, see Theorem 5.68. The trivial type specification tp_0 corresponds to the minimal proper curve with type $\tilde{\mathbf{A}}$ singular fibers only.

The positions of the other points constituting D_{ex} are irrelevant as long as these points do not collide with one another or with singular vertices of $\mathfrak{D}$. Hence, these points can be encoded by the set Ex of their multiplicities. Summarizing, a generic (in its equisingular deformation class) proper trigonal curve C over a base B can be encoded by a triple $(\mathfrak{D}, \operatorname{tp}, \operatorname{Ex})$, where $\mathfrak{D} \subset B$ is a reduced dessin, $\operatorname{tp}\colon \operatorname{Sing}\mathfrak{D} \to \mathbb{N}$ is a function satisfying (4.29), and Ex is a finite set of positive integers. Two such triples $(\mathfrak{D}_i, \operatorname{tp}_i, \operatorname{Ex}_i)$, $i = 1, 2$, are considered equivalent if $\operatorname{Ex}_1 = \operatorname{Ex}_2$ and there is an equivalence between $\mathfrak{D}_1$ and $\mathfrak{D}_2$ taking tp_1 to tp_2. In these terms, Theorem 4.25 takes the following generalized form.

Theorem 4.30. *There is a one-to-one correspondence between the set of equisingular deformation classes of non-isotrivial proper trigonal curves $C \subset \Sigma \to B$ and the set of equivalence classes of triples $(\mathfrak{D}, \operatorname{tp}, \operatorname{Ex})$ as above.* $\triangleleft$

Certainly, a statement similar to Theorem 4.30 holds for real curves as well. As usual, in the real case the situation is slightly more involved. Since the points of D_{ex} are not allowed to collide with one another or with singular vertices of $\mathfrak{D}$, one needs to keep track of the mutual position of the real points of D_{ex} and real singular vertices in the real part $\partial(B/c)$; hence, at least partially, D_{ex} should be treated as a divisor. We leave the precise definitions and the statement to the reader.

The degree of a dessin

Let $\Sigma \to B$ be a compact ruled surface with an exceptional section E. Denote $d := -E^2 \geqslant 0$. Following [60], we define the *degree* of a proper trigonal curve $C \subset \Sigma$ as $\deg C := 3d$. This number can be interpreted as the intersection $C \circ Z$, where Z is the zero section. The *degree* $\deg \mathfrak{D}$ of a dessin $\mathfrak{D}$ is the degree of any minimal (*i.e.*, with type $\tilde{\mathbf{A}}$ singular fibers only) trigonal curve C whose dessin is equal to $\mathfrak{D}$.

Let $\mathfrak{D} \subset B$ be a reduced dessin. For $* = \bullet, \circ, \times$, let $\#_*(\mathfrak{D})$ be the number of $*$-vertices of $\mathfrak{D}$, and let $\#^i_*(\mathfrak{D})$ be the number of $*$-vertices of index $i \in \mathbb{N}$. When $\mathfrak{D}$ is understood, we abbreviate $\#^i_*(\mathfrak{D})$ to $\#^i_*$. Let, further, j_C be the j-invariant of any trigonal curve C as above. By definition, one has

$$\sum_{i\in\mathbb{N}} i \cdot \#^i_\times(\mathfrak{D}) = \sum_{j\in\mathbb{N}} j \cdot \#^j_\bullet(\mathfrak{D}) = \sum_{k\in\mathbb{N}} k \cdot \#^k_\circ(\mathfrak{D}) = \deg j_C. \tag{4.31}$$

From Table 3.1 (the column $\Delta \deg j_C$), we have $\deg j_C = 2\deg\mathfrak{D} - 2\#^1_\bullet - 3\#^3_\circ - 4\#^2_\bullet$; then, taking into account the identities $\#_\bullet = \#^1_\bullet + \#^2_\bullet + \#^3_\bullet$ and $\#_\circ = \#^1_\circ + \#^2_\circ$ (since $\mathfrak{D}$ is reduced), we obtain the following vertex counts:

$$\#_\bullet + \#^2_\bullet + \#^1_\circ = \frac{2}{3}\deg\mathfrak{D}, \qquad \#_\circ + \#^1_\circ + \#^1_\bullet + 2\#^2_\bullet = \deg\mathfrak{D}. \tag{4.32}$$

The total Milnor number

Since we are mainly interested in stable simple maximal trigonal curves, we introduce the following 'correction terms': for a proper trigonal curve C, let

- t_{u} be the number of unstable singular fibers of C (types $\tilde{\mathbf{A}}_0^{**}$, $\tilde{\mathbf{A}}_1^*$, or $\tilde{\mathbf{A}}_2^*$),
- t_0 be the number of singular fibers of type $\tilde{\mathbf{J}}_{r,0}$, $r \geqslant 1$, *i.e.*, those not detected by the j-invariant of the curve, see Table 3.1, and
- t_{ns} be the number of positive Nagata transformations beyond a curve with simple singularities; more precisely, $t_{\mathrm{ns}} = \sum(r-2)$, the summation running over all singular fibers of types $\tilde{\mathbf{J}}_{r,p}$ or $\tilde{\mathbf{E}}_{6r+\epsilon}$ with $r > 2$, $p \geqslant 0$, $\epsilon = 0, 1, 2$.

Theorem 4.33. *Let $\Sigma \to B$ be a compact ruled surface with an exceptional section E. Denote $d := -E^2 \geqslant 0$, and let $C \subset \Sigma$ be a non-isotrivial proper trigonal curve. Then, in the notation above, one has*

$$\mu(C) + t_{\mathrm{u}} + t_0 \leqslant 5d + 2g(B) - 2 + t_{\mathrm{ns}},$$

where $g(B)$ is the genus of B. This inequality turns into an equality if and only if a reduced dessin $\mathfrak{D}$ equivalent to $\operatorname{Dssn} C$ *is maximal.*

Proof. After a small equisingular deformation we can assume that the dessin $\mathfrak{D} := \operatorname{Dssn} C$ is reduced, so that each edge connects two essential vertices. Denote by v, e, and r the numbers of vertices, edges, and regions of $\mathfrak{D}$, respectively. Then $v - e + \sum_R \chi(R) = \chi(B) = 2 - 2g(B)$, where the summation runs over all regions of $\mathfrak{D}$. For each region R we have $\chi(R) \leqslant 1$, the equality holding if and only if $R \cong \bar{D}^2$. Furthermore, since each region is at least a triangle, $r \leqslant \frac{2}{3}e$ and, substituting and expanding $v = \#_\bullet + \#_\circ + \#_\times$, we obtain

$$\#_\bullet + \#_\circ + \#_\times - \frac{1}{3}e \geqslant 2 - 2g(B);$$

the equality holds if and only if each region of $\mathfrak{D}$ is a triangle (then, since $\mathfrak{D}$ is a dessin, each region is also homeomorphic to a disk). As usual, $2e = \sum_v 2 \operatorname{ind} v$ and, due to (4.31), we have $e = 3 \deg j_C$ and $\#_\times - \frac{1}{3}e = -\sum_{i \in \mathbb{N}}(i-1)\#_\times^i$. If C has type $\tilde{\mathbf{A}}$ singular fibers only, the counts $\#_\bullet$ and $\#_\circ$ are found from (4.32), where $\deg \mathfrak{D} = 3d$. Substituting and using the fact that $\mu(\mathbf{A}_p) = p$, we obtain the required inequality. For the general case, note that each positive Nagata transformation increases d by one, and each transformation resulting in a non-simple singular point increases μ by six while increasing t_{nx} by one. Other transformations modify the quantities involved as follows:

- Kodaira type I to type I* ($\mathbf{A}_p \mapsto \mathbf{D}_{p+5}$): $\mu \mapsto \mu + 5$;
- an unstable fiber to a simple type $\tilde{\mathbf{E}}$ fiber: $\mu \mapsto \mu + 6$ and $t_{\mathrm{u}} \mapsto t_{\mathrm{u}} - 1$;
- a nonsingular fiber to a type $\tilde{\mathbf{D}}_4 = \tilde{\mathbf{J}}_{1,0}$ fiber: $\mu \mapsto \mu + 4$ and $t_0 \mapsto t_0 + 1$.

Hence, in all cases, the difference between the two sides of the inequality remains unchanged. □

Corollary 4.34. *For a simple non-isotrivial trigonal curve $C \subset \Sigma \to B$ one has $\mu(C) \leqslant 5d + 2g(B) - 2$, where d and $g(B)$ are as in Theorem* 4.33. ◁

Remark 4.35. The conclusions of Theorem 4.33 and Corollary 4.34 may not hold for isotrivial curves. For such a curve C one always has $\mu(C) + \operatorname{Card}(\operatorname{Sing} C) = 6d$; this statement follows from the description of isotrivial curves found on page 75.

Maximal curves *via* skeletons

Analyzing the definition of $\mathfrak{D}(j)$, see page 113, and the proof of Theorem 4.11, one can easily see that a nonconstant meromorphic function $j\colon B \to \mathbb{P}^1$ has extremal branching behavior if and only if its dessin $\mathfrak{D}(j)$ is maximal. Hence, Definition 3.39 and part of Theorem 4.33 can be rephrased as follows.

Proposition 4.36. *A non-isotrivial trigonal curve C without singular fibers of type $\tilde{\mathbf{J}}_{r,0}$, $r \geqslant 1$, is maximal if and only if its dessin* Dssn C *is maximal.* ◁

Corollary 4.37. *A non-isotrivial simple trigonal curve $C \subset \Sigma \to B$ maximizes the total Milnor number, i.e., $\mu(C) = 5d + 2g(B) - 2$ in the notation of Theorem* 4.33, *if and only if C is maximal and without unstable singular fibers.* ◁

Given a dessin $\mathfrak{D}$, define its *skeleton* $\operatorname{Sk}\mathfrak{D}$ as the union of the $\bullet$- and $\circ$-vertices of $\mathfrak{D}$, its bold edges, and bold monochrome vertices. The *skeleton* $\operatorname{Sk} C$ of a non-isotrivial trigonal curve C is the skeleton of its dessin. If $\mathfrak{D}$ is reduced and $\operatorname{Sk}\mathfrak{D}$ is connected, then $\operatorname{Sk}\mathfrak{D}$ is indeed a finite skeleton of type $(6, 2)$ and without $\bullet$-vertices of valency six, see Definition 1.14; in other words, the valency of each $\bullet$-vertex of $\operatorname{Sk}\mathfrak{D}$ is at most three, and the valency of each $\circ$-vertex is at most two. Skeletons with this property are called *geometric*. (In fact, it is this construction that explains the term 'skeleton'.) A *singular vertex* of a geometric skeleton is a $\bullet$-vertex of valency one or two or a $\circ$-vertex of valency one.

Note that $\operatorname{Sk} C$ is the *dessin d'enfant*, in the original sense of Grothendieck, of the j-invariant j_C. We use the term 'skeleton', reserving 'dessin' for trichotomic graphs, which capture more information in the case of non-maximal curves.

The passage from a dessin to its skeleton is partially invertible. Namely, let $\mathfrak{S}$ be a skeleton and let $B = \operatorname{Supp}\mathfrak{S}$. Place a $\times$-vertex at the center c_R of each region R of $\mathfrak{S}$, and connect this vertex to the $\bullet$- and $\circ$-vertices in the boundary ∂R by solid and dotted edges, respectively. (Recall that the geometric realization of R in B is the closed disk $\bar{D}^2$ and c_R is the center of this disk.) Denote the resulting trichotomic graph $\mathfrak{D} \subset B$ by $\operatorname{Dssn}\mathfrak{S}$. It is immediate that $\operatorname{Dssn}\mathfrak{S}$ is a dessin; if $\mathfrak{S}$ is geometric, then $\operatorname{Dssn}\mathfrak{S}$ is maximal.

Proposition 4.38. *The maps* $\mathfrak{D} \mapsto \operatorname{Sk}\mathfrak{D}$ *and* $\mathfrak{S} \mapsto \operatorname{Dssn}\mathfrak{S}$ *establish a bijection between the set of equivalence classes of maximal dessins and the set of isomorphism classes of geometric skeletons. Furthermore, both sets are in a canonical one-to-one correspondence with the set of analytic isomorphism classes of meromorphic functions* $j\colon S \to \mathbb{P}^1$ *having extremal branching behavior; one has* $S \cong \operatorname{Supp}\mathfrak{S}$.

Proof. It is clear that the two maps establish a bijection between isomorphism classes of skeletons as in the statement and *orientation preserving homeomorphism* classes of maximal dessins. (Note that the skeleton of a maximal dessin is always connected.) On the other hand, a maximal dessin does not admit a nontrivial degeneration, as any nontrivial perturbation produces regions with more than three essential vertices. Hence, a dessin $\mathfrak{D}' \subset B'$ is equivalent to a maximal dessin $\mathfrak{D} \subset B$ if and only if the pairs $(B', \mathfrak{D}')$ and $(B, \mathfrak{D})$ are homeomorphic. The last correspondence is given by the map $j \mapsto \mathfrak{D}(j)$. If $\mathfrak{D}$ is maximal, the ramified covering $j\colon S \to \mathbb{P}^1_{\mathbb{R}}$ given by Theorem 4.11 is unique up to *homeomorphism* and the Riemann existence theorem gives a unique analytic structure on $S = \operatorname{Supp}\mathfrak{S}$ that makes j meromorphic. □

Remark 4.39. The class of geometric skeletons does not quite coincide with that considered in Section 2.1.2 in relation to the modular group: in addition, we need to allow bivalent $\bullet$-vertices, which correspond to the type $\tilde{\mathbf{E}}_{6r+2}$, $r \geqslant 0$, singular fibers of trigonal curves. For this reason, some of the results found below in the book do not extend directly to curves with such fibers.

Let $\mathfrak{D}$ be a maximal dessin and $\mathfrak{S} = \operatorname{Sk}\mathfrak{D}$. The •- and ◦-vertices of $\mathfrak{D}$ coincide with those of $\mathfrak{S}$ (and so do the valencies, *cf.* Convention 4.7), whereas the ×-vertices of $\mathfrak{D}$ are in a one-to-one correspondence with the regions of $\mathfrak{S}$ (the index of a vertex being equal to the width of the corresponding region). Hence, the type specification of a maximal trigonal curve C can be regarded as an $\mathbb{N}$-valued function tp_C defined on the set of singular vertices and regions of the skeleton $\operatorname{Sk} C$, *cf.* Definition 2.53. This function tp_C is subject to the following restrictions, *cf.* Definition 4.28:

- $\operatorname{tp}_C(v) \geqslant 2\deg v$ and $\operatorname{tp}_C(v) = 2\deg v \bmod 6$ for a singular •-vertex v;
- $\operatorname{tp}_C(v) \geqslant 3$ and $\operatorname{tp}_C(v) = 3 \bmod 6$ for a singular ◦-vertex v;
- $\operatorname{tp}_C(R) \geqslant \operatorname{wd} R$ and $\operatorname{tp}_C(R) = \operatorname{wd} R \bmod 6$ for a region R.

A homeomorphism of pairs $(\mathfrak{D}, \operatorname{tp})$ is an orientation preserving homeomorphism of the underlying surfaces taking the dessin to the dessin and preserving the type specification. Since, for a maximal curve, the set Ex of extra multiplicities is empty (there are no singular fibers of type $\tilde{\mathbf{J}}_{r,0}$, $r \geqslant 1$), Theorem 4.30 and Proposition 4.38 give rise to the following classification statement. (For the last part, see also Corollary 3.41.)

Corollary 4.40. *Two maximal trigonal curves C_1, C_2 are equisingular deformation equivalent if and only if the pairs $(\operatorname{Sk} C_i, \operatorname{tp}_{C_i})$, $i = 1, 2$, are homeomorphic. In this case, the curves are also analytically isomorphic.* ◁

Maximal curves of low degree

The *degree* $\deg \mathfrak{S}$ of a skeleton $\mathfrak{S}$ is the degree of $\operatorname{Dssn}\mathfrak{S}$. If $\mathfrak{S}$ is geometric, its degree in terms of the vertex counts is given by (4.32). We are mainly interested in stable maximal curves in the Hirzebruch surface Σ_2. The skeletons of such curves can easily be enumerated manually. The skeleton $\mathfrak{S} := \operatorname{Sk} C$ is of genus zero, and either

- $\deg \mathfrak{S} = 6$ and $\mathfrak{S}$ is $(3, 2)$-regular, see Figure 4.2 (the curve C has type $\tilde{\mathbf{A}}$ singular fibers only), or
- $\deg \mathfrak{S} = 3$ and $\mathfrak{S}$ is $(3, 2)$-regular, see Figure 4.3, top (C has one singular fiber of type $\tilde{\mathbf{J}}_{1,p}$, $p \geqslant 1$, over one of the regions of $\mathfrak{S}$, which should be indicated), or
- $\deg \mathfrak{S} = 3$ and $\mathfrak{S}$ has one singular vertex, see Figure 4.3, bottom (C has one singular fiber of type $\tilde{\mathbf{E}}_{6+\epsilon}$, $\epsilon = 0, 1, 2$, over the singular vertex of $\mathfrak{S}$).

Up to rigid (not necessarily fiberwise) isotopy, all stable proper trigonal curves in Σ_2 can be classified in terms of certain sublattices of $\mathbf{E}_8$, see Theorem 7.1.

A complete list of $(3, 2)$-regular skeletons of degree 12 is found in [20] or at

`http://www.staff.science.uu.nl/~beuke106/mirandapersson/Dessins.html.`

Altogether, there are 191 skeletons.

Observation 4.41. The list in [20] can easily be downgraded to any lower degree: any skeleton $\mathfrak{S}$ of degree $3d$ can be obtained from an appropriate skeleton $\mathfrak{S}'$ of degree $(3d + 3)$ by removing a monogonal region and the 'stem' connecting this

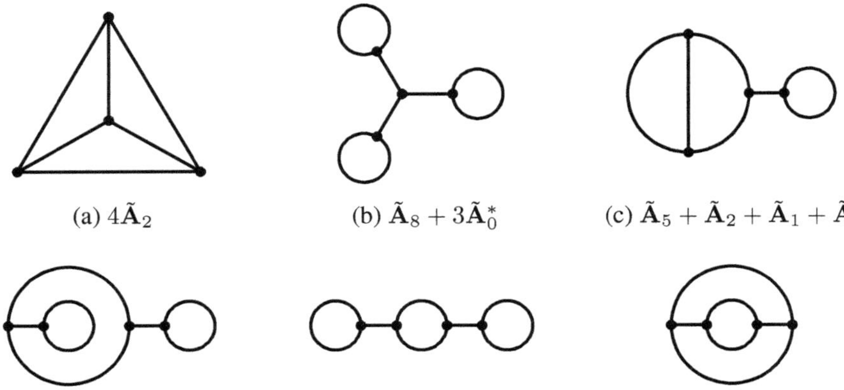

(a) $4\tilde{\mathbf{A}}_2$ (b) $\tilde{\mathbf{A}}_8 + 3\tilde{\mathbf{A}}_0^*$ (c) $\tilde{\mathbf{A}}_5 + \tilde{\mathbf{A}}_2 + \tilde{\mathbf{A}}_1 + \tilde{\mathbf{A}}_0^*$

(d) $2\tilde{\mathbf{A}}_4 + 2\tilde{\mathbf{A}}_0^*$ (e) $\tilde{\mathbf{A}}_7 + \tilde{\mathbf{A}}_1 + 2\tilde{\mathbf{A}}_0^*$ (f) $2\tilde{\mathbf{A}}_3 + 2\tilde{\mathbf{A}}_1$

Figure 4.2. Regular geometric skeletons of genus zero and degree six.

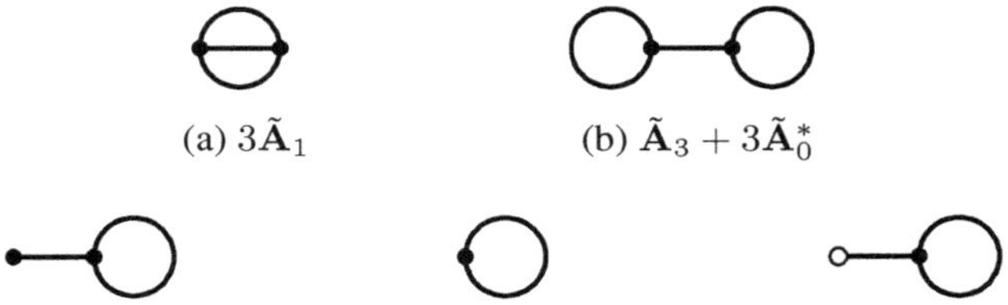

(a) $3\tilde{\mathbf{A}}_1$ (b) $\tilde{\mathbf{A}}_3 + 3\tilde{\mathbf{A}}_0^*$

(c) $\tilde{\mathbf{A}}_0^{**} + \tilde{\mathbf{A}}_2 + \tilde{\mathbf{A}}_0^*$ (d) $\tilde{\mathbf{A}}_2^* + 2\tilde{\mathbf{A}}_0^*$ (e) $\tilde{\mathbf{A}}_1^* + \tilde{\mathbf{A}}_1 + \tilde{\mathbf{A}}_0^*$

Figure 4.3. Geometric skeletons of genus zero and degree three.

region to the rest of $\mathfrak{S}'$. For example, the skeleton shown in Figure 4.3 (a) is obtained in this way from Figure 4.2 (c), and Figure 4.3 (b) is obtained from Figure 4.2 (b), (d), or (e). The lists can also be used to enumerate all skeletons with singular vertices. Formally, a monovalent •-vertex can be obtained by contracting a monogonal region to its only •-vertex, *cf.* Figure 4.3 (c) *vs.* (b), whereas a monovalent ∘-vertex can be obtained by removing a monovalent •-vertex and the incident edge, Figure 4.3 (e) *vs.* (c). Similarly, a bivalent •-vertex is obtained by contracting a bigonal region, *cf.* Figure 4.3 (d) *vs.* (a). It follows from (4.32) that these operations do not change the degree of the skeleton.

A number of similar tricks, producing new skeletons with desired fragments, are discussed and used further in Chapter 8. (The original skeletons are mostly those shown in Figures 4.2 and 4.3.) I do not always know the geometric meaning of these operations in terms of the trigonal curves.

Further properties of maximal curves

We discuss briefly a few other extremal properties of maximal trigonal curves.

Proposition 4.42. *A maximal trigonal curve C does not admit a nontrivial degeneration to a non-isotrivial trigonal curve.*

Proof. Assume that C degenerates to a non-isotrivial curve C_0, which is necessarily proper. Then, up to isotopy, the dessin $\operatorname{Dssn} C$ is obtained from $\operatorname{Dssn} C_0$ by removing disjoint regular neighborhoods of some of its vertices and replacing them with new decorated graphs. (Since the degree of the j-invariant may change, it is no longer required that each of the new graphs should contain essential vertices of at most one kind. We do not discuss the realizability of any such modification by a degeneration of curves.) If this procedure is nontrivial, it results in a graph $\mathfrak{D}$ with at least one non-triangular region. □

Over the rational base $B \cong \mathbb{P}^1$, Proposition 4.42 admits a partial converse: with respect to degenerations, there is another class of extremal curves, *viz.* those whose skeleton is a forest (*i.e.*, disjoint union of trees) and whose dessin has a single $\times$-vertex. (In [52], it is erroneously stated that any curve degenerates to a maximal one.)

We start with a simple lemma.

Lemma 4.43. *Any reduced dessin $\mathfrak{D}$ in the sphere $S \cong S^2$ is equivalent to a reduced dessin whose skeleton is either connected or a forest.*

Proof. Assume that the skeleton $\mathfrak{S} := \operatorname{Sk} \mathfrak{D}$ is neither connected nor a forest. Then there is a simple arc ζ in S, transversal to the edges of $\mathfrak{D}$ and disjoint from its vertices, such that

- the interior of ζ is disjoint from $\mathfrak{S}$;
- the endpoints of ζ are at the centers of two edges e_0, e_1 of $\mathfrak{S}$ that belong to two distinct connected components $\mathfrak{S}_0$, $\mathfrak{S}_1$, respectively;
- the component $\mathfrak{S}_0$ remains connected after removing e_0.

Let s be the last (closest to e_1) point of intersection of $\mathfrak{D}$ and the interior of ζ, and let e be the edge of $\mathfrak{D}$ through s. Assume that it is solid; the case of a dotted edge is treated similarly. Applying, if necessary, a solid monochrome modification, we can assume further that e is incident to the $\bullet$-end v of e_1. Then, replacing the portion of ζ between s and e_1 with a parallel copy of the arc $[p, v]$ of e, we obtain a new simple arc ζ' connecting e_0 and $e_i' = e_1 \uparrow \mathbf{x}^{\pm 1}$ and having fewer points of intersection with the new dessin $\mathfrak{D}'$. Proceeding by induction, we obtain a dessin $\mathfrak{D}'' \sim \mathfrak{D}$ with the same skeleton $\mathfrak{S}$ and a simple arc ζ'' connecting e_0 to an edge e_1'' in another component of $\mathfrak{S}$ and otherwise disjoint from $\mathfrak{D}''$. The bold monochrome modification along ζ'' reduces the number of connected components of $\mathfrak{S}$. □

Corollary 4.44. *Let C be a non-isotrivial trigonal curve in a Hirzebruch surface Σ_d such that $d + u \leqslant 4$, where u is the number of unstable singular fibers of C. Then C is equisingular deformation equivalent to a curve with connected skeleton.*

Proof. If $\operatorname{Sk} C$ is a disconnected forrest, then C has at least four exceptional singular fibers and $\deg C + 3u > 12$. □

Proposition 4.45. *Assume that a non-isotrivial trigonal curve $C \subset \Sigma_d$ admits no nontrivial degeneration to a non-isotrivial curve. Then either C is maximal or the dessin* $\operatorname{Dssn} C$ *has a single $\times$-vertex (and then the skeleton* $\operatorname{Sk} C$ *is a forest).*

Proof. Due to Lemma 4.43, one can assume that the skeleton $\mathfrak{S} := \operatorname{Sk} C$ is either connected or a forest. Then one can degenerate $\operatorname{Dssn} C$ so that all $\times$-vertices within the same region of $\mathfrak{S}$ collide to a single $\times$-vertex. If $\mathfrak{S}$ is connected, the new curve is maximal; otherwise, its dessin has a single $\times$-vertex, which is located in the single region of its skeleton. □

Remark 4.46. In fact, a slightly stronger statement has been proved: one can relax the hypotheses of Proposition 4.45 and consider only degenerations resulting in a continuous change of the j-invariant, *i.e.*, such that only $\tilde{\mathbf{J}}$ type singular fibers are allowed to collide. Such degenerations will be called *simple*. The second class of extremal curves, those with a single $\times$-vertex, deserves a further study; at present, I do not know how 'standard' they can be made if collisions of $\tilde{\mathbf{E}}$ type singular fibers are also allowed.

Corollary 4.47 (of Corollary 4.44 and Proposition 4.45). *Any trigonal curve C as in Corollary* 4.44 *admits a simple degeneration to a maximal curve.* ◁

Dessins *vs.* skeletons

The following statement generalizes the passage from a skeleton to the dessin of a maximal curve. It can be reinterpreted as yet another version of Hurwitz theorem, see Theorem 10.14.

Proposition 4.48. *Let $\mathfrak{S} \subset S$ be an embedded bipartite ribbon graph, and let R be a region of $\mathfrak{S}$ homeomorphic to a disk. Then, up to equivalence identical on ∂R, an extension of $\mathfrak{S}$ to a dessin $\mathfrak{D}$ in R is uniquely determined by the set $\{m_1, \ldots, m_c\}$ of indices of the $\times$-vertices of $\mathfrak{D}$ in R. Any set with $m_i > 0$ and $\sum_i m_i = \operatorname{wd} R$ is realizable.*

Proof. We will show that any extension $\mathfrak{D}$ with a given set of indices of $\times$-vertices is equivalent to a certain standard one. Consider the set $\mathcal{V}$ of all $\bullet$- and $\circ$-vertices in the boundary ∂R, and let $\prec$ be the cyclic order on $\mathcal{V}$ induced by the positive orientation of the boundary. Let, further, $v_1, \ldots, v_c$ be the set of $\times$-vertices of $\mathfrak{D}$, ordered in some

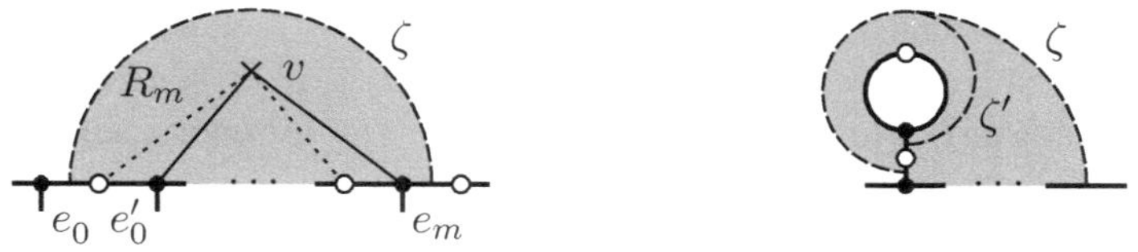

(a) Successful (b) Unsuccessful ($e_0 = e'_2$)

Figure 4.4. Cutting off an m-gonal subregion.

standard way, and denote by $\mathcal{V}_i \subset \mathcal{V}$ the set of the •- and ∘-ends of the solid and dotted edges incident to a vertex v_i, $i = 1, \ldots, c$. Since R is a disk, up to isotopy identical on ∂R the dessin $\mathfrak{D}$ is determined by the partition $\mathcal{V} = \bigcup_i \mathcal{V}_i$, and it suffices to make the latter standard.

Fix a vertex $u_1 \in \mathcal{V}_1$. Let $u \in \mathcal{V}_1$ be the first successor of u_1 (maybe u_1 itself) such that $u' := \operatorname{succ} u$ is not in $\mathcal{V}_1$, and let u'' be the first successor of u' that is again in $\mathcal{V}_1$. Then the inner (with respect to R) edges incident to u' and u'' are of the same kind and in the boundary of the same region of $\mathfrak{D}$; hence we can apply a monochrome modification so that, in the new dessin, one has $u' \in \mathcal{V}_1$ and $u'' \notin \mathcal{V}_1$. (We retain the same notation for all new dessins and related objects.) Proceeding by induction until $u'' = u_1$, we obtain an equivalent dessin in which the set $\mathcal{V}_1$ is a contiguous segment $[u_1, \ldots, u_{2m}]$, where $m = \operatorname{ind} v_1$. Continuing the process, we can move this segment as a whole to any desired position in the boundary ∂R.

Now, starting from the vertex $u_{2m+1} := \operatorname{succ} u_{2m}$, we can make contiguous the set $\mathcal{V}_i$ containing u_{2m+1}, then the set $\mathcal{V}_j$ containing the immediate successor of the last vertex of $\mathcal{V}_i$, and so on. Finally, whenever two sets $\mathcal{V}_i = [\ldots, u]$ and $\mathcal{V}_j = [\operatorname{succ} u, \ldots]$ are adjacent to each other, starting from the first element of $\mathcal{V}_i$ and applying the above procedure $\operatorname{Card} \mathcal{V}_j$ times, we can transpose $\mathcal{V}_i$ and $\mathcal{V}_j$. Hence, the segments $\mathcal{V}_i$ can be ordered in ∂R according to the original order of the ×-vertices v_i of $\mathfrak{D}$. □

Consider a geometric skeleton $\mathfrak{S} := \operatorname{Sk} \mathfrak{D}$ and fix an n-gonal region R of $\mathfrak{S}$. Let $e_0, e'_0, e_1, \ldots$ be the sequence of edges constituting the boundary ∂R, *cf.* Figure 1.2, numbered cyclically starting at any given edge $e_0 \in R$; thus, we have $e_i = e_0 \uparrow (\mathbf{xy})^{-i}$ and $e'_i = e_i \uparrow \mathbf{y}^{-1}$ for all $i \in \mathbb{Z}$. Fix a positive integer $m < n$ and denote by $\mathfrak{S}_m$ the bipartite ribbon graph obtained from $\mathfrak{S}$ by the following modification of the action of $\mathbf{y} \in \mathbb{G}$:

$$e_0 \leftrightarrow e'_m \text{ if } e_m \neq e'_m, \qquad e_0 \leftrightarrow e_0 \text{ if } e_m = e'_m,$$
$$e_m \leftrightarrow e'_0 \text{ if } e_0 \neq e'_0, \qquad e_m \leftrightarrow e_m \text{ if } e_0 = e'_0.$$

(The action of $\mathbf{x}$ and the action of $\mathbf{y}$ on the edges other than e_0, e'_0, e_m, and e'_m remain unchanged. We use the identity $\mathbf{y}^2 = 1$ in Γ and do not attempt a formal description of this construction in the more general setting of arbitrary $\mathbb{G}$-sets.)

The geometric meaning of this construction is a bold monochrome modification along a simple arc ζ connecting the centers of e_0 and e_m in R, see Figure 4.4, and its purpose is cutting off an m-gonal subregion of R. The following statement follows immediately from the definitions in Chapter 1.

Lemma 4.49. *If a triple $(\mathfrak{S}, e_0, m)$ as above satisfies the condition*

1. $e_0 \neq e'_i$ *and* $e_m \neq e'_i$ *for any* $i = 0, \dots, m-1$,

then the edges $e_1, \dots, e_m$ constitute an m-gonal region R_m of $\mathfrak{S}_m$. Independently, if the triple $(\mathfrak{S}, e_0, m)$ satisfies the condition

2. $\operatorname{reg} e'_0 \neq \operatorname{reg} e'_m$,

then $\mathfrak{S}_m$ is connected. ◁

A sequence $e_0, \dots, e'_m$ of edges as above satisfying both conditions in Lemma 4.49 is called a *cutting sequence* of width m. Condition 1 can be rewritten in the equivalent form $e'_0 \neq e_i$ and $e'_m \neq e_i$, $i = 0, \dots, m-1$. Note also that, if $(\mathfrak{S}, e_0, m)$ satisfies condition 1 and $\mathfrak{S}$ is a skeleton of genus zero, then condition 2 is also necessary for the connectedness of $\mathfrak{S}_m$.

Remark 4.50. In the drawings, with bivalent ○-vertices omitted, it is easier to speak about a contiguous sequence $\bar{e}_0, \dots, \bar{e}_m$ of unordered pairs of edges $\bar{e}_i = (e_i, e'_i)$. In these terms, condition 1 in Lemma 4.49 means that $\bar{e}_0$ and $\bar{e}_m$ should be distinct from each other and from all other pairs, and condition 2 means that the regions R_0 and R_m from which R is separated by $\bar{e}_0$ and $\bar{e}_m$, respectively, should be distinct (although one of them may coincide with R itself.)

Lemma 4.51. *Let $\mathfrak{D}$ be a reduced dessin in the sphere $S \cong S^2$ and let $\mathfrak{S} := \operatorname{Sk} \mathfrak{D}$, $R := \operatorname{reg} e_0$, and $m < n := \operatorname{wd} R$ be as above; we assume that $\mathfrak{S}$ is connected, hence R is a disk. Assume that R contains a ×-vertex v of $\mathfrak{D}$ of index m. Then $\mathfrak{D}$ is equivalent to a reduced dessin $\mathfrak{D}_m$ such that:*

- *the solid and dotted edges incident to v end, respectively, at the •- and ○-ends of the edges e'_i, $i = 0, \dots, m-1$, see Figure 4.4 (a);*
- *one has $\operatorname{Sk} \mathfrak{D}_m \cong \mathfrak{S}_m$.*

If, in addition, the triple $(\mathfrak{S}, e_0, m)$ satisfies condition 1 in Lemma 4.49, then v is the only ×-vertex of $\mathfrak{D}_m$ in the m-gonal region R_m given by Lemma 4.49.

Proof. In view of Proposition 4.48, we can assume that the solid and dotted edges incident to v end at the •- and ○-ends of the edges e'_i, $i = 0, \dots, m-1$. Then, the centers of e_0 and e_m can be connected by a simple arc ζ in the interior of R which is disjoint from $\mathfrak{D}$ (except at the ends), see Figure 4.4 (a), and $\mathfrak{D}_m$ is obtained from $\mathfrak{D}$ by a bold monochrome modification along ζ. □

The following lemma is used in Chapter 8.

Lemma 4.52. *Let $\mathfrak{D}$ be a reduced dessin in the sphere $S \cong S^2$ with connected regular skeleton $\mathfrak{S}$, and let v be a bivalent $\times$-vertex of $\mathfrak{D}$ contained in a region R of $\mathfrak{S}$. Then there is a dessin $\mathfrak{D}_2 \sim \mathfrak{D}$ with connected regular skeleton $\mathfrak{S}_2$ and such that v is the only $\times$-vertex of $\mathfrak{D}_2$ in a bigonal region R_2 of $\mathfrak{S}_2$. Furthermore, given a $\circ$-vertex u in the boundary ∂R, the dessin $\mathfrak{D}_2$ can be chosen so that u is in the boundary ∂R_2.*

Proof. Consider the sequence $s = (e_0, e_0', e_1, e_1', e_2, e_2')$ of six edges in ∂R such that u is the common $\circ$-end of e_1 and e_1'. Since $\mathfrak{S}$ is regular, the only possibility for s to fail to satisfy condition 1 in Lemma 4.49 is the case where $e_0 = e_2'$ and $\operatorname{reg} e_1'$ is a monogon, see the arc ζ' in Figure 4.4 (b). Hence, either s or $s_+ := s \uparrow (\mathbf{xy})^{-1}$ satisfy condition 1 and, using Lemma 4.51, we obtain a dessin $\mathfrak{D}'$ and skeleton $\mathfrak{S}'$ with a bigonal region R' containing v; in either case, u is in the boundary $\partial R'$.

If $\mathfrak{S}'$ is disconnected, it consists of two components $\mathfrak{S}_1'$, $\mathfrak{S}_2'$; let $\mathfrak{S}_1'$ be the one containing $\partial R'$, and let R_{12} be the only region of $\mathfrak{S}'$ that is not homeomorphic to a disk. Since $\mathfrak{S}'$ is regular, the component $\mathfrak{S}_1'$ is not a tree, even after shrinking a neighborhood of R' to a single $\circ$-vertex. Hence, there is an edge e_0 of $\mathfrak{S}_1'$ which is in the boundary of R_{12} but *not* in the boundary of R' and such that $\mathfrak{S}_1'$ remains connected after removing e_0. Starting with this edge e_0 and proceeding as in the proof of Lemma 4.43, we obtain a dessin $\mathfrak{D}_2$ as in the statement. □

4.2.3 Generic real curves (see [60])

In this section, trigonal curves are real and almost generic. After a small equisingular deformation, the real dessin $\mathfrak{D} := \operatorname{Dssn}_c C$ of such a curve can be assumed generic. Below, we discuss briefly the relation between the topology of a real trigonal curve and its real dessin and a few important counts and inequalities, of which the most important is Theorem 4.60. The principal reference is [60].

Real fragments

Let $\mathfrak{D} \subset S$ be a real dessin. Recall that points, edges, vertices, *etc.* situated in the boundary ∂S are called real; the connected components S_i of ∂S are called the *real components* of $\mathfrak{D}$.

Fix a color $* \in \{\text{solid}, \text{bold}, \text{dotted}\}$ and let $(\partial S)_*$ be the *closure* of the intersection $\partial S \cap \mathfrak{D}_*$. We emphasize that $(\partial S)_*$ is allowed to contain essential vertices. If a connected component M_j of $(\partial S)_*$ coincides with a real component S_i, the latter is called a *monochrome component* (of the corresponding color). Otherwise, M_j is called a *maximal monochrome segment*.

Let M_j be a component of $(\partial S)_*$. If $\mathfrak{D}$ is generic, the full index of each $\bullet$- or $\times$-vertex is odd and the full index of each $\circ$-vertex is even. Hence, the color of real edges of $\mathfrak{D}$ changes when passing through a $\bullet$- or $\times$-vertex and does not change when passing though a $\circ$-vertex, *cf.* Figure 4.5. It follows that the boundary ∂M_j is formed by $\bullet$- and/or $\times$-vertices, whereas $\circ$-vertices are inside M_j. In particular, the numbers of $\bullet$- and $\times$-vertices within each real component S_i of $\mathfrak{D}$ are even: pairs of such vertices bound maximal bold or, respectively, dotted segments. On the contrary, the number of $\circ$-vertices in S_i or M_j can be even or odd; according to this number, the component S_i or maximal monochrome segment M_j is called *even* or *odd.*

Definition 4.53. A dotted monochrome component of a generic real dessin $\mathfrak{D}$ is called a *hyperbolic component*; such a component can be *even* or *odd.* A maximal even (odd) dotted segment is called an *oval* (respectively, *zigzag*) of $\mathfrak{D}$. A dessin $\mathfrak{D}$ is called *hyperbolic* if all its real components are hyperbolic.

One can easily see that equivalence of generic dessins preserves real components, including their parity, hyperbolic components, ovals, and zigzags. More precisely, we have the following statement.

Proposition 4.54. *Any equivalence of generic real dessins $\mathfrak{D}_i \subset S_i$, $i = 1, 2$, induces a homeomorphism $\partial S_1 \to \partial S_2$, which takes $(\partial S_1)_{\text{dotted}}$ onto $(\partial S_2)_{\text{dotted}}$, preserving the parity of each connected component of $(\partial S_1)_{\text{dotted}}$ and each connected component of the complement $\partial S_1 \smallsetminus (\partial S_1)_{\text{dotted}}$.* $\triangleleft$

Warning 4.55. It is worth emphasizing that equivalence of dessins does *not* necessarily preserve the other monochrome parts $(\partial S)_{\text{solid}}$ or $(\partial S)_{\text{bold}}$, as the $\bullet$-in/out operations may mix the colors, see Figure 4.1.

Let $\mathfrak{D} = \operatorname{Dssn}_c C$ for a real trigonal curve $C \subset \Sigma \to B$. Then $B_{\mathbb{R}} = \partial S$ and the real components of $\mathfrak{D}$ are naturally identified with the real components of B. A component $S_i = B_i$ is even if and only if the corresponding component Σ_i of $\Sigma_{\mathbb{R}}$ is orientable. Indeed, recall that $\mathcal{Y} = \mathcal{O}_B(G_\circ - G_\bullet)$, see the proof of Theorem 3.20, and, as explained above, the restriction of $G_\bullet$ to B_i is even. Furthermore, analyzing Figure 3.1, one can easily see that the identity $\partial S = B_{\mathbb{R}}$ maps $(\partial S)_{\text{dotted}}$ onto the set $\{b \in B_i \mid \operatorname{Card} q_i^{-1}(b) \geqslant 2\}$, see Definition 3.82, taking the hyperbolic components, ovals, and zigzags of $\mathfrak{D}$ to, respectively, hyperbolic components, ovals, and zigzags of C. To distinguish ovals from zigzags, one should notice that the $\circ$-vertices of the dessin are precisely the fibers where C crosses the zero section Z. The same observation shows that two consecutive ovals (possibly separated by zigzags) form a chain if and only if they are separated by an even number of $\circ$-vertices.

Degree and counts

As in the complex case, the *degree* of a real trigonal curve $C \subset \Sigma$ is $\deg C := 3d$, where $d = -E^2$. Dessins of degree three in the disk are called *cubics*; corresponding trigonal curves are indeed plane cubics or, more precisely, blow down to plane cubics after the contraction $\Sigma_1 \to \mathbb{P}^2$ of the exceptional section. Up to equivalence, there are eight cubics; they are listed in Figure 4.5. A formal proof in terms of dessins is found

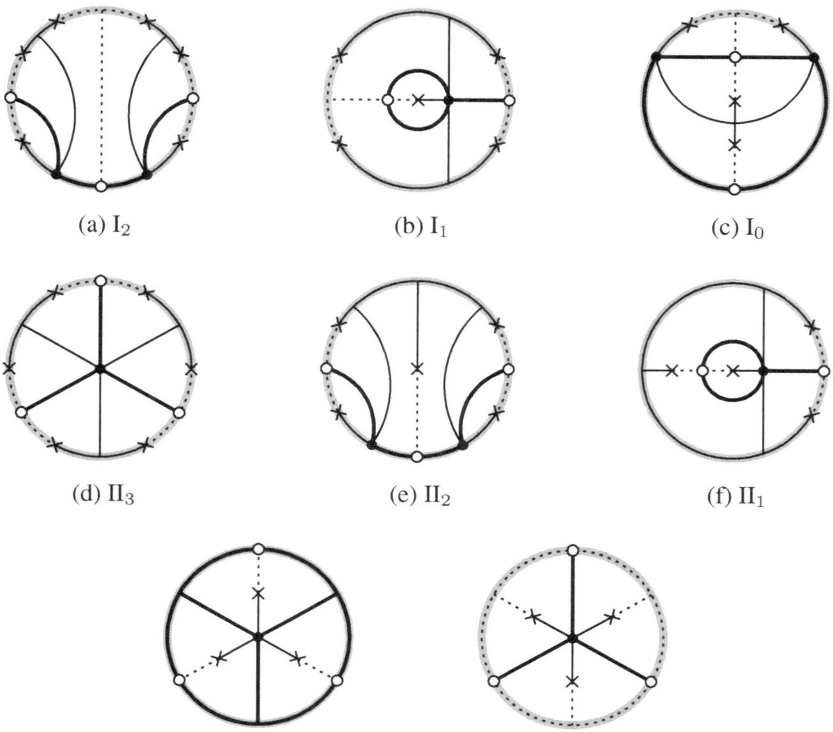

(a) I_2 (b) I_1 (c) I_0

(d) II_3 (e) II_2 (f) II_1

(g) II_0 (h) H

Figure 4.5. Generic real cubic dessins.

in [60]; geometrically, it is probably easier to classify pairs (D, P), where $D \subset \mathbb{P}^2$ is a real cubic and $P \in \mathbb{P}^2_{\mathbb{R}} \smallsetminus D_{\mathbb{R}}$ is a point to be blown up, taking into account the position of P with respect to the inflection tangents of D.

The last dessin in Figure 4.5 is hyperbolic; it is denoted by H. For the others, the notation is T_z, where T is the type (I or II, corresponding to one or none ovals, respectively) and z is the number of zigzags. For the sake of symmetry, the dessin in Figure 4.5 (c) is not reduced: it has an inner solid vertex.

Let $\mathfrak{D} \subset S$ be a generic real dessin of degree $\deg \mathfrak{D} = 3d$, and let $C \subset \Sigma \to B$ be a corresponding real trigonal curve. By the adjunction formula (or by the Riemann–Hurwitz formula), the genus of C is

$$g(C) = \deg \mathfrak{D} - 3\chi(S) + 1. \tag{4.56}$$

Introduce the following notation:

- ℓ_{even}, ℓ_{odd}: the numbers of even/odd hyperbolic components;

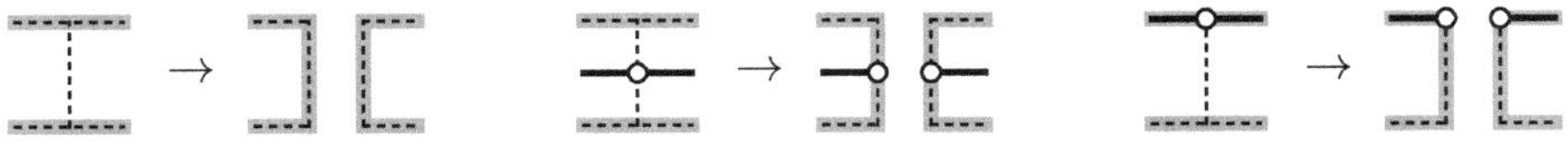

(a) A dotted cut (b) A dotted cut (c) A break

Figure 4.6. Dotted cuts and a break.

- ℓ_{nh}: the number of non-hyperbolic real components;
- n_{o}, n_{z}, n_{i}: the numbers of ovals, zigzags, and inner $\times$-vertices, respectively;
- $\delta := 2 - (\ell_{\mathrm{even}} + \ell_{\mathrm{odd}} + \ell_{\mathrm{nh}}) - \chi(S)$: the 'excessive' Euler characteristic.

Note that all quantities introduced are nonnegative, $\delta > 0$ unless S is a disk with holes, and $\ell_{\mathrm{nh}} > 0$ unless $\mathfrak{D}$ is hyperbolic. Then we have the following vertex counts

$$\#_{\bullet}(\mathfrak{D}) = \frac{2}{3}\deg\mathfrak{D}, \quad \#_{\circ}(\mathfrak{D}) = \deg\mathfrak{D}, \quad \#_{\times}(\mathfrak{D}) = 2(n_{\mathrm{o}}+n_{\mathrm{z}}+n_{\mathrm{i}}) = 2\deg\mathfrak{D}, \tag{4.57}$$

and the number of real components of C is $3\ell_{\mathrm{even}}+2\ell_{\mathrm{odd}}+\ell_{\mathrm{nh}}+n_{\mathrm{o}}$. Combining these identities, we arrive at the following statement.

Proposition 4.58. *If C is an $(M-s)$-curve, then*

$$2\ell_{\mathrm{nh}} + \ell_{\mathrm{odd}} + n_{\mathrm{z}} + n_{\mathrm{i}} + 3\delta = s + 4. \qquad \triangleleft$$

Remark 4.59. Most statements concerning $(M-s)$-dessins to not refer explicitly to the number of *even* hyperbolic components. In fact, after a sequence of $\circ$-in operation, one can assume that such a component S_i contains monochrome vertices only, and then S_i can formally be contracted to a single monochrome vertex without affecting the other invariants involved in the statements. (For more details on this treatment of monochrome components, see [60].) For this reason, in most proofs we silently assume that the dessin has no even hyperbolic components.

Dotted cuts

Most known results concerning the classification of reduced real dessins (*cf.* Theorem 4.69) state that any dessin with certain properties can be cut into simple pieces and, in some circumstances, this procedure is unique. Define a *dotted cut* in a reduced real dessin $\mathfrak{D}$ as an inner dotted edge connecting two real monochrome vertices, as in Figure 4.6 (a), or a pair of inner dotter edges incident to real monochrome vertices on the one hand and to a common inner $\circ$-vertex on the other hand, as in Figure 4.6 (b).

As shown in the figures, a dessin $\mathfrak{D} \subset S$ containing a dotted cut can be broken to produce a simpler dessin (if the element represented by the cut in the group $H^1(S, \partial S)$ is nontrivial) or a pair of dessins. In both cases, the result is somewhat 'simpler' than the original dessin $\mathfrak{D}$. Our principal result in this direction is the following theorem, roughly stating that a dessin with sufficiently many ovals does have a dotted cut.

Theorem 4.60 (see [60]). *If a generic real $(M-s)$-dessin $\mathfrak{D}$ has no dotted cuts, then* $2\deg\mathfrak{D} \leqslant 3(n_z+n_i)+3s-3\delta$.

We precede the proof with a few lemmas.

Define a *break* as an inner dotted edge connecting a real monochrome vertex and a real $\circ$-vertex, see Figure 4.6 (c). As shown in the figure, a dessin can be cut along a break into a simpler dessin (or a pair of dessins), which is generic except that it is allowed to have *real* monovalent $\circ$-vertices. Such a dessin is called a *scrap*. (In the terminology of [60], scraps are not valid dessins.) Conversely, given a scrap (or a pair of scraps) with two marked monovalent $\circ$-vertices, these vertices can be glued together (along some portions of the incident dotted edges) to produce a scrap with fewer monovalent vertices.

Given a scrap $\varsigma \subset S$, let $\kappa(\varsigma)=\chi(S)-\frac{1}{2}\#^1_\circ(\varsigma)$ and $d(\varsigma)=\#_\circ(\varsigma)-\frac{1}{2}\#^1_\circ(\varsigma)$. (We avoid the term *degree* used in [60] as, from our point of view, scraps *are* dessins and hence do already have a well defined degree. Note however that, according to (4.57), one has $d(\varsigma)=\deg(\varsigma)$ whenever ς is a generic dessin, *i.e.*, $\#^1_\circ(\varsigma)=0$.) We extend κ and d to disjoint unions of scraps; the extensions are obviously additive. Furthermore, one has

$$\kappa(\varsigma')=\kappa(\varsigma), \qquad d(\varsigma')=d(\varsigma) \tag{4.61}$$

if the scrap (union of scraps) ς' is obtained from ς by a cut at a break, and

$$\kappa(\varsigma')\geqslant\kappa(\varsigma), \qquad d(\varsigma')=d(\varsigma) \tag{4.62}$$

if ς' is obtained from ς by a cut at a dotted cut.

Always $d(\varsigma)=\#^2_\circ(\varsigma)+\frac{1}{2}\#^1_\circ(\varsigma)>0$. On the contrary, $\kappa(\varsigma)$ is usually negative: assuming that $\partial S\neq\varnothing$, one has $\kappa(\varsigma)>0$ if and only if $S\cong\bar{D}^2$ and $\#^1_\circ(\varsigma)\leqslant 1$, and $\kappa(\varsigma)=0$ if and only if $S=\bar{D}^2$ and $\#^1_\circ(\varsigma)=2$ or S is an annulus or a Möbius band and $\#^1_\circ(\varsigma)=0$.

The counts ℓ_* and n_* introduced above for generic real dessins extend to (disjoint unions of) scraps in the obvious way.

Lemma 4.63. *For a scrap ς, one has $\#_\times(\varsigma)=2d(\varsigma)$ and $\#_\bullet(\varsigma)=\frac{2}{3}d(\varsigma)$. Furthermore,* $d(\varsigma)+\frac{3}{2}\#^1_\circ(\varsigma)=0 \bmod 3\mathbb{Z}$.

Proof. It suffices to complete ς to a generic dessin by patching each monovalent $\circ$-vertex with a half of a cubic, *e.g.*, the one shown in Figure 4.5 (a) (cut at the vertical break), and to use the additivity, (4.61), and known congruences and identities for generic dessins, see (4.57). □

Corollary 4.64. *If $\#^1_\circ(\varsigma)=1$ (respectively, 2), then $d(\varsigma)\geqslant\frac{3}{2}$ (respectively, $d(\varsigma)\geqslant 3$). Hence, for such a scrap ς one has $\#^2_\circ(\varsigma)\geqslant\#^1_\circ(\varsigma)$.* ◁

Lemma 4.65. *Any scrap contains at least $\ell_{\mathrm{odd}}+n_{\mathrm{o}}-n_{\mathrm{i}}$ breaks or dotted cuts.*

Proof. Each oval or odd dotted component contains a monochrome vertex v and, starting from the inner dotted edge incident to v and extending this edge through an inner $\circ$-vertex, if present, one would reach the boundary again, thus obtaining a dotted cut or a break, or end up in an inner $\times$-vertex. Since all $\times$-vertices are monovalent, at most n_{i} chains of dotted edges will not reach the boundary. □

Lemma 4.66. *A scrap ς with $\kappa(\varsigma) > 0$ contains a zigzag or an inner $\times$-vertex.*

Proof. Assume that ς has no inner $\times$-vertices and cut it into smaller scrap(s) along all breaks and dotted cuts present in ς. According to Lemma 4.65, the pieces have no ovals. By additivity and (4.61), (4.62), at least one of the pieces has $\kappa > 0$, and such a piece ς' is a scrap on a disk and has $\#_\circ^1(\varsigma') \leqslant 1$. Hence, due to Corollary 4.64 and Lemma 4.63, ς' has at least three $\times$-vertices; at most one of them is incident to the only monovalent $\circ$-vertex, and hence two others form a zigzag. □

Proof of Theorem 4.60. Let $\deg \mathfrak{D} = 3d$. Cut $\mathfrak{D}$ into scraps along all breaks present in $\mathfrak{D}$. Due to Lemma 4.65, the total number of monovalent $\circ$-vertices obtained is $2b$, where

$$b \geqslant b_0 := \ell_{\mathrm{odd}} + n_{\mathrm{o}} - n_{\mathrm{i}} = \ell_{\mathrm{odd}} + 3d - n_{\mathrm{z}} - 2n_{\mathrm{i}}. \tag{4.67}$$

First, assume that $\mathfrak{D}$ does have at least one break. Then, since the underlying surface S is connected and $\mathfrak{D}$ has no dotted cuts, each of the scraps obtained has at least one monovalent $\circ$-vertex.

Let m_+ be the number of scraps ς with $\kappa(\varsigma) > 0$. Using Lemma 4.63, one can break $m_+ = m'_+ + m''_+$, where m'_+ is the number of scraps ς with $d(\varsigma) = \frac{3}{2}$ and m''_+ is the number of scraps ς with $d(\varsigma) \geqslant \frac{9}{2}$. (Under the assumption $\#_\circ^1(\varsigma) > 0$ the inequality $\kappa(\varsigma) > 0$ implies that $\#_\circ^1(\varsigma) = 1$ and hence $d(\varsigma) + \frac{3}{2} = 0 \bmod 3\mathbb{Z}$.) Due to Lemma 4.66, at least $m'_+ - n_{\mathrm{z}}$ inner $\times$-vertices of $\mathfrak{D}$ are separated by breaks from the ovals and hyperbolic components, and inequality (4.67) can be sharpened to $b \geqslant b_0 + m'_+ - n_{\mathrm{z}}$.

Let b_- be the total number of monovalent $\circ$-vertices in the scraps ς satisfying $\#_\circ^1(\varsigma) \geqslant 3$. Then one has

$$3d = \#_\circ(\mathfrak{D}) \geqslant b + (2b - b_-) + 3m''_+ \geqslant 3\ell_{\mathrm{odd}} + 9d - 6(n_{\mathrm{z}} + n_{\mathrm{i}}) - b_- + 3m_+.$$

Indeed, b is the number of pairs of monovalent $\circ$-vertices, each pair merging to a bivalent $\circ$-vertex of $\mathfrak{D}$, the term $2b - b_-$ is the estimate on the number of bivalent $\circ$-vertices in the scraps given by Corollary 4.64, and the extra term $3m''_+$ is due to the fact that each scrap ς counted by m''_+ has at least four bivalent $\circ$-vertices while contributing only $\#_\circ^1(\varsigma) = 1$ to $2b - b_-$. The second inequality follows from the bound $b \geqslant b_0 + m'_+ - n_{\mathrm{z}}$ and (4.67).

Thus, we have $6d \leqslant 6(n_{\mathrm{z}} + n_{\mathrm{i}}) - 3\ell_{\mathrm{odd}} + b_- - 3m_+$. On the other hand, by the definition of κ, one has $\kappa(\varsigma)/\#_\circ^1(\varsigma) \leqslant -\frac{1}{6}$, hence $\#_\circ^1(\varsigma) \leqslant -6\kappa(\varsigma)$, for any scrap ς with $\#_\circ^1(\varsigma) \geqslant 3$. Using the additivity of κ, we have the bound $b_- \leqslant 3m_+ - 6\chi(S)$.

(The contribution to $\kappa(\mathfrak{D}) = \chi(S)$ of the scraps not counted in b_- does not exceed $\frac{1}{2}m_+$; recall that we assume $\#^1_\circ(\varsigma) \geqslant 1$ and hence $\kappa(\varsigma) \leqslant \frac{1}{2}$ for any scrap in question; for the upper bound we only count the scraps with $\kappa(\varsigma) > 0$.) Thus,

$$6d \leqslant 6(n_z + n_i) - 3\ell_{odd} - 6\chi(S),$$

and it remains to substitute $\chi(S) = 2 - (\ell_{odd} + \ell_{nh}) - \delta$, see Remark 4.59, and use Proposition 4.58.

Now, consider the case when $\mathfrak{D}$ has no breaks. If $\ell_{nh} = 0$, then also $n_z = n_o = 0$ and $n_i = 3d$, and, in view of Proposition 4.58, the inequality in the statement takes the form $4d - 4 + \ell_{odd} + 2\delta \geqslant 0$, which holds automatically since $d \geqslant 1$. In the general case, Lemma 4.65 implies that $n_i \geqslant \ell_{odd} + n_o$ and Proposition 4.58 becomes $s+4 \geqslant 2(\ell_{nh}+\ell_{odd})+n_o+n_z+3\delta$. Assume that the inequality in the statement does *not* hold, *i.e.*, that $2d > n_z + n_i + s - \delta$. Adding the two inequalities and substituting $n_o+n_z+n_i = 3d$, we obtain $4 > 2(\ell_{nh}+\ell_{odd}+\delta)+n_z+d$. Assuming that $\ell_{nh} > 0$, this inequality implies $\ell_{odd} = \delta = 0$ and $\ell_{nh} = d = 1$, *i.e.*, $\mathfrak{D}$ is a cubic. On the other hand, for any cubic dessin one has $n_o \leqslant 1$; hence $n_z+n_i \geqslant 2 = 2d$ and the inequality in the statement of the theorem holds. □

4.3 First applications

In this section, we cite a few further classification results concerning nonsingular real trigonal curves and real Lefschetz fibrations, mainly those statements to which we will contribute in Part II of the book.

4.3.1 Ribbon curves (see [60])

Let $\mathfrak{D}$ be a generic real dessin or a disjoint union of several such dessins. Pick a pair z_1, z_2 of zigzags of $\mathfrak{D}$ and define the *junction* of $\mathfrak{D}$ at z_1, z_2 as the dessin $\mathfrak{D}'$ obtained from $\mathfrak{D}$ by identifying a pair of segments $I_i \subset \operatorname{Int} z_i$, $i = 1, 2$, *via* a homeomorphism $\varphi\colon I_1 \to I_2$ preserving the decorations and such that the resulting dotted cut in $\mathfrak{D}'$ (the common image of I_1 and I_2) connects two ovals, see Figure 4.7 (a). (After the junction, any inner monochrome vertices that may form can be perturbed to obtain a reduced dessin.) Clearly, the equivalence class of $\mathfrak{D}'$ depends only on the pair z_1, z_2 and the mutual orientation of these two zigzags.

Similarly, given a zigzag z of $\mathfrak{D}$, define the *self-junction* at z as the dessin $\mathfrak{D}'$ that is obtained from $\mathfrak{D}$ by identifying a pair of segments $I_1, I_2 \subset \operatorname{Int} z$ *via* a homeomorphism $\varphi\colon I_1 \to I_2$ preserving the decorations and such that the resulting dotted cut in $\mathfrak{D}'$ (the common image of I_1 and I_2) connects an oval and an odd dotted component, see Figure 4.7 (b). The equivalence class of $\mathfrak{D}'$ depends on z only.

Observe that a (self-)junction consumes the zigzag(s) involved. It follows that any two such operations commute. The dotted cuts resulting from the junction operations are also called (self-)junctions.

(a) A junction (b) A self-junction

Figure 4.7. A (self-)junction.

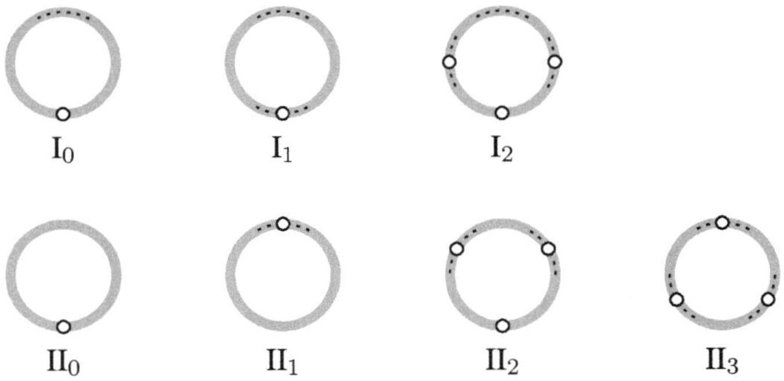

Figure 4.8. Ribbon boxes.

The structure theorems stated below deal with (self-)junctions of cubics. From the point of view of the junction operation, a cubic can be regarded as a 'black box' with a certain decoration of its boundary. More precisely, define a *ribbon box* as a disk $\bar{D}^2$ with a few pairwise disjoint segments, called *dotted*, marked in the boundary $\partial\bar{D}^2$ and a parity assigned to each dotted segment and each complementary segment (*i.e.*, connected component of the complement of the union of the dotted segments). As part of the definition, a ribbon box is required to be one of those listed in Figure 4.8, where odd segments are marked with ∘-vertices.

Each non-hyperbolic cubic dessin $\mathfrak{D} \subset S$, see Figure 4.5, gives rise to a ribbon box (for which we keep the notation used for $\mathfrak{D}$): the disk is the underlying surface S, the dotted segments are the ovals and zigzags of $\mathfrak{D}$, and the parity of a segment is that of the number of ∘-vertices of $\mathfrak{D}$ contained in the segment. Conversely, due to the classification of cubics, see page 132 and Figure 4.5, each ribbon box is obtained in this way from a generic cubic dessin, which is unique up to equivalence. In view of this correspondence, we will often refer to the even (odd) dotted segments of a ribbon box as its *oval* (respectively, *zigzag*) segments.

A *ribbon curve structure* is a collection of ribbon boxes in which some of the boxes are glued *via* identifying certain pairs of zigzag segments, so that the result is a connected surface, and some of the remaining zigzag segments are selected for future self-junction. The selected segments are called *vanishing*. An *orientation* of a ribbon curve structure is an orientation of the underlying surface. An *isomorphism* of two ribbon curve structures is a homeomorphism of the underlying surfaces preserving the decorations, *i.e.*, taking boxes to boxes and dotted segments to dotted segments, preserving all parities, and taking vanishing segments to vanishing segments.

The *junction graph* of a ribbon curve structure is the graph obtained by replacing each box with a vertex and connecting each pair of glued boxes by an edge, one for each pair of zigzags identified. The junction graphs of isomorphic ribbon curve structures are isomorphic (but, in general, not *vice versa*, as some zigzag segments have preferred orientations). Clearly, the valency of a vertex does not exceed the number of zigzag segments in the box represented by the vertex. In particular, each valency is at most three, the trivalent vertices corresponding to the boxes of type II_3.

A ribbon curve structure gives rise to an equivalence class of generic dessins: one replaces each ribbon box with a corresponding cubic, performs a junction at each pair of identified zigzags, and performs a self-junction at each vanishing segment. A dessin obtained in this way is called a *ribbon dessin*. A *ribbon curve* is a generic real trigonal curve whose dessin is equivalent to a ribbon dessin.

Remark 4.68. Since cubics have no cuts, each particular dessin $\mathfrak{D}$ can be represented as a ribbon dessin in at most one way. In fact, $\mathfrak{D}$ is a ribbon dessin if and only if *all* dotted cuts present in $\mathfrak{D}$ are (self-)junctions and, after performing all these cuts, $\mathfrak{D}$ breaks into a union of cubics. Note, however, that in general two ribbon dessins obtained from non-isomorphic ribbon curve structures may be equivalent, and a dessin equivalent to a ribbon dessin does not need to be a ribbon dessin itself. Examples are $(M-1)$-curves, see the discussion after Theorem 4.73. The uniqueness of a ribbon curve structure representing a given curve is addressed in Section 10.3.2.

The classification of M-curves

To simplify the further exposition, we confine ourselves to generic real trigonal curves without even hyperbolic components, *cf.* Remark 4.59. A complete treatment of the general case is given in [60].

Theorem 4.69 (see [60] or [131] for the case of the rational base). *Any generic non-hyperbolic real M-dessin $\mathfrak{D}$ without even hyperbolic components is a ribbon dessin, and the corresponding ribbon curve structure has the following properties*:

- *each box is of type* I, *i.e., represents an M-cubic*;
- *the underlying surface is orientable.*

Conversely, any ribbon curve structure with above two properties defines a generic non-hyperbolic M-dessin without even hyperbolic components; *two such dessins are equivalent if and only if the corresponding ribbon curve structures are isomorphic.*

Proof. Let $\mathfrak{D} \subset S$ be a dessin as in the statement. Then one has $s = 0$ and $\ell_{\mathrm{nh}} \geqslant 1$ in Proposition 4.58; hence $\delta = 0$ (and thus S is a disk with holes; in particular, it

Figure 4.9. Junction graphs of M-dessins.

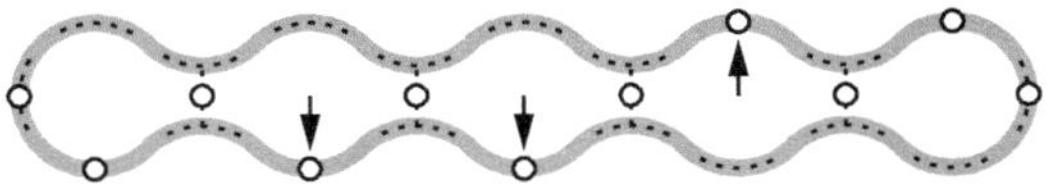

Figure 4.10. An M-curve of degree 15 (vertical arrows marking the chain breaks).

is orientable) and $n_z + n_i = 2$ if S is a disk and $n_z + n_i \leqslant 1$ otherwise (*i.e.*, if ∂S is disconnected). Thus, due to Theorem 4.60, either $\mathfrak{D}$ is a cubic or $\mathfrak{D}$ has a dotted cut c. Using Proposition 4.58 again, one can see that this cut is a (self-)junction and the result of cutting $\mathfrak{D}$ at c is an M-dessin. (Roughly speaking, in order to keep the maximal number of components, one should start with a maximal number, each gluing must create at least two ovals, and each self-gluing must create at least three components.) Proceeding by induction, one concludes that $\mathfrak{D}$ is a (self-)junction of M-cubics, *cf.* Remark 4.68.

The converse statement is straightforward, and the proof of the uniqueness part of the theorem is postponed till Chapter 10, see Theorem 10.73: in fact, ribbon curve structures without boxes of types II_1 or II_2 result in *topologically* distinct curves. □

Remark 4.70. Since each box representing an M-cubic has valency at most two, the junction graph of an M-curve is either a linear tree or a cycle, see Figure 4.9. In the former case, the underlying surface is a disk; in the latter case, it is an annulus.

The real part of an M-curve is easily recovered from its ribbon curve structure. In particular, one can easily describe the maximal chains of ovals: the chain breaks are the odd complementary segments of the boxes of type I_2 *represented by bivalent vertices of the junction graph* (*cf.* Figure 4.10, where the chain breaks are marked with arrows; the curve shown in the figure has three maximal chains of ovals, which are of length 7, 1, and 5). The odd complementary segment of a type I_2 box represented by a monovalent vertex of the junction graph is a chain break if and only if the 'free' zigzag segment of the box is vanishing, as in this case the zigzag turns to an oval. As a simple consequence of this description, we have the following corollary.

Corollary 4.71 (see [60]). *The ovals of a nonsingular trigonal M-curve on a real* rational *ruled surface Σ_k, $k \geqslant 3$, form $(k-2)$ maximal chains, each maximal chain being of odd length.* ▷

Conversely, under certain assumptions the ribbon curve structure, and hence the deformation class, of an M-curve C over B can be recovered from its real part. For example, this is the case if B is an elliptic curve without hyperbolic components or if C has at least one zigzag. In the former case, the sequence of maximal chains of ovals in one of the two components of $B_{\mathbb{R}}$ clearly determines the box decomposition. In the case of the rational base, the junction graph is a linear tree and $C_{\mathbb{R}}$ has two distinguished ovals, those located in the cubics corresponding to the extreme vertices of the tree. If at least one of these extreme ovals is known, the rest of the ribbon curve structure is recovered uniquely, and it remains to observe that a zigzag of an M-curve always points to an extreme oval.

If $B = \mathbb{P}^1$ and $C_{\mathbb{R}}$ has no zigzags, the fiberwise topology of $C_{\mathbb{R}}$ can be encoded by the sequence of lengths of the maximal chains of ovals. There are sequences that can be obtained from two non-isomorphic ribbon curve structures, for example

$$\underline{5}\ \underline{\underline{5}}\ 1\ 3\ \underline{5}\ \underline{\underline{5}}\ 1\ 3 \tag{4.72}$$

(see [60] and Remark 10.84 below): the two chains containing the extreme ovals are either those underlined or those double underlined. In Chapter 10 we obtain a certain characterization of such ambivalent sequences (Corollary 10.83) and show that a given topological type of the real part can be shared by *at most* two non-equivalent M-curves (Corollary 10.82).

The classification of $(M-1)$-curves

The classification of generic trigonal $(M-1)$ curves follows the same lines as that of M-curves and can be stated as follows.

Theorem 4.73 (see [60]). *Any generic non-hyperbolic real $(M-1)$-dessin $\mathfrak{D}$ without even hyperbolic components is equivalent to a dessin admitting a ribbon curve structure such that either*:

- *each box is of type* I *and has valency two (so that the junction graph is a single cycle) and the underlying surface is a Möbius band, or*
- *exactly one box is of type* II, *all other boxes being of type* I, *and the underlying surface is orientable.*

Conversely, any ribbon curve structure as above gives rise to a generic non-hyperbolic $(M-1)$-dessin. ▷ [60]

The junction graph of an $(M-1)$ curve is either a linear tree, or a single cycle, see Figure 4.9, or one of the two graphs shown in Figure 4.11, the trivalent vertex representing a ribbon box of type II_3. In the second graph in Figure 4.11, the cycle may as well consist of one or two vertices, *cf.* Figure 4.9.

In spite of a certain similarity of the statements, the cases of M- and $(M-1)$-curves are essentially different. First, we do not assert that any $(M-1)$-dessin *is* a ribbon

Figure 4.11. Additional junction graphs of $(M-1)$-dessins.

dessin. Using Proposition 4.58 and Theorem 4.60, as in the proof of Theorem 4.69, one can show that the dotted cuts present in $\mathfrak{D}$ break it into a union of M-cubics and, possibly, one $(M-1)$-dessin $\mathfrak{D}'$, which is either of degree three (not necessarily on a disk) or a sextic (*i.e.*, a dessin of degree six on a disk). Such dessins $\mathfrak{D}'$ can be classified and one can show that $\mathfrak{D}'$ is *equivalent* to a (self-)junction of cubics. Second, a ribbon curve structure resulting in an $(M-1)$-curve is not always unique: it is defined up to the moves $\mathrm{II}_2 + \mathrm{I}_2 \leftrightarrow \mathrm{I}_2 + \mathrm{II}_2$ (under certain assumptions on the junction), $\mathrm{II}_2 + \mathrm{I}_1 \leftrightarrow \mathrm{I}_2 + \mathrm{II}_1$, $\mathrm{II}_1 + \mathrm{I}_2 \leftrightarrow \mathrm{I}_1 + \mathrm{II}_2$, or $\mathrm{II}_1 + \mathrm{I}_1 \leftrightarrow \mathrm{I}_1 + \mathrm{II}_1$.

Note that all moves above involve ribbon boxes of types II_1 or II_2. In Chapter 10, see Theorem 10.73, we show that a ribbon curve structure without such boxes is uniquely determined by the *topology* of the curve.

Remark 4.74. In view of Corollary 3.87, Theorems 4.69 and 4.73 also provide a deformation classification of real elliptic M- and $(M-1)$-fibrations: up to deformation, any such fibration is Jacobian and pairs of opposite fibrations are classified by the isomorphism classes of ribbon curve structures as in the theorems. (In the case of $(M-1)$-fibrations, one should also take into account the moves listed above.)

Theorems 4.69 and 4.73 do not extend directly to $(M-2)$-dessins and further, as both key ingredients of their proofs fail: first, Theorem 4.60 does not break a dessin into sufficiently small pieces and, second, it is no longer true that any dotted cut must be a junction. For example, there is an $(M-2)$-sextic that is not equivalent to a junction of two cubics. It appears that general real trigonal curves may have rather large 'unstructured' pieces. Note that Corollary 3.87 does not extend to $(M-2)$-fibrations either: such a fibration does not need to be Jacobian.

4.3.2 Elliptic Lefschetz fibrations revisited (see [142, 141])

As another application of the techniques of real dessins, we follow [142, 141] and discuss the problem of realizability of a directed fibration, see Definition 3.91, by a Jacobian elliptic surface. A directed fibration $p\colon X \to B$ is called *algebraic* if X and B admit analytic structures with respect to which the projection and the section are holomorphic and the topological real structures on X and B are anti-holomorphic; in other words, $X \to B$ is algebraic if it is isomorphic (in the category of directed fibrations) to a real Jacobian elliptic surface.

Define an *essential necklace interval* as a simple segment of an uncoated necklace diagram connecting two vertices of distinct types, *i.e.*, one of the form $\circ\!\!-\!\!\!\times$ or $\times\!\!-\!\!\circ$. It is readily seen that essential are the segments of necklace diagrams connecting the following pairs of stones:

Proposition 4.75 (Salepci [142, 141]). *If a necklace diagram $\mathcal{N}$ with $6n$ stones represents an algebraic directed fibration, it is subject to the inequalities*

$$\#_{\text{ess}} \leqslant 2n, \qquad \#_{\text{ess}} + \#_{<} + \#_{>} \leqslant 6n,$$

where $\#_{\text{ess}}$ is the number of essential necklace intervals in $\mathcal{N}$.

Proof. Let $J \to B \cong \mathbb{P}^1$ be a real Jacobian elliptic fibration represented by $\mathcal{N}$. Then J is a double covering of the Hirzebruch surface Σ_{2n} ramified at the exceptional section and a nonsingular proper real trigonal curve $C \subset \Sigma_{2n}$; up to deformation, the latter can be assumed generic. Let $\mathfrak{D} \subset S = B/c$ be the real dessin of C, so that $\partial S = B_{\mathbb{R}}$. The stones of $\mathcal{N}$ correspond to the maximal dotted segments of ∂S ($\bigcirc$ and $\square$ being ovals and the arrow type stones being zigzags), whereas each essential segment of $\mathcal{N}$ corresponds to a connected component of the complement $\partial S \smallsetminus \mathfrak{D}_{\text{dotted}}$ containing an odd number (hence at least one) of $\circ$-vertices of $\mathfrak{D}$. Thus, the second inequality in the statement follows from the count $\#_{\circ}(\mathfrak{D}) = \deg \mathfrak{D} = 6n$, see (4.57). For the first inequality, observe that each essential segment must also contain at least two $\bullet$-vertices (separating the $\circ$-vertices in the segment from the $\times$-vertices at its boundary) and use the count $\#_{\bullet}(\mathfrak{D}) = \frac{2}{3} \deg \mathfrak{D} = 4n$. $\square$

Warning 4.76. Note that the $\times$- and $\circ$-vertices of uncoated necklace diagrams both correspond to the $\times$-vertices of dessins, whereas the $\circ$- and $\bullet$-vertices of dessins are not directly seen in the necklace diagrams. Since both notation schemes are well established, we leave them intact and just warn the reader.

Corollary 4.77 (Salepci [142, 141]). *The directed fibrations with* 12 *singular fibers represented by the following eight necklace diagrams*

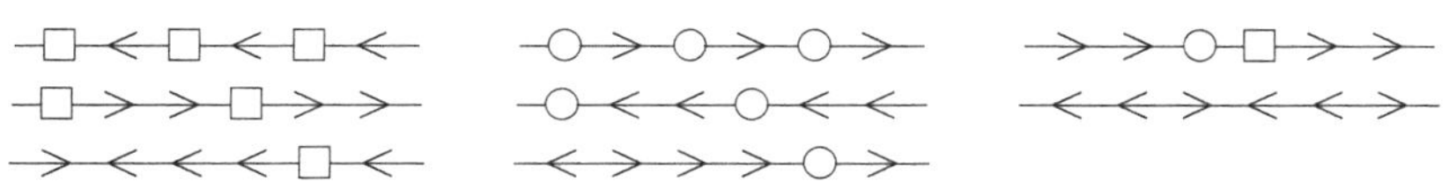

are not algebraic. $\lhd$

Corollary 4.78. *There are at most* 3596 (*undirected*) *equivalence classes of algebraic totally real directed fibrations with* 24 *singular fibers.*

Proof. Analyzing the (computer aided) list of the 8421 equivalence classes, one can see that 4825 of them violate one of the inequalities in Proposition 4.75. □

At present, it is unknown whether all 3596 remaining classes are indeed algebraic.

We conclude with a few positive results. An explicit construction in [142, 141] shows that, among the 25 undirected equivalence classes of totally real elliptic Lefschetz fibrations with twelve singular fibers, see Table 3.4, the seventeen that are not listed in Corollary 4.77 are algebraic. As another realizability statement, a necklace diagram $\mathcal{N}$ represents an algebraic totally real directed fibration if and only if so does $\mathcal{N}^*$: the two diagrams correspond to the two opposite real structures on the same Jacobian elliptic surface. Finally, all maximal directed fibrations, totally real or not, are algebraic. Below, we consider the totally real ones; the general case is postponed till Chapter 10, see Theorem 10.88, where we also show that a given necklace diagram can be shared by at most two maximal fibrations.

Theorem 4.79 (Salepci [141]). *Any maximal totally real directed fibration splits into harsh sums at pairs of contra-oriented arrow type stones, see* (3.97), *of maximal totally real directed fibrations with* 12 *singular fibers each.*

Proof. According to Theorem 3.93 and Corollary 3.95, the fibration is represented by a necklace diagram with total monodromy $1 \in \Gamma$ and with exactly two arrow type stones. Hence, we have the following identity in Γ:

$$\cdots P_{-2}P_{-1}QP_1P_2\cdots = (Q')^{-1},$$

where we can assume that $Q = \mathbb{X}\mathbb{Y}$, the element Q' is either $\mathbb{X}\mathbb{Y}$ or $\mathbb{Y}\mathbb{X}$, and P_i are the local monodromies of ○ and □ type stones, see Table 3.3. Thus, the long word on the left hand side of this identity must cancel to the two letter word on the right hand side. It is easy to see that, in the word corresponding to a chain of ○ and □ type stones, the cancellation never reaches the middle $\mathbb{Y}$ terms of the local monodromies. Hence, a short word can be obtained only if the cancellation process starts at Q and does reach the middle terms of the neighbors of Q. (If the cancellation starting at Q does not reach the middle $\mathbb{Y}$ terms of either of two neighbors P_{-i} and P_j for some $i, j \geqslant 0$, these terms will never get canceled and the reduced form of the right hand side of the above identity will have at least two copies of $\mathbb{Y}$.) Thus, either $P_{-1} = \mathbb{X}^2\mathbb{Y}\mathbb{X}^2$ or $P_1 = \mathbb{Y}\mathbb{X}^2\mathbb{Y}\mathbb{X}^2\mathbb{Y}$, and, analyzing the further cancellation, one can see that the $>$ type stone has a neighborhood

—□—□—>—□— or —○—>—○—○—,

respectively. Passing to the dual inverse diagram, the latter case can be reduced to the former and it suffices to consider —□—□—>—□— only. The monodromy of this fragment cancels down to $\mathbb{X}$ and, to be able to proceed, we have to assume that either P_{-3} or P_2 is $\mathbb{X}^2\mathbb{Y}\mathbb{X}^2$; then, analyzing next few steps, we arrive at

—□—□—□—□—>—□—○— or —○—□—□—>—□—□—□—,

respectively, with the total monodromy $\mathbb{XY}$ or $\mathbb{YX}$. It remains to observe that each fragment can be cut off at its two outermost stones, see (3.97), splitting a maximal diagram with six stones, see Table 3.4, and use induction. □

Corollary 4.80 (Salepci [141]). *A maximal totally real directed fibration is algebraic.*

Proof. Each maximal fibration with six singular fibers, see Table 3.4, is algebraic; in fact, the four fibrations are the four real forms of the two real Jacobian fibrations corresponding to the real trigonal M-curves in Σ_2, see Theorem 4.69 (there are two isomorphism classes of ribbon curve structures with two ribbon boxes of type I_2 each). On the other hand, a harsh sum as in (3.97) of algebraic directed fibrations can be realized by a junction of the corresponding trigonal curves; hence, the statement of the corollary follows from Theorem 4.79. □

As one of the principal results, we extend Corollary 4.80 to all maximal directed fibrations, see Theorem 10.88 in Chapter 10. An alternative proof of Corollary 4.80 is given at the end of Section 10.3.3.

Remark 4.81. The list of non-algebraic fibrations given by Corollary 4.77 shows that Corollary 4.80 does not extend literally to $(M-1)$- or $(M-2)$-fibrations.

Problem 4.82. Does Corollary 4.80 extend to totally real directed fibrations of type I, *cf.* Definition 3.69?

In the algebraic setting, a classification of totally real Jacobian elliptic fibrations of type I over $\mathbb{P}^1$ was recently obtained in [61]. Up to deformation, such a fibration is determined by the restriction $p\colon X_{\mathbb{R}} \to \mathbb{P}^1_{\mathbb{R}}$; the proof uses the techniques of dessins described in this chapter. As a general rule, real algebraic varieties of type I are also known, like M-varieties, to be 'sufficiently rigid' topologically. As a first step towards Problem 4.82, it would be interesting to obtain a characterization of totally real directed fibrations of type I in terms of their necklace diagrams.

Remark 4.83. As another consequence of Theorem 4.79 and Theorem 4.69, see also Remark 4.74, one concludes that the topological classification of *maximal* totally real directed fibrations coincides with the equivariant equisingular deformation classification of maximal real elliptic fibrations. A similar phenomenon takes place in the case of extremal elliptic surfaces, see Section 10.3.1.

Chapter 5

The braid monodromy

The concept of braid monodromy as a means of studying the topology of embedded algebraic curves was introduced by O. Chisini [34, 33] and O. Zariski [167], and the term is probably due to B. Moishezon [120]. For a contemporary introduction to the notion of braid monodromy and its relation to the fundamental group, the reader may consult [65]; the particular case of curves in ruled surfaces is also considered in [9].

5.1 The Zariski–van Kampen theorem

The Zariski–van Kampen theorem computes the fundamental group of the complement of an algebraic curve in terms of the braid monodromy of the curve. We describe the braid monodromy of a proper curve in a ruled surface (making most choices in the definition almost canonical), discuss its relation to the fundamental group, and, for improper curves, introduce the notion of slopes to compensate for the improper fibers.

5.1.1 The monodromy of a proper n-gonal curve

Let $p\colon \Sigma \to B$ be a compact geometrically ruled surface and $E \subset \Sigma$ the exceptional section with $d := -E^2 \geqslant 0$. As a direct generalization of the notion of trigonal curve, define an *n-gonal curve*, $n \in \mathbb{N}$, as a reduced curve $C \subset \Sigma$ not containing E or a fiber of Σ as a component and such that the restriction $p\colon C \to B$ is a map of degree n. An n-gonal curve C is said to be *proper* if it is disjoint from E. A *singular fiber* of C is a fiber F of the ruling intersecting $C + E$ geometrically at fewer than $(n+1)$ points. The set of nonsingular fibers of C is denoted by $B^\sharp$.

One has $H_2(\Sigma) = \mathbb{Z}[E] + \mathbb{Z}[F]$, where F is a fiber, and $[C] = n[E] + r[F]$ for some integer $r \geqslant nd$. In fact, $C \circ E = r - nd$; hence, C is proper if and only if $r = nd$. Let $C_1, \ldots, C_m$ be the irreducible components of C, and let $[C_i] = n_i[E] + r_i[F]$, $r_i \geqslant n_i d$, $i = 1, \ldots, m$, so that $n_1 + \cdots + n_m = n$ and $r_1 + \cdots + r_m = r$. The group $H^2(C)$ is freely generated by the classes c_i Poincaré dual to the standard generators of $H_0(C_i)$, $i = 1, \ldots, m$. Since $H^3(\Sigma) = H_1(\Sigma) = H_1(B)$, from Poincaré–Lefschetz duality and the exact sequences of appropriate pairs one has canonical exact sequences

$$0 \longrightarrow H^2(C) \longrightarrow H_1(\Sigma \smallsetminus (C \cup E \cup F)) \xrightarrow{p_*} H_1(B) \longrightarrow 0 \tag{5.1}$$

and

$$\mathbb{Z} \xrightarrow{\beta} H^2(C) \longrightarrow H_1(\Sigma \smallsetminus (C \cup E)) \xrightarrow{p_*} H_1(B) \longrightarrow 0, \tag{5.2}$$

where β is the homomorphism $\beta\colon 1 \mapsto \sum_i r_i c_i$.

Proper sections and braid monodromy

Fix a compact ruled surface $p\colon \Sigma \to B$ and a *proper* n-gonal curve $C \subset \Sigma$. Define a *monodromy domain* as a subset $\Omega \subset B$ of the form $\Omega = B \smallsetminus U$, where U is the interior of an embedded closed disk. Recall that Ω has homotopy type of a CW-complex of dimension one and the inclusion $\Omega \hookrightarrow B$ induces an isomorphism $H_1(\Omega) = H_1(B)$.

Definition 5.3. A continuous section $s\colon \Omega \to \Sigma$ of the restriction $p\colon p^{-1}(\Omega) \to \Omega$ of the ruling is said to be *proper* (with respect to the given proper curve C) if the image of s is disjoint from both the exceptional section E and convex hull $\operatorname{conv} C$.

Lemma 5.4. *The set of homotopy classes of proper sections over a fixed monodromy domain $\Omega \subset B$ is a principal homogeneous space over $H^1(\Omega) = H^1(B)$. In particular, proper sections always exist.*

Proof. The lemma is a simple application of the obstruction theory. The restriction of the ruling p to $\Sigma \smallsetminus (E \cup \operatorname{conv} C)$ is a locally trivial fibration with a typical fiber $\mathbb{C}^1 \smallsetminus \{\text{disk}\} \sim S^1 = K(\mathbb{Z}, 1)$. Such fibrations are classified by the group $H^2(\Omega) = 0$ (*e.g.*, one can think of complex line bundles and their first Chern classes). Hence, the fibration in question is trivial and a choice of a trivialization identifies homotopy classes of sections with those of maps $\Omega \to K(\mathbb{Z}, 1)$, *i.e.*, with $H^1(\Omega)$. □

Remark 5.5. If the base $B \cong \mathbb{P}^1$ is rational, then Ω is also a disk and a proper section s is unique up to homotopy. In this case, for most computations, one can take for s a 'constant section' constructed as follows: pick an affine coordinate system (x, t) in Σ so that $E = \{x = \infty\}$ and Ω does not contain the point $t = \infty$, and let $s\colon t \mapsto c = \text{const}$, $|c| \gg 0$. Since the intersection $p^{-1}(\Omega) \cap \operatorname{conv} C$ is a compact subset of $\Sigma \smallsetminus E$, such a section is indeed proper whenever $|c|$ is sufficiently large.

Remark 5.6. The Euler class $e \in H^2(B) = \operatorname{Hom}(H_2(B), \mathbb{Z})$ of the (homotopy) S^1-bundle $\Sigma \smallsetminus (E \cup \operatorname{conv} C) \to B$ over the whole base B is the homomorphism $[B] \mapsto d$. In most interesting cases this class is nontrivial and the bundle has no section. (Any section of Σ disjoint from E would necessarily intersect C.) It is this reason why we have to remove a disk and restrict the bundle to a monodromy domain Ω in order to define the braid monodromy. Another manifestation of the Euler class is the nontrivial monodromy at infinity, see Proposition 5.16.

Fix a monodromy domain $\Omega \subset \Sigma$ and let $F_1, \dots, F_k$ be all singular and, possibly, some nonsingular fibers of C over Ω. Denote $b_i = p(F_i)$, $i = 1, \dots, k$, and consider the set $\Omega^\sharp := \Omega \smallsetminus \{b_1, \dots, b_k\}$; without loss of generality, we can assume that all points b_i are in the interior of Ω. Finally, pick a proper section s over Ω and a point $b \in \Omega^\sharp$ and let $F := p^{-1}(b)$ and $\pi_F := \pi_1(F \smallsetminus (C \cup E), s(b))$. The restriction

$p\colon p^{-1}(\Omega^\sharp) \smallsetminus (C \cup F) \to \Omega^\sharp$ is a locally trivial fibration, and the choices made give rise to the *braid monodromy* homomorphism

$$\mathfrak{m}_s\colon \pi_1(\Omega^\sharp, b) \to \operatorname{Aut} \pi_F.$$

As explained in Section D.1.2, see page 361, formally one considers the associated principal bundle with discrete fibers $\operatorname{Aut} \pi_1(p^{-1}(x) \smallsetminus (E \cup C), s(x))$, $x \in \Omega^\sharp$. This bundle is a covering, and each loop $\delta\colon I \to \Omega^\sharp$ with $\delta(0) = b$ has a unique lift $\tilde{\delta}$ with $\tilde{\delta}(0) = \mathrm{id} \in \operatorname{Aut} \pi_F$; by definition, $\mathfrak{m}_s[\delta] = \tilde{\delta}(1)$.

It is important to emphasize that, even though any section of a fibration can be used to construct an $\operatorname{Aut} \pi_1$-valued monodromy, we reserve the term 'braid monodromy' for the homomorphism obtained using a *proper* section s in the sense of Definition 5.3, *i.e.*, disjoint from $\operatorname{conv} C$ and extending through the punctures $b_1, \dots, b_k$ to the whole monodromy domain Ω. When the section is understood or irrelevant, the subscript s will be omitted.

Occasionally, we extend $\mathfrak{m}_s$ to the homotopy classes of paths in $\Omega^\sharp$. Thus, given a class $\zeta \in \pi_1(\Omega^\sharp; b', b'')$, we will speak about the monodromy

$$\mathfrak{m}_s(\zeta)\colon \pi_1(p^{-1}(b') \smallsetminus (C \cup E), s(b')) \to \pi_1(p^{-1}(b'') \smallsetminus (C \cup E), s(b'')).$$

The monodromy groups

There is a strict deformation retraction $F \smallsetminus (C \cup E) = F^\circ \smallsetminus C \to \bar{D}^\circ_n$, where $\bar{D}^\circ_n$ is an embedded closed disk $\bar{D}^2 \subset F^\circ$ containing $\operatorname{conv} C$ in its interior and $s(b)$ in its boundary and punctured at the n points $F \cap C$. A geometric basis $\{\alpha_1, \dots, \alpha_n\}$ for the group $\pi_F = \pi_1(\bar{D}^\circ_n, s(b))$ is called *proper*. A proper geometric basis gives rise to an identification $\pi_F = \mathbb{F}_n$ and an isomorphism $\operatorname{Aut} \pi_F = \operatorname{Aut} \mathbb{F}_n$. The image under this isomorphism of the braid group $\mathbb{B}_n \subset \operatorname{Aut} \mathbb{F}_n$ is denoted by $\mathbb{B}(F) \subset \operatorname{Aut} \pi_F$; this image does not depend on the basis used (provided that the latter is proper).

For a point $x \in \Omega$, denote by $\rho_x \in \pi_1(F^\circ_x \smallsetminus C, s(x))$ the image under the inclusion homomorphism of the 'counterclockwise' (with respect to the complex orientation of the affine line F°_x) generator of the cyclic group $\pi_1(F^\circ_x \smallsetminus \operatorname{conv} C, s(x))$. Since the S^1-bundle $p^{-1}(\Omega) \smallsetminus (E \cup \operatorname{conv} C) \to \Omega$ is obviously orientable, hence 1-simple, all these elements are naturally identified and we usually omit the subscript x. Over the basepoint $x = b$, one has $\rho = \alpha_1 \cdots \alpha_n$ (in fact, this property characterizes proper geometric bases), and this element ρ is invariant under the braid monodromy.

Remark 5.7. The affine action of $H^1(B)$ on the set of homotopy classes of proper sections given by the obstruction theory, see Lemma 5.4, has a transparent geometric description. Let s and $s' := s + h$, $h \in H^1(B) = \operatorname{Hom}(H_1(B), \mathbb{Z})$ be two sections, and assume that $s(b) = s'(b)$. Then, for any loop γ in Ω, the loop $s'_*(\gamma) \cdot s_*(\gamma^{-1})$ in $p^{-1}(\Omega) \smallsetminus (E \cup \operatorname{conv} C)$ is homotopic to $\rho^{h[\gamma]}$.

As explained above, the element $\rho \in \pi_F$ is invariant under the braid monodromy. Since also the braid monodromy $\mathfrak{m}(\gamma)$ along any loop γ takes a geometric generator to a conjugate of a generator, Theorem 2.28 implies the following statement, which explains the term 'braid monodromy'.

Proposition 5.8. *For any proper section s, the braid monodromy $\mathfrak{m}_s$ of a proper n-gonal curve C takes values in the braid group $\mathbb{B}(F) \subset \operatorname{Aut} \pi_F$.* ◁

Any element $h \in H^1(\Omega^\sharp) = \operatorname{Hom}(H_1(\Omega^\sharp), \mathbb{Z})$ gives rise to a homomorphism

$$\Delta^2 h \colon \pi_1(\Omega^\sharp, b) \to H_1(\Omega^\sharp) \to \mathbb{Z} = \mathcal{Z}(\mathbb{B}_n),$$

the identification $\mathbb{Z} = \mathcal{Z}(\mathbb{B}_n)$ sending 1 to Δ^2, which does not depend on the choice of a proper geometric basis. Hence, the set of homomorphisms $\mathfrak{m} \colon \pi_1(\Omega^\sharp, b) \to \mathbb{B}_n$ has a natural affine $H^1(\Omega^\sharp)$-action given by

$$\mathfrak{m} \uparrow h = \mathfrak{m} + \Delta^2 h \colon \gamma \mapsto \mathfrak{m}(\gamma) \cdot \Delta^2 h(\gamma). \tag{5.9}$$

This action can be composed with the restriction homomorphism $H^1(B) \to H^1(\Omega^\sharp)$, and the next statement is an immediate consequence of the definition of the braid monodromy (see also Remark 5.7).

Lemma 5.10. *The braid monodromies defined using two proper sections s and $s + h$, $h \in H^1(B)$, see Lemma 5.4, are related via $\mathfrak{m}_{s+h} = \mathfrak{m}_s + \Delta^2 h$.* ◁

For a fixed proper curve C and nonsingular fiber F, define the *monodromy group* $MG(C, F)$ and *restricted monodromy group* $MG_+(C, F)$ of C with respect to F as the images

$$MG(C, F) := \operatorname{Im} \mathfrak{m}_s \quad \text{and} \quad MG_+(C, F) := \mathfrak{m}_s \operatorname{Ker}[\pi_1(\Omega^\sharp, b) \to \pi_1(\Omega, b)],$$

respectively. *A priori*, these subgroups depend on the choice of a monodromy domain Ω and a proper section s over Ω. Note that $MG_+(C, F)$ is a normal subgroup of $MG(C, F)$ but, in general, neither of the two subgroups is normal in $\mathbb{B}(F)$.

Theorem 5.11. *Assume that the monodromy domain Ω used contains (in its interior) all singular fibers of C. Then the monodromy groups $MG(C, F)/\Delta^2 \subset \mathbb{B}(F)/\Delta^2$ and $MG_+(C, F) \subset \mathbb{B}(F)$ do not depend on Ω or on the proper section s.*

Proof. Due to Lemma 5.10, the subgroups in the statement do not depend on the choice of a section s. Given two monodromy domains Ω_1, Ω_2, both containing the basepoint $b := p(F)$ and all singular fibers of C, one can find a third domain Ω with the same properties and such that $\Omega \subset \Omega_i$, $i = 1, 2$. Then the inclusions induce isomorphisms $\pi_1(\Omega, b) = \pi_1(\Omega_i, b)$ and $\pi_1(\Omega^\sharp, b) = \pi_1(\Omega_i^\sharp, b)$, $i = 1, 2$, and the braid monodromies can be defined using a proper section s over Ω and its extensions to Ω_i, $i = 1, 2$. Under this assumption, the resulting homomorphisms are equal (in the obvious sense), and so are their images. □

The identifications $\pi_F \to \mathbb{F}_n$ and $\mathbb{B}(F) \to \mathbb{B}_n$ depend on the choice of a geometric basis $\alpha = \{\alpha_1, \dots, \alpha_n\}$ for π_F. Since any two proper geometric bases are related by a braid, see Theorem 2.28, the resulting identifications $\mathbb{B}(F) \to \mathbb{B}_n$ differ by an inner automorphism of $\mathbb{B}_n$. Furthermore, proper geometric bases in two distinct nonsingular fibers F', F'' also differ by a braid, namely, the monodromy $\mathfrak{m}_s(\zeta)$ along any path $\zeta \in \pi_1(\Omega^\sharp; p(F'), p(F''))$. Hence, up to conjugation, the images of the groups $MG(C, F)$ and $MG_+(C, F)$ in the abstract braid group $\mathbb{B}_n$ do not depend on the fiber F or a proper geometric basis for π_F. The conjugacy classes $MG(C) := [\![MG(C, F)]\!]$ and $MG_+(C) := [\![MG_+(C, F)]\!]$ are also referred to as the *monodromy group* and *restricted monodromy group* of C, respectively. Since $\Delta^2 \in \mathbb{B}_n$ is a central element, for any $r \in \mathbb{N}$ one can also speak about the conjugacy class $MG(C)/\Delta^{2r} := [\![MG(C, F)/\Delta^{2r}]\!]$ in the group $\mathbb{B}_n/\Delta^{2r}$.

Corollary 5.12. *The conjugacy classes $MG(C)/\Delta^2$ in $\mathbb{B}_n/\Delta^2$ and $MG_+(C)$ in $\mathbb{B}_n$ depend on the curve C only (assuming that the monodromy domain used contains in its interior all singular fibers of C).* ◁

Remark 5.13. If the base B of the ruling is rational, then any monodromy domain Ω is a disk, one has $\pi_1(\Omega) = 0$, and hence the subgroups MG and MG_+ coincide. Thus, in this case the subgroup $MG(C, F) \subset \mathbb{B}(F)$ and the conjugacy class $MG(C)$ in $\mathbb{B}_n$ are also independent of the choices made.

Proposition 5.14 (*cf.* Zariski [167]). *Assume that an n-gonal curve $C' \subset \Sigma' \to B'$ is obtained from a proper n-gonal curve $C \subset \Sigma \to B$ by a small perturbation. Then one has $MG_+(C) \prec MG_+(C')$ and $MG(C)/\Delta^2 \prec MG(C')/\Delta^2$.*

Proof. Topologically, one can identify $\Sigma' \to B'$ and $\Sigma \to B$ and choose a common monodromy domain $\Omega' = \Omega$, reference fiber $F' = F$, and proper section $s' = s$. Furthermore, one can assume that the two curves coincide outside the pull-back of a regular neighborhood U of the union of the singular fibers b_j of C. Then the braid monodromy of C is induced from that of C' by the inclusion $\Omega \smallsetminus U \hookrightarrow (\Omega')^\sharp$ and strict deformation retraction $\Omega^\sharp \to \Omega \smallsetminus U$. □

The monodromy at infinity

Consider a monodromy domain $\Omega = B \smallsetminus U$, where U is the interior of an embedded closed disk $\bar{U} \subset B$. Similar to Definition 5.3, one can introduce the notion of proper section s_∞ over $\bar{U}$ (as a continuous section of the restricted locally trivial fibration $p^{-1}(\bar{U}) \smallsetminus (E \cup \operatorname{conv} C) \to \bar{U}$); since the disk $\bar{U}$ is contractible, such a section is unique up to homotopy in the class of proper sections, *cf.* Lemma 5.4.

Consider a pair of proper sections s over Ω and s_∞ over $\bar{U}$ and assume that $s(b) = s_\infty(b)$ for some point $b \in \partial\Omega = \partial\bar{U}$. Let $F = p^{-1}(b)$ and $\tilde{b} = s(b) \in F$. We have the following Serre exact sequence

$$0 \to \pi_1(F \smallsetminus (E \cup \operatorname{conv} C), \tilde{b}) \to \pi_1(p^{-1}(\partial\Omega) \smallsetminus (E \cup \operatorname{conv} C), \tilde{b}) \to \pi_1(\partial\Omega, b) \to 0.$$

Regard $\partial\Omega$ as a loop starting at b and traversing the boundary of Ω once in the positive direction (with respect to the standard complex orientation of Ω) and consider the classes $s_*[\partial\Omega]$ and $(s_\infty)_*[\partial\Omega]$. Their images in $\pi_1(\partial\Omega, b)$ coincide; hence they differ by a certain element of the group $\pi_1(F \smallsetminus (E \cup \operatorname{conv} C), \tilde{b})$. The following lemma is essentially the obstruction theoretical definition of the Euler class and the fact that the Euler number of the S^1-bundle $\Sigma \smallsetminus (E \cup \operatorname{conv} C) \to B$ (with the counterclockwise orientation induced from the affine fibers F°) equals $d := \deg \mathcal{Y} = -E^2$, see page 65.

Lemma 5.15. *In the notation above, one has* $s_*[\partial\Omega] \cdot (s_\infty)_*[\partial\Omega]^{-1} = \rho^{-d}$. ◁

Fix a monodromy domain Ω and a basepoint $b \in \Omega^\sharp$ and let $F = p^{-1}(b)$. Pick a path γ in $\Omega^\sharp$ connecting b to a point $b' \in \partial\Omega$ and consider the class in $\pi_1(\Omega^\sharp, b)$ represented by the loop $\gamma \cdot \partial\Omega \cdot \gamma^{-1}$. Disregarding the dependence on γ, denote this class by $[\partial\Omega]$. The image $\mathfrak{m}_s[\partial\Omega] \in \mathbb{B}(F)$ is called the *monodromy at infinity.*

Proposition 5.16. *If the monodromy domain Ω contains in its interior all singular fibers of C, then one has $\mathfrak{m}_s[\partial\Omega] = \Delta^{2d}$, where $d = -E^2$. In particular, $\mathfrak{m}_s[\partial\Omega]$ is independent of the choices made in the construction.*

Proof. Let $\bar{U}$ be the closure of $B \smallsetminus \Omega$. Under the assumptions, the restricted fibration $p^{-1}(\bar{U}) \smallsetminus (C \cup E) \to \bar{U}$ is locally trivial (as $\bar{U}$ contains no singular fibers of C), hence trivial (as $\bar{U}$ is contractible), and so is its braid monodromy *defined using a proper section s_∞ over $\bar{U}$*. Thus, if the basepoint b is in the boundary $\partial\Omega$ and the auxiliary path γ in the definition of $[\partial\Omega]$ is constant, the statement follows from Lemma 5.15, *cf.* Lemma 5.10. To show that the result does not depend on the point b or path γ, it suffices to apply Proposition 5.8 and the fact that $\Delta^{2d} \in \mathcal{Z}(\mathbb{B}(F))$. □

Nagata transformations and induced curves

As in Section 3.1.1, for n-gonal curves one can define the notions of Nagata transformation and induced curve. Fix a proper n-gonal curve $C \subset \Sigma \to B$, a monodromy domain Ω, and a proper section s over Ω, and let $\mathfrak{m}_s\colon \pi_1(\Omega^\sharp, b) \to \mathbb{B}_n$ be the braid monodromy of C.

Let ζ be a simple arc connecting a point in the boundary $\partial\Omega$ and a puncture b_i. It defines a certain class $[\zeta] \in H_1(\Omega, \partial\Omega \cup \{b_1, \ldots, b_k\})$ and, *via* Poincaré duality, a class $\mathrm{D}^{-1}[\zeta] \in H^1(\Omega^\sharp)$.

Lemma 5.17. *Let C' be the curve obtained from C by a positive Nagata transformation at a fiber F_i, and let ζ be a simple arc as above. Then a section s' over Ω proper with respect to C' can be chosen so that the braid monodromy $\mathfrak{m}'_{s'}$ of C' is given by $\mathfrak{m}'_{s'} = \mathfrak{m}_s + \Delta^2\,\mathrm{D}^{-1}[\zeta]$, see* (5.9).

Proof. Over $\Omega^\sharp$, the two curves are diffeomorphic and their monodromies coincide. However, the original section s is not proper with respect to C', as it intersects the zero section at F_i. Assuming ζ disjoint from the other punctures, we can modify s in a neighborhood of ζ (by cutting, twisting, and regluing) to make it proper. This new section results in the monodromy $\mathfrak{m}'_{s'}$ as stated, *cf.* Lemma 5.10. □

Corollary 5.18. *The conjugacy classes $MG(C)/\Delta^2$ and $MG_+(C)/\Delta^2$ are invariant under Nagata equivalence of proper curves.* ◁

For the next statement, assume that a curve $C' \subset \Sigma' \to B'$ is induced from C by a nonconstant base change $\varphi\colon B' \to B$ and let $\Omega' = \varphi^{-1}(\Omega)$. Consider a monodromy domain $\bar{\Omega}'$ containing Ω', pick a basepoint b' over b, and let F' be the fiber over b'. By definition, the fibers F and F', and hence the groups π_F and $\pi_{F'}$, are canonically identified. Denote by $\mathfrak{m}'\colon \pi_1((\bar{\Omega}')^\sharp, b') \to \mathbb{B}(F') = \mathbb{B}(F)$ the braid monodromy of C' (defined using some proper section $\bar{s}'$ over $\bar{\Omega}'$), and let $\mathfrak{m}'|_{\Omega'} := \mathfrak{m}' \circ \mathrm{in}_*$ be its restriction, where $\mathrm{in}\colon (\Omega')^\sharp \hookrightarrow (\bar{\Omega}')^\sharp$ is the inclusion.

Lemma 5.19. *In the notation above, one has $\mathfrak{m}'|_{\Omega'} = (\mathfrak{m}_s \circ \varphi_*) + \Delta^2 h$ for some class $h \in H^1(\Omega')$, cf.* (5.9).

Proof. Let s' be the pull-back $\varphi^* s$. It is a proper section over Ω' and, since the construction of monodromy is natural, the monodromy $\mathfrak{m}_{s'}\colon \pi_1((\Omega')^\sharp, b') \to \mathbb{B}(F')$ defined using s' equals $\mathfrak{m}_s \circ \varphi_*$. On the other hand, up to homotopy s' differs from the restriction $\bar{s}'|_{\Omega'}$ by a certain cohomology class $h \in H^1(\Omega')$, *cf.* Lemma 5.4, and hence $\mathfrak{m}'|_{\Omega'} = \mathfrak{m}_{s'} + \Delta^2 h$, *cf.* Lemma 5.10. □

Warning 5.20. Typically, the pull-back s' used in the proof of Lemma 5.19 does not extend to a proper section over $\bar{\Omega}'$ and one unavoidably has $h \neq 0$. The homomorphism $\mathfrak{m}_s \circ \varphi_*$ does not need to vanish on the kernel $\mathrm{Ker}\, \mathrm{in}_*$.

Corollary 5.21. *If a curve C' is induced from a curve C, then $MG_+(C') \prec MG_+(C)$ and $MG(C')/\Delta^2 \prec MG(C)/\Delta^2$. The latter inclusion also holds if C' is Nagata equivalent to a curve induced from C.* ◁

5.1.2 The fundamental groups

For most of this section, we assume that $C \subset \Sigma \to B$ is a proper n-gonal curve and keep the notation introduced in Section 5.1.1: $\Omega \subset B$ is a fixed monodromy domain without singular fibers of C in the boundary, $s\colon \Omega \to \Sigma$ is a proper section, $b_1, \ldots, b_k$ are all singular and, possibly, some nonsingular fibers of C in the interior of Ω, and $\Omega^\sharp = \Omega \smallsetminus \{b_1, \ldots, b_k\}$. Theorem 5.22 and Proposition 5.26 appeared in [161] and [167] and are commonly referred to as the *Zariski–van Kampen method* of computation of the fundamental group of the complement of an algebraic curve.

Theorem 5.22 (Zariski–van Kampen). *In the notation above, the fundamental group* $\pi_1(p^{-1}(\Omega^\sharp) \smallsetminus (C \cup E))$ *is a semidirect product* $\pi_F \rtimes \pi_1(\Omega^\sharp)$. *More precisely, a choice of a proper section* s *gives rise to an isomorphism*

$$\pi_1(p^{-1}(\Omega^\sharp) \smallsetminus (C \cup E), s(b)) = \pi_F * \pi_1(\Omega^\sharp, b)/\{\gamma^{-1}\alpha\gamma = \alpha \uparrow \mathfrak{m}_s(\gamma)\},$$

the relations running over all elements $\alpha \in \pi_F$ *and* $\gamma \in \pi_1(\Omega^\sharp, b)$.

Proof. Since the fiber $F \smallsetminus (C \cup E) = F^\circ \smallsetminus C$ is homotopy equivalent to a wedge of circles, the Serre exact sequence of the restricted fibration reduces to a short exact sequence

$$0 \longrightarrow \pi_F \xrightarrow{i_*} \pi_1(p^{-1}(\Omega^\sharp) \smallsetminus (C \cup E), s(b)) \xrightarrow{p_*} \pi_1(\Omega^\sharp, b) \longrightarrow 0,$$

where i is the inclusion. This sequence splits *via* s_*, representing the middle term as a semidirect product, and the conjugation action of $\pi_1(\Omega^\sharp, b)$ on π_F is essentially the definition of monodromy. □

Theorem 5.22 gives us an explicit presentation of the fundamental group. Represent the monodromy domain Ω as a connected sum of a surface B' of genus $g = g(B)$ and a punctured disk D'. Let $\{\mu_1, \nu_1, \dots, \mu_g, \nu_g\}$ be a standard symplectic basis for the group $\pi_1(B')$, and let $\{\gamma_1, \dots, \gamma_k\}$ be a geometric basis for $\pi_1(D')$; together, these elements form a free basis for $\pi_1(\Omega^\sharp, b)$. For the future computations, we can assume that

$$[\mu_1, \nu_1] \cdots [\mu_g, \nu_g]\gamma_1 \cdots \gamma_k = [\partial\Omega]. \tag{5.23}$$

Pick a proper section s and identify the elements of $\pi_1(\Omega^\sharp, b)$ with their images under the induced homomorphism s_*. (Since $p_* \circ s_*$ is the identity, s_* is a monomorphism.) Let, further, $\{\alpha_1, \dots, \alpha_n\}$ be a proper geometric basis for $\pi_F = \pi_1(F, s(b))$. Then the group $\pi_1(p^{-1}(\Omega^\sharp) \smallsetminus (C \cup E), s(b))$ is generated by the elements

$$\alpha_1, \dots, \alpha_n, \mu_1, \nu_1, \dots, \mu_g, \nu_g, \gamma_1, \dots, \gamma_k, \tag{5.24}$$

and the defining relations are

$$\gamma^{-1}\alpha_i\gamma = \alpha_i \uparrow \mathfrak{m}_s(\gamma), \quad i = 1, \dots, n,\ \gamma \in \{\mu_j, \nu_j, \gamma_r\}. \tag{5.25}$$

In order to pass to the group of the complement of $C \cup E$ and, possibly, some (but not necessarily all) singular or nonsingular fibers of C, we need to patch back in (some of) the fibers over the punctures of $\Omega^\sharp$. The fibers can be patched one by one, and the result is described by the following proposition, which holds for any, non necessarily proper, n-gonal curve C.

Proposition 5.26 (van Kampen [161]). *Let* $C \subset \Sigma \to B$ *be an* n*-gonal curve and let* $F_1, \dots, F_k$ *be a finite collection of fibers of* Σ. *Consider a small embedded disk*

$\Phi \subset \Sigma$ transversal to F_1 at a single point $0 \in \Phi$ and disjoint from C, E, and the other fibers F_i, $i \geqslant 2$. Then the inclusion homomorphism

$$\pi_1(\Sigma \smallsetminus (C \cup E \cup \bigcup_{i\geqslant 1} F_i)) \to \pi_1(\Sigma \smallsetminus (C \cup E \cup \bigcup_{i\geqslant 2} F_i))$$

is an epimorphism and its kernel is normally generated by the class $[\partial\Phi]$.

Proof. Let $S_1, S_2, \ldots$ be the points of intersection of $(C+E)$ and F_1, and pick some Milnor balls $M_1, M_2, \ldots$ for the curve $C+E+F_1$ at these points, disjoint from one another and from the other fibers F_i, $i \geqslant 2$. Then, for each $r = 1, 2$, there is a strict deformation retraction

$$\Sigma \smallsetminus \Bigl(C \cup E \cup \bigcup\nolimits_{i\geqslant r} F_i\Bigr) \to X_r := \Sigma \smallsetminus \Bigl(\bigcup\nolimits_j M_j \cup C \cup E \cup \bigcup\nolimits_{i\geqslant r} F_i\Bigr).$$

On the other hand, X_2 is the union of X_1 and a sufficiently small tubular neighborhood T of $F_1' := F_1 \smallsetminus \bigcup_j M_j$ in X_2. Since the normal bundle of F_1 in Σ is trivial, one can adjust T and Φ so that $T = F_1' \times \Phi$ and, hence, $T \cap X_1 = F_1' \times (\Phi \smallsetminus 0)$. Then the inclusion $T \cap X_1 \hookrightarrow T$ induces an epimorphism

$$\pi_1(T \cap X_1) = \pi_1(F_1') \times \pi_1(\partial\Phi) \twoheadrightarrow \pi_1(T) = \pi_1(F_1')$$

and its kernel is the cyclic group $\pi_1(\partial\Phi) = \mathbb{Z}[\partial\Phi]$. Hence, the statement of the proposition follows from the Seifert–van Kampen theorem. □

Remark 5.27. Assume that the monodromy domain Ω contains in its interior all singular fibers of C. Pick a point $b_\infty \in B \smallsetminus \Omega$ and denote $B^{\sharp\sharp} = B \smallsetminus \{b_1, \ldots, b_k, b_\infty\}$. Then the spaces $\Omega^\sharp$ and $p^{-1}(\Omega^\sharp) \smallsetminus (C \cup E)$ are strict deformation retracts of $B^{\sharp\sharp}$ and $p^{-1}(B^{\sharp\sharp}) \smallsetminus (C \cup E)$, respectively, and Theorem 5.22 computes the fundamental group $\pi_1(p^{-1}(B^{\sharp\sharp}) \smallsetminus (C \cup E), s(b))$. On the other hand, the latter space is obtained from $\Sigma \smallsetminus (C \cup E)$ by removing finitely many fibers, and these fibers can be patched, one by one, using Proposition 5.26.

For any proper section s_∞ over the closure $\bar{U}$ of the complement $B \smallsetminus \Omega$, see page 150, the image $s_\infty(\bar{U})$ is a disk transversal to the fiber $F_\infty = p^{-1}(b_\infty)$. Hence, according to Proposition 5.26, Lemma 5.15, and the choice of the basis, see (5.23), patching the fiber F_∞ gives the so-called *relation at infinity*

$$[\mu_1, \nu_1] \cdots [\mu_g, \nu_g]\gamma_1 \cdots \gamma_k = \rho^{-d}, \tag{5.28}$$

where $d = -E^2$. For each 'finite' fiber F_j, $j = 1, \ldots, k$, the generator γ_j can be represented by a loop bounding a disk in the image $s(\Omega)$; hence, patching this fiber results in a relation $\gamma_j = 1$. In other words, one can eliminate γ_j from the generating set (5.24) and relation (5.28) (if present) and replace the corresponding relations in (5.25) with the relations

$$\alpha_i = \alpha_i \uparrow \mathfrak{m}_s(\gamma_j), \quad i = 1, \ldots, n, \tag{5.29}$$

often called *braid relations* and abbreviated to $\mathfrak{m}_s(\gamma_j) = \mathrm{id}$, see Section A.2.1. Note that the value $\mathfrak{m}_s(\gamma_j)$ does not depend on s, see Lemma 5.10. Note also that, since the automorphism $\mathfrak{m}_s(\gamma_j)$ of π_F is a braid, see Proposition 5.8, and hence $\rho \uparrow \mathfrak{m}_s(\gamma_j) = \rho$, any one of the n relations (5.29) can be omitted: it would follow from the others. On the other hand, when simplifying the presentation, it is often convenient to move in the opposite direction and *add* as many relations as possible. Thus, relations (5.29) imply that

$$\alpha = \alpha \uparrow \mathfrak{m}_s(\gamma_j) \quad \text{for any } \alpha \in \pi_F = \langle \alpha_1, \ldots, \alpha_n \rangle. \tag{5.30}$$

Furthermore, if all singular fibers F_j of C are patched, one has the relations

$$\alpha = \alpha \uparrow \mathfrak{m} \quad \text{for all } \alpha \in \pi_F \text{ and } \mathfrak{m} \in MG_+(C, F), \tag{5.31}$$

see Corollary A.22. Both statements are easy exercises in elementary group theory.

Each braid relation (5.29) can be written in the form $(\alpha \uparrow \beta) \cdot \alpha^{-1} = 1$, where α is an element of a certain group G and β is a braid acting on G. A list of these relations (for a fixed braid β and a given list of elements $\alpha \in G$) is returned by the function "`BraidRelations`" in Listing 2.5.

In the sequel, we are primarily interested in the following four groups:

$$\begin{aligned}
\pi^{\mathrm{aff}}(C, \tilde{b}) &:= \pi_1(\Sigma \smallsetminus (C \cup E \cup F_\infty), \tilde{b}),\\
\pi^{\mathrm{aff}}_+(C, \tilde{b}) &:= \mathrm{Ker}[\mathrm{inclusion}_* \colon \pi^{\mathrm{aff}}(C, \tilde{b}) \to \pi_1(\Sigma \smallsetminus F_\infty, \tilde{b})],\\
\pi^{\mathrm{proj}}(C, \tilde{b}) &:= \pi_1(\Sigma \smallsetminus (C \cup E), \tilde{b}),\\
\pi^{\mathrm{proj}}_+(C, \tilde{b}) &:= \mathrm{Ker}[\mathrm{inclusion}_* \colon \pi^{\mathrm{proj}}(C, \tilde{b}) \to \pi_1(\Sigma, \tilde{b})],
\end{aligned}$$

where F_∞ is a nonsingular fiber of C and the basepoint $\tilde{b}$ is assumed outside the union $E \cup \operatorname{conv} C$; the basepoint is usually omitted in the notation. The first two groups are referred to as the *affine fundamental groups* of C, the last two, its *projective fundamental groups*. Summarizing the discussion above, one arrives at the following description of these groups.

Theorem 5.32. *If the base B is a compact curve of positive genus $g \geqslant 1$, then, in the notation of Section* A.2.1,

$$\pi^{\mathrm{aff}}_+(C, \tilde{b}) = \pi^{\mathrm{proj}}_+(C, \tilde{b}) = \pi_F / MG_+(C, F),$$

where F is the fiber containing $\tilde{b}$ and $\pi_F := \pi_1(F^\circ \smallsetminus C, \tilde{b})$. Furthermore, one has

$$\pi^{\mathrm{aff}}(C) = \pi^{\mathrm{aff}}_+(C) \rtimes \pi_1(B \smallsetminus b_\infty),$$

with $\pi_1(B \smallsetminus b_\infty)$ acting on $\pi^{\mathrm{aff}}_+(C)$ via the braid monodromy $\mathfrak{m}_s$, and

$$\pi^{\mathrm{proj}}(C) = \pi^{\mathrm{aff}}(C) / [\mu_1, \nu_1] \cdots [\mu_g, \nu_g] \rho^d,$$

where $d = -E^2$. If the base is rational, then $\pi^{\mathrm{aff}}_+(C)$ is as above and one has $\pi^{\mathrm{aff}}(C) = \pi^{\mathrm{aff}}_+(C)$ and $\pi^{\mathrm{proj}}(C) = \pi^{\mathrm{proj}}_+(C) = \pi^{\mathrm{aff}}(C)/\rho^d$. ◁

For the statement of Theorem 5.32, observe that, since $MG_+(C,F) \subset MG(C,F)$ is a normal subgroup, the braid monodromy does factor to a well defined homomorphism $\mathfrak{m}_s\colon \pi_1(B \smallsetminus b_\infty) \to \operatorname{Aut} \pi_+^{\mathrm{aff}}(C)$, see Corollary A.25. For the relation at infinity $[\mu_1,\nu_1]\cdots[\mu_g,\nu_g] = \rho^{-d}$, we assume that the standard symplectic basis $\{\mu_i,\nu_i\}$ for the group $\pi_1(B \smallsetminus b_\infty) = \pi_1(\Omega)$ is chosen so that $[\mu_1,\nu_1]\cdots[\mu_g,\nu_g] = [\partial\Omega]$, *cf.* (5.23).

According to Theorem 5.32, the affine group $\pi_+^{\mathrm{aff}}(C)$ of an n-gonal curve is naturally equipped with an epimorphism $\pi_F \twoheadrightarrow \pi_+^{\mathrm{aff}}(C,\tilde{b})$, which is induced by the inclusion of the fiber $F \ni \tilde{b}$. Hence, a choice of a proper geometric basis $\{\alpha_1,\ldots,\alpha_n\}$ for π_F gives rise to an epimorphism $\mathbb{F}_n \twoheadrightarrow \pi_+^{\mathrm{aff}}(C)$, called a *geometric presentation* of $\pi_+^{\mathrm{aff}}(C)$. Since any two proper geometric bases in any two nonsingular fibers differ by a braid, a geometric presentation is well defined up to the action of the braid group $\mathbb{B}_n$ on $\mathbb{F}_n$. In particular, the homomorphism $\deg\colon \mathbb{F}_n \to \mathbb{Z}$ descends to a well defined *degree homomorphism* $\deg\colon \pi_+^{\mathrm{aff}}(C) \to \mathbb{Z}$. One has $\deg\alpha_i = 1$, $i = 1,\ldots,n$.

A homomorphism $\varphi\colon \pi_+^{\mathrm{aff}}(C_1) \to \pi_+^{\mathrm{aff}}(C_2)$ between the affine fundamental groups of two proper n-gonal curves $C_i \subset \Sigma_i \to B_i$, $i = 1,2$, is called *geometric*, or *induced by a braid*, if there is a braid $\beta \in \mathbb{B}_n$ that makes the following diagram commutative (where the vertical arrows are geometric presentations of the groups):

$$\begin{array}{ccc} \mathbb{F}_n & \xrightarrow{\ \beta\ } & \mathbb{F}_n \\ \downarrow & & \downarrow \\ \pi_+^{\mathrm{aff}}(C_1) & \xrightarrow{\ \varphi\ } & \pi_+^{\mathrm{aff}}(C_2). \end{array}$$

Clearly, a geometric homomorphism is necessarily onto. If such a homomorphism exists, $\pi_+^{\mathrm{aff}}(C_2)$ is said to be a *geometric quotient* of $\pi_+^{\mathrm{aff}}(C_1)$. More generally, an epimorphism $\mathbb{F}_n \twoheadrightarrow G$ is a *geometric quotient* of $\pi_+^{\mathrm{aff}}(C_1)$ if, after applying a braid, it factors through a geometric presentation $\mathbb{F}_n \twoheadrightarrow \pi_+^{\mathrm{aff}}(C_1)$. This property is equivalent to the existence of a braid β such that $K_1 \uparrow \beta \subset K_2$, where $K_1 = \operatorname{Ker}[\mathbb{F}_n \twoheadrightarrow \pi_+^{\mathrm{aff}}(C_1)]$ and $K_2 = \operatorname{Ker}[\mathbb{F}_n \twoheadrightarrow G]$ are the kernels. In terms of the monodromy groups, $\pi_+^{\mathrm{aff}}(C_2)$ is a geometric quotient of $\pi_+^{\mathrm{aff}}(C_1)$ if and only if $MG_+(C_1) \prec MG_+(C_2)$.

The next observation follows from Theorem 5.32 and Proposition 5.14. Originally it was stated in [167] for plane curves, which can be regarded as proper curves in the blown up plane $\mathbb{P}^2(P) = \Sigma_1$.

Proposition 5.33 (Zariski [167]). *Assume that an n-gonal curve $C' \subset \Sigma' \to B'$ is obtained from a proper n-gonal curve $C \subset \Sigma \to B$ by a small perturbation. Then the group $\pi_+^{\mathrm{aff}}(C')$ is a geometric quotient of $\pi_+^{\mathrm{aff}}(C)$.* ◁

The case of rational base

If the base B of the ruling is rational, $B \cong \mathbb{P}^1$, some of the concepts introduced above can be simplified. According to Theorem 5.32, the affine fundamental group

$\pi^{\mathrm{aff}}(C) = \pi_+^{\mathrm{aff}}(C)$ of a proper n-gonal curve C in a Hirzebruch surface Σ_d, $d \geqslant 0$, has a geometric presentation of the form

$$\pi^{\mathrm{aff}}(C) = \langle \alpha_1, \dots, \alpha_n \mid \mathfrak{m}(\gamma_j) = \mathrm{id},\ j = 1, \dots, k \rangle, \tag{5.34}$$

where $\{\alpha_1, \dots, \alpha_n\}$ is a proper geometric basis for the fundamental group π_F of a nonsingular fiber and $\{\gamma_1, \dots, \gamma_k\}$ is a geometric basis for the group $\pi_1(\Omega^\sharp)$ of a monodromy domain punctured at all singular fibers of C. The projective group $\pi^{\mathrm{proj}}(C) = \pi_+^{\mathrm{proj}}(C)$ is obtained from (5.34) by adding the *relation at infinity*

$$\rho^d := (\alpha_1 \cdots \alpha_n)^d = 1. \tag{5.35}$$

It follows that, in the case of rational base, the degree homomorphism $\pi^{\mathrm{aff}}(C) \to \mathbb{Z}$ descends to an epimorphism $\deg\colon \pi^{\mathrm{proj}}(C) \twoheadrightarrow \mathbb{Z}_{nd}$.

Recall that the central element $\Delta^2 \in \mathbb{B}_n$ acts on $\mathbb{F}_n$ *via* the conjugation by ρ^{-1}. Hence, the following statement is an immediate consequence of presentation (5.34) and Proposition 5.16; see also (5.31) and, for the commutants, Corollary A.30.

Proposition 5.36. *The element $\rho^d \in \pi^{\mathrm{aff}}(C)$ is central and the quotient projection $\pi^{\mathrm{aff}}(C) \to \pi^{\mathrm{proj}}(C)$ restricts to an isomorphism of the commutants.* $\triangleleft$

The importance of Proposition 5.36 is in the fact that, due to Poincaré–Lefschetz duality, the abelianizations $H_1(\Sigma_d \smallsetminus (C \cup E \cup F_\infty))$ of $\pi^{\mathrm{aff}}(C)$ and $H_1(\Sigma_d \smallsetminus (C \cup E))$ of $\pi^{\mathrm{proj}}(C)$ are easily computable and depend on the irreducible components of C only, *cf.* (5.1) and (5.2), respectively. In the notation introduced prior to (5.1) (with $r_j = n_j$ for all $j = 1, \dots, m$), one has

$$H_1(\Sigma_d \smallsetminus (C \cup E \cup F_\infty)) = H^2(C) = \bigoplus \mathbb{Z} c_j,$$

and $H_1(\Sigma_d \smallsetminus (C \cup E))$ is the quotient of this group by $\sum_j n_j c_j$. Each geometric generator α_i of π_F corresponds to a certain point of intersection of C and F, which in turn belongs to a certain component C_j of C. In these terms, the orbits of the induced (*via* the permutation representation $\mathbb{B}_n \twoheadrightarrow \mathbb{S}_n$) action of *MG*$(C, F)$ on the set of conjugacy classes of geometric generators are in an obvious one-to-one correspondence with the irreducible components of C, and the abelianization epimorphism is given by $\alpha_i \mapsto c_j$. The degree homomorphism $\deg\colon \pi^{\mathrm{aff}}(C) \to \mathbb{Z}$ is the further composition of this map $\alpha_i \mapsto c_j$ with the map $c_j \mapsto 1 \in \mathbb{Z}$, $j = 1, \dots, m$.

The above description of the abelianization does not generalize to the group $\pi_+^{\mathrm{aff}}(C)$ of a curve C over a base B of positive genus. In this general case, the orbits of the permutation representation of the monodromy group *MG*(C, F) are still enumerated by the irreducible components of C, but this is no longer true for the restricted group *MG*$_+(C, F)$. For example, the isotrivial trigonal curve C with $j_C \equiv 0$ constructed in Remark 3.32 is irreducible, but one has *MG*$_+(C, F) = 0$ (as C has no singular fibers) and hence $\pi_+^{\mathrm{aff}}(C) = \pi_F \cong \mathbb{F}_3$.

We conclude this section with a simple lemma, which facilitates the computation of the fundamental groups.

Lemma 5.37. *In the presence of the relation* (5.35) *or, more generally, the relations* $[\rho^d, \alpha_i] = 1$ *for all* $i = 1, \dots, n$, *any one of the braid relations* $\mathfrak{m}(\gamma_j) = \mathrm{id}$ *in the presentation* (5.34) *can be omitted*: *it would follow from the others.*

Proof. In $\pi_1(\Omega^\sharp)$ one has $\gamma_1 \dots \gamma_k = [\partial\Omega]$. Hence, the group $MG(C, F)$ is generated by the images under $\mathfrak{m}$ of the class $[\partial\Omega]$ and all but any one generators γ_j. On the other hand, due to Proposition 5.16, the braid relation $\mathfrak{m}[\partial\Omega] = \mathrm{id}$ is equivalent to the relations $[\rho^d, \alpha_i] = 1$ for all $i = 1, \dots, n$, which are assumed present in the presentation. □

5.1.3 Improper curves: slopes

Now, let $\tilde{C}$ be an arbitrary, not necessarily proper, n-gonal curve in a ruled surface $\tilde{p}\colon \tilde{\Sigma} \to B$ with exceptional section $\tilde{E}$. As in the case of trigonal curves, the curve $\tilde{C}$ is Nagata equivalent to a certain proper n-gonal curve C in a ruled surface $p\colon \Sigma \to B$ with exceptional section E. Fix a monodromy domain $\Omega \subset B$ and let $b_1, \dots, b_k \in \Omega$ be the images of all singular fibers of $\tilde{C}$ and all singular fibers of C; we assume that all these points are in the interior of Ω. Denote $\tilde{F}_j = \tilde{p}^{-1}(b_j)$ and $F_j = p^{-1}(b_j)$, $j = 1, \dots, k$. Consider the punctured domain $\Omega^\sharp := \Omega \smallsetminus \{b_1, \dots, b_k\}$ and fix a basepoint $b \in \Omega^\sharp$. Let $\tilde{F} = \tilde{p}^{-1}(b)$ and $F = p^{-1}(b)$.

Fix a section $s\colon \Omega \to \Sigma$ proper *with respect to* C and consider the corresponding braid monodromy $\mathfrak{m} := \mathfrak{m}_s\colon \pi_1(\Omega^\sharp, b) \to \mathbb{B}(F)$ of C. The birational transformation converting $\tilde{C}$ to C establishes a diffeomorphism

$$\varphi\colon \tilde{p}^{-1}(\Omega^\sharp) \smallsetminus (\tilde{C} \cup \tilde{E}) \xrightarrow{\cong} p^{-1}(\Omega^\sharp) \smallsetminus (C \cup E) \tag{5.38}$$

and an isomorphism of the fundamental groups. In fact, φ also restricts to a diffeomorphism $\tilde{F}^\circ \smallsetminus \tilde{C} \to F^\circ \smallsetminus C$ and gives rise to a section $\tilde{s}\colon \Omega^\sharp \to \tilde{\Sigma} \smallsetminus (\tilde{C} \cup \tilde{E})$, which can be used to define the monodromy $\pi_1(\Omega^\sharp, b) \to \operatorname{Aut} \pi_1(\tilde{F}^\circ \smallsetminus \tilde{C}, \tilde{s}(b))$ of the original curve $\tilde{C}$; under the identifications made, the latter monodromy obviously coincides with $\mathfrak{m}$.

As an immediate consequence of the above discussion, we conclude that the group $\pi_1(\tilde{p}^{-1}(\Omega^\sharp) \smallsetminus (\tilde{C} \cup \tilde{E}))$ is still computed by Theorem 5.22 and the fibers $\tilde{F}_j$ (and the fiber $\tilde{F}_\infty$ at infinity) can be patched using Proposition 5.26. The difference from the previous case is the fact that, in general, the section $\tilde{s}$ does not extend through the punctures of $\Omega^\sharp$ (to a section disjoint from $\tilde{C}$ and $\tilde{E}$) and, hence, the lifts $\tilde{s}_*(\gamma_j)$ of the geometric generators of the group $\pi_1(\Omega^\sharp)$ do not bound disks; thus, the extra relations arising from patching the fibers are *not* $\gamma_j = 1$. In order to describe these new relations, we introduce the concept of slopes.

Local slopes

In the local setting, we fix one of the fibers $\tilde{F}_i$ and consider a small closed analytic disk $\tilde{\Phi} \subset \tilde{\Sigma}$ transversal to $\tilde{F}_i$ at a single point $0 \in \tilde{\Phi}$ and disjoint from $\tilde{C}$, $\tilde{E}$, and the other fibers $\tilde{F}_j$, $j \neq i$, *cf.* Proposition 5.26. Let $\Phi \subset \Sigma$ be the transform of $\tilde{\Phi}$; we can assume that the restriction of p to Φ is a one-to-one map into Ω. Let $\Phi' := p(\Phi)$, pick a basepoint $b'_i \in \partial\Phi'$, and denote $F'_i = p^{-1}(b'_i)$ and $\gamma'_i = [\partial\Phi'] \in \pi_1(\partial\Phi', b'_i)$. Let

$$\pi'_i := \pi_1((F'_i)^\circ \smallsetminus C, s(b'_i)) \quad \text{and} \quad \pi''_i := \pi_1((F'_i)^\circ \smallsetminus C; s(b'_i), F'_i \cap \Phi).$$

Consider the 'cylinders'

$$\tilde{X}_i := \tilde{p}^{-1}(\Phi') \smallsetminus (\tilde{C} \cup \tilde{E}), \qquad X_i := p^{-1}(\Phi') \smallsetminus (C \cup E)$$

and their diffeomorphic boundaries

$$\partial\tilde{X}_i \cong \partial X_i := p^{-1}(\partial\Phi') \smallsetminus (C \cup E).$$

The restriction $p\colon \partial X_i \to \partial\Phi'$ is a locally trivial fibration. It has two distinguished sections: the restriction of s and the section $s_i := (p|_{\partial\Phi})^{-1}$ whose image is $\partial\Phi$, and, similar to the absolute case, we can use this pair of sections to define the *relative monodromy* $\mathfrak{m}_i^{\mathrm{rel}} \in \operatorname{Aut}\pi''_i$ along γ'_i. Since the path product makes π''_i a principal left homogeneous space over π'_i, for any $\xi_i \in \pi''_i$ one can form the 'difference' $\varkappa'_i := (\xi_i \uparrow \mathfrak{m}_i^{\mathrm{rel}}) \cdot \xi_i^{-1} \in \pi'_i$, which is the only element of π'_i such that $\varkappa'_i \downarrow \xi_i = \xi_i \uparrow \mathfrak{m}_i^{\mathrm{rel}}$. This difference is called a *local slope* of $\tilde{C}$ at its fiber $\tilde{F}_i$.

For the next few statements, we assume fixed the curve $\tilde{C}$, its proper model C, a fiber $\tilde{F}_i$ over a point $b_i \in B$, a basepoint $b'_i \in B$ close to b_i, and a proper (with respect to C) section s defined in a closed regular neighborhood $\Phi' \ni b_i$ containing b'_i. Under these assumptions, we also have the local monodromy $\mathfrak{m}'_i \in \operatorname{Aut}\pi'_i$ along γ'_i, which is a braid. Similar to Theorem 5.22, one can see that the group $\pi_1(\partial X_i, s(b'_i))$ is the semidirect product $\pi'_i \rtimes \pi_1(\partial\Phi', b'_i)$, with $\pi_1(\partial\Phi', b'_i) = \langle\gamma'_i\rangle$ acting on π'_i *via* $\mathfrak{m}'_i$:

$$\pi_1(\partial X_i, s(b'_i)) = \pi'_i * \langle\gamma'_i\rangle / \{(\gamma'_i)^{-1}\alpha\gamma'_i = \alpha \uparrow \mathfrak{m}'_i\}, \tag{5.39}$$

the relations running over all $\alpha \in \pi'_i$. (We identify γ'_i and its image under s_*.)

Introduce the *twisted local monodromy* $\tilde{\mathfrak{m}}'_i := \mathfrak{m}'_i\varkappa'_i\colon \alpha \mapsto (\varkappa'_i)^{-1}(\alpha \uparrow \mathfrak{m}'_i)\varkappa'_i$ and the element $\tilde{\gamma}'_i := \gamma'_i\varkappa'_i \in \pi_1(\partial X_i, s(b'_i))$. In these terms, presentation (5.39) can be rewritten in the form

$$\pi_1(\partial X_i, s(b'_i)) = \pi'_i * \langle\tilde{\gamma}'_i\rangle / \{(\tilde{\gamma}'_i)^{-1}\alpha\tilde{\gamma}'_i = \alpha \uparrow \tilde{\mathfrak{m}}'_i\}. \tag{5.40}$$

Remark 5.41. According to Proposition 5.8, the twisted local monodromy $\tilde{\mathfrak{m}}'_i$ is an element of the extended group $\mathbb{B}(F'_i) \cdot \operatorname{Inn}\pi'_i \subset \operatorname{Aut}\pi'_i$, see page 55. Hence, due to Lemma 2.33, both $\mathfrak{m}'_i$ and $\varkappa'_i$ are recovered from $\tilde{\mathfrak{m}}'_i$ uniquely up to the transformation $(\mathfrak{m}'_i, \varkappa'_i) \mapsto (\Delta^2\mathfrak{m}'_i, \rho\varkappa'_i)$, which is induced by a positive Nagata transformation of C, *cf.* Lemma 5.17. Note also that $\tilde{\mathfrak{m}}'_i$ itself is independent of C.

Lemma 5.42. *Under the assumptions above, the local slope $\varkappa'_i \in \pi'_i$ is well defined up to the transformation $\varkappa'_i \mapsto (\alpha \uparrow \mathfrak{m}'_i) \cdot \varkappa'_i \cdot \alpha^{-1}$, $\alpha \in \pi'_i$.*

Proof. The ambiguity in the definition of $\varkappa'_i$ is in the choice of a transversal disk $\tilde{\Phi}$ over Φ' and an element $\xi_i \in \pi''_i$. Any two transversal disks are homotopic in the class of such disks, resulting in homotopic sections s_i and equal relative monodromies $\mathfrak{m}_i^{\mathrm{rel}}$. Any other element $\xi'_i \in \pi''_i$ has the form $\alpha \downarrow \xi_i$ for some $\alpha \in \pi'_i$. Since the π'_i-action on π''_i commutes with the monodromy, $(\alpha \downarrow \xi_i) \uparrow \mathfrak{m}_i^{\mathrm{rel}} = (\alpha \uparrow \mathfrak{m}'_i) \downarrow (\xi_i \uparrow \mathfrak{m}_i^{\mathrm{rel}})$, the slope defined *via* ξ'_i is $(\alpha \uparrow \mathfrak{m}'_i) \cdot \varkappa'_i \cdot \alpha^{-1}$, as stated. □

Corollary 5.43. *The twisted monodromy $\tilde{\mathfrak{m}}'_i$ is well defined up to the conjugation $\tilde{\mathfrak{m}}'_i \mapsto \alpha^{-1}\tilde{\mathfrak{m}}'_i\alpha$, $\alpha \in \operatorname{Inn}\pi'_i = \pi'_i$, and the image of $\tilde{\mathfrak{m}}'_i$ in the Burau group $\mathrm{Bu}_n(F'_i)$ does not depend on the choice of a proper model C or slope $\varkappa'_i$.* ◁

Corollary 5.44. *The relator subgroup $\operatorname{Rel}\tilde{\mathfrak{m}}'_i \subset \pi'_i$ does not depend on the choice of a slope $\varkappa'_i$. All local slopes project to a single element in the quotient $\pi'_i/\tilde{\mathfrak{m}}'_i$, and this element commutes with the image of ρ or any other element fixed by $\mathfrak{m}'_i$.*

Proof. The only statement that needs proof is the fact that the images of $\varkappa'_i$ and ρ commute. This follows from the identity $\rho \uparrow \mathfrak{m}'_i = \rho$. In fact, for any element $\alpha \in \pi'_i$ fixed by $\mathfrak{m}'_i$ one has $(\varkappa'_i)^{-1}\alpha\varkappa'_i = \alpha \uparrow \tilde{\mathfrak{m}}'_i = \alpha$ in $\pi'_i/\tilde{\mathfrak{m}}'_i$. □

For any transversal disk $\tilde{\Phi}$ over Φ' as above, the boundary $\partial\Phi$ of its transform Φ gives rise to a well defined conjugacy class in the group $\pi_1(\partial X_i, s(b'_i))$.

Lemma 5.45. *The conjugacy class in $\pi_1(\partial X_i, s(b'_i))$ defined by the boundaries of the transforms of transversal disks $\tilde{\Phi}$ over Φ' is $\{\gamma'_i\varkappa'_i\}$, where $\varkappa'_i$ runs over the set of all local slopes at $\tilde{F}_i$, see Lemma* 5.42.

Proof. The statement is essentially the definition of monodromy. Pick an element $\xi_i \in \pi''_i$. Then, representing homotopy classes by paths, the product

$$\xi_i \cdot [\partial\Phi] \cdot (\xi_i \uparrow \mathfrak{m}_i^{\mathrm{rel}})^{-1} \cdot (\gamma'_i)^{-1} \sim \xi_i[\partial\Phi]\xi_i^{-1} \cdot (\gamma'_i\varkappa'_i)^{-1}$$

is a null homotopic loop at $s(b'_i)$. The product $\xi_i[\partial\Phi]\xi_i^{-1}$ represents the conjugacy class in question. On the other hand, due to Lemma 5.42 and (5.39), all elements $\gamma'_i\varkappa'_i$, where $\varkappa'_i$ is a local slope, also constitute a whole conjugacy class. □

Lemma 5.46. *The inclusion homomorphism $\pi_1(\partial X_i, s(b'_i)) \to \pi_1(\tilde{X}_i, s(b'_i))$ is an epimorphism, and its kernel is normally generated by $\gamma'_i\varkappa'_i$, where $\varkappa'_i$ is any local slope. Hence, $\pi_1(\tilde{X}_i, s(b'_i)) = \pi'_i/\tilde{\mathfrak{m}}'_i$.*

Proof. By a local version of Proposition 5.26, the inclusion induces an epimorphism and its kernel is normally generated by the conjugacy class $[\partial\Phi]$. This class is given by Lemma 5.45, and the presentation of $\pi_1(\tilde{X}_i, s(b'_i))$ is obtained from (5.39) by introducing an extra relation $\gamma'_i\varkappa'_i = 1$ and eliminating γ'_i. □

Global slopes and fundamental groups

To make the construction of slopes global, fix a monodromy domain Ω, a common basepoint $b \in \Omega^\sharp$, and a geometric basis $\{\mu_1, \nu_1, \dots, \gamma_1, \dots, \gamma_k\}$ for the group $\pi_1(\Omega^\sharp, b)$ as in (5.23), (5.24). In the notation of the previous subsection, each basis element γ_i has the form $\zeta_i^{-1}\gamma_i'\zeta_i$, where $\zeta_i \in \pi_1(\Omega^\sharp; b_i', b_i)$.

Definition 5.47. The *(global) slope* of an n-gonal curve $\tilde{C}$ at its fiber $\tilde{F}_i$ is the element $\varkappa_i := \varkappa_i' \uparrow \mathfrak{m}(\zeta_i) \in \pi_F$, where $\varkappa_i'$ is a local slope and ζ_i is as above. The *twisted braid monodromy* is the homomorphism

$$\tilde{\mathfrak{m}} \colon \pi_1(\Omega^\sharp, b) \to \mathbb{B}(F) \cdot \operatorname{Inn} \pi_F \subset \operatorname{Aut} \pi_F$$

defined *via* $\tilde{\mathfrak{m}}(\mu_i) = \mathfrak{m}(\mu_i)$, $\tilde{\mathfrak{m}}(\nu_i) = \mathfrak{m}(\nu_i)$, and $\tilde{\mathfrak{m}}(\gamma_i) = \mathfrak{m}(\gamma_i) \cdot \operatorname{inn} \varkappa_i$.

Both the slopes $\varkappa_i$ and twisted monodromy $\tilde{\mathfrak{m}}$ depend on a number of choices, most important being the paths ξ_i in the definition of local slopes and paths ζ_i in Definition 5.47. We summarize this dependence in the following proposition.

Proposition 5.48. *Fix a proper model C and a geometric generator $\gamma_i \in \pi_1(\Omega^\sharp, b)$. Let $\mathfrak{m}_i := \mathfrak{m}(\gamma_i)$; note that this value does not depend on the choice of a proper section. Then:*

1. the slope $\varkappa_i$ is defined up to the move $\varkappa_i \mapsto (\alpha \uparrow \mathfrak{m}_i)\varkappa_i\alpha^{-1}$, $\alpha \in \pi_F$;
2. the value $\tilde{\mathfrak{m}}_i := \tilde{\mathfrak{m}}(\gamma_i)$ is defined up to the conjugation $\varkappa_i \mapsto \alpha\tilde{\mathfrak{m}}_i\alpha^{-1}$, $\alpha \in \operatorname{Inn} \pi_F$; this value does not depend on the choice of a proper model C;
3. the relator subgroup $\operatorname{Rel} \tilde{\mathfrak{m}}_i \subset \pi_F$ and the quotient group $\pi_F/\tilde{\mathfrak{m}}_i$ depend on the generator γ_i only; they are independent of C or the other choices;
4. the image of $\varkappa_i$ in $\pi_F/\tilde{\mathfrak{m}}_i$ depends on C and γ_i only; this image commutes with the image of ρ or any other element fixed by $\mathfrak{m}_i$.

Proof. As in the previous section, we use the prime notation for the local objects. The first two statements follow from Lemma 5.42 and Corollary 5.43. One should also observe that, since the group $\pi_1(\Omega^\sharp)$ is free, the path ζ_i in Definition 5.47 is determined by γ_i uniquely up to right multiplication by γ_i' and, on the other hand, the transformation $\varkappa_i' \mapsto \varkappa_i' \uparrow \mathfrak{m}_i' = (\varkappa_i' \uparrow \mathfrak{m}_i')\varkappa_i'(\varkappa_i')^{-1}$ is a special case of the transformation in Lemma 5.42. For the independence of C, note that a positive Nagata transformation of C at F_i acts *via* $(\mathfrak{m}_i, \varkappa_i) \mapsto (\Delta^2\mathfrak{m}_i, \rho\varkappa_i)$, *cf.* Remark 5.41, leaving $\tilde{\mathfrak{m}}_i$ intact. The last two statements follow from the first two, *cf.* Corollary 5.44. □

Assume that the monodromy domain Ω contains all singular fibers of $\tilde{C}$ and C and define the *twisted monodromy groups* as the images

$$MG^{\sim}(\tilde{C}, \tilde{F}) := \operatorname{Im} \tilde{\mathfrak{m}} \quad \text{and} \quad MG^{\sim}_{+}(\tilde{C}, \tilde{F}) := \tilde{\mathfrak{m}} \operatorname{Ker}[\pi_1(\Omega^\sharp, b) \to \pi_1(\Omega, b)].$$

The notation is slightly misleading as both groups depend on a number of choices. However, both the saturation Sat $MG^{\sim}_{+}(\tilde{C},\tilde{F})$ and the projection of $MG^{\sim}_{+}(\tilde{C},\tilde{F})$ to the Burau group $\mathrm{Bu}_n(\tilde{F})$ depend on the pair $(\tilde{C},\tilde{F})$ only, see Theorem 5.50 below and Proposition 5.48, respectively.

Pick a point $b_\infty \in B \smallsetminus \Omega$, *cf.* Remark 5.27. Due to the diffeomorphism (5.38), the group $\pi_1(\tilde{\Sigma} \smallsetminus (\tilde{C} \cup \tilde{E} \cup \bigcup_i \tilde{F}_i))$, $i = 1, \dots, k, \infty$, is still given by (5.24) and (5.25), and patching the fiber $\tilde{F}_\infty$ results in the relation at infinity (5.28). However, patching a finite fiber $\tilde{F}_j$, $j = 1, \dots, k$, gives a relation $\gamma_j = \varkappa_j^{-1}$, see Lemma 5.46, or, after eliminating γ_j, the *braid relations*

$$\alpha_i = \alpha_i \uparrow \tilde{\mathfrak{m}}(\gamma_j), \quad i = 1, \dots, n, \quad \text{or just} \quad \tilde{\mathfrak{m}}_j = \mathrm{id}. \tag{5.49}$$

Summarizing, we obtain the following version of Theorem 5.32 for improper curves.

Theorem 5.50. *If the base B is a compact curve of positive genus $g \geqslant 1$, then*

$$\pi_+^{\mathrm{aff}}(\tilde{C},\tilde{b}) = \pi_+^{\mathrm{proj}}(\tilde{C},\tilde{b}) = \pi_{\tilde{F}}/MG^{\sim}_{+}(\tilde{C},\tilde{F}),$$

where $\tilde{F}$ is the fiber containing $\tilde{b}$ and $\pi_{\tilde{F}} := \pi_1(\tilde{F}^\circ \smallsetminus \tilde{C},\tilde{b})$. Furthermore, one has

$$\pi^{\mathrm{aff}}(\tilde{C}) = \pi_+^{\mathrm{aff}}(\tilde{C}) \rtimes \pi_1(B \smallsetminus b_\infty),$$

with $\pi_1(B \smallsetminus b_\infty)$ acting on $\pi_+^{\mathrm{aff}}(\tilde{C})$ via the twisted braid monodromy $\tilde{\mathfrak{m}}$, and

$$\pi^{\mathrm{proj}}(\tilde{C}) = \pi^{\mathrm{aff}}(\tilde{C})/[\mu_1,\nu_1]\cdots[\mu_g,\nu_g]\rho^d\varkappa_1^{-1}\cdots\varkappa_k^{-1},$$

where $d = -E^2$. If the base is rational, then $\pi_+^{\mathrm{aff}}(\tilde{C})$ is as above and one has $\pi^{\mathrm{aff}}(\tilde{C}) = \pi_+^{\mathrm{aff}}(\tilde{C})$ and $\pi^{\mathrm{proj}}(\tilde{C}) = \pi_+^{\mathrm{proj}}(\tilde{C}) = \pi^{\mathrm{aff}}(\tilde{C})/\rho^d\varkappa_1^{-1}\cdots\varkappa_k^{-1}$. $\lhd$

Warning 5.51. In the relation at infinity in Theorem 5.50 one has $d = -E^2$, *not* $-\tilde{E}^2$, *i.e.*, the ruled surface containing the proper model C of $\tilde{C}$ is considered. According to Proposition 5.48, it is the product $\rho^d\varkappa_1^{-1}\cdots\varkappa_k^{-1}$ that is independent of C.

Remark 5.52. For the computations, it is usually more convenient to keep track of the ordinary braid monodromy $\mathfrak{m}$ and the slopes $\varkappa_i$ separately. Conceptually, all four groups can be expressed in terms of the twisted monodromy $\tilde{\mathfrak{m}}$, independent of a proper model. According to Proposition 5.16, the monodromy $\tilde{\mathfrak{m}}[\partial\Omega]$ is an inner automorphism of $\pi_{\tilde{F}}$; modulo the braid relations (which are expressed in terms of $\tilde{\mathfrak{m}}$ only), this automorphism is the conjugation by the element $\rho^{-d}\varkappa_k\cdots\varkappa_1$ appearing in the relation at infinity.

The presentation $\pi_+^{\mathrm{aff}}(\tilde{C},\tilde{b}) = \pi_{\tilde{F}}/MG^{\sim}_{+}(\tilde{C},\tilde{F})$ given by Theorem 5.50 (sometimes composed with the isomorphism $\pi_{\tilde{F}} = \mathbb{F}_n$ given by a proper geometric basis) is called a *geometric presentation* of $\pi_+^{\mathrm{aff}}(\tilde{C})$. As in the case of proper curves, it gives rise to a well defined *degree homomorphism* $\deg\colon \pi_+^{\mathrm{aff}}(\tilde{C}) \to \mathbb{Z}$.

Proposition 5.53 (Zariski [167]). *Assume that an n-gonal curve $\tilde{C}' \subset \tilde{\Sigma}' \to B'$ is obtained from an n-gonal curve $\tilde{C} \subset \tilde{\Sigma} \to B$ by a small perturbation. Then the slopes can be chosen so that $MG^{\sim}(\tilde{C}) \prec MG^{\sim}(\tilde{C}')$ and $MG^{\sim}_{+}(\tilde{C}) \prec MG^{\sim}_{+}(\tilde{C}')$. In particular, the group $\pi^{\mathrm{aff}}_{+}(\tilde{C}')$ is a geometric quotient of $\pi^{\mathrm{aff}}_{+}(\tilde{C})$.*

Proof. Fix a monodromy domain Ω for $\tilde{C}$ and a geometric basis $\{\mu_1, \nu_1, \dots, \gamma_1, \dots\}$ for the group $\pi_1(\Omega^\sharp)$. Realize each basis element μ_i, ν_i by an embedded circle, and each element γ_i, by a lasso about the corresponding singular fiber. These loops can be chosen pairwise disjoint except for the common basepoint. Let Ω^1 be the union of the images of the loops. Then Ω^1 is a strict deformation retract of $\Omega^\sharp$, and the twisted braid monodromy $\tilde{\mathfrak{m}}$ of $\tilde{C}$ can be defined as the ordinary braid monodromy, with respect to an appropriate section $\tilde{s}$, of the restriction of the bundle to Ω^1. This section $\tilde{s}$ can be constructed as follows: over the loop l_i of the lasso realizing an element γ_i, let $\tilde{s} = s_i$, where s_i is the section used in the definition of the local slope, see page 158, and extend the result to the rest of Ω^1 *via* a proper section s for the proper model C of $\tilde{C}$. (We can assume that s and s_i match over the common points.) We can use a similar construction for the perturbed curve $\tilde{C}'$, replacing each loop l_i with the disk $\bar{l}_i$ bounded by l_i, puncturing $\bar{l}_i$ at the singular fibers of $\tilde{C}'$ emerging from the corresponding singular fiber b_i of $\tilde{C}$, and restricting s_i to this punctured disk. Then, the rest of the proof repeats almost literally the proof of Proposition 5.14. □

The case of rational base

As in the proper case, we discuss a few properties of the groups $\pi^{\mathrm{aff}}(\tilde{C})$ and $\pi^{\mathrm{proj}}(\tilde{C})$ specific to curves over the rational base $B = \mathbb{P}^1$. Consider a curve $\tilde{C} \subset \Sigma_{\tilde{d}}$ and fix a proper model $C \subset \Sigma_d$. Consider a proper geometric basis $\{\alpha_1, \dots, \alpha_n\}$ for the group $\pi_F = \pi_{\tilde{F}}$ of a nonsingular fiber and a geometric basis $\{\gamma_1, \dots, \gamma_k\}$ for the group $\pi_1(\Omega^\sharp)$ of a monodromy domain punctured at all singular fibers of $\tilde{C}$ and C. Use the proper model C to define the braid monodromy $\mathfrak{m}$, slopes $\varkappa_1, \dots, \varkappa_k$, and twisted braid monodromy $\tilde{\mathfrak{m}}$. Then, according to Theorem 5.50, the affine fundamental group $\pi^{\mathrm{aff}}(\tilde{C}) = \pi^{\mathrm{aff}}_{+}(\tilde{C})$ has a geometric presentation of the form

$$\pi^{\mathrm{aff}}(\tilde{C}) = \left\langle \alpha_1, \dots, \alpha_n \;\middle|\; \tilde{\mathfrak{m}}(\gamma_j) = \mathrm{id},\ j = 1, \dots, k \right\rangle; \tag{5.54}$$

the projective group $\pi^{\mathrm{proj}}(\tilde{C}) = \pi^{\mathrm{proj}}_{+}(\tilde{C})$ is obtained from (5.54) by adding the *relation at infinity*

$$\rho^d = \varkappa_k \cdots \varkappa_1. \tag{5.55}$$

(Note the reverse order in the product of the slopes and the fact that d is the index of the Hirzebruch surface containing the proper model C, not the original curve $\tilde{C}$.)

Proposition 5.56. *The following statements hold:*

1. *the projection $\varkappa_i \in \pi^{\mathrm{aff}}(\tilde{C})$ does not depend on the choices in the construction; it only depends on the curves $\tilde{C}$ and C and the basis element γ_i;*

2. *in* $\pi^{\text{aff}}(\tilde{C})$ *one has* $[\varkappa_i, \rho] = 1$, $i = 1, \dots, k$;
3. *the element* $\rho^d \varkappa_1^{-1} \cdots \varkappa_k^{-1} \in \pi^{\text{aff}}(\tilde{C})$ *is central; it does not depend on* C;
4. *in* $\pi^{\text{aff}}(\tilde{C})$, *each slope* $\varkappa_i$ *commutes with the product* $\varkappa_k \cdots \varkappa_1$.

Proof. The first two statements follow from Proposition 5.48 (4). For the third one, let $\mathfrak{m}_i := \mathfrak{m}(\gamma_i)$ and $\tilde{\mathfrak{m}}_i := \tilde{\mathfrak{m}}(\gamma_i) = \mathfrak{m}_i \varkappa_i$. Each $\tilde{\mathfrak{m}}_i$ belongs to the monodromy group $M := MG\tilde{}(\tilde{C}, \tilde{F})$, hence so does the product $\tilde{\mathfrak{m}}_1 \dots \tilde{\mathfrak{m}}_k$. Due to Corollary A.26, one can move the $\varkappa_i$ factors, one by one, to the right; then, using Proposition 5.16, one concludes that $\Delta^{2d} \varkappa_k \cdots \varkappa_1 \in \operatorname{Sat} M$, which implies the statement of Item 3. (The independence of C is explained in Remark 5.41.) The last statement follows from Items 2 and 3. □

Remark 5.57. In view of Proposition 5.56 (4), the images of the slopes in the group $\pi^{\text{aff}}(\tilde{C})$ cyclically commute. Hence, if there are at most two nontrivial slopes $\varkappa_1$, $\varkappa_2$, the order of the factors in the relation at infinity (5.55) is irrelevant.

Corollary 5.58 (of Proposition 5.56 (3) and Corollary A.30). *The quotient projection* $\pi^{\text{aff}}(\tilde{C}) \to \pi^{\text{proj}}(\tilde{C})$ *restricts to an isomorphism of the commutants.* ◁

Lemma 5.59. *In the presence of the relation* (5.55) *or, more generally, the relations* $[\rho^d \varkappa_1^{-1} \cdots \varkappa_k^{-1}, \alpha_i] = 1$ *for all* $i = 1, \dots, n$, *one of the braid relations* $\tilde{\mathfrak{m}}(\gamma_j) = \text{id}$ *in the presentation* (5.54) *can be omitted: it would follow from the others.*

Proof. Let $\tilde{\mathfrak{m}}(\gamma_j) = \text{id}$ be the omitted relation. Denote $\mathfrak{m}_i := \mathfrak{m}(\gamma_i)$, $\tilde{\mathfrak{m}}_i := \tilde{\mathfrak{m}}(\gamma_i)$, and $M := \operatorname{Zar} \operatorname{Ker}[\pi_{\tilde{F}} \to \pi^{\text{aff}}(\tilde{C})]$. By the assumptions, $\tilde{\mathfrak{m}}_i = \mathfrak{m}_i \varkappa_i \in M$ for all $i \neq j$, hence also $\varkappa_i \mathfrak{m}_i \in M$, see Corollary A.26, and $\mathfrak{m}_1 \cdots \mathfrak{m}_k \varkappa_k \cdots \varkappa_1 \in M$, see Proposition 5.16. Using Corollary A.26 again, the last statement can be rewritten in the form $(\varkappa_1 \mathfrak{m}_1) \cdots (\varkappa_{j-1} \mathfrak{m}_{j-1}) \cdot \tilde{\mathfrak{m}}_j \cdot \tilde{\mathfrak{m}}_{j+1} \cdots \tilde{\mathfrak{m}}_k \in M$; hence $\tilde{\mathfrak{m}}_j \in M$. □

5.2 The case of trigonal curves

The Zariski–van Kampen method gives one a complete description of the fundamental groups related to an algebraic curve C in a geometrically ruled surface in terms of the braid monodromy of C. However, the computation of the latter is usually a difficult task, as one needs to trace the n roots of the defining equation $f(x,t) = 0$ of the curve varying the parameter t along a loop in the base. The purpose of this section is to show that, if C is a trigonal curve, the braid monodromy can easily be computed combinatorially in terms of the dessin of C.

5.2.1 Monodromy *via* skeletons

Let $C \subset \Sigma \to B$ be a non-isotrivial proper trigonal curve. According to Theorem 3.19 and Lemma 5.19, the Γ-valued braid monodromy $\mathfrak{m}\colon \pi_1(B^\sharp) \to \Gamma = \mathbb{B}_3/\Delta^2$ is induced from that of the universal cubic and hence is completely determined by the

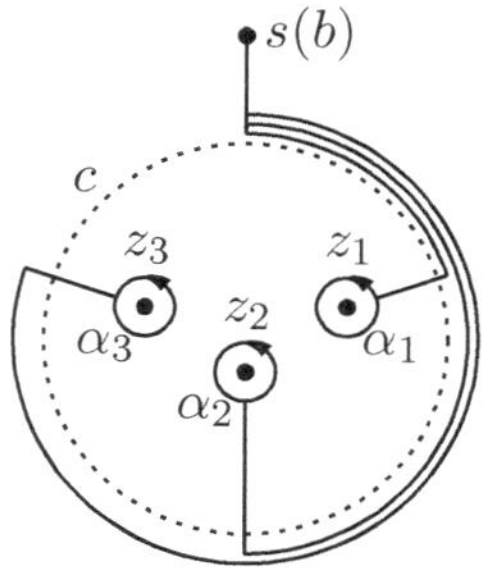

Figure 5.1. A canonical basis.

j-invariant $j := j_C$. (Recall that the Γ-valued braid monodromy is independent of the choice of a proper section, see Lemma 5.10, and factors through the fundamental group $\pi_1(B^\sharp)$, see Proposition 5.16.)

Let $\mathfrak{D} := \operatorname{Dssn} C$ and $\mathfrak{S} := \operatorname{Sk} C$. Throughout this section, *we assume that $\mathfrak{D}$ has no bold monochrome vertices, so that $\mathfrak{S}$ is a bipartite ribbon graph.* The inflation $|\operatorname{Inf}\mathfrak{S}|$ is naturally, up to isotopy, embedded into $B^\sharp$; hence, any chain in $\mathfrak{S}$ represents a homotopy class of paths in $B^\sharp$. Pick a point b inside an edge of $\mathfrak{S}$ and let $F := p^{-1}(b)$. According to Proposition 3.13, the three points z_1, z_2, z_3 of the intersection $C \cap F$ form an isosceles triangle T with an angle greater than $\pi/3$ at the vertex; important is the fact that T is not equilateral and hence has a clearly distinguished vertex. These points can be numbered canonically, so that the orientation z_1, z_2, z_3 of T is *clockwise* and z_2 is the vertex of T, as shown in Figure 5.1.

Consider a circle $c \subset F$ centered at the barycenter of T and containing T inside. We can assume that the basepoint $s(b)$ is outside of c. A *canonical basis* is a proper geometric basis $\{\alpha_1, \alpha_2, \alpha_3\}$ for $\pi_F := \pi_1(F^\circ \smallsetminus \{z_1, z_2, z_3\}, s(b))$ defined as follows: the element α_i is represented by the lasso formed by a small counterclockwise circle about z_i, $i = 1, 2, 3$, connected to $s(b)$ by a radial segment, an arc l_i of c, and another radial segment common to all three elements, see Figure 5.1. We require that each consecutive arc l_i, $i = 2, 3$, is longer than its predecessor l_{i-1} by a positive angle that is less than 2π. However, we do not make any assumption about the first arc l_1; thus, a canonical basis is well defined up to simultaneous conjugation by ρ, *i.e.*, up to the action of the central element $\Delta^2 \in \mathbb{B}_3$.

Convention 5.60. From now on, when speaking about the braid monodromy of a non-isotrivial trigonal curve C, we always assume that the basepoint b is chosen inside a bold edge e of the dessin $\operatorname{Dssn} C$ and that the basis for the free group π_F is canonical. The edge e is called the *reference edge*.

Remark 5.61. Originally, in [52], a base point b was chosen at a nonsingular •-vertex of the dessin of the curve, where the three points z_1, z_2, z_3 form an equilateral triangle.

Under this convention, the definition of canonical basis required an extra structure, called *marking*, which was essentially a choice of one of the bold edges incident to the vertex. Thus, the approach adopted above is equivalent to that of [52], with the extra advantage that no additional structure is required and one does not need to avoid singular •-vertices.

By means of a canonical basis, we can identify the group π_F with the free group $\mathbb{F}_3$; up to the action of Δ^2, this identification is determined by the base point b. Hence, given two points b and b', each inside an edge of $\mathfrak{S}$, and a path ζ connecting b to b' in $B^\sharp$, the monodromy $\mathfrak{m}(\zeta)\colon \pi_F \to \pi_{F'}$ composed with these identifications gives rise to a well defined element $\mathfrak{m}(\zeta) \bmod \Delta^2 \in \mathbb{B}_3/\Delta^2 = \Gamma$. In particular, this construction can be applied to a chain in $\mathfrak{S}$. Our principal result is the following theorem, computing $\mathfrak{m}(\zeta) \bmod \Delta^2$ in terms of $\mathfrak{S}$.

Theorem 5.62. *Let C be a proper non-isotrivial trigonal curve, and let ζ be a chain in the skeleton* $\operatorname{Sk} C$. *Then the monodromy* $\mathfrak{m}(\zeta) \bmod \Delta^2 \in \Gamma$ *is the image of* $\operatorname{val}\zeta$ *under the quotient projection* $\mathbb{G} \to \mathbb{G}_{(3,2)} = \Gamma$.

Proof. The statement can easily be proved using the properties of the j-invariant given by Propositions 3.13 and 3.16. Alternatively, in view of Theorem 3.19 and Lemma 5.19, it suffices to compute the monodromy for the universal cubic $\bar{C}$. Since the skeleton $\operatorname{Sk}\bar{C}$ has a single edge $[0,1] \in \mathbb{P}^1_{\mathbb{R}}$, we only need the monodromy along the two loops $\gamma_1\colon t \mapsto (1/2)\exp(2\pi i t)$ and $\gamma_2\colon t \mapsto 1 - (1/2)\exp(2\pi i t)$, $t \in I$. A simple computation shows that $\mathfrak{m}(\gamma_1) = \sigma_1\sigma_2$ and $\mathfrak{m}(\gamma_2) = \sigma_1\sigma_2\sigma_1$, and under the identification $\mathbb{B}_3/\Delta^2 = \Gamma$ fixed in Section 2.2.3, see (2.39), these elements project to $\mathbb{X}$ and $\mathbb{Y}$, respectively. □

A certain lift to $\mathbb{B}_3$ of the monodromy $\mathfrak{m}(\zeta)$ along a chain ζ is computed by function "`Path`" in Listing C.1. The argument can be given either as an element of "`Gamma`", *e.g.*, `x^-1*y*x`, or, for easier input, as a list with entries ± 1 (for $\mathbb{X}^{\pm 1}$) or 0 (for $\mathbb{Y}$).

Corollary 5.63. *Let $C \subset \Sigma \to B$ be a maximal trigonal curve without singular fibers of type $\tilde{\mathbf{E}}_{6k+2}$, $k \geqslant 0$. Then one has $MG(C)/\Delta^2 = \operatorname{Stab}_{(3,2)} \operatorname{Sk} C$.*

Proof. Under the assumptions, $\mathfrak{S} := \operatorname{Sk} C$ is a skeleton of type $(3,2)$ and there is a homeomorphism $B \cong \operatorname{Supp}\mathfrak{S}$. Let $B^{\sharp\sharp}$ be the base B punctured at all vertices of the dessin $\operatorname{Dssn} C$. Then the inclusions $|\operatorname{Inf}\mathfrak{S}| \hookrightarrow B^{\sharp\sharp} \hookrightarrow B^\sharp$, *cf.* Lemma 1.31, induce an epimorphism $\pi_1(|\operatorname{Inf}\mathfrak{S}|) \twoheadrightarrow \pi_1(B^\sharp)$, and the statement of the corollary follows from Theorems 1.36 and 5.62. □

Extension to the regions

In general, the inclusion homomorphism $\pi_1(|\operatorname{Inf}\mathfrak{S}|) \to \pi_1(B^\sharp)$ is not onto. In order to extend the braid monodromy to the whole group $\pi_1(B^\sharp)$, consider the cut $\operatorname{Cut}\mathfrak{S}$

and, for each region R of the cut, let $R^\sharp$ be R with the essential vertices of $\mathfrak{D}$ removed. Consider a basepoint b inside an edge e of $\mathfrak{S}$. The interior of e has two copies $e_\pm$ in the boundary $\partial \operatorname{Cut} \mathfrak{S}$, which are contained in two regions $R_\pm$, not necessarily distinct. Hence, we also have two copies $b_\pm \in \partial R_\pm$. To distinguish between the copies, we will use the following convention.

Definition 5.64. The basepoint (edge, region) $b_+ \in e_+ \subset R_+ \subset \operatorname{Cut} \mathfrak{S}$ are said to be *positive* with respect to e or b if, for a sufficiently small neighborhood U of b_+ in R_+, the j-invariant j_C has positive imaginary part on $U \smallsetminus \partial R_+$. Otherwise, the basepoint (edge, region) $b_- \in e_- \subset R_-$ are said to be *negative*. The positive (negative) region is also said to lie *to the left* (respectively, *to the right*) from e or b. For the regions, we will use the notation $R_\pm = \operatorname{reg}_\pm e$.

We emphasize once again that the regions R_+ and R_- may coincide, whereas the edges $e_\pm$ and basepoints $b_\pm$ are always distinct. With respect to the complex orientation of the surface B and the canonical orientation $\bullet\!\!\rightarrow\!\!\circ$ of the edge e, the positive region is indeed to the left, as one can easily check considering the universal cubic. In the special case when $\mathfrak{S}$ is a skeleton (*i.e.*, connected) and $B \cong \operatorname{Supp} \mathfrak{S}$ (*i.e.*, all regions are disks), one has $R_+ = \operatorname{reg} e$, see Figure 1.2.

Given a region R, we denote by $\partial_+ R$ and $\partial_- R$ the unions of the open bold edges in the boundary ∂R with respect to which R is positive (respectively, negative). Let, further, $\bar\partial_\pm R$ be the closure of $\partial_\pm R$, *i.e.*, the union of corresponding closed edges.

Consider a region R of the cut $\operatorname{Cut} \mathfrak{S}$ and let $R_\times$ be the union of the $\times$-vertices of $\mathfrak{D}$ that are in R. The closure of the solid part $\mathfrak{D}_{\text{solid}}$, intersected with R, defines a certain class $[\mathfrak{D}_{\text{solid}}] \in H_1(R, R_\times \cup \partial R) = H^1(R^\sharp)$ and, hence, a homomorphism $H_1(R^\sharp) \to \mathbb{Z}$ given by $\gamma \mapsto \gamma \circ [\mathfrak{D}_{\text{solid}}]$. Similarly, the dotted part of $\mathfrak{D}$ gives rise to a class $[\mathfrak{D}_{\text{dotted}}]$ and homomorphism $\gamma \mapsto \gamma \circ [\mathfrak{D}_{\text{dotted}}]$. According to Item 3 in Definition 4.1, one has $[\mathfrak{D}_{\text{solid}}] + [\mathfrak{D}_{\text{dotted}}] = 0$ in $H_1(R, R_\times \cup \partial R)$.

Theorem 5.65. *The Γ-valued braid monodromy* $\mathfrak{m}\colon \pi_1(B^\sharp, b) \to \Gamma$ *restricted to the group* $\pi_1(R_+^\sharp, b_+)$ *or* $\pi_1(R_-^\sharp, b_-)$ *is given by* $\gamma \mapsto (\mathbb{X}\mathbb{Y})^{i(\gamma)}$ *or* $\gamma \mapsto (\mathbb{Y}\mathbb{X})^{i(\gamma)}$, *respectively, where* $i(\gamma) := [\gamma] \circ [\mathfrak{D}_{\text{solid}}] = -[\gamma] \circ [\mathfrak{D}_{\text{dotted}}]$. *In particular, the restriction takes values in the cyclic subgroup* $\langle \mathbb{X}\mathbb{Y} \rangle \subset \Gamma$ *(respectively, in the cyclic subgroup* $\langle \mathbb{Y}\mathbb{X} \rangle \subset \Gamma$*).*

Proof. We will prove the statement for $R := R_+$. According to Propositions 3.13 and 3.16, inside any *inner* point $b' \in R^\sharp \smallsetminus \partial R$, one of the three points z_1', z_2', z_3' of the intersection $C \cap p^{-1}(b')$ is distinguished by the condition that it opposes the shortest edge of the triangle formed by the points. (Even if the triangle degenerates to a segment, the condition still makes sense, and it is the closure of $\mathfrak{D}_{\text{bold}}$ where the shortest edge may change.) Comparing Definition 5.64 and the definition of a canonical basis (and using Propositions 3.13 and 3.16 again), one concludes that this distinguished point tends to z_1 over b when b' approaches b_+. (It is this part of the proof that makes the difference between R_+ and R_-: the distinguished point tends to z_3 when b' approaches b_-.)

Consider a loop γ at b_+. Step away from the boundary and assume that γ is in the interior of the region. Then, up to an affine shift of the fiber, the three points of C over a point b' on γ are of the form $\pm z'$ and $z'_1 \neq 0$. Making another change of coordinates (and thus disregarding the possible action of Δ^2), one can assume that $z'_1 = 1$. Then $2|z'| < |z' \pm 1|$, *i.e.*, the points $\pm z'$ belong to the intersection of the disks $|z' \pm \frac{1}{3}| < \frac{2}{3}$, and the three points $1, \pm z'$ line up whenever γ crosses the dotted part of $\mathfrak{D}$, see Proposition 3.13; the direction of the rotation is given by Proposition 3.16. Now, the statement of the theorem follows from the geometric definition of a braid. (Recall that, according to our convention (2.39), the element $(\mathbb{X}\mathbb{Y})^{-1} \in \Gamma$ is the image of the Artin generator $\sigma_2 \in \mathbb{B}_3$ and, since the point z_1 remains distinguished, it is z_2 and z_3 that get intertwined by the braid. For the other region R_-, the distinguished point is z_3 and the 'acting braid' is $\sigma_1 \mapsto (\mathbb{Y}\mathbb{X})^{-1}$.) □

The restriction of the monodromy to the kernel of the canonical homomorphism $\pi_1(R^\sharp) \to H_1(R^\sharp) \to H_1(R)$ depends on the set of the $\times$-vertices of $\mathfrak{D}$ only. Let $D_\times$ be the divisor of the $\times$-vertices of $\mathfrak{D}$ restricted to R, *i.e.*, $D_\times = \sum_v (\operatorname{ind} v) v$, the summation running over all $\times$-vertices of $\mathfrak{D}$ that are in R. Then $D_\times = \partial[\mathfrak{D}_{\text{dotted}}]$ modulo ∂R and hence, for any loop γ in $R^\sharp$ that is null-homologous in R, one has $[\gamma] \circ [\mathfrak{D}_{\text{dotted}}] = \operatorname{lk}(\gamma, D_\times)$, essentially by the definition of the linking coefficient lk. If R is a disk and $p \in R$ is a single point, then $\operatorname{lk}(\gamma, p)$ is what is usually called the *winding number* of γ about p; we can extend this definition to divisors by additivity. The most important special case is covered by the following corollary.

Corollary 5.66. *For the class $\gamma \in \pi_1(R^\sharp_\pm, b_\pm)$ realized by a lasso about a $\times$-vertex v, the reduced monodromy $\mathfrak{m}(\gamma) \bmod \Delta^2 \in \Gamma$ is equal to $(\mathbb{X}\mathbb{Y})^{-\operatorname{ind} v}$ (for the positive region R_+, b_+) or $(\mathbb{Y}\mathbb{X})^{-\operatorname{ind} v}$ (for the negative region R_-, b_-).* ◁

Disconnected boundary

For the sake of completeness, we consider the case of a region with disconnected boundary as well. Due to Poincaré duality, there are isomorphisms

$$H_1(R, R_\times \cup \bar{\partial}_\mp R) = H^1(R^\sharp, \partial_\pm R) = \operatorname{Hom}(H_1(R^\sharp, \partial_\pm R), \mathbb{Z})$$

and we have classes $[\mathfrak{D}_{\text{solid}}] = -[\mathfrak{D}_{\text{dotted}}]$ in $H_1(R, R_\times \cup \bar{\partial}_\mp R)$ and homomorphisms $H_1(R^\sharp, \partial_\pm R) \to \mathbb{Z}$, $\gamma \mapsto \gamma \circ [\mathfrak{D}_*]$. The following statement is an immediate generalization of Theorem 5.65. We consider the case of positive region only; for a negative region, one should replace $\mathbb{X}\mathbb{Y}$ with $\mathbb{Y}\mathbb{X}$ in the statement.

Proposition 5.67. *Let R be a region, and let ζ be a path in $R^\sharp$ with the endpoints in $\partial_+ R$. Then $\mathfrak{m}(\zeta) \bmod \Delta^2 = (\mathbb{X}\mathbb{Y})^{i(\zeta)}$, where $i(\zeta) := [\zeta] \circ [\mathfrak{D}_{\text{solid}}]$.* ◁

Due to the Seifert–van Kampen theorem, any loop in $B^\sharp$ is homotopic to a product of paths as in Theorem 5.62 and Proposition 5.67; hence, these two statements give one a complete description of the Γ-valued monodromy.

By deforming the curve and introducing bold monochrome vertices (roughly, by contracting appropriate paths ζ as in Proposition 5.67), one can make the skeleton $\operatorname{Sk} C$ connected and incorporate Proposition 5.67 into Theorem 5.62. However, this approach would require an extension of the notions of inflation, chain, evaluation homomorphism, *etc.* to the more general skeletons with monochrome vertices. For this reason, we leave this simple exercise to the reader.

The lift to $\mathbb{B}_3$

To complete the computation of the braid monodromy, we need to describe its lift from Γ to $\mathbb{B}_3$. Recall that this lift is only well defined if a nonsingular fiber $b_\infty \in B$ is specified, and it is only meaningful on the kernel of the natural homomorphism

$$\pi_1(B^{\sharp\sharp}, b) \to H_1(B^{\sharp\sharp}) \to H_1(B \smallsetminus b_\infty) = H_1(B),$$

as otherwise the lift depends on the choice of a proper section. (Here, $B^{\sharp\sharp} = B^\sharp \smallsetminus b_\infty$.) Since a braid $\beta \in \mathbb{B}_3$ is uniquely determined by its projection to Γ and its degree $\operatorname{dg} \beta \in \mathbb{Z}$, it suffices to describe the homomorphism $\operatorname{dg} \mathfrak{m} \colon K \to \mathbb{Z}$, where

$$K := \operatorname{Ker}[H_1(B^{\sharp\sharp}) \to H_1(B \smallsetminus b_\infty)] = H_2(B \smallsetminus b_\infty, B^{\sharp\sharp}) = H^0(\{b_1, \ldots, b_k\})$$

and $b_1, \ldots, b_k$ are the singular fibers of C. (For the identifications, we use the exact sequence of pair $(B \smallsetminus b_\infty, B^{\sharp\sharp})$ and Poincaré–Lefschetz duality.) In other words, $\operatorname{dg} \mathfrak{m}$ can be regarded as an element of the dual group $H_0(\{b_1, \ldots, b_k\})$.

Theorem 5.68. *One has* $\operatorname{dg} \mathfrak{m} = \sum_i (\operatorname{mult} b_i) b_i$ *in* $H_0(\{b_1, \ldots, b_k\})$, *where* $\operatorname{mult} b_i$ *is the multiplicity of the singular fiber* b_i *of* C, *see Table* 3.1.

In other words, for a lasso γ about b_i one has $\operatorname{dg} \mathfrak{m}(\gamma) = \operatorname{mult} b_i$. Since b_i can be perturbed to $\operatorname{mult} b_i$ type $\tilde{\mathbf{A}}_0^*$ singular fibers and the monodromy about a type $\tilde{\mathbf{A}}_0^*$ fiber is conjugate to σ_1, this observation constitutes a proof of Theorem 5.68. □

Isotrivial curves

As usual, we consider separately the special case of an isotrivial trigonal curve C. In this case, all triples $\{z_1, z_2, z_3\} = C \cap F^\circ$ in nonsingular fibers are analytically isomorphic and, hence, the reduced braid monodromy $\mathfrak{m} \bmod \Delta^2$ takes values in the group of automorphisms of any such triple, which is $\mathbb{Z}_3$ generated by $\mathbb{X}$ (if $j_C \equiv 0$), $\mathbb{Z}_2$ generated by $\mathbb{Y}$ (if $j_C \equiv 1$), or 0 (if $j_C \not\equiv 0, 1$). The restriction to $MG(C)/\Delta^2$ of the permutation representation $s \colon \Gamma \to \mathbb{S}_3$ is a monomorphism, and the composition $s \circ \mathfrak{m}$ is the ordinary monodromy of the ramified covering $C \to B$.

For the lift to $\mathbb{B}_3$, one can use Theorem 5.68, which still applies to isotrivial curves. A basis $\{\alpha_1, \alpha_2, \alpha_3\}$ in (any) nonsingular fiber F can be constructed as in Figure 5.1, with the only difference that, if $j_C \equiv 0$ or 1, the numbering of the points is not quite well defined (and then is irrelevant). Then $MG_+(C, F)$ lies in the cyclic subgroup generated, respectively, by $\sigma_1\sigma_2$, $\sigma_1\sigma_2\sigma_1$, or Δ^2 if $j_C = 0$, 1, or otherwise. Furthermore, the divisor $\sum_i(\operatorname{mult} b_i)b_i$ in Theorem 5.68 equals $2G_3 = 3G_2 = 6G$, where G_3, G_2, and G are the divisors (whichever makes sense for the particular value of the j-invariant) introduced in the classification of isotrivial curves on page 75.

The homological invariant of an elliptic surface

Let $\tilde{F}^\circ$ be the double covering of a fiber F° ramified at $F \cap C$; it is a punctured torus. A canonical basis for π_F gives rise to a basis $\mathbf{a} := [\alpha_2\alpha_1^{-1}]$, $\mathbf{b} := [\alpha_3\alpha_2^{-1}]$ for $H_1(\tilde{F}^\circ)$ (*cf.* Section 2.2.2), identifying the latter group with $\mathcal{H}$ and the induced $\mathbb{B}_3$-action on $H_1(\tilde{F}^\circ)$ with the specialization of the Burau representation at $t = -1$. The conjugation by Δ^2 multiplies $\{\mathbf{a}, \mathbf{b}\}$ by (-1) and, comparing the definitions, we arrive at the following statement, relating the homological invariant of an elliptic surface and the monodromy of its ramification curve.

Proposition 5.69. *Let $J \to B$ be a Jacobian elliptic surface and $C \subset \Sigma \to B$ its ramification curve. Then, in the bases $\{\alpha_1, \alpha_2, \alpha_3\}$ and $\{\mathbf{a}, \mathbf{b}\}$ related as explained above, one has $\mathfrak{h}_J \bmod \pm\mathrm{id} = \mathfrak{m} \bmod \Delta^2$, where $\mathfrak{m}$ is the braid monodromy of C with respect to any proper section. If B is rational, then also $\mathfrak{h}_J = \mathfrak{m} \bmod \Delta^4$.* $\lhd$

Remark 5.70. Since the reduction $\mathfrak{m} \bmod \Delta^2$ is completely determined by $j_C = j_J$, Proposition 5.69 gives us another interpretation of the fact that the homological invariant of an elliptic surface belongs to its j-invariant.

5.2.2 Slopes

Slopes of improper trigonal curves also admit a simple combinatorial description. For a simple way to describe the topological type of an improper singular fiber $\tilde{F}$, we refer to the type of its proper model F (as explained below, F can always be assumed of type $\tilde{\mathbf{A}}$) and the sequence of the blow-up centers of the inverse Nagata transformations converting F back to $\tilde{F}$.

We start with two simple observations. Consider a trigonal curve C and let C' be the curve obtained from C by one negative Nagata transformation centered at a point P in a fiber F. Then

- if $P \notin C$, then the slopes of C and C' at F are equal;
- if P is a triple point of C, then both C and C' are proper and, hence, C' can be used as a proper model of the original improper curve, *cf.* Remark 5.41.

For this reason, we always assume that $P \in C$ and that this point has multiplicity at most two. In particular, we consider proper models with $\tilde{\mathbf{A}}$ type singular fibers only and we do not continue blow-ups after the curve becomes 'totally improper', *i.e.*, intersects the fiber at its only intersection point with the exceptional section.

The former observation deserves a separate statement.

Proposition 5.71. *Assume that a trigonal curve $\tilde{C}$ has a triple singular point P at infinity, and let C be the curve obtained from $\tilde{C}$ by one positive Nagata transformation centered at P. Then $\pi_+^{\mathrm{aff}}(\tilde{C}) = \pi_+^{\mathrm{aff}}(C)$ and $\pi_+^{\mathrm{proj}}(\tilde{C}) = \pi_+^{\mathrm{proj}}(C)$.* ◁

Thus, we fix a proper trigonal curve C with a type $\tilde{\mathbf{A}}$ (Kodaira's I–IV) singular fiber F and consider a curve $\tilde{C}$ obtained from C by a sequence of negative Nagata transformations with all blow-up centers $P_1 \leftarrow P_2 \leftarrow \cdots \leftarrow P_n$ in (the consecutive transforms of) C. For the blow-up centers, we use the classical language of infinitely near points; for example, we say that P_i lies in a branch $\mathfrak{b}$ of C if this point belongs to the appropriate transform of $\mathfrak{b}$. For each type of F and sequence of blow-ups, we describe an appropriate canonical basis for the local fundamental group and, in this basis, the slope $\varkappa$ of $\tilde{C}$ and the braid relations $\tilde{\mathfrak{m}} = \mathrm{id}$, *cf.* (5.49). With Corollary A.30 in mind, we also try to indicate some central elements in the quotient. In each case, we merely state the final result, leaving technical details to the reader; all computations can easily be done using the local normal forms of C, see (3.26) and (3.27), and of an analytic branch intersecting C at the blow-up centers, including infinitely near.

Non-exceptional singular fibers (Kodaira's type I_m)

A type I_m singular fiber is located over a $\times$-vertex v of the dessin $\mathfrak{D} := \operatorname{Dssn} C$. (In the special case $m = 0$, the fiber is 'freely movable' and we assume it inside a region R_v describe below.) Let R_v be any negative, with respect to the standard chessboard coloring, region of the cut $\operatorname{Cut}\mathfrak{D}$ containing v in its boundary. Then R_v is to the right from any bold edge $e \in \partial R_v$, and we take for $\{\alpha_1, \alpha_2, \alpha_3\}$ a canonical basis over an inner point of such an edge. Over R_v, the curve C has three separate branches $\mathfrak{b}_i$ corresponding to the basis elements α_i, $i = 1, 2, 3$, see Section 5.2.1, so that $\mathfrak{b}_3$ remains a nonsingular branch over v whereas $\mathfrak{b}_1$ and $\mathfrak{b}_2$ collide at a singular point (if $m > 0$). If m is even, $\mathfrak{b}_1$ and $\mathfrak{b}_2$ remain separate analytic branches over v; if m is odd, $\mathfrak{b}_1$ and $\mathfrak{b}_2$ belong to a single analytic branch.

The local monodromy $\mathfrak{m}$ about v is σ_1^m, see Theorems 5.65 and 5.68. It follows that, modulo the braid relation $\tilde{\mathfrak{m}} = \mathrm{id}$, one has

$$[\varkappa, \rho] = [\varkappa, \alpha_1\alpha_2] = [\varkappa, \alpha_3] = 1,$$

see Proposition 5.48 (4). We consider separately the following four cases.

Case 1: $m \geqslant 0$ and $P_1 \in \mathfrak{b}_3$ In this case, all blow-up centers P_i belong to $\mathfrak{b}_3$ and we have

$$\varkappa = \alpha_3^n \quad \text{and} \quad \alpha_3^n \alpha_i \alpha_3^{-n} = \alpha_i \uparrow \sigma_1^m,\ i = 1, 2. \tag{5.72}$$

Letting $n = 0$ or -1, we formally obtain the braid relations for a proper singular fiber of type $\tilde{\mathbf{A}}_{m-1}$ or $\tilde{\mathbf{D}}_{m+2}$, respectively (with $\varkappa = 1$ in both cases). If $n = \pm 1$, the product $(\alpha_1\alpha_2)^m\alpha_3^{-2n}$ is a central element and, if $m = 2k$ is even, then $(\alpha_1\alpha_2)^k\alpha_3^{-n}$ is also a central element.

Case 2: $m \geqslant 2n \geqslant 0$ and $P_1 \in \mathfrak{b}_1$ If $n = 0$, the fiber is proper. All blow-up centers are double points of C (shared by $\mathfrak{b}_1$ and $\mathfrak{b}_2$), and we have

$$\varkappa = (\alpha_1\alpha_2)^n \quad \text{and} \quad \{\alpha_1, \alpha_2\}_{m-2n} = [(\alpha_1\alpha_2)^n, \alpha_3] = 1. \tag{5.73}$$

If $m > 2n > 0$, then $(\alpha_1\alpha_2)^s$ is a central element, where $s := \text{l.c.m.}(m - 2n, n)$.

Case 3: $m = 2k + 1 > 0$, $n = k + 1$, and $P_1 \in \mathfrak{b}_1$ The blow-up centers $P_1 \leftarrow \cdots \leftarrow P_k$ are double points of C, and P_{k+1} is a simple point in the branch $\mathfrak{b}_1 = \mathfrak{b}_2$. Note that no further negative Nagata transformation is meaningful, as the curve becomes totally improper. After simplifications, we have

$$\varkappa = \alpha_1^m \quad \text{and} \quad \alpha_1 = \alpha_2, \quad [\alpha_1^m, \alpha_3] = 1. \tag{5.74}$$

Hence, α_1^m is a central element.

Case 4: $m = 2k \geqslant 0$, $n > k$, and $P_n \in \mathfrak{b}_j$, $j = 1, 2$ The first k blow-up centers are double points of C, and the remaining $n - k$ ones are simple points in the same branch, $\mathfrak{b}_1$ or $\mathfrak{b}_2$. We have

$$\varkappa = (\alpha_1\alpha_2)^k\alpha_j^{n-k} \quad \text{and} \quad [\alpha_j^{n-k}, \alpha_i] = [\varkappa, \alpha_3] = 1, \ i = 1, 2. \tag{5.75}$$

If $k = 0$, then α_j^{n-k} is a central element. If $n - k = 1$, then $[\alpha_1, \alpha_2] = 1$ and $\varkappa$ is a central element. In general, the group does not seem to have an obvious center.

Remark 5.76. Let R be a region of $\text{Cut}(\text{Sk}\, C)$, and let e be an edge in $\partial_- R$. Then e can be used as a common reference edge for all singular fibers inside R, and one does not need to consider separate regions R_v. Indeed, due to Theorem 5.65, over R the braid monodromy takes values in the abelian subgroup generated by σ_1 and Δ^2, and all relations except (5.75) are invariant under this subgroup. If R contains a $\times$-vertex of odd index, then one can take $j = 1$ in all relations (5.75); otherwise, the curve has three separate analytic branches over R and the parameter $j = 1$ or 2 at each fiber as in Case 4 is well defined.

Remark 5.77. Choosing a common reference edge e in $\partial_- R$ as in Remark 5.76, one can combine and simplify the relations arising from all singular fibers inside R. Groups obtained in this way are studied in details in Appendix B. As a very simple observation, assume that R contains a number of singular fibers as in Case 2, and define the *braid index* of R as $s := \text{g.c.d.}(m_i - 2n_i)$, where $(m_1, n_1), \ldots$ are the corresponding parameters. (If R contains a fiber as in Case 3, we can let $s = 1$.) Then, amongst others, the fundamental group has a relation $\{\alpha_1, \alpha_2\}_s = 1$, see (A.27). We mainly use this observation when studying perturbations, when a single type $\tilde{\mathbf{A}}_{m-1}$ proper singular fiber inside R splits into fibers of types $\tilde{\mathbf{A}}_{m_1-1}$, $\tilde{\mathbf{A}}_{m_2-1}$, *etc.* In this case, the new braid index of R is $s = \text{g.c.d.}(m_i)$.

In the case of a simple maximal curve C, each region contains a unique singular fiber of C inside. If this fiber is proper, then, depending on its type, we speak about *regions of type* $\tilde{\mathbf{A}}$ *or* $\tilde{\mathbf{D}}$. The braid index of an $\tilde{\mathbf{A}}$ type region R equals $\operatorname{wd} R$; the braid index of a $\tilde{\mathbf{D}}$ type region is 0.

Exceptional singular fibers (Kodaira's types II, III, IV)

An exceptional singular fiber is located over a singular vertex v of the skeleton $\mathfrak{S} := \operatorname{Sk} C$, and we take for $\{\alpha_1, \alpha_2, \alpha_3\}$ a canonical basis over an inner point of an edge e of $\mathfrak{S}$ incident to v. For a fiber of type II, III, or IV, the local braid monodromy is $\sigma_1\sigma_2$, $\sigma_1\sigma_2\sigma_1$, and $(\sigma_1\sigma_2)^2$, respectively. The statements found in this subsection, see (5.78)–(5.80), apply equally well to isotrivial curves, with the appropriate modification of the notion of canonical basis, as explained on page 169.

Case 5: The type II $=$ $\tilde{\mathbf{A}}_0^{**}$ Only one negative Nagata transformation makes sense (the transform having a type $\tilde{\mathbf{A}}_2^*$ singularity at infinity), and we have

$$\varkappa = \alpha_1 \quad \text{and} \quad \alpha_1 = \alpha_2 = \alpha_3. \tag{5.78}$$

Hence, the group is automatically cyclic.

Case 6: The type IV $=$ $\tilde{\mathbf{A}}_2^*$ As in the previous case, only one negative Nagata transformation is meaningful (the transform having a type $\tilde{\mathbf{A}}_0^{**}$ singularity at infinity), and we have

$$\varkappa = \alpha_1\alpha_2 \quad \text{and} \quad \alpha_1 = \alpha_2 = \alpha_3. \tag{5.79}$$

Hence, the group is cyclic.

Case 7: The type III $=$ $\tilde{\mathbf{A}}_1^*$ blown-up $n > 0$ times The first blow-up center is a double point of C, and the other $n-1$ ones are simple points in the branch of C transversal to the fiber (this branch corresponds to the generator α_2). We have

$$\varkappa = \alpha_1\alpha_2^n \quad \text{and} \quad \alpha_3 = \alpha_2^{-n}\alpha_1\alpha_2^n, \quad [\alpha_1, \alpha_2^{2n-1}] = 1. \tag{5.80}$$

Hence, α_2^{2n-1} is a central element.

5.2.3 The strategy

The results of this section give one a relatively simple way to compute the groups $\pi_+^{\mathrm{aff}}(C)$ and $\pi_+^{\mathrm{proj}}(C)$ of any trigonal curve C. Below, we outline an algorithm that will be used, with appropriate modifications, in the subsequent chapters. The curve is assumed to be represented by its dessin $\mathfrak{D} := \operatorname{Dssn} C$ and its type specification, but we mainly make use of the skeleton $\mathfrak{S} := \operatorname{Sk} C$ only.

1. Choose a reference edge e of $\mathfrak{S}$ and a canonical basis $\{\alpha_1, \alpha_2, \alpha_3\}$ over e.
2. For each singular $\bullet$- or $\circ$-vertex v of $\mathfrak{S}$, choose an edge $e(v)$ incident to v; for each region R of $\operatorname{Cut} \mathfrak{S}$, choose an edge $e(R)$ in $\partial_- R$.

```
# Relations arising from a region/singular vertex of a skeleton, see Section 5.2.3
# "p" is the path from the basepoint to the region/vertex in question
# "t" is the type specification of the region/vertex
# "Rels*" functions return relators in "G" (see Listing C.1), so that the group π1 should be defined via
    "g := G/Union(RelsA(...), RelsD(...), ...);"
```

```
Reread("braid.txt");   # (see Listing 2.5)
Reread("common.txt");  # (see Listing C.1)
# Braid monodromy: translate elements in "list" back to e
bm := function(p, list) return Braid(Path(p)^-1, G, list); end;
# This function implements all others (see "BraidRelations" in Listing 2.5)
_rels := function(p, b, t, gens)
 return bm(p, BraidRelations(b*(S.1*S.2)^(3*t), G, gens));
end;
# Type A* ("t=0"), D* ("t=1"), or worse ("t>1") "n"-gonal regions
RelsD := function(p, n, t) return _rels(p, S.1^n, t, [G.2, G.3]); end;
# Simpler relations for type A* "n"-gonal regions: one is enough
RelsA := function(p, n) return _rels(p, S.1^n, 0, [G.2]); end;
# Type E* ("t=1") and worse ("t>1") vertices
RelsE6 := function(p, t) return _rels(p, Path(x), t, [G.2, G.3]); end;
RelsE7 := function(p, t) return _rels(p, Path(y^-1), t, [G.2, G.3]); end;
RelsE8 := function(p, t) return _rels(p, Path(x^2), t, [G.2, G.3]); end;
# Useful shortcuts ("t=0" or "t=1")
A  := RelsA;     # Type A* region
RA := function(p, n) return A(p, n)[1]; end;          # (as a single relator)
D  := function(p, n) return RelsD(p, n, 1); end;      # Type D* region
E6 := p -> RelsE6(p, 1);     # Type E6: monovalent •-vertex
E7 := p -> RelsE7(p, 1);     # Type E7: monovalent ○-vertex
E8 := p -> RelsE8(p, 1);     # Type E8: bivalent •-vertex
```

Listing 5.2. Computing π_1 with `GAP` ("`pi1.txt`").

3. For each singular vertex v, write the corresponding braid relations, slopes, *etc.* in a canonical basis over $e(v)$, see Section 5.2.2.
4. For each region R and each singular fiber inside R, write the corresponding braid relations, slopes, *etc.* in a canonical basis over $e(R)$, see Section 5.2.2.
5. Translate all relations/elements obtained to the common reference edge e *via* the monodromy $\mathfrak{m}(\zeta)^{-1} = (\operatorname{val}\zeta)^{-1}$, see Theorem 5.62, where ζ is a chain in $\mathfrak{S}$ connecting e to $e(v)$ or $e(R)$.
6. Write down the appropriate presentation (5.54), (5.55) and analyze the group.

Since $\Delta^2 \in \mathbb{B}_3$ is a central element and $[\varkappa_i, \rho] = 1$, see Proposition 5.48 (4), the group thus obtained does not depend on the particular lifts of the braid monodromy to $\mathbb{B}_3$ used in Step 5. In Step 4, one would usually pre-simplify the relations as explained in Remark 5.77 before applying Step 5.

This algorithm is partially implemented in Listing 5.2. (Human intervention is needed to identify the regions and select chains; an attempt of a fully automated brute force computation is discussed in Section C.1.2.) I did not implement nontrivial slopes

```
# A few shortcuts and service functions used in the computations, cf. Chapter 8

Reread("pi1.txt");    # (see Listing 5.2)
# A service function: stores "g := G/[arg];" and its commutant "h" and returns "h"
DS := function(arg)
 g := G/Union(arg); h := DerivedSubgroup(g); return h;
end;
# Simplifying a presentation of a normal subgroup (for groups suspected to be B3 or Γ)
simplify := function(arg)
 P := PresentationNormalClosure(g, Subgroup(g, arg)); SimplifyPresentation(P);
end;
# Analyzing various quotients (to prove that a group is not B3 or Γ)
qgroup := function(arg) return FactorGroupFpGroupByRels(g, arg); end;
qsize := function(arg) return Size(CallFuncList(qgroup, arg)); end;

# This function returns a list of A type relations, united with a predefined list "fixed"
# The chains to the regions are passed in "chains" (usually predefined), and their widths (braid
    indices) are in "wd", which may be shorter than "chains" (e.g., "wd" can be passed as "arg" from
    a function with a variable number of arguments)

fixed := [];     # usually, relations at infinity
AA := function(chains, wd)
 return Union(List([1..Length(wd)], i -> RA(chains[i], wd[i])), fixed);
end;
```

Listing 5.3. Shortcuts for computing π_1 ("pi1sc.txt").

as they do not appear very often. A typical usage of this program is as follows:

```
g := G/Union(A(p1,n1),..., D(p2,n2),..., fixed);
```

where n1, n2, ..., are the widths of regions $\operatorname{reg}_-(e \uparrow \zeta)$, $\zeta =$ p1, p2, ..., and fixed is a fixed set of relations depending on a particular problem (*e.g.*, relation at infinity, some central elements, 'special' relations at a singular fiber close to e, *etc*). Then, the group g can be analyzed by GAP, *cf.* Chapter 8. A few shortcuts commonly used in the sequel are collected in Listing 5.3.

Convention 5.81. In the figures, the common reference edge e is marked by a grey diamond $\diamond$ next to the $\bullet$-end of e.

A few observations

In practice, we usually use the braid relations arising from all but one singular fibers, see Lemma 5.59, or even just as many relations as needed to show that the group is abelian or, more generally, not larger than expected. (For example, for sextics of torus type in Chapter 8 we will usually try to prove that the group is a quotient of $\mathbb{B}_3$.) For future references, we state a few simple results that, in some cases, give us an upper bound on the size of the fundamental group and a supply of central elements.

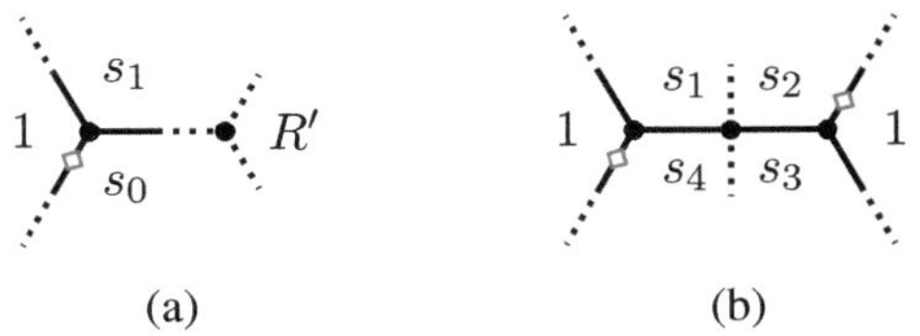

Figure 5.4. Two special fragments.

In Lemmas 5.82 and 5.83, we refer to Figure 5.4, where two special fragments of a skeleton are shown. The numbers inside regions indicate their braid indices.

Lemma 5.82. *Assume that the skeleton* Sk C *has a fragment shown in Figure* 5.4 (a), *and let* $s = \text{g.c.d.}(s_0, s_1, \dots)$ *be the greatest common divisor of the braid indices of all regions adjacent to the region* R *marked with* 1. *Then* $\pi_+^{\text{aff}}(C)$ *is generated by* $\alpha_1 = \alpha_2$ *and* α_3 *and has relation* $\{\alpha_2, \alpha_3\}_s = 1$. *In particular,* $(\alpha_2\alpha_3)^s$ *is a central element and the following are sufficient conditions for the group to be abelian*:

- $s = 1$ *or* 2;
- *one of the regions adjacent to* R *contains a* $\tilde{\mathbf{D}}$ *type fiber*;
- *the horizontal edge in Figure* 5.4 *(a) extends towards the boundary of a region* R' *(shown in dotted lines) with braid index* 1.

Proof. From the region R we have $\alpha_1 = \alpha_2$, and the other regions adjacent to R give us the relations $\{\alpha_1 \uparrow \sigma_2^{-i}, \alpha_3\}_{s_i} = 1$, $i = 0, 1, \dots$, (*via* the chains $(\mathbf{x}^2\mathbf{y})^i\mathbf{x}$), which, in view of $\alpha_1 = \alpha_2$, are equivalent to $\{\alpha_2, \alpha_3\}_s = 1$, see (A.27). Hence, $(\alpha_2\alpha_3)^s$ is a central element, see (A.28), and the group is abelian whenever $s = 1$ or 2. If one of the regions, *e.g.*, the one marked with s_0, contains a $\tilde{\mathbf{D}}$ type fiber, we have $[\alpha_1, \alpha_2\alpha_3] = 1$ which, in view of $\alpha_1 = \alpha_2$, implies that $[\alpha_2, \alpha_3] = 1$. Finally, if the region R' with braid index 1 is present, it gives us the relation $\alpha_2 = (\alpha_2\alpha_3)^{-1}\alpha_1(\alpha_2\alpha_3)$, which also implies $[\alpha_2, \alpha_3] = 1$. □

Lemma 5.83. *Assume that the skeleton* Sk C *has a fragment shown in Figure* 5.4 *(b), where the* •*-vertex in the middle may be of any index at least two. Then the map* $\sigma_1 \mapsto \alpha_1 = \alpha_2$, $\sigma_2 \mapsto \alpha_3$ *establishes an epimorphism* $\mathbb{B}_3 \twoheadrightarrow \pi_+^{\text{aff}}(C)$. *If at least one of the braid indices* s_1, s_2, s_3, s_4 *is prime to* 3, *the group is abelian.*

Proof. From the leftmost and rightmost regions marked with 1 we have $\alpha_1 = \alpha_2$ and

$$\alpha_2 = \rho^{-1}\alpha_2\alpha_3\alpha_2^{-1}\rho \quad \text{or} \quad \alpha_2\alpha_3\alpha_2^{-1} = \rho^{-1}\alpha_1\rho,$$

respectively. (Depending on the number of edges at the •-vertex in the middle, we use the chain $\mathbf{xyx}^2\mathbf{yx}^2$ or $\mathbf{xyxyx}^2$, respectively.) In view of $\alpha_1 = \alpha_2$, both are equivalent to the braid relation $\alpha_2\alpha_3\alpha_2 = \alpha_3\alpha_2\alpha_3$. In particular, $(\alpha_2\alpha_3)^3$ is a central element. Hence, if s_1 or s_4 is prime to 3, the group is abelian due to Lemma 5.82. If s_2 or s_3 is prime to 3, we can repeat the same argument taking for the basepoint the right grey diamond in the figure. □

5.3 Universal curves

One of the most important manifestations of the fact that the braid monodromy of a trigonal curve is essentially induced by its j-invariant is the existence of the so-called universal curves, which, in some situations, lets one deduce various geometric properties of a curve from algebraic properties of its monodromy group.

5.3.1 Universal curves

The results of this chapter let one compute, in a more or less algorithmic way, various fundamental groups related to a trigonal curve. Here, we try to address the inverse problem: what can be said about a curve C if its group $\pi^{\text{aff}}(C)$ is known.

Fix an orientation preserving homeomorphism $\operatorname{Supp}(\operatorname{Sk}\Gamma) \to \mathbb{C} \cup \{\infty\}$ so that the only edge $\bullet\!\!-\!\!\circ$ of the skeleton $\operatorname{Sk}\Gamma$ be mapped to the segment $[0, 1]$ and the center of its only region be mapped to ∞. Consider a subgroup $H \subset \Gamma$. Let $\mathfrak{S} := \operatorname{Sk} H$ and consider the extension $j\colon \operatorname{Supp}\mathfrak{S} \to \operatorname{Supp}(\operatorname{Sk}\Gamma)$ of the natural morphism $\mathfrak{S} \to \operatorname{Sk}\Gamma$, see Definition 1.27 and the construction thereafter. Denote by $\mathit{UB}(H)$ the surface $\operatorname{Supp}\mathfrak{S}$ with the complex structure induced by j. Up to isomorphism, the curve $\mathit{UB}(H)$ and the meromorphic function $j\colon \mathit{UB}(H) \to \mathbb{P}^1$ are determined by H. If H is a subgroup of finite index, then $\mathit{UB}(H)$ is the modular curve $H\backslash\mathfrak{H}^*$ and j is its modular j-invariant, see Lemma 2.22.

If H is a subgroup of finite index, and hence $\mathit{UB}(H)$ is compact, Theorem 3.20 implies that there is a ruled surface $\mathit{U}\Sigma(H) \to \mathit{UB}(H)$ and a proper trigonal curve $\mathit{UC}(H) \subset \mathit{U}\Sigma(H)$ such that j is the j-invariant of $\mathit{UC}(H)$. Both $\mathit{U}\Sigma(H)$ and $\mathit{UC}(H)$ are unique up to Nagata equivalence. Clearly, the curve $\mathit{UC}(H)$ depends only on the conjugacy class $[\![H]\!]$.

Definition 5.84. The curve $\mathit{UC}(H) \subset \mathit{U}\Sigma(H) \to \mathit{UB}(H)$ just constructed is called the *universal trigonal curve* corresponding to the finite index subgroup $H \subset \Gamma$ or to the conjugacy class $[\![H]\!]$.

For example, the universal cubic $\bar{C}$ introduced prior to Theorem 3.19 is the ultimate universal trigonal curve $\mathit{UC}(\Gamma)$. In general, $\mathit{UC}(H)$ is the trigonal curve constructed from the maximal dessin $\operatorname{Dssn}\mathfrak{S}$, see Proposition 4.38.

Proposition 5.85. *Let C be a non-isotrivial trigonal curve over a compact base B, and assume that the reduced monodromy group $M := \mathit{MG}(C)/\Delta^2$ is subconjugate to a subgroup $H \subset \Gamma$. Then the j-invariant $j_C\colon B \to \mathbb{P}^1$ of C factors through the j-invariant $j_H\colon \mathit{UB}(H) \to \mathbb{P}^1$ of the universal curve $\mathit{UC}(H)$.*

Proof. Denote by $(\mathbb{P}^1)^{\sharp\sharp}$ the Riemann sphere punctured at 0, 1, ∞, and all critical values of j_C. Consider the surfaces $\mathit{UB}(H)^{\sharp\sharp} := j_H^{-1}(\mathbb{P}^1)^{\sharp\sharp}$ and $B^{\sharp\sharp} := j_C^{-1}(\mathbb{P}^1)^{\sharp\sharp}$, so that the restrictions of j_H and j_C are topological coverings. There is a natural epimorphism $\mathfrak{m}_\Gamma \colon \pi_1((\mathbb{P}^1)^{\sharp\sharp}) \twoheadrightarrow \pi_1(\mathbb{C} \smallsetminus \{0,1\}) \twoheadrightarrow \Gamma$ (the braid monodromy of the universal cubic) and, by the definition of $\mathit{UC}(H)$, the image of the induced homomorphism $(j_H)_* \colon \pi_1(\mathit{UB}(H)^{\sharp\sharp}) \to \pi_1((\mathbb{P}^1)^{\sharp\sharp})$ is the pull-back $\mathfrak{m}_\Gamma^{-1}(H)$. On the other hand, since the Γ-valued monodromy $\mathfrak{m}_C \bmod \Delta^2$ of C is induced *via* j_C from that of $\mathit{UC}(\Gamma)$, see Theorem 3.19 and Lemma 5.19, the basepoints can be chosen so that $\operatorname{Im}(j_C)_* \subset \operatorname{Im}(j_H)_*$. From the elementary theory of covering spaces it follows that there is a covering $\bar{\jmath} \colon B^{\sharp\sharp} \to \mathit{UB}(H)^{\sharp\sharp}$ such that $j_C = j_H \circ \bar{\jmath}$. Since the degree $\deg j_C$ is finite, so is $\deg j_H$; hence $\mathit{UB}(H)$ is compact and $\bar{\jmath}$ extends to a ramified covering $\bar{\jmath} \colon B \to \mathit{UB}(H)$, which is obviously holomorphic. □

Corollary 5.86. *The reduced monodromy group $\mathit{MG}(C)/\Delta^2 \subset \Gamma$ of a non-isotrivial trigonal curve C over a compact base B is a subgroup of finite index; its genus does not exceed the genus of the base B.* ◁

Corollary 5.87 (of Proposition 5.85 and Theorem 3.19). *Up to Nagata equivalence, a trigonal curve C as in Proposition* 5.85 *is induced from the universal curve $\mathit{UC}(H)$ by a non-constant base change $\bar{\jmath} \colon B \to \mathit{UB}(H)$.* ◁

According to Corollary 5.86, the monodromy group of a non-isotrivial trigonal curve C over the rational base $B \cong \mathbb{P}^1$ is a subgroup of genus zero. On the other hand, in the case of the rational base, one has a well defined conjugacy class $\mathit{MG}(C)$ in $\mathbb{B}_3$, see Remark 5.13, and the universal curve can be defined for a subgroup of $\mathbb{B}_3$ or $\tilde{\Gamma}$. (Recall that, according to our convention, subgroups of $\tilde{\Gamma}$ can be regarded as those of $\mathbb{B}_3$ of depth 6 or 12, see Remark 2.61.) Namely, for a subgroup $H \subset \mathbb{B}_3$ of genus zero, a *universal curve* $\mathit{UC}_{\mathbb{B}}(H) \subset U\Sigma_{\mathbb{B}}(H)$ is defined as a universal curve $\mathit{UC}(H/\Delta^2)$ with the additional property that its type specification modulo $\operatorname{dp} H$ equals the type specification of H. (It is this construction that explains the term *type specification* for subgroups.) This universal curve $\mathit{UC}_{\mathbb{B}}(H) \subset U\Sigma_{\mathbb{B}}(H)$ is well defined up to m-Nagata equivalence, where $6m = \operatorname{dp} H$, and H represents the conjugacy class $\mathit{MG}(\mathit{UC}_{\mathbb{B}}(H))/\Delta^{2m}$. By introducing an extra type $\tilde{\mathbf{J}}_{m,0}$ singular fiber, one can make sure that $H \in \mathit{MG}(\mathit{UC}_{\mathbb{B}}(H))$.

Corollary 5.87 takes the following refined form.

Corollary 5.88. *Let $C \subset \Sigma_d$ be a non-isotrivial proper trigonal curve in a Hirzebruch surface, and assume that the monodromy group $\mathit{MG}(C)$ is subconjugate to a subgroup $H \subset \mathbb{B}_3$ (which is necessarily of genus zero). Let $\operatorname{dp} H = 6m$. Then C is m-Nagata equivalent to a curve induced from $\mathit{UC}_{\mathbb{B}}(H)$.* ◁

The notion of universal curve and Corollary 5.88 can be restated in terms of the fundamental group. Given an epimorphism $\varphi \colon \mathbb{F}_3 \twoheadrightarrow G$, define the *universal curve* $\mathit{UC}_{\mathbb{B}}(\varphi)$ as the curve $\mathit{UC}_{\mathbb{B}}(\operatorname{Zar}_{\mathbb{B}_3} \operatorname{Ker} \varphi)$. Then we have the following corollary.

Corollary 5.89. *An epimorphism* $\varphi\colon \mathbb{F}_3 \twoheadrightarrow G$ *can appear as a geometric quotient of the fundamental group* $\pi^{\mathrm{aff}}(C)$ *of a non-isotrivial proper trigonal curve* $C \subset \Sigma_d$ *if and only if* $\mathrm{Zar}_{\mathbb{B}_3}(\mathrm{Ker}\,\varphi) \subset \mathbb{B}_3$ *is a subgroup of genus zero. If this is the case, any such curve* C *is Nagata equivalent to a curve induced from* $\mathit{UC}_{\mathbb{B}}(\varphi)$. ◁

Speculation 5.90. Combining the results of this subsection, one can speculate that

1. only 'few' epimorphisms $\varphi\colon \mathbb{F}_3 \twoheadrightarrow G$ may appear as geometric quotients of the fundamental groups $\pi^{\mathrm{aff}}(C)$ of non-isotrivial trigonal curves $C \subset \Sigma_d$;
2. the existence of a geometric quotient $\pi^{\mathrm{aff}}(C) \twoheadrightarrow G$ may imply certain geometric properties of C, *viz.* those induced from $\mathit{UC}_{\mathbb{B}}(\mathbb{F}_3 \twoheadrightarrow G)$;
3. in particular, the existence of a quotient $\pi^{\mathrm{aff}}(C) \twoheadrightarrow G$ may imply the existence of a larger quotient $\pi^{\mathrm{aff}}(C) \twoheadrightarrow \tilde{G} \twoheadrightarrow G$.

We illustrate these speculations in subsequent chapters, see Theorems 6.1, 6.9, and 6.16 and geometric applications in Section 6.1.2. Some results extend to the case of improper curves, which is slightly more delicate as slopes should be taken into account.

5.3.2 The irreducibility criteria

As a simple illustration of the concept of universal curve, we discuss reducible trigonal curves. Consider the permutation representation $s\colon \Gamma \twoheadrightarrow \mathbb{S}_3$ and the standard action of $\mathbb{S}_3$ on $\{1,2,3\}$. Clearly, the irreducible components of a trigonal curve C can be identified with the orbits of the subgroup $s(\mathit{MG}(C)) \subset \mathbb{S}_3$. Hence, C is reducible (splits into three linear components) if and only if $\mathit{MG}(C)/\Delta^2 \prec \Gamma_1(2) = s^{-1}\langle(1\,3)\rangle$ (respectively, $\mathit{MG}(C)/\Delta^2 \subset \Gamma(2) = \mathrm{Ker}\, s$). In view of Corollary 5.87, this can be restated as follows.

Proposition 5.91. *A non-isotrivial trigonal curve* C *is reducible or splits into three linear components if and only if, up to Nagata equivalence, it is induced from the universal curve* $\mathit{UC}(\Gamma_1(2))$ *or* $\mathit{UC}(\Gamma(2))$, *respectively.* ◁

The universal reducible curves $\mathit{UC}(\Gamma_1(2))$ and $\mathit{UC}(\Gamma(2))$ can be realized in Σ_1; their skeletons are shown in Figures 4.3 (e) and (a), respectively.

In the rest of this section, we characterize reducible curves in terms of their dessins. Define a *splitting marking* of the skeleton $\mathfrak{S} := \mathrm{Sk}\, C$ as a function $\mu\colon \mathrm{Edg}\,\mathfrak{S} \to \mathbb{Z}_3$ satisfying the following three conditions:

1. for each edge e, one has $\mu(e \uparrow \mathbf{x}) = \mu(e) + 1$ and $\mu(e \uparrow \mathbf{y}) = -\mu(e)$;
2. if the boundary of a region R of $\mathrm{Cut}\,\mathfrak{S}$ contains an edge e with $\mu(e) = 0$, then $[\mathfrak{D}_{\mathrm{solid}}] = 0$ in $H_1(R, R_\times \cup \partial R; \mathbb{Z}_2)$, see page 166.

If a region R satisfies condition (2), one has a well defined linking coefficient form $l\colon \tilde{H}_0(\partial_+ R) \to \mathbb{Z}_2$, $a \mapsto \mathrm{lk}(a, [\mathfrak{D}_{\mathrm{solid}}])$. Hence, for a pair of edges e_1, e_2 in $\partial_+ R$ one can speak about the linking coefficient $l(e_2 - e_1) \in \mathbb{Z}_2$, and the third condition is

3. if the boundary $\partial_+ R$ of a region R contains an edge e with $\mu(e) = 0$, then any other edge e' in $\partial_+ R$ is marked with 0, if $l(e' - e) = 0$, or 2, if $l(e' - e) \neq 0$.

In most interesting cases, this definition can be simplified. Condition (1) means that, at each $\bullet$-vertex of the skeleton, the edges are marked with $0, 1, 2$ cyclically in the counterclockwise direction, whereas at each $\circ$-vertex, either all edges are marked with 0 or the markings alternate between 1 and 2. Hence, $\mathbf{xy}$ transposes 0 and 2 and condition (3) holds trivially within each connected component of the boundary ∂R, as for $e' = e \uparrow (\mathbf{xy})^m$ one has $l(e' - e) = m \bmod 2$. Thus, (3) is only meaningful for regions with disconnected boundary. Condition (2) takes a simpler form if R is homeomorphic to a disk:

2′. if the boundary of a region $R \cong \bar{D}^2$ contains an edge e with $\mu(e) = 0$, then all $\times$-vertices in R are of even index.

Finally, due to (1), the boundary of each region that contains an edge marked with 0 is of even length. Hence, if C is a maximal curve, all three conditions follow from (1) and a splitting marking is easily defined in terms of the skeleton $\mathfrak{S}$ only.

If $\mathfrak{S}$ is connected, a splitting marking is uniquely determined by its value on any one of the edges, but not any such value extends to a splitting marking. For example, if $\mathfrak{S}$ has a $\bullet$-vertex v with $\operatorname{ind} v \neq 0 \bmod 3$, it does not admit a splitting marking.

Theorem 5.92. *There is a canonical bijection between the linear components of a trigonal curve $C \subset \Sigma$ (i.e., irreducible components that are sections of Σ) and splitting markings of the skeleton* $\operatorname{Sk} C$. *In particular, C is reducible if and only if its skeleton admits a splitting marking.*

Proof. Recall that the map $s\colon \Gamma \twoheadrightarrow \mathbb{S}_3$ is given by $\mathbb{X} \mapsto (3\,2\,1)$, $\mathbb{Y} \mapsto (1\,3)$. Using Theorem 5.62 and Proposition 5.67, one can easily see that the existence of a splitting marking is equivalent to the requirement that the monodromy $s \circ \mathfrak{m}$ takes values in the cyclic subgroup generated by $(1\,3)$ (with respect to a canonical basis over any edge e with $\mu(e) = 0$), *i.e.*, that the vertices z_2 over the edges marked with 0 belong to a separate component of C, which is a section of Σ. □

Another characterization of reducible trigonal curves in Hirzebruch surfaces, *i.e.*, over the rational base is discussed in Chapter 6, see Proposition 6.2.

Part II

Applications

Chapter 6

The metabelian invariants

In this chapter, we illustrate Speculation 5.90 on the example of various metabelian invariants (uniform dihedral quotients and the Alexander module) of the fundamental group of a trigonal curve.

6.1 Dihedral quotients

6.1.1 Uniform dihedral quotients

Consider a trigonal curve $\tilde{C} \subset \Sigma_d$, $d \geqslant 0$. As explained in Section 5.1.3, the group $\pi^{\text{aff}}(\tilde{C})$ is equipped with a canonical homomorphism $\deg\colon \pi^{\text{aff}}(\tilde{C}) \twoheadrightarrow \mathbb{Z}$ and one can speak about its uniform dihedral quotients, see Section A.2.3. If $\tilde{C}$ is irreducible, any dihedral quotient of $\pi^{\text{aff}}(\tilde{C})$ is uniform. The following theorem asserts that only finitely many groups can appear as such quotients, *cf.* Speculation 5.90 (1).

Theorem 6.1. *If the group $\pi^{\text{aff}}(\tilde{C})$ of a nontrivial trigonal curve $\tilde{C} \subset \Sigma_d$ admits a uniform dihedral quotient $\mathbb{D}[\mathcal{Q}]$, then $\mathcal{Q}$ is a quotient of one of the following groups*:

$$\mathbb{Z}_2 \oplus \mathbb{Z}_8, \quad \mathbb{Z}_4 \oplus \mathbb{Z}_4, \quad \mathbb{Z}_2 \oplus \mathbb{Z}_6, \quad \mathbb{Z}_3 \oplus \mathbb{Z}_6, \quad \mathbb{Z}_9, \quad \mathbb{Z}_5 \oplus \mathbb{Z}_5, \quad \mathbb{Z}_{10}, \quad \mathbb{Z}_7.$$

Conversely, any quotient $\mathcal{Q}$ of any of these group does appear in a uniform quotient $\pi^{\text{aff}}(C) \twoheadrightarrow \mathbb{D}[\mathcal{Q}]$ for some nontrivial proper simple trigonal curve C.

Proof. Consider the geometric presentation $\pi^{\text{aff}}(\tilde{C}) = \pi_{\tilde{F}}/MG^{\sim}_{+}(\tilde{C}, \tilde{F})$ given by Theorem 5.50. There is an identification $\mathcal{H} = \mathrm{A}_3/(t+1)$, $\mathbf{e}_1 \mapsto \mathbf{a}$, $\mathbf{e}_2 \mapsto \mathbf{b}$, and $MG^{\sim}_{+}(\tilde{C}, \tilde{F})$ acts on $\mathcal{H}$ *via* the specialization of the Burau group Bu_3 at $t = -1$, *cf.* (2.39). Denote by $H \subset \tilde{\Gamma}$ the image of $MG^{\sim}_{+}(\tilde{C}, \tilde{F})$ under this representation; then, due to Lemma A.37, it suffices to consider the quotient $\mathcal{H}/H$. Computing the Zariski kernels $\operatorname{Zar} K \subset \tilde{\Gamma}$ for various subgroups $K \subset \mathcal{H}$, we conclude that the maximal dihedral quotient of $\pi^{\text{aff}}(\tilde{C})$ is of the form $\mathbb{D}[\mathcal{Q}]$, where either

- $\mathcal{Q} = \mathcal{H}$, and then $H = 0$, or
- $\mathcal{Q} \cong \mathbb{Z} \oplus \mathbb{Z}_n$, and then $H \prec \langle (\mathbb{X}\mathbb{Y})^n \rangle$, or
- $\mathcal{Q} \cong \mathbb{Z}_m \oplus \mathbb{Z}_n$, $m \mid n$, and then $H \prec \tilde{\Gamma}_m(n)$.

Here, $\tilde{\Gamma}_m(n)$ is the congruence subgroup

$$\tilde{\Gamma}_m(n) := \left\{ \begin{bmatrix} a & b \\ c & d \end{bmatrix} \in \tilde{\Gamma} \;\middle|\; a = d = 1 \bmod n,\ b = 0 \bmod n,\ c = 0 \bmod m \right\}.$$

If $\tilde{C}$ is not isotrivial, then H is a subgroup of genus zero, see Corollary 5.86; hence, we are in the third case and the pair (m, n) is such that $\tilde{\Gamma}_m(n) \subset \tilde{\Gamma}$ is of genus zero. Such pairs are precisely those listed in the statement, see [37] or the list of congruence subgroups of genus zero found in [40].

If $\tilde{C}$ is isotrivial, then H is the finite cyclic subgroup generated by $\pm\,\mathrm{id}$, $\pm\mathbb{X}$, or $\mathbb{Y}$, see page 75. If $H = \langle \mathrm{id} \rangle = 0$, the curve is trivial; in all other cases, $\mathcal{Q} = 0$, $\mathbb{Z}_2$, $\mathbb{Z}_3$, or $\mathbb{Z}_2 \oplus \mathbb{Z}_2$, *i.e.*, a group as in the statement. □

Proposition 6.2. *A trigonal curve $\tilde{C} \subset \Sigma_d$ is reducible (respectively, splits into three components) if and only if the group $\pi^{\mathrm{aff}}(\tilde{C})$ admits a quotient to $\mathbb{D}_4 = \mathbb{Z}_2 \oplus \mathbb{Z}_2$ (respectively, to $\mathbb{D}[\mathbb{Z}_2 \oplus \mathbb{Z}_2] = \mathbb{Z}_2 \oplus \mathbb{Z}_2 \oplus \mathbb{Z}_2$).*

Proof. Since both dihedral groups in the statement are abelian, the quotient maps in question factor through the abelianization $H_1(\Sigma_d \smallsetminus (\tilde{C} \cup E \cup F_\infty))$, and the assertion is an immediate consequence of Poincaré–Lefschetz duality. □

Corollary 6.3. *If an irreducible trigonal curve $\tilde{C} \subset \Sigma_d$ admits a $\mathbb{D}[\mathcal{Q}]$-covering, then $\mathcal{Q}$ is a quotient of $\mathbb{Z}_3 \oplus \mathbb{Z}_3$, $\mathbb{Z}_9$, $\mathbb{Z}_5 \oplus \mathbb{Z}_5$, or $\mathbb{Z}_7$.* ◁

Proper curves

From now on, we assume that $C \subset \Sigma_d$ is a *proper* trigonal curve. Denote by $\mathbb{D}[\mathcal{Q}_C]$ the maximal uniform dihedral quotient of $\pi^{\mathrm{aff}}(C)$. The next statement follows from Lemma A.37 and the fact that $\rho \in \mathbb{F}_3$ is fixed by $\mathbb{B}_3$.

Lemma 6.4. *For a proper trigonal curve $C \subset \Sigma_d$, one has $\mathcal{Q}_C = \mathcal{H}/H$, where H is the image of $MG_+(C, F)$ under the representation* (2.39). ◁

Consider one of the subgroups $H = \tilde{\Gamma}_m(n)$ introduced in the proof of Theorem 6.1 and assume that it is of genus zero. According to our general convention, we can regard H as a subgroup of $\mathbb{B}_3$ and hence consider the corresponding universal curve $UC_{\mathbb{B}}(H)$, which we denote by $C(\mathcal{Q})$, where $\mathcal{Q} := \mathcal{H}/H$. If $\mathcal{Q} = \mathbb{Z}_2$ or $\mathbb{Z}_2 \oplus \mathbb{Z}_2$, *i.e.*, $H = \tilde{\Gamma}_m(2)$, $m = 1, 2$, one has $H \ni -\mathrm{id}$ and the corresponding *universal reducible curves* are defined up to Nagata equivalence; assuming all singular fibers of type $\tilde{\mathbf{A}}$, we obtain curves $C(\mathcal{Q}) \subset \Sigma_1$, see Figure 6.2 (a) and (b). In all other cases, $H \not\ni -\mathrm{id}$, the curve is defined up to *even* Nagata equivalence, see Corollary 5.88, and there is a unique *simple* curve $C(\mathcal{Q})$. The skeletons of these curves are shown in Figures 6.1 and 6.2. (In Figure 6.1 (a)–(c) and Figure 6.2 (a)–(e), (i), the reader may recognize some of the skeletons shown in Figures 4.2, 4.3, redrawn from a slightly different perspective showing the symmetries. The grey dotted circle in the figures is contracted to a point to form a sphere.) The type specification is uniquely (up to isomorphism) determined by the requirement $\operatorname{dp} H = 12$, *i.e.*, $C(\mathcal{Q}) \subset \Sigma_{2k}$. In fact, the sets of singular fibers are $\tilde{\mathbf{E}}_6 + \tilde{\mathbf{A}}_2 + \tilde{\mathbf{A}}_0^*$ for $C(\mathbb{Z}_3) \subset \Sigma_2$, see Figure 6.1 (a), and

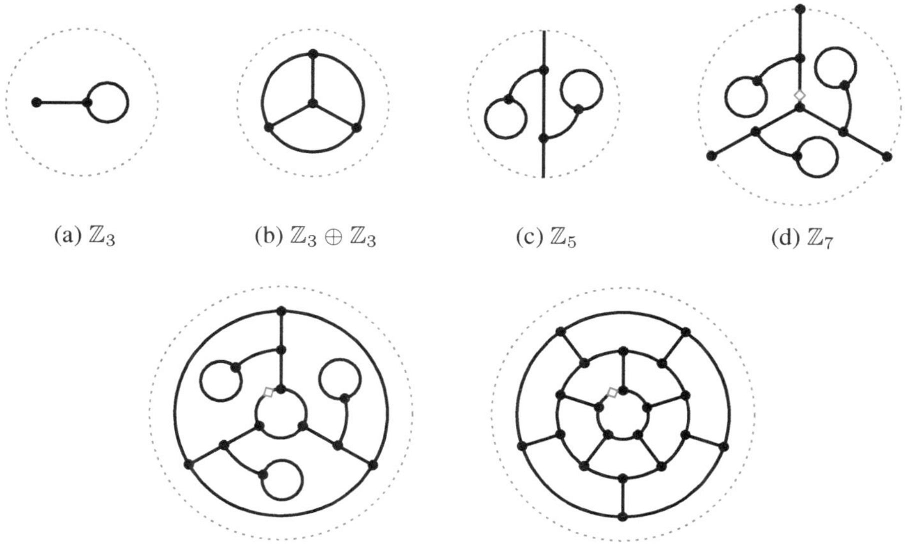

(a) $\mathbb{Z}_3$ (b) $\mathbb{Z}_3 \oplus \mathbb{Z}_3$ (c) $\mathbb{Z}_5$ (d) $\mathbb{Z}_7$

(e) $\mathbb{Z}_9$ (f) $\mathbb{Z}_5 \oplus \mathbb{Z}_5$

Figure 6.1. Irreducible universal curves admitting dihedral coverings.

$\tilde{\mathbf{D}}_5+\tilde{\mathbf{A}}_3+\tilde{\mathbf{A}}_0^*$ for $C(\mathbb{Z}_4) \subset \Sigma_2$, see Figure 6.2 (c); in all other cases all singular fibers are of type $\tilde{\mathbf{A}}$. According to Corollary 5.88, we have the following statement.

Corollary 6.5. *The group $\pi^{\text{aff}}(C)$ of a non-isotrivial proper trigonal curve $C \subset \Sigma_d$ admits a uniform dihedral quotient $\mathbb{D}[\mathcal{Q}]$ if and only if C is r-Nagata equivalent to a curve induced from $C(\mathcal{Q})$, where $r = 1$ for $\mathcal{Q} \subset \mathbb{Z}_2 \oplus \mathbb{Z}_2$ and $r = 2$ otherwise.* ◁

Corollary 6.6 (of Corollary 6.5 and Lemma 3.31). *The group $\pi^{\text{aff}}(C)$ of a non-isotrivial proper trigonal curve $C \subset \Sigma_d$ with Kodaira type* I *singular fibers only admits a uniform dihedral quotient $\mathbb{D}[\mathcal{Q}]$, $\mathcal{Q} \neq 0$, $\mathbb{Z}_2$, $\mathbb{Z}_3$, or $\mathbb{Z}_4$, if and only if C is induced from the universal curve $C(\mathcal{Q})$.* ◁

In order to illustrate Speculation 5.90 (3), we compute the fundamental groups of all irreducible universal curves. (The computation for the reducible curves is similar, but the groups are necessarily infinite and more difficult to be analyzed with `GAP`.) The groups of the curves $C(\mathbb{Z}_3)$, $C(\mathbb{Z}_3 \oplus \mathbb{Z}_3)$, and $C(\mathbb{Z}_5)$ in Σ_2 are computed in Chapter 7, see Theorems 7.7 and 7.8. The other three groups are easily computed as explained in Section 5.2.3, with the reference edge chosen as shown in Figure 6.1; the `GAP` input is found in "`misc/dihedral.txt`".

The curve $C(\mathbb{Z}_7)$ does not lead to any surprises: the commutant of its group is $\mathbb{Z}_7$.

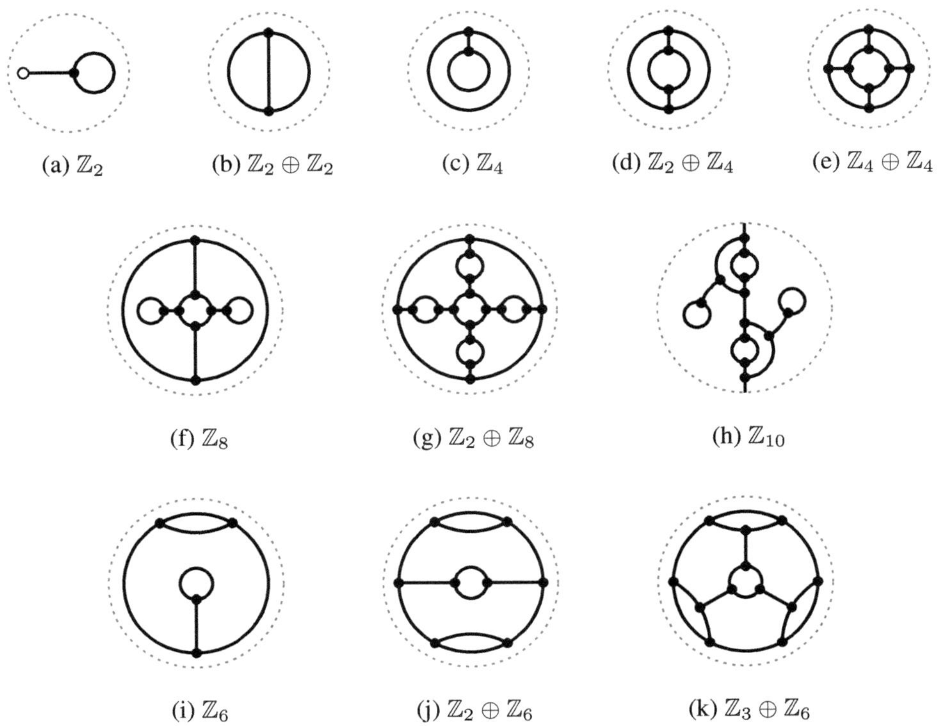

(a) $\mathbb{Z}_2$ (b) $\mathbb{Z}_2 \oplus \mathbb{Z}_2$ (c) $\mathbb{Z}_4$ (d) $\mathbb{Z}_2 \oplus \mathbb{Z}_4$ (e) $\mathbb{Z}_4 \oplus \mathbb{Z}_4$

(f) $\mathbb{Z}_8$ (g) $\mathbb{Z}_2 \oplus \mathbb{Z}_8$ (h) $\mathbb{Z}_{10}$

(i) $\mathbb{Z}_6$ (j) $\mathbb{Z}_2 \oplus \mathbb{Z}_6$ (k) $\mathbb{Z}_3 \oplus \mathbb{Z}_6$

Figure 6.2. Reducible universal curves admitting dihedral coverings.

For the group π_1 of the curve $C(\mathbb{Z}_9)$, we obtain the relations

$$\begin{gathered}\{\alpha_1, \alpha_2\}_3 \uparrow \sigma_2^{-i} = \{\alpha_3, (\alpha_2\alpha_1\alpha_2)^{-1}\alpha_1(\alpha_2\alpha_1\alpha_2)\}_1 \uparrow \sigma_2^{-i} = 1, \quad i = 0, 1, 2, \\ \{\alpha_2, \alpha_3\}_3 = 1, \quad \rho^6 = 1.\end{gathered} \tag{6.7}$$

The call `simplify(g.1/g.3, g.2/g.3)` returns three generators $\kappa_9, \kappa_{12}, \kappa_{14}$ that are subject to the relations (copying *verbatim*)

$$\begin{aligned}\kappa_{12}^{-1}\kappa_9\kappa_{12}\kappa_9^{-1} &= \kappa_9^{-3}\kappa_{12}^3 = \kappa_9^{-1}\kappa_{14}\kappa_{12}^{-1}\kappa_9\kappa_{14}^{-1}\kappa_{12} \\ &= \kappa_{12}^{-1}\kappa_{14}^{-1}\kappa_{12}^2\kappa_{14}^{-1}\kappa_9^{-1}\kappa_{12}\kappa_{14}\kappa_{12}^{-2}\kappa_{14}\kappa_9 = 1,\end{aligned}$$

which simplify to

$$(\kappa_9^{-1}\kappa_{12})^3 = [\kappa_9, \kappa_{12}] = [\kappa_{14}, \kappa_9^{-1}\kappa_{12}] = 1.$$

Hence, $[\pi_1, \pi_1] = \langle\kappa_9, \kappa_{14}\rangle \times \mathbb{Z}_3$, with the $\mathbb{Z}_3$ factor generated by $\kappa_9^{-1}\kappa_{12}$. A similar computation for the group $\bar{G} := \pi_1/\rho^2$ shows that $[\bar{G}, \bar{G}] \cong [\pi_1, \pi_1]$; since the groups are Hopfian, the canonical epimorphism $[\bar{G}, \bar{G}] \twoheadrightarrow [\pi_1, \pi_1]$ is an isomorphism.

Finally, for $C(\mathbb{Z}_5 \oplus \mathbb{Z}_5)$ we have the relations

$$\begin{gathered}\{\alpha_1, \alpha_2\}_5 \uparrow \sigma_2^{-i} = \{\alpha_3, (\alpha_1\alpha_2)^{-1}\alpha_2(\alpha_1\alpha_2)\}_5 \uparrow \sigma_2^{-i} = 1, \quad i = 0, \dots, 4, \\ \{\alpha_2, \alpha_3\}_5 = 1, \quad \rho^{10} = 1.\end{gathered} \tag{6.8}$$

This group is infinite; its Alexander module is $(\Lambda \oplus \Lambda)/\tilde{\varphi}_5(-t)$. For the quotient group $\bar{G} := \pi_1/\rho^2$, we have $\bar{G}'/\bar{G}'' = \mathbb{Z}_5^2$ (as expected) and $\bar{G}''/\bar{G}''' = \mathbb{Z}_5^6$, where $'$ temporarily stands for the commutant of a group. This group also appears infinite, but my laptop fails to compute the other consecutive commutants.

Summarizing, we have the following theorem.

Theorem 6.9. *Assume that the fundamental group $\pi_1 := \pi^{\text{proj}}(C)$ of a proper trigonal curve $C \subset \Sigma_d$ admits a uniform dihedral quotient $\mathbb{D}[\mathcal{Q}]$. Then:*

- *if $\mathcal{Q} = \mathbb{Z}_3$, then π_1 factors to Γ;*
- *if $\mathcal{Q} = \mathbb{Z}_3 \oplus \mathbb{Z}_3$, then π_1 factors to the group given by* (7.10);
- *if $\mathcal{Q} = \mathbb{Z}_9$, then π_1 factors to G/ρ^2, where G is the group given by* (6.7); *hence, $[\pi_1, \pi_1]$ factors to $\mathbb{F}_2 \times \mathbb{Z}_3$;*
- *if $\mathcal{Q} = \mathbb{Z}_5 \oplus \mathbb{Z}_5$, then π_1 factors to G/ρ^2, where G is the group given by* (6.8).

In the last case, if C has Kodaira type I *singular fibers only, then π_1 factors to G.*

Proof. If C is not isotrivial, the statement of the theorem follows from Corollaries 6.5 and 6.6, the computation of the groups of the universal curves, and the obvious fact that even Nagata equivalence preserves the group $\pi^{\text{proj}}(C)/\rho^2$. If C is isotrivial, dihedral quotients as in the statement may exist only if the monodromy group of C is generated by a power $(\sigma_1\sigma_2)^k$, see page 75, and all assertions are straightforward. (Since the group π_1 is computed modulo ρ^2, we can assume that the monodromy group contains $(\sigma_1\sigma_2)^6$, and it suffices to consider the cases $k = 0, 1, 2, 3$.) □

6.1.2 Geometric implications

In this subsection we illustrate Speculation 5.90 (2), exploring the relation between the existence of uniform dihedral quotients and various geometric properties of the curve.

Trigonal curves of torus type

Remarkably, the following two theorems are almost literal translation of the extended version of *Oka's conjecture* for plane sextics, see Theorems 7.33 and 7.34.

Theorem 6.10. *For an irreducible proper trigonal curve $C \subset \Sigma_d$, the following four statements are equivalent:*

1. *the curve C is of torus type;*

2. $(t^2 - t + 1)$ *divides the Alexander polynomial* $\Delta_C(t)$;
3. *the group* $\pi^{\text{proj}}(C)$ *factors to* Γ;
4. *the group* $\pi^{\text{proj}}(C)$ *factors to* $\mathbb{S}_3 = \mathbb{D}_6$.

Theorem 6.11. *Let* $C \subset \Sigma_d$ *be an irreducible proper trigonal curve, and let* $\mathbb{D}[\mathcal{Q}_C]$ *be the maximal dihedral quotient of its fundamental group. Then there is a bijection between the set of torus structures on* C *and the dual projective space* $\check{\mathbb{P}}(\mathcal{Q}_C \otimes \Bbbk_3)$. *If* C *is simple, both sets are isomorphic to* $\mathbb{P}(\mathcal{K}_C \otimes \Bbbk_3)$, *see Theorem* 6.12.

Proof of Theorem 6.10. The implications (3) $\Longrightarrow$ (2) and (3) $\Longrightarrow$ (4) are obvious, and the implication (2) $\Longrightarrow$ (4) is given by Lemma A.35. The converse (4) $\Longrightarrow$ (3) is part of Theorem 6.9, and the equivalence (4) $\Longleftrightarrow$ (1) follows from Theorem 6.11. □

Proof of Theorem 6.11. An injective map from the set of torus structures to the space $\check{\mathbb{P}}(\mathcal{Q}_C \otimes \Bbbk_3)$ parameterizing the quotients $\pi^{\text{aff}}(C) \twoheadrightarrow \mathbb{S}_3$ is given by Corollary 3.36. If C is not isotrivial, the surjectivity is given by Proposition 3.34, Corollary 6.5, and the fact that the universal curve $C(\mathbb{Z}_3)$ is of torus type: it is given by the affine equation $x^3 + (x+t)^2 = 0$. If C is isotrivial and irreducible, $\mathcal{Q}_C$ has 3-torsion if and only if the monodromy group is generated by $(\sigma_1\sigma_2)^{2 \bmod 6}$. In this case, $\mathcal{Q}_C = \mathbb{Z}_3$ and the curve is given by an equation of the form $x^3 + (\bar{g}_3)^2 = 0$ for some polynomial $\bar{g}_3(t)$, *i.e.*, it is of torus type. □

Simple curves

Consider a proper *simple* trigonal curve $C \subset \Sigma_{2k}$ and let X be the covering Jacobian elliptic surface. Identify $E \subset \Sigma_{2k}$ with the section of X and let $\tilde{F}_\infty \subset X$ be the pull-back of the fiber at infinity F_∞. Let, further, $E_* = \sum_i E_i$, where E_i, $i = 1, \ldots, r$, are the exceptional divisors in X that contract to the singular points of C. The lattice $\mathbf{S} := H_2(E_*)$ can be identified with the set of singularities of C, see Convention D.1.

The classes $[E]$ and $[\tilde{F}_\infty]$ span a unimodular sublattice of $H_2(X)$. Hence, in view of Corollary 3.58, the orthogonal complement $\mathbf{P}_k := (\mathbb{Z}[E] \oplus \mathbb{Z}[\tilde{F}_\infty])^\perp$ in $H_2(X)$ is the (only) even unimodular lattice of rank $12n - 4$ and signature $-8n$. (The only definite lattice obtained in this way is $\mathbf{P}_1 \cong \mathbf{E}_8$.) The lattice extension $\mathbf{S} \subset \mathbf{P}_k$, considered up to isomorphism, is called the *homological type* of C.

An important invariant of a homological type is the kernel $\mathcal{K}_C \subset \operatorname{discr} \mathbf{S}$ of the finite index extension $\mathbf{S} \subset (\mathbf{S} \otimes \mathbb{Q}) \cap \mathbf{P}_k$.

Theorem 6.12. *Given a simple proper trigonal curve* $C \subset \Sigma_{2k}$, $k > 0$, *there is a canonical isomorphism* $\mathcal{Q}_C = \operatorname{Ext}(\mathcal{K}_C, \mathbb{Z}) = \operatorname{Hom}(\mathcal{K}_C, \mathbb{Q}/\mathbb{Z})$.

Proof. Identify C with its proper transform in X and consider the unramified double covering $X^\circ := X \smallsetminus (C \cup E \cup E_* \cup \tilde{F}_\infty) \to \Sigma_{2k} \smallsetminus (C \cup E \cup F_\infty)$. The homology $H_1(X^\circ)$ is the group K_G introduced in Corollary A.33, where $G := \pi^{\text{aff}}(C)$. Due to Corollary A.33, we have $\mathcal{Q}_C = K_G / \langle [\alpha_1^2], [\alpha_2^2], [\alpha_3^2] \rangle$. On the other hand, the

Table 6.1. Universal Z-splitting sections of class order $n \leqslant 6$.

n	The curve $C(\mathbb{Z}_n) \subset \Sigma_2$	Sections
3	$x^3 + (x+t)^2 = 0$	$x = 0$
4	$(x^2 - t)(2x - t - 1) = 0$	$x = 1$
5	$4x^3 - 3xp(t) + q(t) = 0$ [1]	$2x = (t^2+1) \pm 6$
6	$(x^2 - t)(16x + 16t^2 - 24t - 3) = 0$	$4x = -4t + 3$

[1] $p(t) = t^4 - 12t^3 + 14t^2 + 12t + 1, \quad q(t) = (t^2+1)(t^4 - 18t^3 + 74t^2 + 18t + 1)$

elements $[\alpha_i^2]$ are represented by meridians of a tubular neighborhood of C in X and the passage to the quotient is realized geometrically by patching C. Thus, we have $\mathcal{Q}_C = H_1(X \smallsetminus (E \cup E_* \cup \tilde{F}_\infty))$ and, using Poincaré–Lefschetz duality and the exact sequence of pair, the latter group is the cokernel of the inclusion homomorphism $H^2(X) \to H^2(E \cup E_* \cup \tilde{F}_\infty)$. This homomorphism is dual to $H_2(E \cup E_* \cup \tilde{F}_\infty) \to H_2(X)$, which, up to a free summand, is a projective resolution of $\mathcal{K}_C$. By definition, the cokernel above equals $\operatorname{Ext}(\mathcal{K}_C, \mathbb{Z})$. Since $\mathcal{K}_C$ is a finite abelian group, applying $\operatorname{Hom}(\mathcal{K}_C, \,\cdot\,)$ to the injective resolution $0 \to \mathbb{Z} \to \mathbb{Q} \to \mathbb{Q}/\mathbb{Z} \to 0$, we also have $\operatorname{Ext}(\mathcal{K}_C, \mathbb{Z}) = \operatorname{Hom}(\mathcal{K}_C, \mathbb{Q}/\mathbb{Z})$. □

Corollary 6.13 (of Theorem 6.12 and Corollary 3.61). *Let $C \subset \Sigma_{2k}$, $k > 0$, be a proper simple trigonal curve, and let X be the covering Jacobian elliptic surface. Then there is a canonical isomorphism* $\operatorname{Tors} MW(X) = \operatorname{Ext}(\mathcal{Q}_C, \mathbb{Z})$. ◁

Corollary 6.14 (Cox, Parry [37]). *The torsion* $\operatorname{Tors} MW(X)$ *of the Mordell–Weil group of a Jacobian elliptic surface X over the rational base $\mathbb{P}^1$ is a quotient of one of the groups listed in Theorem* 6.1. ◁

As a last geometric application, combining Corollary 6.13, Theorem 3.38, and Corollary 6.5, we obtain the following complete description of Z-splitting sections.

Theorem 6.15. *Let $C \subset \Sigma_{2k}$ be a proper non-isotrivial trigonal curve. Then, the class order n of a Z-splitting section for C can take values between* 3 *and* 9. *The Z-splitting sections of class order n are in a one-to-one correspondence with the pairs of opposite epimorphisms $\mathcal{Q}_C \twoheadrightarrow \mathbb{Z}_n$; up to even Nagata equivalence, any such section is induced (together with C) from a* universal Z-splitting section *for $C(\mathbb{Z}_n)$.* ◁

An irreducible curve may only have Z-splitting section of odd class order $n = 3$, 5, 7, or 9. The universal Z-splitting sections of class order $n \leqslant 6$ are known; these sections and corresponding universal curves $C(\mathbb{Z}_n) \subset \Sigma_2$ are listed in Table 6.1.

For other applications of dihedral quotients, see [7, 11, 13, 158, 159].

6.2 The Alexander module

The concept of *Alexander module* (also called *Alexander invariant*) and *Alexander polynomial* in the context of algebraic curves in surfaces was originally introduced by O. Zariski [168] and later developed by A. Libgober [105, 106, 107, 108]. Since then, it has been a subject of intensive study by a number of authors, see recent surveys [58, 109, 110, 128]. One of the best *topological* estimates, due to Libgober [105], states that the Alexander module A_C of a curve C is a quotient of the sum of the local Alexander modules at all singular points of C. (For a brief explanation of this fact, observe that the fundamental group of a nonsingular curve is abelian; hence, all 'non-abelian features' come from the singular points.) Other estimates, making use of algebraic geometry, still depend on the singularities of the curve. The principal result of this section, Theorem 6.16, states that, in agreement with Speculation 5.90 (1), the module A_C of a *trigonal curve* C can essentially take but finitely many values, no matter how complicated the singularities of C are. (An exception is the specialization at a cubic root of unity, see Section 6.2.4.)

6.2.1 Statements

Let $\tilde{C} \subset \Sigma_d$ be a trigonal curve, which we do *not* assume proper. The group $\pi^{\mathrm{aff}}(\tilde{C})$ is equipped with a distinguished epimorphism $\deg\colon \pi^{\mathrm{aff}}(\tilde{C}) \twoheadrightarrow \mathbb{Z}$, and we define the *Alexander module* $\mathrm{A}_{\tilde{C}}$ of $\tilde{C}$ as that of deg, see Section A.2.3. Recall that $\pi^{\mathrm{aff}}(\tilde{C})$ is a Zariski quotient $\pi_{\tilde{F}}/MG_+^{\sim}(\tilde{C}, \tilde{F})$, see Theorem 5.50, and, in order to simplify the statements and make them more precise, we choose a geometric basis in $\pi_{\tilde{F}}$ and consider the *extended Alexander module* $\bar{\mathrm{A}}_{\tilde{C}} := \mathrm{A}_3/MG_+^{\sim}(\tilde{C}, \tilde{F})$, the monodromy group acting on A_3 *via* the Burau representation. The module $\bar{\mathrm{A}}_{\tilde{C}}$ is an invariant of the monodromy group of $\tilde{C}$ rather than of its fundamental group. Its advantage is in the fact that it depends only on the image of the monodromy group in Bu_3. There is a natural epimorphism $\bar{\mathrm{A}}_{\tilde{C}} \twoheadrightarrow \mathrm{A}_{\tilde{C}}$, which is often a isomorphism, see Corollary 6.20.

For a further simplification, fix a value p, prime or 0, and an algebraic number ξ over $\Bbbk_p$, denote by $\psi_\xi \in \Lambda \otimes \Bbbk_p$ the minimal polynomial of ξ, and consider the field $\bar{\Bbbk} := \Lambda(\xi) := (\Lambda \otimes \Bbbk_p)/\psi_\xi$ and the $\bar{\Bbbk}$-vector spaces

$$\mathrm{A}_{\tilde{C}}(\xi) := (\mathrm{A}_{\tilde{C}} \otimes \Bbbk_p)/\psi_\xi = \mathrm{A}_{\tilde{C}} \otimes_\Lambda \bar{\Bbbk}, \quad \bar{\mathrm{A}}_{\tilde{C}}(\xi) := (\bar{\mathrm{A}}_{\tilde{C}} \otimes \Bbbk_p)/\psi_\xi = \bar{\mathrm{A}}_{\tilde{C}} \otimes_\Lambda \bar{\Bbbk}.$$

(Occasionally, to avoid repetition, we will use this specialization notation with $\xi = t$, referring to the original modules.) Our goal is the description of the pairs (p, ξ) for which $\bar{\mathrm{A}}_{\tilde{C}}(\xi)$ may not vanish. In other words, we disregard the possible integral torsion of the form $\mathbb{Z}_{p^r}$, $r > 1$, as well as the torsion $(\Lambda \otimes \Bbbk_p)/\psi_\xi^r$. These more subtle properties are to be the subject of a further study. Note that both phenomena do occur: for the former, see Theorem 6.1, Lines 1–8 in Table 6.3, Theorem 6.42, or Line 2 in Table 6.5; for the latter, see Line 7 in Table 6.5 or the case $m = 3$ in Addendum 6.46. A description of the integral modules $\bar{\mathrm{A}}_{\tilde{C}}/\Phi_N(-t)$ for the values

$N = 1$, $N = 2, 3, 4, 5$, and $N = 6$ is outlined in Sections 6.1.1, 6.2.3, and 6.2.4, respectively.

Throughout this section, whenever p and an algebraic number ξ over $\Bbbk_p$ are fixed, we reserve the notation $N := \operatorname{ord}(-\xi)$ and $M := \operatorname{ord}\xi$ for the multiplicative orders of, respectively, $-\xi$ and ξ in $\bar{\Bbbk}^*$. One has $M = \mathrm{e}_p(N)$, $N = \mathrm{e}_p(M)$, where $\mathrm{e}_2(N) = N$ and, for $p \neq 2$,

$$\mathrm{e}_p(N) = \begin{cases} 2N, & \text{if } N = 1 \bmod 2, \\ \frac{1}{2}N, & \text{if } N = 2 \bmod 4, \\ N, & \text{if } N = 0 \bmod 4. \end{cases}$$

The monodromy groups of isotrivial curves are abelian, see page 169, and their Alexander modules are easily computed. Hence, in this section *we consider non-isotrivial curves only.* The principal results are summarized by the following theorem.

Theorem 6.16. *Let $\tilde{C} \subset \Sigma_d$ be a non-isotrivial trigonal curve.*

- *For $1 \leqslant N \leqslant 5$, the module $\bar{\mathrm{A}}_{\tilde{C}}/\tilde{\varphi}_N(-t)$ can take but finitely many values, see Section 6.2.3 and Theorem 6.32 for details.*
- *For $N = 6$, the module $\bar{\mathrm{A}}_{\tilde{C}}/\tilde{\varphi}_3$ is given by Theorems 6.42, 6.44, and 6.45; it is finite whenever $\tilde{C}$ is proper.*
- *For $N \geqslant 7$, the module $\bar{\mathrm{A}}_{\tilde{C}}(\xi)$ may be nontrivial only if (p, ψ_ξ) is one of the pairs listed in Table 6.2; in this case, one has $\dim_{\bar{\Bbbk}} \bar{\mathrm{A}}_{\tilde{C}}(\xi) = 1$.*

In view of the correspondence explained on page 78, Theorem 6.16 also gives an upper bound to the Alexander module of a plane curve $D \subset \mathbb{P}^2$ with a singular point P of multiplicity $\deg D - 3$ and without linear components through P (assuming that the trigonal model of D is not isotrivial).

Addendum 6.17. *In all cases listed in Table 6.2, the extended Alexander module admits a geometric presentation of the form $\mathrm{A}_3(\xi)/\bar{\Bbbk}\mathbf{e}_1$. Furthermore, at most one pair (p, ψ_ξ) as in the table may appear in the Alexander module of any given curve.*

Remark 6.18. Each pair (p, ψ_ξ) marked with a * in Table 6.2 may appear in the Alexander module of a proper trigonal curve, and we assert that such curves do exist; according to Corollary 5.89, there are universal curves with this property.

The data in the last column do not determine a skeleton uniquely; isomorphic are only the skeletons corresponding to the polynomials ψ_ξ separated by a comma rather than a semicolon in the table.

The universal subgroups $(H/\Delta^2) \subset \Gamma$ corresponding to the sporadic pairs listed in Table 6.2 are *not* congruence subgroups (as they are not present in [40]), even though they satisfy the restrictions on the cusp widths, see Remark 2.25.

The fundamental groups of the proper universal curves corresponding to the pairs (p, ψ_ξ) listed in Table 6.2 are discussed at the end of Section C.1.2.

Table 6.2. Exceptional Alexander modules ($N \geqslant 7$), see Theorem 6.16.

#	p	N	$\psi_\xi \in \Bbbk_p[t]$ or $\xi \in \Bbbk_p$	Sk H
1	2	*7	$t^3+t+1,\ t^3+t^2+1$	$(9;1,0;1^2 7^1;1)$
2		*15	$t^4+t+1,\ t^4+t^3+1$	$(17;1,2;1^2 15^1;3)$
3	3	*8	$t^2+t+2,\ t^2+2t+2$	$(10;0,1;1^2 8^1;1)$
4	5	*8	$t^2+2,\ t^2+3$	$(78;0,0;1^6 8^9;3)$
5		12	$t^2+2t+4,\ t^2+3t+4$	$(52;0,4;1^4 12^4;2)$
6	11	*10	9; 5; 4; 3	$(24;2,0;1^2 2^1 10^2;2)$
7	13	*12	11, 6; 7, 2	$(14;0,2;1^2 12^1;3)$
8	17	*8	15, 8; 9, 2	$(36;0,0;1^4 8^4;2)$
9	19	9	15, 14; 13, 3; 10, 2	$(20;0,2;1^2 9^2;1)$
10		18	17; 16; 9; 6; 5; 4	$(40;2,4;1^2 2^1 18^2;2)$
11	29	*7	22, 4; 13, 9; 6, 5	$(60;0,0;1^4 7^8;2)$
12	37	9	30, 21; 28, 4; 25, 3	$(76;0,4;1^4 9^8;2)$
13	43	*7	39, 32; 27, 8; 22, 2	$(132;0,0;1^6 7^{18};3)$

Listed are the values of p, N, and ψ_ξ and certain information $(e; c_2, c_3; r; a)$ about the skeleton of the universal subgroup $H \subset \mathrm{Bu}_3$; here, e is the number of edges, c_2 and c_3 are the numbers of monovalent $\circ$- and $\bullet$-vertices, respectively, r is the set of the region widths in the partition notation, and a is the order of the automorphism group. Marked with a * are the pairs that may appear in the Alexander module of a proper trigonal curve, see Remark 6.18.

Proof of Theorem 6.16 is based on the fact that, according to Corollary 5.86, the monodromy group $H := MG^{\sim}_{+}(\tilde{C}, \tilde{F}) \subset \mathbb{B}_3 \cdot \operatorname{Inn} \mathbb{F}_3$ is a subgroup of genus zero. In fact, both Theorem 6.16 and all its refinements hold as well for the module $\bar{\mathrm{A}}_{\mathbb{F}_n/H}$, where $H \subset \mathbb{B}_3 \cdot \operatorname{Inn} \mathbb{F}_3$ is any subgroup of genus zero, with references to proper trigonal curves replaced with the requirement that $H \subset \mathbb{B}_3$, and in these settings we also assert the existence. For proper curves, the existence of a subgroup $H \subset \mathbb{B}_3$ resulting in a given Alexander module implies also the existence of the corresponding universal curve, see Corollary 5.88.

Thus, fix a subgroup $H \subset \mathbb{B}_3 \cdot \operatorname{Inn} \mathbb{F}_3$ and, for $\eta = t$ or $\eta = \xi \in \bar{\Bbbk}^*$, introduce the submodules

$$\mathrm{V}_H(\eta) := \operatorname{Ker}[\mathrm{A}_n(\eta) \twoheadrightarrow \mathrm{A}_{\mathbb{F}_n/H}(\eta)], \quad \bar{\mathrm{V}}_H(\eta) := \operatorname{Ker}[\mathrm{A}_n(\eta) \twoheadrightarrow \bar{\mathrm{A}}_{\mathbb{F}_n/H}(\eta)].$$

The submodules $\mathrm{V}_H(\eta)$ depend on H, whereas $\bar{\mathrm{V}}_H(\eta)$ depend on the image of H in the Burau group Bu_3 only. Obviously, one always has $\bar{\mathrm{V}}_H(\eta) \subset \mathrm{V}_H(\eta)$.

Lemma 6.19. *If H is a subgroup of $\mathbb{B}_3$, then for any $\eta = t$ or $\eta = \xi \in \bar{\Bbbk}^*$ one has the inclusion $\tilde{\varphi}_3(\eta)\mathrm{V}_H(\eta) \subset \bar{\mathrm{V}}_H(\eta)$.*

Proof. For any element $\alpha \in \mathbb{F}_3$ of degree 1, any element $h \in H$, and any integer n, one has $[(\alpha^n \uparrow h)\alpha^{-n}] = \tilde{\varphi}_n(t)[(\alpha \uparrow h)\alpha^{-1}]$ in $\mathrm{A}_3(\eta)$. Since $\deg \rho = 3$, the difference

$[(\alpha^3 \uparrow h)\alpha^{-3}] - [(\rho \uparrow h)\rho^{-1}]$ belongs to $\mathrm{V}_H(\eta)$. On the other hand, ρ is $\mathbb{B}_3$-invariant and $[(\rho \uparrow h)\rho^{-1}] = 0$. Hence, the statement follows from Lemma A.36. □

Corollary 6.20. *If $H \subset \mathbb{B}_3$ and $\xi \in \bar{\Bbbk}^*$ is such that $\bar{\varphi}_3(\xi) \neq 0$, then the natural epimorphism $\bar{\mathrm{A}}_{\mathbb{F}_n/H}(\xi) \twoheadrightarrow \mathrm{A}_{\mathbb{F}_n/H}(\xi)$ is an isomorphism.* ◁

6.2.2 Proof of Theorem 6.16: the case $N \geqslant 7$

Fix a subgroup $H \subset \mathrm{Bu}_3$ and consider the skeleton $\mathfrak{S} := \operatorname{Sk} H$ with the distinguished edge e_0 (the orbit of $1 \in \Gamma$) and its type specification tp. Fix, further, a value p, prime or 0, and an algebraic number ξ over $\Bbbk_p$. *We assume that $\xi \neq \pm 1$ and $\xi^2 + \xi + 1 \neq 0$,* so that $M := \operatorname{ord} \xi > 3$. Let $N := \operatorname{ord}(-\xi)$. (We will show that $N < \infty$.) A region R of $\mathfrak{S}$ is called *trivial* if $N \mid \operatorname{wd} R$; otherwise R is called *essential.*

Since $\dim_{\bar{\Bbbk}} \mathrm{A}_3 = 2$, we are interested in subgroups H with $\dim_{\bar{\Bbbk}} \bar{\mathrm{V}}_H \leqslant 1$. We start with an analysis of the local contributions to $\bar{\mathrm{V}}_H$. Given an edge e of $\mathfrak{S}$, connect it to e_0 by a chain $\zeta = (e_0, g)$, $g \in \Gamma$, lift g to an element $g' \in \mathbb{B}_3$, and refer to the basis $(\mathbf{e}_1', \mathbf{e}_2') := (\mathbf{e}_1, \mathbf{e}_2) \uparrow g'$ as a *local basis* over e. For a monovalent vertex v or region R, denote by $\bar{\mathrm{V}}_v(\xi)$ or $\bar{\mathrm{V}}_R(\xi)$ the image $\operatorname{Im}(\tilde{\mathfrak{m}} - 1)$, where $\tilde{\mathfrak{m}}$ is the monodromy about, respectively, v or the boundary ∂R. We compute these spaces in a local basis over the edge e incident to v or, respectively, any edge e in the boundary $\partial_- R$.

Lemma 6.21. *If $\dim_{\bar{\Bbbk}} \bar{\mathrm{V}}_H(\xi) \leqslant 1$, then $N, M < \infty$ and the following holds:*

1. *for a trivial region R, one has $\operatorname{tp}(R) = \operatorname{wd} R \bmod 2M$ and $\bar{\mathrm{V}}_R(\xi) = 0$;*
2. *essential regions can be subdivided into two types,* I *and* II*, as in Items* 3 *and* 4 *below;*
3. *for a type* I *region R, one has $\operatorname{tp}(R) = \operatorname{wd} R \bmod 2M$ and $\bar{\mathrm{V}}_R(\xi) = \bar{\Bbbk}\mathbf{e}_1'$;*
4. *for a type* II *region R of width $n := \operatorname{wd}(R)$, one has: if n is even or $p = 2$, then $\operatorname{tp}(R) = -n \bmod 2M$; otherwise, $\operatorname{tp}(R) = M - n \bmod 2M$ and M is even; in both cases, $\bar{\mathrm{V}}_R(\xi) = \bar{\Bbbk}(\mathbf{e}_1' + (\xi + 1)\mathbf{e}_2')$;*
5. *for a monovalent $\bullet$-vertex v, one has: if $p \neq 3$, then $M = 0 \bmod 3$ and $\operatorname{tp}(v) = \pm\frac{2}{3}M \bmod 2M$; otherwise, $M \neq 0 \bmod 3$ and $\operatorname{tp}(v) = 0 \bmod 2M$; in both cases, $\bar{\mathrm{V}}_v(\xi) = \bar{\Bbbk}(\mathbf{e}_1' - \xi^{-s}\mathbf{e}_2')$, where $s = \frac{1}{2}\operatorname{tp}(v) - 1$;*
6. *if $\mathfrak{S}$ has a monovalent $\circ$-vertex v, then M must be odd, $\operatorname{tp}(v) = M \bmod 2M$, and $\bar{\mathrm{V}}_v(\xi) = \bar{\Bbbk}(\mathbf{e}_1' + \xi^s\mathbf{e}_2')$, where $s = \frac{1}{2}(M + 1)$;*
7. *at most one of the three regions incident to a trivalent $\bullet$-vertex is essential;*
8. *the region incident to a monovalent $\bullet$- or $\circ$-vertex is trivial;*
9. *two monovalent vertices cannot be incident to a common edge.*

Proof. In a local basis, one has $\tilde{\mathfrak{m}} = t^s\sigma_1^n$ for a region R with $\operatorname{wd} R = n$ and $\operatorname{tp}(R) = n + 2s$, $\tilde{\mathfrak{m}} = t^s\sigma_1\sigma_2$ for a $\bullet$-vertex v with $\operatorname{tp}(v) = 2 + 2s$, and $\tilde{\mathfrak{m}} = t^s\sigma_1\sigma_2\sigma_1$ for a $\circ$-vertex v with $\operatorname{tp}(v) = 3 + 2s$, see Theorem 5.62, Definition 2.53, and (2.39). Hence, Items 1–6 are easily proved using (2.42) and (2.41) and analyzing the rank of the matrix $\tilde{\mathfrak{m}} - 1$. For Item 7, assume that, for some edge e, the regions $R = \operatorname{reg}_- e$

and $R' = \operatorname{reg}_-(e\uparrow\mathbb{X})$ are both essential and consider the space $\bar{\mathrm{V}}_{R'}(\xi)+\bar{\mathrm{V}}_R(\xi)\uparrow\sigma_1\sigma_2$. Unless $\xi^2+\xi+1=0$ (and both regions are of type II), this space has dimension 2. Remaining Items 8 and 9 are proved similarly, by comparing two subspaces. □

Corollary 6.22. *If* $\bar{\mathrm{V}}_H(\xi)=0$*, then* $\mathfrak{S}$ *is a regular skeleton and each region* R *of* $\mathfrak{S}$ *is trivial and satisfies* $\operatorname{tp}(R)=\operatorname{wd} R \bmod 2M$*, cf.* 6.21 (1)*. If* H *is a subgroup of genus zero, the converse also holds.* ◁

An edge e of $\mathfrak{S}$ is said to be of type I, II, III$_\epsilon$ ($\epsilon=\pm$ if $p\neq 3$ or $\epsilon=0$ if $p=3$), or IV if, respectively, e is in the boundary ∂_-R of an essential region R of type I or II, is incident to a monovalent •-vertex v, or is incident to a monovalent ∘-vertex v. (If $p\neq 3$, the two types III$_\pm$ are distinguished according to the sign in the equality $\operatorname{tp}(v)=\pm\frac{2}{3}M \bmod 2M$.) In all other cases, e is said to be of type 0. According to Items 8 and 9 in Lemma 6.21, the types are indeed mutually exclusive, and we can consider the space $\bar{\mathrm{V}}_e(\xi)$, referring to the appropriate space ($\bar{\mathrm{V}}_R$ or $\bar{\mathrm{V}}_v$, depending on the type T of e) in a local basis over e. This space is generated by $\mathbf{e}_1'+a_T(\xi)\mathbf{e}_2'$, where

$$a_T(t)=0,\quad t+1,\quad -t^{-s},\quad \text{or}\quad t^{\frac{1}{2}(M+1)} \tag{6.23}$$

for $T=$ I, II, III, or IV, respectively, see Lemma 6.21. Here, $s=\pm\frac{1}{3}M-1$ for $T=\mathrm{III}_\pm$ and $s=-1$ for $T=\mathrm{III}_0$.

Lemma 6.24. *If* $\dim_{\bar{\Bbbk}}\bar{\mathrm{V}}_H\leqslant 1$*, then, within each trivial region* R *of* $\mathfrak{S}$*, the distance between any two edges of the same type other than* 0 *is divisible by* N*.*

Proof. Consider two edges $e',e''\in R$ of types $T',T''\neq 0$, and let $e''=e'\uparrow\sigma_2^d$. For $\dim_{\bar{\Bbbk}}\bar{\mathrm{V}}_H\leqslant 1$, the generators of $\bar{\mathrm{V}}_{e''}(\xi)$ and $\bar{\mathrm{V}}_{e'}(\xi)\uparrow\sigma_2^d$ must be linearly dependent; using (2.42) and equating to 0 the corresponding determinant, we arrive at

$$\tilde{\varphi}_d(-\xi)[\xi-a_{T'}(\xi)(\xi+1)]=a_{T''}(\xi)-a_{T'}(\xi), \tag{6.25}$$

see (6.23). If $T'=T''$, then the right hand side vanishes and either $\tilde{\varphi}_d(-\xi)=0$, and then $d\,|\,N$ as stated, or $\xi=a_{T'}(\xi)(\xi+1)$. Using (6.23) again, one can see that the last equation and $\xi^M=1$ either have no common solutions or imply $\xi^2+\xi+1=0$. □

From now on, assume that H is a subgroup of genus zero.

Let R_i and S_j be, respectively, the trivial and essential regions of $\mathfrak{S}$ (where i and j run over certain index sets). Introduce the following counts:

- v_1 and v_3 are the numbers of, respectively, mono- and trivalent •-vertices;
- u_1 and u_2 are the numbers of, respectively, mono- and bivalent ∘-vertices;
- Nm_i is the width of the trivial region R_i; let $m:=\sum_i m_i$;
- n_j is the width of the essential region S_j; let $n:=\sum_j n_j$.

The number of edges is $e = v_1 + 3v_3 = u_1 + 2u_2 = Nm + n$. Since the total number of regions does not exceed $m + n$, we can eliminate v_3 and u_2 and rewrite Euler's identity in the form $(6 - N)m + 5n + 4v_1 + 3u_1 \geqslant 12$.

For a trivial region R_i and a type $T \in \{\mathrm{I}, \mathrm{II}, \mathrm{III}, \mathrm{IV}\}$, let $k_i^T m_i$ be the number of edges of type T in R_i. Denote $k_i^{\mathrm{all}} := 5k_i^{\mathrm{I}} + 5k_i^{\mathrm{II}} + 4k_i^{\mathrm{III}} + 3k_i^{\mathrm{IV}}$ and $k^{\mathrm{all}} := \max_i k_i^{\mathrm{all}}$. According to Items 7 and 8 in Lemma 6.21, each edge of type $T \neq 0$ is in a trivial region; hence $5n + 4v_1 + 3u_1 = \sum_i k_i^{\mathrm{all}} m_i \leqslant k^{\mathrm{all}} m$ and Euler's inequality implies

$$N < 6 + k^{\mathrm{all}}. \tag{6.26}$$

Due to Lemma 6.24, we have $k_i^{\mathrm{III}} \leqslant 2$ and $k_i^T \leqslant 1$ for $T \neq \mathrm{III}$. Hence, $k^{\mathrm{all}} \leqslant 21$ and we obtain the first estimate $N \leqslant 26$.

Step 1: reduction to a finite list The rest of the proof is heavily computer aided. For each $N = 7, \dots, 26$, we will try to show that, with finitely many explicit exceptions, $r(n + v_1 + u_1) \leqslant m$, where $r := \lceil 5/(N-6) \rceil = 5, 3, 2, 2$ or 1 for $N = 7, 8, 9, 10$ or $N \geqslant 11$, respectively. Then, Euler's inequality above would imply $(5/r) + 6 - N > 0$, which is a contradiction.

Since $\sigma_1^N = \mathrm{id} \bmod \tilde{\varphi}_N(-t)$, see (2.42), we can assume that $[\![\sigma_1^N]\!] \subset H$; then the widths of all regions of $\mathfrak{G}$ divide N.

Fix a value N as above and consider r elements $g_1, \dots, g_r \in \mathbb{B}_3$. For the inequality $r(n + v_1 + u_1) \leqslant m$, it would suffice to show that, for any edge e of type $T \neq 0$, all members of the set $E_e := \{e \uparrow g_i(\mathbb{X}\mathbb{Y})^{-s}\}$, $i = 1, \dots, r$, $s = 0, \dots, N-1$, are pairwise distinct and, moreover, the sets E_e, $E_{e'}$ corresponding to two distinct edges are disjoint (assuming that $\dim_{\bar{\Bbbk}} \bar{\mathrm{V}}_H(\xi) \leqslant 1$). To this end, we consider a pair of types $T, T' \neq 0$, a pair of indices $i, i' = 1, \dots, r$, and an index $s = 0, \dots, N-1$ and try to show that, whenever $s \neq 0$ or $i \neq i'$ or $T \neq T'$, the vectors $(\mathbf{e}_1 + a_T \mathbf{e}_2) \uparrow g_i \sigma_2^s$ and $(\mathbf{e}_1 + a_{T'} \mathbf{e}_2) \uparrow g_{i'}$ specialized at $t = \xi$ are linearly independent. As a sufficient condition, the determinant $D(t)$ of the matrix formed by the two vectors should have no common roots with the cyclotomic polynomial $\Phi_M(t)$, *i.e.*, the resultant $\mathcal{R}$ of the two polynomials should not be zero.

For each N, it is possible to choose elements $g_1, \dots, g_r$ so that none of the above resultants is $0 \in \mathbb{Z}$, and this fact eliminates all values $N \geqslant 7$ for $p = 0$. Whenever one of the resultants $\mathcal{R}$ has a prime divisor p, we record all irreducible common divisors ψ_ξ of $D(t)$ and $\Phi_M(t)$ over $\Bbbk_p$; these pairs (p, ψ_ξ) constitute the finite list of exceptions mentioned above. For some values of N, we try several sets $\{g_1, \dots, g_r\}$, retaining only those pairs (p, ψ_ξ) that fail all tests.

Step 2: computing the genus Now, for each pair (p, ψ_ξ) found at Step 1, compute the corresponding maximal subgroups $H' \subset \mathrm{Bu}_3$ and $H \subset \mathbb{B}_3$ and evaluate their genera. Consider the representation $r_\xi \colon \mathrm{Bu}_3 \to GL(2, \bar{\Bbbk})$ obtained by specializing matrices at $t = \xi$. Since $p > 0$, the latter group is finite. Let $G' := r_\xi(\mathrm{Bu}_3)$ and $G := r_\xi(\mathbb{B}_3)$. Then, for each type $T \neq 0$ allowed for the given pair (p, ψ_ξ), compute the subgroups $Z'_T(\xi) := \mathrm{Zar}_{G'} \bar{\Bbbk}\mathbf{u}_T(\xi)$ and $Z_T(\xi) := \mathrm{Zar}_G \bar{\Bbbk}\mathbf{u}_T(\xi)$, where $\mathbf{u}_T(\xi) :=$

$\mathbf{e}_1 + a_T(\xi)\mathbf{e}_2$, see (6.23). The subgroups $H'_T(\xi) \subset \mathrm{Bu}_3$ and $H_T(\xi) \subset \mathbb{B}_3$ in question are the pull-backs of, respectively, $Z'_T(\xi)$ and $Z_T(\xi)$ under r_ξ.

In practice, we compute the group $Z_T(\xi)$ as the stabilizer of $\mathbf{u}_T^\perp := [a_t(\xi), -1]^{\mathrm{t}}$ under the left action of G by the left multiplication. (Obviously, in dimension two, $\mathrm{Im}(\tilde{\mathfrak{m}} - 1) \in \bar{\Bbbk}\mathbf{u}$ if and only if $(\tilde{\mathfrak{m}} - 1)\mathbf{u}^\perp = 0$. To comply with the `GAP` right group action convention, we transpose all matrices and use the right multiplication.) Then the skeleton $\mathfrak{S}_T := \mathrm{Sk}\, H_T(\xi)$ is found as $(Z_T(\xi) \cdot \langle \xi^3 \,\mathrm{id}\rangle)\backslash G$. This skeleton has an automorphism U of order 3 or 1 (the multiplication by $\xi\,\mathrm{id}$), and $\mathfrak{S}'_T := \mathrm{Sk}\, H'_T(\xi) = \mathfrak{S}/U$, see Lemma 2.57. Selecting the pairs (p, ψ_ξ) for which $\mathfrak{S}_T$ or $\mathfrak{S}'_T$ has genus zero, we arrive at Table 6.2. Technical details are explained in Section C.1.2. □

Proof of Addendum 6.17. The first statement is established in Step 2 of the previous proof: each time, the subgroup $H_T(\xi)$ is conjugate to $H_{\mathrm{I}}(\xi)$ (see Listing C.4 and Section C.1.2) and, hence, a canonical basis can be chosen so that $\bar{\mathrm{V}}_H(\xi) = \bar{\Bbbk}\mathbf{e}_1$. The second statement is proved by computing the intersections of the corresponding universal subgroups. More precisely, for each pair of subgroups, we compute the product of their skeletons and show that none of its components has genus zero, see page 15. This computation is carried out in "`Burau2.txt`". □

6.2.3 Congruence subgroups (the case $N \leqslant 5$)

For an integer $N \geqslant 2$, consider the *Burau congruence subgroups*

$$\mathrm{Bu}(N) := \{g \in \mathrm{Bu}_3 \mid g = \mathrm{id} \bmod \tilde{\varphi}_N(-t)\}, \quad \bar{\mathbb{B}}(N) := \mathrm{Bu}(N) \cap \mathbb{B}_3,$$

which can be defined as the kernels of the induced representations $\mathrm{Bu}_3, \mathbb{B}_3 \to \mathrm{Aut}\,\mathrm{A}'$, where $\mathrm{A}' := \mathrm{A}_3/\tilde{\varphi}_N(-t)$. Let $\mathbb{B}(N) \subset \bar{\mathbb{B}}(N)$ be the kernel of the $\mathbb{B}_3$-action on the semidirect product $\mathrm{A}' \rtimes \mathbb{Z}$, $\mathbb{Z}$ acting on A' *via* t.

Consider also the normal closures $\mathit{NC}(\sigma_1^N) \subset \mathbb{B}_3$. For $2 \leqslant N \leqslant 5$, these subgroups are of finite index: one has $\mathit{NC}(\sigma_1^N)/\Delta^2 = \Gamma(N)$ and $\mathrm{dp}\,\mathit{NC}(\sigma_1^N) = 6$, 12, 24, and 60 for $N = 2$, 3, 4, and 5, respectively.

Lemma 6.27. *For any integer $N \geqslant 2$, one has $\mathit{NC}(\sigma_1^N) \subset \mathbb{B}(N) \subset \bar{\mathbb{B}}(N) \subset \mathrm{Bu}(N)$.*

Proof. The inclusion $\mathit{NC}(\sigma_1^N) \subset \bar{\mathbb{B}}(N)$ follows from (2.42). For $\mathit{NC}(\sigma_1^N) \subset \mathbb{B}(N)$, observe that σ_1^N acts trivially on α_3, and hence, for each $\beta \in \mathbb{B}_3$, one has

$$\big(\alpha_3 \uparrow (\beta^{-1}\sigma_1^N\beta)\big)\alpha_3^{-1} = \big((\alpha_3 \uparrow \sigma_1^N)\alpha_3^{-1} \cdot \alpha_3 \cdot (h \uparrow \sigma_1^N)h^{-1} \cdot \alpha_3^{-1}\big) \uparrow \beta,$$

where $h := \alpha_3^{-1}(\alpha_3 \uparrow \beta) \in \mathrm{Ker}\,\mathrm{deg}$. This element projects to $0 \in \mathrm{A}'$. □

Corollary 6.28. *All cusp widths of $\mathbb{B}(N)$ divide N. If $N \leqslant 5$, then $\mathbb{B}(N)/\Delta^2 \subset \Gamma$ is a congruence subgroup of level N.* ◁

Conjecture 6.29. For each $N \geqslant 2$, one has $\mathbb{B}(N) = \mathit{NC}(\sigma_1^N)$.

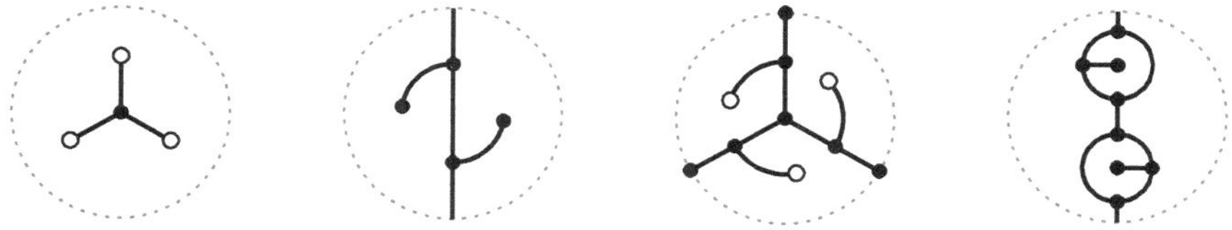

(a) $3A^0 = \Gamma^3$ (b) $4D^0$ (c) $5E^0$ (d) $5F^0$

Figure 6.3. Some universal subgroups of level $l \leqslant 5$.

Conjecture 6.30. For each $N \geqslant 6$, the subgroup $\bar{\mathbb{B}}(N) \subset \mathbb{B}_3$ is of infinite index.

We establish Conjecture 6.29 for all values $N \leqslant 6$, see Lemmas 6.31 and 6.40. Conjecture 6.30 would follow from Conjecture 6.29; in particular, it holds for $N = 6$. I know a computer aided proof of Conjecture 6.30 for $N = 7$ and 9: it suffices to show that the eigenvalues of the matrix $\sigma_2\sigma_1^{-1}$ specialized at a primitive N-th root of unity are *not* roots of unity, which can be done by computing the resultants of the characteristic polynomial and a number of cyclotomic polynomials. The same approach should work for any fixed value of N, but I do not know a general proof.

Lemma 6.31. *For* $N = 2, 3, 4, 5$, *one has* $\bar{\mathbb{B}}(N) = \mathbb{B}(N) = NC(\sigma_1^N)$.

Proof. Computing powers of $(\sigma_1\sigma_2)^3 = t^3 \operatorname{id}$, we see that $\operatorname{dp}\bar{\mathbb{B}}(N) = \operatorname{dp} NC(\sigma_1^N)$. Furthermore, since $NC(\sigma_1^N)/\Delta^4 = \tilde{\Gamma}(N)$ for $N \leqslant 5$ and $\tilde{\Gamma}/\tilde{\Gamma}(N) = GL(2, \mathbb{Z}_N)$ acts faithfully on $\mathcal{H} \otimes \mathbb{Z}_N = \mathrm{A}'(-1)$, we also have $\bar{\mathbb{B}}(N)/\Delta^2 = NC(\sigma_1^N)/\Delta^2$. □

There are 23 congruence subgroups of level $l \leqslant 5$, see [40], and trying various type specifications modulo $\operatorname{dp}\mathbb{B}(N)$, one can easily enumerate all submodules of the form $\bar{\mathrm{V}}_H \subset \mathrm{V}_H$ in $\mathrm{A}_3/\tilde{\varphi}_N(-t)$, $2 \leqslant N \leqslant 5$. (The case $N = 1$ should be treated separately as $t + 1$ is not a divisor of $\tilde{\varphi}_1(-t)$. This case corresponds to the uniform dihedral quotients of the fundamental group and thus is completely covered by Theorem 6.1.) We leave this tedious task to the reader and merely state a few consequences in terms of the specializations $\bar{\mathrm{A}}_{\tilde{C}}(\xi)$ at roots of unity over fields.

Theorem 6.32. *Let* $\tilde{C} \subset \Sigma_d$ *be a non-isotrivial trigonal curve, and let* $\xi \in \bar{\Bbbk}^*$ *be a root of unity over* $\Bbbk_p$ *such that* $N := \operatorname{ord}(-\xi) \leqslant 5$. *Then the module* $\bar{\mathrm{A}}_{\tilde{C}}(\xi)$ *may be nontrivial if and only if the pair* (p, ψ_ξ) *is one of those listed in Table* 6.3. *All pairs listed can be realized by proper trigonal curves.* ◁

In the columns $(H/\Delta^2) \subset \Gamma$ and $\operatorname{Sk} H$ in Table 6.3, we assume that H is the universal subgroup of $\mathbb{B}_3$. In all but one cases, the subgroup $H' \subset \mathrm{Bu}_3$ either equals H or differs by the type specification only. The exception is Line 11, the pair $(p, \xi) = (2, t^2 - t + 1)$, where $H' = (\Gamma)^-$. Note that this is also the only case where we cannot assert that $\bar{\mathrm{V}}_H(\xi) = \mathrm{V}_H(\xi)$, see Corollary 6.20.

Table 6.3. Alexander modules $A_{\bar{C}}(\xi)$ with $N := \mathrm{ord}(-\xi) \leqslant 5$, see Theorem 6.32.

#	p	N	$\psi_\xi \in \Bbbk_p[t]$	$\bar{V}_H$	$(H/\Delta^2) \subset \Gamma$	Sk H	Remarks
1	0	2	$t-1$	I	$2B^0 = \Gamma_1(2)$	6.2 (a)	$\Leftrightarrow 9$
2				0	$2C^0 = \Gamma(2)$	6.2 (b)	$\Leftrightarrow 10$
3		3	t^2-t+1	I	$3B^0 = \Gamma_1(3)$	6.1 (a)	$\Leftrightarrow 13$
4				0	$3D^0 = \Gamma(3)$	6.1 (b)	$\Leftrightarrow 14$
5		4	t^2+1	I	$4B^0 = \Gamma_1(4)$	6.2 (c)	$\Rightarrow 1, 9$
6				0	$4G^0 = \Gamma(4)$	6.2 (e)	$\Rightarrow 2, 10$
7		5	$\tilde{\varphi}_5(-t)$	I	$5D^0 = \Gamma_1(5)$	6.1 (c)	$\Rightarrow 18$
8				0	$5H^0 = \Gamma(5)$	6.1 (f)	$\Rightarrow 19$
9	2	1	$t+1$	I	$2B^0 = \Gamma_1(2)$	6.2 (a)	$\Leftrightarrow 1$
10				0	$2C^0 = \Gamma(2)$	6.2 (b)	$\Leftrightarrow 2$
11		3	t^2-t+1	II	$3A^0 = \Gamma^3$	6.3 (a)	
12		5	$\tilde{\varphi}_5(-t)$	IV	$5E^0$	6.3 (c)	
13	3	1	$t+1$	I	$3B^0 = \Gamma_1(3)$	6.1 (a)	$\Leftrightarrow 3$
14				0	$3D^0 = \Gamma(3)$	6.1 (b)	$\Leftrightarrow 4$
15		2	$t-1$	II	$2A^0 = \Gamma^2$	•—•	
16		4	t^2+1	III_0	$4D^0$	6.3 (b)	$\Rightarrow 15$
17		5	$\tilde{\varphi}_5(-t)$	III_0	$5F^0$	6.3 (d)	
18	5	1	$t+1$	I	$5D^0 = \Gamma_1(5)$	6.1 (c)	
19				0	$5H^0 = \Gamma(5)$	6.1 (f)	
20	7	1	$t+1$	I	$7E^0 = \Gamma_1(7)$	6.1 (d)	

Listed are the values of p, N, and ψ_ξ, the subspace $\bar{V}_H(\xi)$, the universal subgroup $H/\Delta^2 \subset \Gamma$ (in the notation of [40] and the conventional notation, if available), the skeleton Sk H, and a list of dependencies. The subspace $\bar{V}_H(\xi)$ is represented by its type T as in Section 6.2.3: $\bar{V}_H(\xi) = 0$ if $T = 0$, otherwise $\bar{V}_H(\xi) = \Bbbk \mathbf{u}_T$, where $\mathbf{u}_T := \mathbf{e}_1 + a_T(\xi)\mathbf{e}_2$, see (6.23).

In the last column of Table 6.3, as well as in Addendums 6.33 and 6.34 below, we also deal with subgroups $H \subset \mathbb{B}_3$, *i.e.*, with proper trigonal curves. The implications and equivalences listed in the table represent, respectively, the inclusions and equalities of the corresponding universal subgroups.

Addendum 6.33. *For a subgroup* $H \subset \mathbb{B}_3$, *integer* $N = 2, 3, 4, 5$, *and* $p \not| N$ *prime or zero, let* $\bar{V}_p := \bar{V}_H(\xi_p)$, *where* ξ_p *is a root of* $\Phi_N(-t)$ *over* $\Bbbk_p$. *Then:*

- *if* $\dim \bar{V}_0 \leqslant 1$, *Lines* 1–8 *in Table* 6.3, *then* $\dim \bar{V}_p = \dim \bar{V}_0$ *for any* $p \not| N$;
- *conversely, if* $p \neq 2$ *and* $\bar{V}_p \prec \Bbbk \mathbf{u}_{\mathrm{I}}(\xi_p)$, *then* $\dim \bar{V}_0 = \dim \bar{V}_p$. ◁

The condition $\bar{V}_p \prec \Bbbk \mathbf{u}_{\mathrm{I}}(\xi_p)$ rather than just $\dim \bar{V}_p \leqslant 1$ in the second part of Addendum 6.33 is essential; in fact, modules $\bar{V}_p$ of other types, see Lines 11, 12, and 15–17 in Table 6.3, are incompatible with $p = 0$. The reason for the restriction $p \neq 2$, which is meaningful for $N = 3$ or 5, is the fact that the type specification of the corresponding universal subgroup H_p is defined modulo $2M = 2\mathrm{e}_p(N)$; hence, one has $[H_2 : H_p] = 2$ for $p \neq 2$. For all other values of p, including $p = 0$, the universal subgroups are equal, and this observation proves the statement of the addendum.

The same pair (p, ψ_ξ) may result in non-conjugate subspaces $\bar{\mathrm{V}}_H(\xi) \subset \mathrm{A}_3(\xi)$, see, *e.g.*, Line 1 *vs.* 15 or Line 7 *vs.* 12 or 17 in the table. In such cases, even though the quotients $\bar{\mathrm{A}}_{\mathbb{F}_3/H}(\xi)$ are isomorphic, the corresponding universal subgroups differ.

Addendum 6.34. *In addition to the implications and equivalences listed in Table* 6.3, *one has the following compatibility relations*:

- *Lines* 1 $\Leftrightarrow$ 9 (*reducible curves, see Proposition* 6.2) *can appear in the same module with Lines* 5, 11, 3 $\Leftrightarrow$ 13, 4 $\Leftrightarrow$ 14, *and* 7 $\Rightarrow$ 18;
- *Lines* 3 $\Leftrightarrow$ 13 (*curves of torus type, see Theorem* 6.10) *can appear in the same module with Lines* 5, 15, 1 $\Leftrightarrow$ 9, *and* 2 $\Leftrightarrow$ 10.

Otherwise, two distinct summands $\Lambda(\xi')$, $\Lambda(\xi'')$ *with* $\psi_{\xi'} \neq \psi_{\xi''}$ *cannot appear in the extended Alexander module of the same proper trigonal curve.* $\triangleleft$

Since the intersection of two congruence subgroups is also a congruence subgroup, all statements are easily proved using the tables found in [40].

6.2.4 The parabolic case $N = 6$

For this subsection, fix the notation $\Lambda' := \Lambda/(t^2+t+1)$ and $\mathrm{A}' = \mathrm{A}_3/(t^2+t+1)$. Consider the vector $\mathbf{v} := -t\mathbf{e}_1 + \mathbf{e}_2 \in \mathrm{A}'$. It is immediate that $\mathbf{v} \uparrow \sigma_1 = \mathbf{v} \uparrow \sigma_2 = \mathbf{v}$, and in the basis $\{\mathbf{v}, \mathbf{e}_2\}$ the induced $\mathbb{B}_3$-action is given by the matrices

$$\sigma_1 \mapsto \begin{bmatrix} 1 & 0 \\ -t^2 & -t \end{bmatrix}, \qquad \sigma_2 \mapsto \begin{bmatrix} 1 & 0 \\ 0 & -t \end{bmatrix}; \tag{6.35}$$

hence,

$$\delta_1 := \sigma_1\sigma_2^{-1} \mapsto \begin{bmatrix} 1 & 0 \\ t & 1 \end{bmatrix}, \qquad \delta_2 := \sigma_2^{-1}\sigma_1 \mapsto \begin{bmatrix} 1 & 0 \\ t+1 & 1 \end{bmatrix}. \tag{6.36}$$

(We fix the notation δ_1, δ_2 till the end of the section.) Thus, the image of $\mathbb{B}_3$ in $GL(\mathrm{A}')$ is the full group of lower triangular matrices with $[1, (-t)^s]$ in the diagonal. Besides, since $\Delta^2 \mapsto \mathrm{id}$, the $\mathbb{B}_3$-action on A' factors through Γ. For a submodule $\bar{\mathrm{V}} \subset \mathrm{A}'$, we denote by $\operatorname{Zar} \bar{\mathrm{V}} \subset \Gamma$ the Zariski kernel of $\bar{\mathrm{V}}$ with respect to this Γ-action.

Let $\Gamma' := [\Gamma, \Gamma] = 6A^1$; recall that it is freely generated by $\delta_i \bmod \Delta^2$, $i = 1, 2$. Let further $\Gamma'' := [\Gamma', \Gamma']$. Next few statements follow directly from (6.35) and (6.36).

Lemma 6.37. *The kernel of the* Γ*-action on* A' *is* Γ'', *and the image of* Γ'/Γ'' *in* $GL(\mathrm{A}')$ *is the subgroup of all unipotent lower triangular matrices.* $\triangleleft$

Lemma 6.38. *Any vector of the form* $\mathbf{e}_2 + f\mathbf{v}$, $f \in \Lambda'$, *is conjugate to* $\mathbf{e}_2$. $\triangleleft$

Table 6.4. Submodules $\bar{\mathrm{V}}'_H \subset \mathrm{A}'$ with $H \subset \mathbb{B}_3$ ($N = 6$), see Theorem 6.42.

#	Generator $\mathbf{u} \notin \Lambda'\mathbf{v}$	H/H'
1	$\mathbf{u}_1 := \mathbf{e}_2$	$(\mathbb{X}\mathbb{Y})^1$
2	$\mathbf{u}_2 := (t^2-1)\mathbf{e}_2 = -(t+2)\mathbf{e}_2$	$(\mathbb{X}\mathbb{Y})^2$
3	$\mathbf{u}_3 := 2\mathbf{e}_2$	$(\mathbb{X}\mathbb{Y})^3$
4	$\mathbf{u}_\circ := 2\mathbf{e}_2 - \mathbf{v} = t\mathbf{e}_1 + \mathbf{e}_2$	$\mathbb{Y}$
5	$\mathbf{u}_\bullet := (t^2-1)\mathbf{e}_2 - t^2\mathbf{v} = \mathbf{e}_1 - \mathbf{e}_2$	$\mathbb{X}$

Listed are an extra generator $\mathbf{u} \notin \Lambda'\mathbf{v}$ and an element $g \in \Gamma$ generating the corresponding universal subgroup $H \subset \Gamma$ over the intersection $H' := H \cap \Gamma'$.

Lemma 6.39. *The map* $\varphi\colon J \mapsto \operatorname{Zar}(J\mathbf{v})$ *establishes a one-to-one correspondence between the set of ideals* $J \subset \Lambda$ *and the set of subgroups* $H \subset \Gamma'$ *containing* Γ'' *and normal in* Γ; *furthermore, one has* $\operatorname{Rel}\varphi(J) = J\mathbf{v}$ *and the map* $g \mapsto (\mathbf{e}_2 \uparrow g) - \mathbf{e}_2$ *is an isomorphism* $\Gamma'/\Gamma'' = \Lambda'\mathbf{v}$ *of abelian groups that takes* $\varphi(J)$ *onto* $J\mathbf{v}$. ◁

Lemma 6.40. *One has* $\mathbb{B}(6) = NC(\sigma_1^6)$ *and* $\bar{\mathbb{B}}(6) = \mathbb{B}(6) \cdot \langle\Delta^2\rangle$.

Proof. The second statement follows from Lemma 6.38 and the fact that $\Gamma'' \subset \Gamma$ is normally generated by the image of $\delta_1\delta_2\delta_1^{-1}\delta_2^{-1} = \sigma_1\sigma_2^{-6}\sigma_1^{-1}\Delta^2$. For the first one, observe that

$$[(\alpha_1 \uparrow \Delta^{2s})\alpha_1^{-1}] = s[(t-1)\mathbf{e}_1 + (t^2-1)\mathbf{e}_2] \in \mathrm{A}' \tag{6.41}$$

for any $s \in \mathbb{Z}$; hence, $\operatorname{dp}\mathbb{B}(6) = 0$. □

Theorem 6.42. *Let* $H \subset \Gamma$ *be a subgroup of genus zero, and let* $\bar{\mathrm{V}}'_H := \operatorname{Rel} H \subset \mathrm{A}'$. *Then* $\bar{\mathrm{V}}'_H$ *is conjugate to a submodule* $\Lambda'\mathbf{u} + J\mathbf{v}$, *where* $\mathbf{u}$ *is one of the vectors listed in Table* 6.4 *and* $J \subset \Lambda'$ *is an ideal of finite index*; *in particular,* $\operatorname{Card}(\mathrm{A}'/\bar{\mathrm{V}}'_H) < \infty$.

Conversely, any submodule $\Lambda'\mathbf{u} + J\mathbf{v} \subset \mathrm{A}'$ *with* $\mathbf{u}$ *and* J *as above is of the form* $\bar{\mathrm{V}}'_H$ *for some subgroup* $H \subset \Gamma$ *of genus zero.*

Proof. We can assume that H is saturated, *i.e.*, $H = \operatorname{Zar}\bar{\mathrm{V}}'_H$; then $H \supset \Gamma''$, see Lemma 6.37, and all its cusp widths divide 6. Let $H' := H \cap \Gamma'$; it is a finite index subgroup of Γ' containing Γ''. Any such subgroup is torsion free and has all cusp widths equal to 6; hence, it is of genus one and H is of genus zero if and only if the covering $\operatorname{Sk} H' \to \operatorname{Sk} H$ is ramified, see Corollary 1.29 (4), *i.e.*, if $\mathfrak{S} := \operatorname{Sk} H$ has a region of widths $w = 1$, 2, or 3 or a monovalent vertex. In the former case, we can assume that $H \ni \sigma_2^w \bmod \Delta^2$ and $\bar{\mathrm{V}}'_H \ni \mathbf{u}_w$, Lines 1–3 in Table 6.4; in the latter case, H contains $\mathbb{Y}$ or $\mathbb{X}$ and $\bar{\mathrm{V}}'_H$ contains $\mathbf{u}_\circ$ or $\mathbf{u}_\bullet$, Lines 4 and 5, respectively.

Note that $(t^2-1)\Lambda' \ni 3$. Hence, if $\bar{\mathrm{V}}'_H$ contains a vector conjugate to $\mathbf{u}_2$ or $\mathbf{u}_\bullet$ *and* a vector conjugate to $\mathbf{u}_3$ or $\mathbf{u}_\circ$, it also contains a vector $\mathbf{e}_2 + f\mathbf{v}$, $f \in \Lambda'$, and, due to Lemma 6.38, we can assume that $H \ni \sigma_2 \bmod \Delta^2$. Thus, since the index $[H : H']$

Table 6.5. Submodules $\bar{\mathrm{V}}'_H \subset \mathrm{A}'$ with $H \not\subset \mathbb{B}_3$ ($N = 6$), see Theorem 6.44.

#	Generators of $\bar{\mathrm{V}}'_H$	$\mathrm{A}'/\bar{\mathrm{V}}'_H$	$H \subset \mathrm{Bu}_3$	$\mathrm{Sk}\,H$
1	$2\mathbf{e}_2, \mathbf{v}$	$\Lambda' \otimes \Bbbk_2$	$(\Gamma)^-$	○—•
2	$\mathbf{v}$	Λ'	$(2A^0)^-$	•—•
3	$(t-1)\mathbf{e}_2, \mathbf{v}$	$\Lambda'/(t-1) \cong \Bbbk_3$	$(2A^0)^{\mathrm{bu}}$	•—•
4	$2\mathbf{e}_2, (t-1)\mathbf{v}$	$(\Lambda' \otimes \Bbbk_2) \oplus \Lambda'/(t-1)$	$(2B^0)^-$	6.2 (a)
5	$\mathbf{e}_2, (t-1)\mathbf{v}$	$\Lambda'/(t-1) \cong \Bbbk_3$	$(2B^0)^{\mathrm{bu}}$	6.2 (a)
6	$(t-1)\mathbf{e}_2, (t-1)\mathbf{v}$	$\mathrm{A}'/(t-1)$	$(2C^0)^{\mathrm{bu}}$	6.2 (b)
7	$(t-1)\mathbf{e}_2 - \mathbf{v}, (t-1)\mathbf{v}$	$\Lambda'/(t-1)^2$	$(6A^0)^0$	

Listed are a set of generators of $\bar{\mathrm{V}}'_H$, the module $\mathrm{A}'/\bar{\mathrm{V}}'_H$, the universal subgroup $H \subset \mathrm{Bu}_3$ (including the type specification), and its skeleton $\mathrm{Sk}\,H$. The skeleton $\mathrm{Sk}\,6A^0$ is similar to Figure 6.3 (a), except that the monovalent vertices are •-.

divides $[\Gamma : \Gamma'] = 6$, up to conjugation H is generated over H' by a single element as shown in the table and, taking for J the ideal $\varphi^{-1}(H')$, see Lemma 6.39, we arrive at the presentation of $\bar{\mathrm{V}}'_H$ as in the statement. (Since obviously $H' = \mathrm{Zar}(\bar{\mathrm{V}}'_H \cap \Lambda'\mathbf{v})$, this subgroup is normal in Γ and $\varphi^{-1}(H')$ is defined.)

The converse statement is given by Lemma 6.39: a submodule $\Lambda'\mathbf{u} + J\mathbf{v}$ with $\mathbf{u}$ as in Table 6.4 is $\bar{\mathrm{V}}'_H$, where $H = \varphi(J) \cdot \langle g \rangle$ and g, depending on the type of $\mathbf{u}$, is the extra generator given in the table. Since $\varphi(J) \subset \Gamma'$ is a subgroup of finite index (as so is $J \subset \Lambda'$) and the covering $\mathrm{Sk}\,\varphi(J) \to \mathrm{Sk}\,H$ is ramified, H is a subgroup of genus zero, see Corollary 1.29 (4). □

Corollary 6.43 (of the proof). *A subgroup $H \subset \Gamma$ equals* $\mathrm{Zar}\,\bar{\mathrm{V}}'$ *for a submodule $\bar{\mathrm{V}}' \subset \mathrm{A}'$ if and only if $H \supset \Gamma''$ and $H \cap \Gamma'$ is normal in Γ.* ◁

As explained in the proof of Theorem 6.42, a subgroup $H \supset \Gamma''$ is of genus zero if and only if H contains an element conjugate to $\mathbb{X}$, $\mathbb{Y}$, or $(\mathbb{X}\mathbb{Y})^w$, $w = 1, 2, 3$. Observe that the quotient Γ'/Γ'' is the Alexander module of Γ, and as such it is isomorphic to $\Lambda/(t^2 - t + 1)$. Hence, a subgroup $H' := H \cap \Gamma'$ is normal in Γ if and only if $H'/\Gamma'' \subset \Gamma'/\Gamma''$ is a submodule, and this is automatically the case whenever H contains an element conjugate to $\mathbb{X}$, $\mathbb{X}\mathbb{Y}$, or $(\mathbb{X}\mathbb{Y})^2$, *i.e.*, the skeleton $\mathrm{Sk}\,H$ has a mono- or bigonal region or a monovalent •-vertex. In the other two cases, the normality condition needs to be checked explicitly.

Theorem 6.44. *Let $H \subset \mathrm{Bu}_3$, $H \not\subset \mathbb{B}_3$, be a subgroup of genus zero. Then the submodule $\bar{\mathrm{V}}'_H :=$* $\mathrm{Rel}\,H \subset \mathrm{A}'$ *is conjugate to one of those listed in Table* 6.5.

Proof. If $s \neq 0 \bmod 3$, the ideal $(t^s - 1)\Lambda'$ contains 3 and $t - 1$; hence, if tp_H is not trivial modulo 6, we have relations $t\mathbf{v} = \mathbf{v} \bmod \bar{\mathrm{V}}'_H$ and $3\mathbf{v} = 0 \bmod \bar{\mathrm{V}}'_H$. Modulo these relations, the action of Γ factors through the finite group $\Gamma/6C^1$, where $6C^1$ is the genus one subgroup normally generated by δ_1^3 and $\delta_1\delta_2$. Now, the statement can be proved by trying various *non-trivial* type specifications for the seven genus zero subgroups containing $6C^1$, see [40]. □

We conclude this section with a description of the specializations $\bar{\mathrm{A}}_{\tilde{C}}(\xi)$, where ξ is a cubic root of unity over $\Bbbk_p$ with $p \neq 2, 3$ prime or zero. The exceptions $p = 2$ or 3 are covered by Theorem 6.32, as in these cases $N = 3$ or 2, respectively.

Theorem 6.45. *Let $H \subset \mathrm{Bu}_3$ be a subgroup of genus zero, and let $\xi \in \bar{\Bbbk}^*$ be a root of $t^2 + t + 1$ over $\Bbbk_p$, with $p \neq 2, 3$ prime or zero, such that* $\dim \bar{\mathrm{V}}_H(\xi) \leqslant 1$. *Then one has one of the following two mutually exclusive cases:*

- $\bar{\mathrm{V}}_H(\xi) = \Lambda(\xi)(-t\mathbf{e}_1 + \mathbf{e}_2)$ *(for all $p \neq 2, 3$ and all ξ) and $H \prec (2A^0)^-$;*
- *$p \geqslant 5$, the module $\bar{\mathrm{V}}_H(\xi)$ is conjugate to $\bar{\Bbbk}\mathbf{e}_2$, and $H \prec (H_\xi)^+ \subset \mathbb{B}_3$ for a certain genus zero subgroup $H_\xi \subset \Gamma$ of index $p^{\deg \psi_\xi}$.*

In the latter case, any finite *number of* distinct *pairs (p, ψ_ξ) may appear in the Alexander module of a particular subgroup.* ◁

The two cases in Theorem 6.45 are mutually exclusive since the maximal subgroup of $2A^0$ on which the slopes $+\,\mathrm{dg}$ and $-\,\mathrm{dg}$ are equal is Γ', which is of genus one. The subgroups H_ξ and their finite intersections can be characterized as follows: the skeleton of the group has exactly two monovalent vertices, one •- and one ○-, and exactly one monogonal region, all other regions being hexagons.

Addendum 6.46. *For any odd integer $m \geqslant 3$, there exists a proper trigonal curve C whose* conventional *Alexander module A_C factors to $(\Lambda \otimes \mathbb{Z}_m)/(t^2 + t + 1)$.*

Proof. Consider the subgroup $H \subset \mathbb{B}_3$ generated by $NC(\sigma_1^6)$, σ_2, δ_1^m, δ_2^m, and Δ^{2m}, and take for C the universal curve $UC_{\mathbb{B}}(H)$. The projection H/Δ^2 is the subgroup $\varphi(m\Lambda') \cdot \langle \mathbb{X}\mathbb{Y} \rangle$, see the proof of Theorem 6.42. Hence, $\bar{\mathrm{V}}'_H = \Lambda'\mathbf{e}_2 + m\Lambda'\mathbf{v}$, and it remains to show that $[(\alpha_1 \uparrow \beta)\alpha_1^{-1}] \in \bar{\mathrm{V}}'_H$ for any $\beta \in H$, see Lemma A.36. This statement holds trivially for $\beta = \sigma_2$ and $\beta \in NC(\sigma_1^6)$, see Lemma 6.27, and it follows from (6.41) for $\beta = \Delta^{2m}$. Finally, using the simple identity

$$[(\alpha_1 \uparrow gh)\alpha_1^{-1}] = [(\alpha_1 \uparrow g)\alpha_1^{-1}] \uparrow h + [(\alpha_1 \uparrow h)\alpha_1^{-1}], \quad g, h \in \mathbb{B}_3,$$

and induction, one can compute the classes

$$[(\alpha_1 \uparrow \delta_1^r \delta_2^s)\alpha_1^{-1}] = -r(t^2\mathbf{e}_2 + \mathbf{v}) + s(\mathbf{e}_2 - \mathbf{v}) - \binom{r}{2}\mathbf{v} - \binom{s}{2}t^2\mathbf{v} + rst\mathbf{v}.$$

These classes belong to $\bar{\mathrm{V}}'_H$ if $r = s = 0 \bmod m$. □

Chapter 7

A few simple computations

In this and next chapter, we apply the machinery developed in Chapter 5 to compute the fundamental groups of some classes of curves. Here, we consider a few simple cases that eventually reduce to proper trigonal curves in Σ_2.

7.1 Trigonal curves in Σ_2

We start with computing the fundamental groups of proper trigonal curves in Σ_2. The projective groups obtained appear later as the fundamental groups of certain non-simple sextics, see Section 7.2.2. The affine groups of some unstable curves are closely related to perturbations of simple singularities, see Section 7.1.2.

7.1.1 Proper curves in Σ_2

Let $C \subset \Sigma_2$ be a simple proper trigonal curve, and let J be the minimal resolution of singularities of the double covering of Σ_2 ramified at $E + C$. Due to Corollary 3.58, one has $H_2(J) \cong \mathbf{E}_8 \oplus \mathbf{U}$; hence, the homological type of C, see Section 6.1.2, is an embedding $\varphi\colon \mathbf{S} \hookrightarrow \mathbf{E}_8$ of root lattices (where $\mathbf{S}$ is the set of singularities of C) and one may expect a certain degree of 'rigidity'. The following theorem resembles Theorem 7.27 below but, unlike Theorem 7.27, I do not know a conceptual proof of this statement: it follows from comparing two independent classifications.

Theorem 7.1. *Up to equisingular (not necessarily fiberwise) deformation, a simple proper trigonal curve $C \subset \Sigma_2$ is determined by its homological type $\varphi\colon \mathbf{S} \hookrightarrow \mathbf{E}_8$. An embedding $\varphi\colon \mathbf{S} \hookrightarrow \mathbf{E}_8$ of a root lattice $\mathbf{S}$ appears as the homological type of a simple proper trigonal curve $C \subset \Sigma_2$ if and only if $\ell(\operatorname{Tors} \mathbf{E}_8/\varphi(\mathbf{S})) \leqslant 2$.*

Remark 7.2. According to Proposition 6.2 and Theorem 6.10, see also Theorem 6.12, the trigonal curve C corresponding to an embedding $\varphi\colon \mathbf{S} \hookrightarrow \mathbf{E}_8$ is reducible if and only if $\ell_2(\mathbf{E}_8/\varphi(\mathbf{S})) > 0$, and it is of torus type if and only if $\ell_3(\mathbf{E}_8/\varphi(\mathbf{S})) > 0$.

Remark 7.3. There is one family of non-simple proper trigonal curves in Σ_2. Each curve C of this family is the union of three sections tangent at a common point, which is a type $\mathbf{J}_{10}$ singular point of C.

Proof of Theorem 7.1. The necessity of the condition $\ell(\operatorname{Tors} \mathbf{E}_8/\varphi(\mathbf{S})) \leqslant 2$ follows from Theorem 6.12 and the fact that the maximal possible uniform dihedral quotient

Table 7.1. Maximal stable proper trigonal curves of degree six.[1]

#	Set of singularities	Figure	Count	π_1	$[\pi_1, \pi_1]$
* 1	$4\mathbf{A}_2$	4.2(a)	$(1,0)$	7.9	
* 2	$(\mathbf{A}_8)$	4.2(b)	$(1,0)$	5.83, 7.11	$\mathbb{F}_2$
* 3	$2\mathbf{A}_4$	4.2(d)	$(1,0)$	7.12	$\mathbb{Z}_5$
* 4	$(\mathbf{E}_6 \oplus \mathbf{A}_2)$	4.3(c)	$(1,0)$	7.11	$\mathbb{F}_2$
5	$\mathbf{E}_8$	4.3(d)	$(1,0)$	5.82	0
* 1′	$(\mathbf{A}_5 \oplus \mathbf{A}_2) \oplus \mathbf{A}_1$	4.2(c)	$(1,0)$	7.13	$\mathbb{F}_2$
* 2′	$\mathbf{A}_7 \oplus \mathbf{A}_1$	4.2(e)	$(1,0)$	7.13	$\mathbb{Z}$
* 3′	$2\mathbf{A}_3 \oplus 2\mathbf{A}_1$	4.2(f)	$(1,0)$	7.14	$\mathbb{Z} \times \mathbb{Z}$
* 4′	$\mathbf{D}_6 \oplus 2\mathbf{A}_1$	4.3(a)	$(1,0)$	7.15	
5′	$\mathbf{D}_8$	4.3(b)	$(1,0)$	5.82	0
* 6′	$\mathbf{D}_5 \oplus \mathbf{A}_3$	4.3(b)	$(1,0)$	7.13	$\mathbb{Z}$
7′	$\mathbf{E}_7 \oplus \mathbf{A}_1$	4.3(e)	$(1,0)$	5.82	0

[1] There are 12 classes realizing 12 sets of singularities

of $\pi^{\text{aff}}(C)$ is $\mathbb{D}[\mathcal{H}]$, *cf.* Theorem 6.1, and $\ell(\mathcal{H}) = 2$. For the other two statements, we compare the classification of lattice extensions $\mathbf{S} \subset \mathbf{E}_8$ given by Propositions A.13 and A.18 and independent geometric classification of trigonal curves. Since we do not require that deformations should be fiberwise, we consider stable curves only.

If C is isotrivial, its set of singularities is $2\mathbf{D}_4$ (all other isotrivial simple curves in Σ_2 are unstable), and such curves form a single deformation family, see page 75.

Any non-isotrivial stable simple curve admits a simple degeneration to a maximal one, see Corollary 4.47, and maximal curves are classified by their skeletons, see Figures 4.2 and 4.3; these curves are listed in Table 7.1, see comments to Table 8.1 on page 235 for the conventions. (For the skeleton in Figure 4.3 (b), one should also distinguish between two distinct type specifications.) All other curves are obtained from the maximal ones by perturbing one or several non-exceptional singular fibers, and their dessins can be compared using Proposition 4.48 and Lemma 4.51. In most cases, the equivalence class of a dessin is determined by the set of singularities of the corresponding curve.

The five exceptional lattices in Proposition A.18 can be realized by both irreducible and reducible curves; the former can be obtained by a perturbation of $(\mathbf{A}_8)$, Line 2 in Table 7.1, the latter, by a perturbation of $\mathbf{A}_7 \oplus \mathbf{A}_1$, Line 2′ to $\mathbf{A}_7$ (the curve remains reducible), followed by further perturbations of the type $\tilde{\mathbf{A}}_7$ fiber. □

As an addendum, we have the following lemma.

Lemma 7.4. *Any irreducible stable simple proper trigonal curve in Σ_2 without triple points admits a simple degeneration to an* irreducible *maximal curve.*

Proof. It suffices to consider the few perturbations of the curves with the skeletons shown in Figure 4.2 (c), (e), (f) breaking a fiber of type I_{2k} into two 'odd' fibers. Using Lemma 4.51, one can show that each dessin can be modified so that its skeleton is one of those shown in Figure 4.2 (a), (b), or (d). □

Consider a pair (C, F), where $C \subset \Sigma_2$ is a simple proper trigonal curve and F is a distinguished stable *singular* fiber of C. Two such pairs (C_i, F_i), $i = 1, 2$, are *deformation equivalent* if the curves $C_i + F_i$ are related by an equisingular, but not necessarily fiberwise, deformation. Define the *refined homological type* of (C, F) as the homological type $\varphi\colon \mathbf{S} \hookrightarrow \mathbf{E}_8$ of C refined with the splitting $\mathbf{S} = \mathbf{S}' \oplus \mathbf{T}$, where $\mathbf{T}$ is the irreducible summand representing the singular point of C in F (or 0 if F is of type $\mathbf{A}_0^*$) and $\mathbf{S}' = \mathbf{T}^\perp$. An *isomorphism* of such refined homological types is an isomorphism of the underlying ordinary homological types preserving the splitting.

Theorem 7.5. *There is a one-to-one correspondence between the set of deformation classes of pairs (C, F) as above and the set of isomorphism classes of refined homological types $\varphi\colon \mathbf{S}' \oplus \mathbf{T} \hookrightarrow \mathbf{E}_8$ such that $\ell(\mathrm{Tors}\, \mathbf{E}_8/\varphi(\mathbf{S}')) \leqslant 1$.*

Proof. If $\mathbf{T} \neq 0$, Proposition A.18 implies that the condition $\ell(\mathrm{Tors}\, \mathbf{E}_8/\varphi(\mathbf{S}')) \leqslant 1$ is equivalent to $\ell(\mathrm{Tors}\, \mathbf{E}_8/\varphi(\mathbf{S})) \leqslant 2$, and the existence of a pair (C, F) is given by Theorem 7.1. If $\mathbf{T} = 0$, the fiber F is of type $\tilde{\mathbf{A}}_0^*$, and the only curves that do *not* have such a fiber are $4\mathbf{A}_2$, see Figure 4.2 (a), and the curves splitting into three sections, *i.e.*, precisely those with $\ell(\mathrm{Tors}\, \mathbf{E}_8/\varphi(\mathbf{S})) = 2$. The fact that the refined homological type determines a pair (C, F) uniquely up to deformation is proved as in Theorem 7.1. □

Similarly, we can define the refined homological type $\varphi\colon \mathbf{S}' \oplus \mathbf{T} \hookrightarrow \mathbf{E}_8$ of a pair (C, F) with F an unstable fiber of C. (Here, $\mathbf{T} \cong 0$, $\mathbf{A}_1$, or $\mathbf{A}_2$ if F is of type $\tilde{\mathbf{A}}_0^{**}$, $\tilde{\mathbf{A}}_1^*$, or $\tilde{\mathbf{A}}_2^*$, respectively.)

Theorem 7.6. *There is a one-to-one correspondence between the set of deformation classes of pairs (C, F) as above with the fiber F unstable and the set of isomorphism classes of refined homological types $\varphi\colon \mathbf{S}' \oplus \mathbf{T} \hookrightarrow \mathbf{E}_8$ such that $\mathrm{Tors}(\mathbf{E}_8/\varphi(\mathbf{S}')) = 0$.*

This theorem is proved in Section 7.1.2, where the three types of unstable fibers are considered one by one: the curves C are constructed as perturbations, 'fixed at infinity', of an $\mathbf{E}$ type singular point of an appropriate isotrivial curve. The point perturbed is of type $\varphi(\mathbf{T})^\perp = \mathbf{E}_8$, $\mathbf{E}_7$, or $\mathbf{E}_6$, and the classification of pairs is reduced to that of primitive embeddings $\mathbf{S}' \hookrightarrow \varphi(\mathbf{T})^\perp$, see Propositions A.13 and A.16–A.18.

Theorem 7.7. *With the exception of the set of singularities*

- $2\mathbf{A}_4$, *Line* 3 *in Table* 7.1, *where* $\pi_1 \cong \mathbb{D}_{10} \times \mathbb{Z}_3$,

the fundamental group $\pi_1 := \pi^{\mathrm{proj}}(C)$ of an irreducible proper trigonal curve $C \subset \Sigma_2$ that is not of torus type is abelian.

Theorem 7.8. *With the exception of the set of singularities*

- $4\mathbf{A}_2$, *Line* 1 *in Table* 7.1, *where* π_1 *is given by* (7.10),

the fundamental group $\pi_1 := \pi^{\mathrm{proj}}(C)$ *of an irreducible proper trigonal curve* $C \subset \Sigma_2$ *of torus type is isomorphic to* Γ.

Proof of Theorems 7.7 *and* 7.8. We start with computing the groups of all maximal stable curves, both irreducible and reducible, using the skeletons shown in Figures 4.2 and 4.3 and the strategy outlined in Section 5.2.3. The relation at infinity (5.35) takes the form $\rho^2 = 1$ and, in view of Lemma 5.37, it suffices to consider the braid relations given by all but one regions of the skeleton.

Most groups are easily constructed and analyzed using GAP, see Listing 5.3; details are found in "misc/Sigma2.txt" and the arguments below.

Argument 7.9 ($4\mathbf{A}_2$, Line 1). The braid relations from the triangles $\mathrm{reg}_-(e \uparrow \zeta)$, $\zeta = \mathbf{y}, \mathbf{yx}, \mathbf{yx}^2$, combined with the relation at infinity, give us the presentation

$$\pi_1 = \langle \alpha_1, \alpha_2, \alpha_3 \mid \{\alpha_1, \alpha_2\}_3 = \{\alpha_1, \alpha_3\}_3 = \{\alpha_2, \alpha_3\}_3 = \rho^2 = 1 \rangle. \qquad (7.10)$$

This curve is of torus type; it has four distinct torus structures and is a universal curve with this property, see Theorem 6.11 and Figure 6.1 (b). According to [49], the group given by (7.10) is the 'minimal' fundamental group of a sextic of weight eight, see G_0 in Section 7.2.3 (*cf.* also Argument 8.45).

Argument 7.11 (other irreducible curves of torus type). There is an epimorphism $\mathbb{B}_3 \twoheadrightarrow \pi_1$: for $(\mathbf{A}_8)$, Line 2, it is given by Lemma 5.83, and for $(\mathbf{E}_6 \oplus \mathbf{A}_2)$, Line 4, it follows immediately from the two braid relations. Hence, in view of Theorem 6.10, we have $\pi_1 \cong \Gamma$, *cf.* Lemma 7.32 below.

Argument 7.12 ($2\mathbf{A}_4$, Line 3). Choosing the reference edge e in the boundary $\partial_- R$ of a monogonal region R, we have $\alpha_1 = \alpha_2$ and, together with the braid relation from the pentagon adjacent to R, this gives us the presentation

$$\pi_1 = \langle \alpha_2, \alpha_3 \mid \{\alpha_2, \alpha_3\}_5 = (\alpha_2^2\alpha_3)^2 = 1 \rangle \cong \mathbb{D}_{10} \times \mathbb{Z}_3.$$

The braid relation from the other pentagon can be omitted as it does not change the size of the group.

Argument 7.13 (reducible curves with a monogonal region). If the skeleton $\mathfrak{S} := \mathrm{Sk}\, C$ has a monogonal region R, then, assuming that the region R' adjacent to R is of type $\tilde{\mathbf{A}}$, we have the relations $\alpha_1 = \alpha_2$ and $\{\alpha_2, \alpha_3\}_w = 1$, where $w := \mathrm{wd}\, R'$ is even (since C is assumed reducible). Hence, $\delta := (\alpha_2\alpha_3)^{w/2}$ is a central element and, due to Corollary A.30, we can use GAP to analyze the commutant of the quotient π_1/δ. For the sets of singularities $\mathbf{A}_7 \oplus \mathbf{A}_1$, Line 2′ and $\mathbf{D}_5 \oplus \mathbf{A}_3$, Line 6′, the groups can easily be identified as semidirect products $(\Lambda/(t^2-1)) \rtimes \mathbb{Z}_2$, with the quotient $\mathbb{Z}_2$ (generated by ρ) acting *via* the multiplication by t.

Argument 7.14 ($2\mathbf{A}_3 \oplus 2\mathbf{A}_1$, Line 3′). Choosing the reference edge e in the boundary $\partial_- R$ of a bigonal region R, see Figure 4.2 (f), from R and reg$_-(e \uparrow \mathbf{xyx}^2)$ we have relations $[\alpha_1, \alpha_2] = [\rho^{-1}\alpha_1\rho, \alpha_2] = 1$. Since $\rho^2 = 1$, it follows that the four elements α_i, $\rho^{-1}\alpha_i\rho$, $i = 1, 2$, pairwise commute, and this implies $[(\alpha_1^{-1}\rho)^2, \alpha_2] = 1$, which is equivalent to the third braid relation from the tetragon reg$_-(e \uparrow \mathbf{x})$. Hence, π_1 is a semidirect product $(\Lambda^2/(t^2 - 1)) \rtimes \mathbb{Z}_2$, with the quotient $\mathbb{Z}_2$, generated by ρ, acting *via* the multiplication by t.

Argument 7.15 ($\mathbf{D}_6 \oplus 2\mathbf{A}_1$, Line 4′). Choosing the reference edge e so that both reg$_-$ e and reg$_-(e \uparrow \mathbf{x})$ are $\tilde{\mathbf{A}}$ type regions, we have $[\alpha_1, \alpha_2] = [\alpha_2, \alpha_3] = \rho^2 = 1$, *i.e.*, $\pi_1 = (\mathbb{Z} * \mathbb{Z}_2) \times \mathbb{Z}$.

Any stable irreducible proper trigonal curve $C \subset \Sigma_2$ is not isotrivial. Hence, C admits a simple degeneration to a maximal curve C', see Corollary 4.47, and the group π_1 of C can be computed together with that of C', by adjusting the braid indices of the regions, *cf.* Observation 8.4. In fact, in all cases one can conclude that π_1 is as expected, *i.e.*, $\mathbb{Z}_6$ or Γ, using Lemmas 5.82 and 5.83. □

Remark 7.16. The fundamental groups of the two isotrivial stable curves can be computed as explained on page 169. The monodromy groups of these curves are generated by Δ^2 (for the set of singularities $2\mathbf{D}_4$) or Δ^4 (for the set of singularities $\mathbf{J}_{10}$); hence, the fundamental groups are $\langle\alpha_1, \alpha_2\rangle \times \mathbb{Z}_2$ or $\langle\alpha_1, \alpha_2\rangle * \mathbb{Z}_2$, respectively. In both cases, the $\mathbb{Z}_2$ factor is generated by ρ.

7.1.2 Perturbations of simple singularities

As another application, we compute the local fundamental groups of perturbations of all simple singularities.

Consider an isolated singular point p of a curve $C \subset S$, and let D_ϵ be a Milnor ball about p. The group $\pi_1(D_\epsilon \smallsetminus C) = \pi_1(\partial D_\epsilon \smallsetminus C)$ is called the *local fundamental group* of C at p. Consider a perturbation C_s of $C = C_0$, and assume that it remains transversal to the boundary ∂D_ϵ. Then all spaces $\partial D_\epsilon \smallsetminus C_s$ are homotopy equivalent and, similar to Proposition 5.33, the inclusion $\partial D_\epsilon \smallsetminus C_1 \hookrightarrow D_\epsilon \smallsetminus C_1$ induces an epimorphism $\pi_1(D_\epsilon \smallsetminus C) \twoheadrightarrow \pi_1(D_\epsilon \smallsetminus C_1)$ of the local fundamental groups. It is called the *perturbation epimorphism*.

Assume that p is a simple singular point of type $\mathbf{T}$. According to Looijenga [113], the deformation classes of the germs of perturbations of p are in a canonical one-to-one correspondence with the isomorphism classes of *primitive* embeddings $\mathbf{S} \hookrightarrow \mathbf{T}$ of root lattices, and a perturbation corresponding to a sublattice $\mathbf{S} \subset \mathbf{T}$ has $\mathbf{S}$ as its set of singularities. Using Proposition A.13, one can deduce that any perturbation factors through a *maximal* one (in the sense of the total Milnor number) and each maximal perturbation is obtained by removing a single vertex from the Dynkin diagram $\mathbf{T}$.

Below we consider, one by one, the types of simple singularities and compute the local fundamental groups of their perturbations.

Perturbations of $\mathbf{E}_8$

Consider the isotrivial curve $C \subset \Sigma_2$ given in affine coordinates (x, t) by the equation $x^3 + t^5 = 0$. It has a type $\mathbf{E}_8$ singular point at $(0, 0)$, a type $\tilde{\mathbf{A}}_0^{**}$ singular fiber F_∞ over $t = \infty$, and no other singular fibers. Since the curve is proper and has two singular fibers only, the pair (D_ϵ, C) is deformation equivalent to the pair $(\mathbb{C}^2, C)$, where $\mathbb{C}^2 = \Sigma_2 \smallsetminus (E \cup F_\infty)$, and the local fundamental group at $(0, 0)$ is equal to the group $\pi_1(\Sigma_2 \smallsetminus (C \cup E \cup F_\infty))$.

Let $C' \subset \Sigma_2$ be another proper trigonal curve given by its Weierstraß equation $f(x, t) = 0$, see (3.4), where g_2 and g_3 are regarded as polynomials in t. The curve C' has a type $\tilde{\mathbf{A}}_0^{**}$ fiber at infinity if and only if $\deg g_2 \leqslant 3$ and $\deg g_3 = 5$. Renormalize the equation to make the leading coefficient of g_3 equal to 1 and consider the family $C_s := \{s^{15} f(x/s^5, t/s^3) = 0\}$, where $s \in \mathbb{C}$ is a parameter. It is immediate that $C_1 = C'$ and the family is a perturbation of $C_0 = C$, as the curve C_s, $s \neq 0$, is obtained from C_1 by the automorphism $(x, t) \mapsto (s^5 x, s^3 t)$. Furthermore, this perturbation preserves the singular fiber at infinity, thus remaining transversal to the boundary of a sufficiently large Milnor ball. Hence, the perturbed local fundamental group can be found as $\pi_1(\Sigma_2 \smallsetminus (C' \cup E \cup F_\infty))$.

In view of [113], there are eight maximal perturbations of a type $\mathbf{E}_8$ singular point, see Proposition A.13, and all these perturbations can be realized by maximal trigonal curves $C' \subset \Sigma_2$, see Figure 7.1, where the fiber F_∞ is represented by the rightmost monovalent •-vertex v. (These skeletons are obtained by contracting a monogonal region in Figures 4.2 and 4.3, see Observation 4.41.)

Analyzing further perturbations as in the proof of Theorem 7.1, we obtain a proof of Theorem 7.6 for the case where $F = F_\infty$ is of type $\tilde{\mathbf{A}}_0^{**}$.

Let $\{\beta_1, \beta_2, \beta_3\}$ be a canonical basis over the edge incident to v. In this basis, the original local group of the type $\mathbf{E}_8$ singular point is

$$\pi_1(D_\epsilon \smallsetminus C) = \langle \beta_1, \beta_2, \beta_3 \mid \beta_1\rho^2 = \rho^2\beta_2,\ \beta_2\rho^2 = \rho^2\beta_3,\ \beta_3\rho = \rho\beta_1 \rangle, \qquad (7.17)$$

where $\rho = \beta_1\beta_2\beta_3$. Observe that ρ^5 is a central element.

Theorem 7.18. *Up to deformation, there are three proper perturbations of a type* $\mathbf{E}_8$ *singularity with nonabelian fundamental group. They are as follows*:

- $\mathbf{A}_4 \oplus \mathbf{A}_3$: $\{\beta_1, \beta_2\}_4 = \{\beta_1, \beta_3\}_5 = 1$ *and* $\beta_1 = \beta_3\beta_1\beta_2\beta_1^{-1}\beta_3^{-1}$;
- $\mathbf{A}_4 \oplus \mathbf{A}_2 \oplus \mathbf{A}_1$: $\{\beta_1, \beta_2\}_5 = \{\beta_1, \beta_3\}_3 = [\beta_1\beta_2\beta_1^{-1}, \beta_3] = 1$;
- $\mathbf{D}_5 \oplus \mathbf{A}_2$: $\{\beta_1, \beta_2\}_3 = [\beta_2, \beta_3\beta_1] = 1$ *and* $\beta_3 = (\beta_1\beta_2\beta_3)\beta_1(\beta_1\beta_2\beta_3)^{-1}$,

where listed are the sets of singularities of the perturbed curves C' *and the relations in the group* $\pi_1(D_\epsilon \smallsetminus C')$ *in the basis* $\{\beta_1, \beta_2, \beta_3\}$ *described above.*

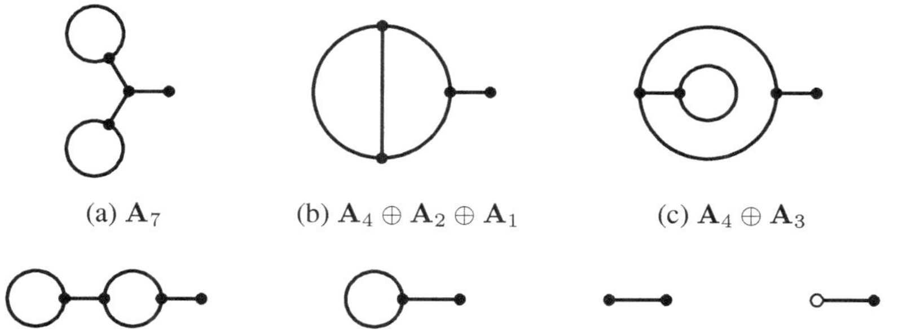

(a) $\mathbf{A}_7$ (b) $\mathbf{A}_4 \oplus \mathbf{A}_2 \oplus \mathbf{A}_1$ (c) $\mathbf{A}_4 \oplus \mathbf{A}_3$

(d) $\mathbf{A}_6 \oplus \mathbf{A}_1$ (e) $\mathbf{D}_7$, $\mathbf{D}_5 \oplus \mathbf{A}_2$ (f) $\mathbf{E}_6 \oplus \mathbf{A}_1$ (g) $\mathbf{E}_7$

Figure 7.1. Maximal perturbations of $\mathbf{E}_8$.

For the three groups listed in the statement, one has, in the order of appearance, $\operatorname{ord}(\pi_1/\beta_1^3) = 360$, $\operatorname{ord}(\pi_1/\beta_1^2) = 120$, and $\operatorname{ord}(\pi_1/\beta_1^5) = 600$; hence, they are not abelian. All three groups have perfect commutants. For the last perturbation $\mathbf{D}_5 \oplus \mathbf{A}_2$, one can use GAP to analyze the central quotient π_1/ρ^5, see Corollary A.30, and show that $\pi_1 \cong \mathbb{Z} \times SL(2, \Bbbk_5)$. I do not know whether the commutants of the other two groups are finite.

Proof. We compute the groups of the eight maximal perturbations using the approach outlined in Section 5.2.3. Since the fiber F_∞ remains removed, one has

$$\pi_1 := \pi_1(\Sigma_2 \smallsetminus (C' \cup E \cup F_\infty)) = \langle \beta_1, \beta_2, \beta_3 \mid \mathfrak{m}_i = \mathrm{id} \rangle,$$

where $\mathfrak{m}_i$ are the braid monodromies about all regions of the skeleton and all its singular vertices other than v.

For the perturbations $\mathbf{A}_7$ and $\mathbf{D}_7$, see Figures 7.1 (a) and (e), respectively, the groups are abelian due to Lemma 5.82.

For the other maximal perturbations, a simple way to conclude that their groups are abelian is to use GAP and show that $\operatorname{ord}(\pi_1/\rho^5) = 15$, see Corollary A.30. The GAP input is found in `"misc/simple.txt"`. For non-maximal perturbations, we use Lemma 4.51 to show that any such perturbation degenerates to a maximal one with abelian fundamental group. Alternatively, adjusting the braid indices of regions, one can use GAP to show that $\operatorname{ord}(\pi_1/\rho^5) = 15$. Details are left to the reader. □

Perturbations of $\mathbf{E}_7$

As in the previous case, any perturbation of a type $\mathbf{E}_7$ singularity can be realized by a family of trigonal curves $C_s := \{s^9 f(x/s^3, t/s^2) = 0\} \subset \Sigma_2$, where $f(x,t) = 0$ is a Weierstraß equation (3.4) with $\deg g_2 = 3$ and $\deg g_3 \leqslant 4$, so that all curves have a type $\tilde{\mathbf{A}}_1^*$ singular fiber at infinity. The limit C_0 is the isotrivial curve $x^3 + xt^3 = 0$

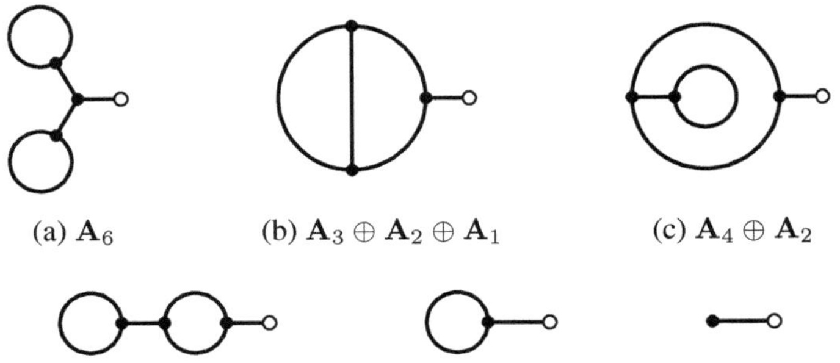

(a) $\mathbf{A}_6$ (b) $\mathbf{A}_3 \oplus \mathbf{A}_2 \oplus \mathbf{A}_1$ (c) $\mathbf{A}_4 \oplus \mathbf{A}_2$

(d) $\mathbf{A}_5 \oplus \mathbf{A}_1$ (e) $\mathbf{D}_6,\ \mathbf{D}_5 \oplus \mathbf{A}_1$ (f) $\mathbf{E}_6$

Figure 7.2. Maximal perturbations of $\mathbf{E}_7$.

with a type $\mathbf{E}_7$ singular point at the origin. There are seven maximal perturbations, see Proposition A.13; their skeletons are shown in Figure 7.2, where the type $\tilde{\mathbf{A}}_1^*$ fiber F_∞ is represented by the monovalent $\circ$-vertex v.

Analyzing further perturbations, we obtain a proof of Theorem 7.6 for the fiber $F = F_\infty$ of type $\tilde{\mathbf{A}}_1^*$.

Let $\{\beta_1, \beta_2, \beta_3\}$ be a canonical basis over the edge incident to v. For the perturbed group, the relations look simpler in the basis $\delta_i = \beta_i \uparrow \sigma_1\sigma_2$, $i = 1, 2, 3$. In this basis, the local fundamental group of C has the form

$$\pi_1(D_\epsilon \smallsetminus C) = \langle \delta_1, \delta_2, \delta_3 \mid \delta_2\rho^2 = \rho^2\delta_3,\ \delta_3\rho = \rho\delta_2 \rangle. \tag{7.19}$$

Note that ρ^3 is a central element.

Theorem 7.20. *Up to deformation, there are five proper perturbations of a type* $\mathbf{E}_7$ *singularity with nonabelian fundamental group. They are as follows*:

- $\mathbf{A}_4 \oplus \mathbf{A}_2$: $\{\delta_1, \delta_2\}_3 = \{\delta_2, \delta_3\}_5 = 1$, $\delta_2 = \delta_3\delta_1\delta_3^{-1}$;
- $\mathbf{A}_3 \oplus \mathbf{A}_2 \oplus \mathbf{A}_1$: $[\delta_1, \delta_3] = \{\delta_1, \delta_2\}_4 = \{\delta_2, \delta_3\}_3 = 1$;
- $\mathbf{A}_5 \oplus \mathbf{A}_1$: $[\delta_2, \delta_3] = \{\delta_1, \delta_2\}_6 = 1$, $\delta_3 = \delta_1\delta_2\delta_1^{-1}$;
- $\mathbf{D}_5 \oplus \mathbf{A}_1$ *and* $\mathbf{A}_2 \oplus 3\mathbf{A}_1$: $[\delta_1, \delta_2] = [\delta_1, \delta_3] = \{\delta_2, \delta_3\}_3 = 1$,

where listed are the sets of singularities of the perturbed curves C' *and the relations in the group* $\pi_1(D_\epsilon \smallsetminus C')$ *in the basis* $\delta_1 = \rho\beta_3\rho^{-1}$, $\delta_2 = \beta_1$, $\delta_3 = \beta_2$.

One can use `GAP` to show that the first group in the statement is isomorphic to $\mathbb{Z} \times SL(2, \Bbbk_5)$. The second group factors to the last one, which is obviously $\mathbb{Z} \times \mathbb{B}_3$. For the third group (the perturbation $\mathbf{A}_5 \oplus \mathbf{A}_1$), eliminate δ_3 and change to the basis δ_2, $\kappa := \delta_1\delta_2$. The new relations are $[\delta_2, \kappa^3] = [\delta_2, \kappa\delta_2\kappa^{-1}]$, and it is easily seen that the group is a semidirect product $(\Lambda/(t^3 - 1)) \rtimes \mathbb{Z}$, where the kernel is generated by δ_2 as a Λ-module and the quotient, generated by κ, acts by the multiplication by t.

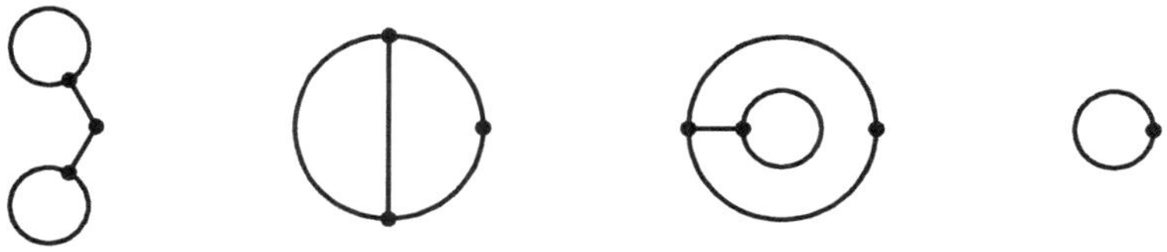

(a) $\mathbf{A}_5$ (b) $2\mathbf{A}_2 \oplus \mathbf{A}_1$ (c) $\mathbf{A}_4 \oplus \mathbf{A}_1$ (d) $\mathbf{D}_5$

Figure 7.3. Maximal perturbations of $\mathbf{E}_6$.

Proof of Theorem 7.20. The groups are computed as outlined in Section 5.2.3. It has already been explained that the four groups listed in the statement are not abelian, whereas the three other maximal groups and most non-maximal ones are abelian due to Lemma 5.82. □

Perturbations of $\mathbf{E}_6$

Any perturbation of a type $\mathbf{E}_6$ singularity can be realized by a family of trigonal curves $C_s := \{s^{12}f(x/s^4, t/s^3) = 0\} \subset \Sigma_2$, where $f(x,t) = 0$ is a Weierstraß equation (3.4) with $\deg g_2 \leqslant 2$ and $\deg g_3 = 4$, so that all curves have a type $\tilde{\mathbf{A}}_2^*$ singular fiber at infinity. The limit C_0 is the isotrivial curve $x^3 + t^4 = 0$ with a type $\mathbf{E}_6$ singular point at the origin. According to [113] and Proposition A.13, there are four maximal perturbations; their skeletons are shown in Figure 7.3, where the type $\tilde{\mathbf{A}}_2^*$ fiber F_∞ is represented by the bivalent •-vertex v. These skeletons can be obtained from Figures 4.2 and 4.3, *cf.* Observation 4.41. Figures 4.2 (d) and (e) result in isomorphic skeletons, see Figure 7.3 (c).

Further analysis of non-maximal perturbations gives us a proof of Theorem 7.6 for the remaining case where $F = F_\infty$ is a fiber of type $\tilde{\mathbf{A}}_2^*$.

In a canonical basis $\{\beta_1, \beta_2, \beta_3\}$ over the upper edge incident to v in the figures, the original local fundamental group is

$$\pi_1(D_\epsilon \smallsetminus C) = \langle \beta_1, \beta_2, \beta_3 \mid \beta_2\rho = \rho\beta_1,\ \beta_3\rho = \rho\beta_2 \rangle. \tag{7.21}$$

Thus, the conjugation by ρ^{-1} acts *via* $\beta_1 \mapsto \beta_2 \mapsto \beta_3 \mapsto \rho^{-1}\beta_1\rho \mapsto \beta_1$ and ρ^4 is a central element.

Theorem 7.22. *Up to deformation, there are three proper perturbations of a type* $\mathbf{E}_6$ *singularity with nonabelian fundamental group. They are as follows*:

- $2\mathbf{A}_2 \oplus \mathbf{A}_1$: $[\beta_1, \beta_3] = \{\beta_1, \beta_2\}_3 = \{\beta_2, \beta_3\}_3 = 1$;
- $2\mathbf{A}_2$ *and* $\mathbf{A}_5$: $\{\beta_1, \beta_2\}_3 = 1$, $\beta_1 = \beta_3$,

where listed are the sets of singularities of the perturbed curves C' *and the relations in the group* $\pi_1(D_\epsilon \smallsetminus C')$ *in the basis* $\{\beta_1, \beta_2, \beta_3\}$ *described above.*

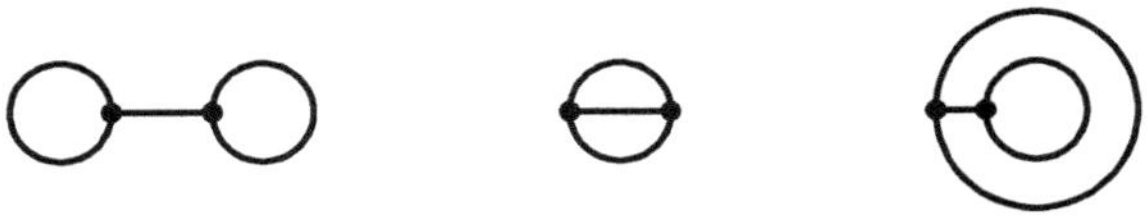

(a) $\mathbf{A}_{m-1}$ (b) $\mathbf{D}_2 \oplus \ldots$ (c) $\mathbf{D}_3 \oplus \ldots$

Figure 7.4. Some perturbations of $\mathbf{D}_m$.

Obviously, the former group is $\mathbb{B}_4$, and the latter is $\mathbb{B}_3$. The three perturbations that have nonabelian fundamental groups are said to be of *torus type*.

Proof of Theorem 7.22. The relations are computed as outlined in Section 5.2.3. For all perturbations not listed in the statement, including the non-maximal ones, the fact that the group is abelian (in fact, cyclic) is immediate; in most cases, this can formally be deduced from Lemma 5.82. □

Perturbations of A and D type singularities

The perturbations of a type $\mathbf{A}_{m-1}$ singularity are well known. Any such perturbation can be realized by a family $C_s := \{x^2 + s^m f(t/s) = 0\}$, where f is a polynomial of degree m. The braid monodromy of the hyperelliptic curve C_s takes values in the cyclic group $\mathbb{B}_2 = \mathbb{Z}$ and is easily computable. For references, we state the result.

Theorem 7.23. *Any perturbation of a type* $\mathbf{A}_{m-1}$ *singularity is of the form*

- $\bigoplus_i \mathbf{A}_{s_i-1}$, *where* $d := m - \sum_i s_i \geqslant 0$.

Let $s = 1$ *if* $d > 0$ *and* $s = \text{g.c.d.}(s_i)$ *if* $d = 0$. *Then the local fundamental group of the perturbed curve is* $\langle \beta_1, \beta_2 \mid \{\beta_1, \beta_2\}_s = 1\rangle$. ◁

Perturbations with $s = 0 \bmod 3$ (since $s \mid m$, one has $m = 0 \bmod 3$ as well) are said to be of *torus type*. They are related to sextics of torus type, see Observation 8.4.

Perturbations of a $\mathbf{D}_m$ type singular point are given by [113] and Proposition A.13. If $m \geqslant 5$, such a point can be realized within the only singular fiber of a proper trigonal curve C in an $(m-4)$-gonal region R of the skeleton $\operatorname{Sk} C$. Let $\{\beta_1, \beta_2, \beta_3\}$ be a canonical basis over any edge e in $\partial_- R$.

Theorem 7.24. *Any proper perturbation of a type* $\mathbf{D}_m$ *singularity is either*

- $\mathbf{A}_{m-1}$ *or a further perturbation thereof, see Theorem* 7.23, *or*
- $\mathbf{D}_p \oplus \bigoplus_i \mathbf{A}_{s_i-1}$ *with* $p \geqslant 2$ *and* $d := m - p - \sum_i s_i \geqslant 0$.

In the former case, the perturbed local fundamental group is abelian. In the latter case, it is

$$\langle \beta_1, \beta_2 \mid \{\beta_1, \beta_2\}_s = 1\rangle \rtimes \langle \beta_3\rangle, \quad \beta_3^{-1}\beta_i\beta_3 = \beta_i \uparrow \sigma_1^{m-2},\ i = 1, 2,$$

where $s = 1$ *if* $d > 0$ *and* $s = \text{g.c.d.}(s_i)$ *if* $d = 0$. *This group is abelian if and only if either* $s = 1$ *or* m *is even and* $s = 2$.

Proof. The exceptional case $\mathbf{D}_4$ can easily be handled by considering plane cubics as perturbations of a triple of lines intersecting at a single point. If $m \geqslant 5$, consider the trigonal curve C mentioned prior to the statement and compute the fundamental group over the region R regarded as a monodromy domain. The braid monodromy about the boundary ∂R is $\sigma_1^{m-4}\Delta^2$, and the original group is easily seen to be

$$\langle \beta_1, \beta_2 \rangle \rtimes \langle \beta_3 \rangle, \quad \beta_3^{-1}\beta_i\beta_3 = \beta_i \uparrow \sigma_1^{m-2}, \; i = 1, 2. \tag{7.25}$$

In terms of Zariski quotients, this relation can be written in the form $\beta_3 = \sigma_1^{m-2}$, where β_3 is identified with $\operatorname{inn}\beta_3 \in \operatorname{Aut}\langle \beta_1, \beta_2, \beta_3 \rangle$.

Consider the dessin of a perturbed curve. For the perturbation $\mathbf{A}_{m-1}$, the original $\times$-vertex u of index $(m-4)$ is replaced with an m-valent $\times$-vertex and a fragment as in Figure 7.4 (a) (with two monovalent $\times$-vertices inside), which are connected in a certain way with solid and dotted edges. (Note that this procedure changes the degree of the j-invariant.) The resulting group is abelian due to Lemma 5.82. For the other perturbations, the original $\times$-vertex u is replaced with

- a number of $\times$-vertices of indices $s_1, \ldots$, one for each $\mathbf{A}$ type point,
- d monovalent $\times$-vertices (if $d > 0$), and
- a fragment shown in Figure 7.4 (b) or (c) (if $p = 2$ or 3, respectively) or a $\times$-vertex of index $(p-4)$ (if $p > 4$).

Thus, the braid index of R changes to s, resulting in an extra relation $\sigma_1^s = \text{id}$, or $\{\beta_1, \beta_2\}_s = 1$. Note that $s \mid (m-p)$, hence $\beta_3 = \sigma_1^{p-2}$ in $\operatorname{Aut}\mathbb{F}_3$. If $p \geqslant 4$, the latter is the braid relation about the remaining type $\tilde{\mathbf{D}}_p$ singular fiber. If $p = 2$ or 3, the additional relations, obtained from the insertions shown in Figure 7.4, are, respectively, $[\beta_1, \beta_3] = [\beta_2, \beta_3] = 1$ or $\{\beta_2, \beta_3\}_4 = 1$ and $\beta_2 = \beta_3^{-1}\beta_2^{-1}\beta_1\beta_2\beta_3$. One can see that these extra relations follow from $\beta_3 = \text{id}$ (for $p = 2$) or $\beta_3 = \sigma_1$ (for $p = 3$), which are already present. □

7.2 Sextics with a non-simple triple point

A *sextic* is a plane curve of degree six. Usually we assume sextics reduced, although formally this requirement is not part of the definition. Given a sextic $D \subset \mathbb{P}^2$, we use the shortcut $\pi_1 := \pi_1(\mathbb{P}^2 \setminus D)$ for its fundamental group.

7.2.1 A gentle introduction to plane sextics

Topology of singular plane sextics is a vast subject deserving a separate monograph. A brief survey of known tools and results and an outline of a few possible directions of the further investigation can be found in [58]. In this introductory section, we merely recall some concepts and basic properties needed in the sequel.

Simple sextics

A *simple sextic* is a sextic with all singular points simple. A set of singularities of such a sextic is identified with a negative definite root lattice $\mathbf{S}$, see Convention D.1.

Given a simple sextic $D \subset \mathbb{P}^2$, the minimal resolution of singularities X of the double covering of $\mathbb{P}^2$ ramified at D is a $K3$-surface. The classes of the exceptional divisors over the singular points of D span a sublattice in $\mathbf{L} := H_2(X)$, which is isomorphic to $\mathbf{S}$. Besides, the pull-back of a generic line in $\mathbb{P}^2$ defines a class $h \in \mathbf{L}$, $h^2 = 2$, orthogonal to $\mathbf{S}$. Thus, associated to a simple sextic D is a lattice extension

$$\mathbf{S} \oplus \mathbb{Z}h \subset \mathbf{L} \cong 2\mathbf{E}_8 \oplus 3\mathbf{U}, \quad h^2 = 2, \tag{7.26}$$

called the *homological type* of D. As a consequence, since $\sigma_-(\mathbf{L}) = 19$, the total Milnor number of D is $\mu(D) \leqslant 19$. A sextic D with $\mu(D) = 19$ is called *maximizing*, see [135]. (Both the inequality and the term apply to simple sextics only.)

The homological type of a simple sextic has the following properties, see [160]:

1. there is no root $r \in \mathbf{S}$ such that $\frac{1}{2}(r + h) \in \mathbf{L}$;
2. each root $r \in (\mathbf{S} \otimes \mathbb{Q}) \cap \mathbf{L}$ belongs to $\mathbf{S}$.

Here, the first condition guarantees that, generically, the linear system corresponding to h defines a degree 2 map $X \to \mathbb{P}^2$ and the ramification locus of this map is a certain plane sextic $D \subset \mathbb{P}^2$, see [140], and the second one makes sure that the set of singularities of this sextic D is precisely $\mathbf{S}$. A lattice extension (7.26) satisfying the above conditions and with $\mathbf{S}$ a negative root lattice is called an *abstract homological type*. A *morphism* between two abstract homological types $\mathbf{S}_i \oplus \mathbb{Z}h_i \subset \mathbf{L}$, $i = 1, 2$, is an automorphism of $\mathbf{L}$ taking h_1 to h_2 and $\mathbf{S}_1$ onto a *primitive* sublattice of $\mathbf{S}_2$.

Given an abstract homological type (7.26), define its *transcendental lattice* as $\mathbf{T} := (\mathbf{S} \oplus \mathbb{Z}h)^\perp \subset \mathbf{L}$. Since $\sigma_+\mathbf{T} = 2$, all positive definite 2-planes in $\mathbf{T} \otimes \mathbb{R}$ can be oriented in a coherent way, so that, for any two such planes ω_1, ω_2, the orthogonal projection $\omega_1 \to \omega_2$ is orientation preserving. A choice $\mathfrak{o}$ of one of these coherent orientations is called an *orientation* of the abstract homological type. The homological type of a plane sextic has a canonical orientation given by $(\operatorname{Re}\omega, \operatorname{Im}\omega)$, where $\omega \in \mathbf{L} \otimes \mathbb{C}$ is the class of a holomorphic 2-form on the covering $K3$-surface.

The next statement is based on the global Torelli theorem [136] and surjectivity of the period map [97] for $K3$-surfaces; its proof is outside of the scope of this book.

Theorem 7.27 (see [46]; the existence part is found in [166]). *The map sending a sextic to its oriented homological type establishes a bijection between the set of equisingular deformation families of simple sextics and that of isomorphism classes of oriented abstract homological types.* ▷ [46]

As a consequence, two simple sextics $D_1, D_2 \subset \mathbb{P}^2$ are deformation equivalent if and only if the pairs $(\mathbb{P}^2, D_1)$ and $(\mathbb{P}^2, D_2)$ are related by an orientation preserving diffeomorphism analytic in a neighborhood of each singular point, see [46].

Theorem 7.27 reduces the deformation classification of simple sextics to a purely arithmetical problem, which can be solved using Nikulin's theory of discriminant forms (see [125] and Section A.1.1) and the results of [117, 118]. According to I. Shimada, there must be about twelve thousand classes. The classification of maximizing sextics is known, see [166] (a list of maximizing sets of singularities) and [148] (sets of singularities extending to more than one homological type). A few other results are cited further in this section.

This approach, based on the global Torelli theorem, is not constructive: although clarifying certain algebro-geometric aspects, it leaves obscured the more topological properties of curves, such as their braid monodromy and fundamental groups.

As yet another consequence of the theory of $K3$-surfaces (the existence of a fine period space, see [17]) we have the following statement concerning perturbations.

Theorem 7.28. *Let D be a simple sextic with the set of singularities $\mathbf{S}$ and let $\mathbf{S}' \subset \mathbf{S}$ be a primitive root sublattice of $\mathbf{S}$. Then the restricted lattice extension $\mathbf{S}' \oplus \mathbb{Z}h \subset \mathbf{L}$ is the homological type of a certain simple sextic D' which can be obtained from D by a small perturbation.* ▷ [49, 149]

Thus, any morphism of homological types is induced by a degeneration of sextics. Alternatively, in view of [113], Theorem 7.28 states that the singular points of a simple sextic can be perturbed arbitrarily and independently. In general, the existence of an independent perturbation of singularities is a difficult problem. The simplest known counterexample of degree seven is the union of a three-cuspidal quartic Q and the three tangents to Q at its cusps. These tangents meet at a single point to form a $\mathbf{D}_4$ singularity, which cannot be perturbed, *e.g.*, to $3\mathbf{A}_1$ independently of the three other singular points of type $\mathbf{E}_7$.

Sextics of torus type

A plane sextic D is said to be of *torus type* if its equation can be represented in the form $f_2^3 + f_3^2 = 0$, where f_2 and f_3 are some homogeneous polynomials of degree 2 and 3, respectively. A representation as above, regarded up to obvious equivalence, is called a *torus structure* on D. With the exception of a few very degenerate curves (which are all unions of lines passing through a common point, thus reducible and non-simple), a torus structure is determined by the conic $Q_2 := \{f_2 = 0\}$. A torus structure represents D as the branch locus of a projection to $\mathbb{P}^2$ of a cubic surface $V \subset \mathbb{P}^3$ from a point $O \in \mathbb{P}^3 \smallsetminus V$; if $O = [0 : 0 : 0 : 1]$, the surface V is given by the equation $z_3^3 + 2z_3 f_2 + 3f_3 = 0$, *cf.* (3.4). Conversely, the branch locus of any such projection is a sextic of torus type, possibly non-reduced.

Fix a torus structure on a sextic D and consider the conic $Q_2 := \{f_2 = 0\}$ and the cubic $Q_3 := \{f_3 = 0\}$. Simple analysis using local normal forms shows that each point $P \in Q_2 \cap Q_3$ is singular for D, and this point is of type

- $\mathbf{A}_{3k-1}$, if Q_3 is nonsingular at P and $(Q_2 \circ Q_3)_P = k$,

- $\mathbf{E}_6$, if Q_3 is singular at P and $(Q_2 \circ Q_3)_P = 2$, or
- non-simple (adjacent to $\mathbf{J}_{10}$) otherwise.

The singular points of D that are in the intersection $Q_2 \cap Q_3$ are called *inner* (with respect to the given torus structure), and the other singular points are called *outer*. If a torus structure is unique or assumed, the inner points are parenthesized in the notation for the set of singularities.

Define the *weight* $w(\mathbf{T})$ of a simple singularity of type $\mathbf{T}$ as $w(\mathbf{A}_{3k-1}) = k$, $k \in \mathbb{N}$, $w(\mathbf{E}_6) = 2$, and $w(\mathbf{T}) = 0$ otherwise. The *weight* of a simple sextic is the sum of the weights of all its singular points. From the description of the inner singularities and the fact that $Q_2 \circ Q_3 = 6$ we obtain the following statement, which implies, in particular, that the total weight of a simple sextic of torus type is at least six.

Lemma 7.29. *Given a torus structure, the total weight of its inner singular points does not exceed six. If all inner points are simple, their total weight equals six.* $\triangleleft$

The property of being of torus type is invariant under equisingular deformations. In fact, this property is of purely topological nature; for irreducible sextics, this assertion follows, *e.g.*, from Theorem 7.33 below.

A Zariski open dense subset in the space of sextics of torus type is formed by the so-called *Zariski sextics*, *i.e.*, the six cuspidal sextics given by the equation $f_2^3 + f_3^2 = 0$ for a sufficiently generic pair f_2, f_3. Note that there also exist six cuspidal sextics that are not of torus type. The existence of two distinct families of irreducible six cuspidal sextics, those of and not of torus type, was first stated by Del Pezzo and then proved by Segre; see, *e.g.*, [154, p. 407]. Zariski [167] later showed that the two families differ by the fundamental groups, which are Γ and $\mathbb{Z}_6$, respectively.

Since any sextic of torus type is a degeneration of a family of Zariski sextics, taking into account Proposition 5.33 we have the following well known lemma.

Lemma 7.30. *For any reduced sextic $D \subset \mathbb{P}^2$ of torus type, there is an epimorphism $\pi_1 \twoheadrightarrow \Gamma$.* $\triangleleft$

Lemma 7.31. *Let $D \subset \mathbb{P}^2$ be an irreducible sextic of torus type, and assume that there is an isomorphism $[\pi_1, \pi_1] \cong \mathbb{F}_2$. Then $\pi_1 \cong \Gamma$.*

This approach to identifying Γ as the fundamental group of a plane sextic works well with `GAP`. It was suggested to me by E. Artal Bartolo.

Proof. The epimorphism $\varphi\colon \pi_1 \twoheadrightarrow \Gamma$ given by Lemma 7.30 induces an isomorphism of the abelianizations of the groups and, hence, an epimorphism $[\pi_1, \pi_1] \twoheadrightarrow [\Gamma, \Gamma] \cong \mathbb{F}_2$ of their commutants. Since $\mathbb{F}_2$ is a Hopfian group, the latter is an isomorphism and, by the 5-lemma, so is φ. $\square$

Corollary 7.32. *Let $D \subset \mathbb{P}^2$ be an irreducible sextic of torus type, and assume that there is an epimorphism $\mathbb{B}_3 \twoheadrightarrow \pi_1$. Then $\pi_1 \cong \Gamma$.* $\triangleleft$

The following statement is known as *Oka's conjecture*; originally, the implication (2) $\Longrightarrow$ (1) was suggested in [69].

Theorem 7.33 (see [45, 50]). *For an irreducible plane sextic D, the following four statements are equivalent*:

1. *the sextic D is of torus type*;
2. *the Alexander polynomial $\Delta_D(t)$ is non-trivial*;
3. *the group π_1 factors to Γ*;
4. *the group π_1 factors to $\mathbb{S}_3 = \mathbb{D}_6$.*

Proof. The implications (3) $\Longrightarrow$ (2) $\Longrightarrow$ (4) are proved as in Theorem 6.10, with the additional observation that the Alexander polynomial of an irreducible sextic has the form $(t^2 - t + 1)^s$ for some $s \geqslant 0$, see Lemma A.34. The implication (1) $\Longrightarrow$ (3) is given by Lemma 7.30, and the last implication (4) $\Longrightarrow$ (1) follows from Theorem 7.34 below, which refines the equivalence (4) $\Longleftrightarrow$ (1). □

Theorem 7.34. *Let D be an irreducible plane sextic, and let $\mathbb{D}[\mathcal{Q}_D]$ be the maximal dihedral quotient of its fundamental group. Then there is an injective map from the set of torus structures on D to the dual projective space $\check{\mathbb{P}}(\mathcal{Q}_D \otimes \Bbbk_3)$. With one exception, this map is bijective. The exception is a nine cuspidal sextic, where the twelve torus structures are mapped into the thirteen points of $\check{\mathbb{P}}(\Bbbk_3^3)$.*

Proof. The space $\check{\mathbb{P}}(\mathcal{Q}_D \otimes \Bbbk_3)$ classifies the $\mathbb{S}_3$-coverings of $\mathbb{P}^2$ ramified at D. Each torus structure (f_2, f_3) defines such a covering, *viz.* the Galois closure of the projection $V \to \mathbb{P}^2$, where V is the cubic surface $\{z_3^3 + 2z_3 f_2 + 3f_3 = 0\}$. Distinct torus structures correspond to distinct coverings: an inner singular point P is characterized by the property that the composed map $\pi_1(U \smallsetminus D) \to \pi_1(\mathbb{P}^2 \smallsetminus D) \twoheadrightarrow \mathbb{S}_3$ should be an epimorphism, where U is a Milnor ball about P.

The proof of the surjectivity depends on the nature of D.

If D has a non-simple triple or quadruple singular point, the proof is given at the end of Section 7.2.2 or in Appendix B (see Theorem B.13), respectively.

In the most interesting case of a simple sextic, the proof is based on the theory of $K3$-surfaces and is beyond the scope of this book. Briefly, from Theorem 7.37 below one concludes that almost each hyperplane in $\mathcal{Q}_D \otimes \Bbbk_3$ defines a pair of numerically effective classes in the Picard group $\operatorname{Pic} X$ of the covering $K3$-surface X. These classes are realized by a pair of rational (-2)-curves (not necessarily irreducible), which project to the same conic $Q_2 \subset \mathbb{P}^2$, and the term f_2 in the torus structure is the defining polynomial of Q_2. (In the exceptional case of a nine cuspidal sextic, there is a covering in which all nine cusps behave as inner points. This covering does not come from a torus structure.) For further details, see [45]. □

As a consequence of Theorem 7.33, we have the following partial converse of the statement that the weight of a simple sextic of torus type is at least six, *cf.* Lemma 7.29.

Proposition 7.35. *Let D be an irreducible simple plane sextic of weight w. If $w \geqslant 7$ or $w = 6$ and at least one of the singular points of D of weight zero is not a node $\mathbf{A}_1$, then D is of torus type.*

Proof. The Alexander polynomial of an irreducible sextic is $(t^2 - t + 1)^s$, where the exponent s can be computed algebro-geometrically as the superabundance of the linear system of conics subject to certain linear conditions at the singular points of D, see [42, 106, 111]. Each point P of weight $w(P) > 0$ imposes $w(P)$ conditions, and each point other than a node imposes at least one condition. Hence, under the hypotheses, the virtual dimension of the system is less than -1 and one has $s > 0$. □

Thus, the only sets of singularities that can be shared by sextics of and not of torus type are of the form

$$\sum_i a_{3i-1}\mathbf{A}_{3i-1} + e_6\mathbf{E}_6 + a_1\mathbf{A}_1, \quad \sum_i i a_{3i-1} + 2e_6 = 6.$$

(This is a direct generalization of original Zariski's example, see [42].) A complete deformation classification of all sextics with these sets of singularities was recently obtained in [2, 3].

Other special sextics

For a plane sextic $D \subset \mathbb{P}^2$, define the *affine fundamental group* as

$$\pi^{\text{aff}}(D) := \pi_1(\mathbb{P}^2 \smallsetminus (D \cup L), b)$$

where L is a generic line and $b \in \mathbb{P}^2 \smallsetminus (D \cup L)$ is a basepoint. By Poincaré–Lefschetz duality, the abelianization of this group is freely generated by the classes of meridians of small tubular neighborhoods of the components of D, and sending each meridian to $1 \in \mathbb{Z}$ defines distinguished epimorphisms $\pi^{\text{aff}}(D) \twoheadrightarrow \mathbb{Z} \twoheadrightarrow \mathbb{Z}_2$. A dihedral quotient $\pi^{\text{aff}}(D) \twoheadrightarrow \mathbb{D}[Q]$ is called *uniform* if it factors this distinguished epimorphism. Let $\pi^{\text{aff}}(D) \twoheadrightarrow \mathbb{D}[\mathcal{Q}_D]$ be the maximal uniform dihedral quotient. If D is irreducible, any dihedral quotient of $\pi^{\text{aff}}(D)$ is uniform and $\mathbb{D}[\mathcal{Q}_D]$ is also the maximal dihedral quotient of $\pi_1(\mathbb{P}^2 \smallsetminus D)$.

For a simple sextic D, we denote by $\mathcal{K}_D \subset \operatorname{discr}(\mathbf{S} \oplus \mathbb{Z}h)$ the 'imprimitivity' of its homological type, see (7.26), *i.e.*, the kernel of the finite index extension $\mathbf{S}' \subset \tilde{\mathbf{S}}'$, where $\mathbf{S}' := \mathbf{S} \oplus \mathbb{Z}h$ and $\tilde{\mathbf{S}}' := (\mathbf{S}' \otimes \mathbb{Q}) \cap \mathbf{L}$.

The next two statements are analogues of Proposition 6.2 and Theorem 6.12 for trigonal curves, and the proofs are literally the same.

Proposition 7.36. *A sextic D is reducible if and only if the group $\mathcal{Q}_D$ has 2-torsion, i.e., if $\pi^{\text{aff}}(D)$ factors to $\mathbb{D}_4 \cong \mathbb{Z}_2 \oplus \mathbb{Z}_2$.* ◁

Theorem 7.37. *For a simple plane sextic $D \subset \mathbb{P}^2$, there are canonical isomorphisms $\mathcal{Q}_D = \text{Ext}(\mathcal{K}_D, \mathbb{Z}) = \text{Hom}(\mathcal{K}_D, \mathbb{Q}/\mathbb{Z})$.* ◁

Corollary 7.38. *If the set of singularities $\mathbf{S}$ of a simple plane sextic D satisfies the inequality $\ell_2(\operatorname{discr} \mathbf{S}) + \mu(\mathbf{S}) > 20$, then D is reducible.* ◁

An irreducible sextic D is called *special* or, more precisely, a $\mathbb{D}_{2n}$*-sextic* if the fundamental group $\pi_1(\mathbb{P}^2 \smallsetminus D)$ admits a dihedral quotient $\mathbb{D}_{2n}$, $n \geqslant 3$. According to Theorem 7.33, the $\mathbb{D}_6$-sextics are precisely those of torus type. All other special sextics are known: we have the following two theorems, which resemble similar statements for trigonal curves, *cf.* Theorem 6.1.

Theorem 7.39 (see [45]). *For an irreducible sextic D, the group $\mathcal{Q}_D$ is a quotient of $\mathbb{Z}_3 \oplus \mathbb{Z}_3 \oplus \mathbb{Z}_3$, $\mathbb{Z}_5$, or $\mathbb{Z}_7$. Conversely, any such quotient appears as $\mathcal{Q}_D$.*

Theorem 7.40 (see [45]). *The $\mathbb{D}_{2n}$-sextics, $n \geqslant 5$, form fifteen equisingular deformation families, one family for each of the following sets of singularities*:

- $n = 5$: $4\mathbf{A}_4 \oplus k\mathbf{A}_1$, $0 \leqslant k \leqslant 2$, $4\mathbf{A}_4 \oplus \mathbf{A}_2$, $\mathbf{A}_9 \oplus 2\mathbf{A}_4 \oplus k\mathbf{A}_1$, $0 \leqslant k \leqslant 1$, $\mathbf{A}_9 \oplus 2\mathbf{A}_4 \oplus \mathbf{A}_2$, $2\mathbf{A}_9$ (*simple*),
  $\mathbf{J}_{10} \oplus 2\mathbf{A}_4$, $\mathbf{J}_{11} \oplus 2\mathbf{A}_4$, $\mathbf{J}_{15} \oplus \mathbf{A}_4$, $\mathbf{Y}^1_{1,1} \oplus 2\mathbf{A}_4$, $\mathbf{W}_{12} \oplus 2\mathbf{A}_4$ (*non-simple*);
- $n = 7$: $3\mathbf{A}_6 \oplus k\mathbf{A}_1$, $k \leqslant 1$.

Remark 7.41. The sets of singularities $4\mathbf{A}_4 \oplus k\mathbf{A}_1$, $k \leqslant 1$, $\mathbf{A}_9 \oplus 2\mathbf{A}_4 \oplus k\mathbf{A}_1$, $k \leqslant 1$, $2\mathbf{A}_9$, and $3\mathbf{A}_6$ are also realized by non-special irreducible sextics. The other ten sets of singularities are only realized by special sextics.

Proof of Theorems 7.39 *and* 7.40. Sextics with a triple (type $\mathbf{J}$) or quadruple (types $\mathbf{Y}$ and $\mathbf{W}$) non-simple singular point are treated in Theorem 7.45 or Theorem B.13, respectively. For simple sextics, the proof is based on Theorem 7.27: in view of Proposition 7.36 and Theorem 7.37, it suffices to enumerate the homological types with the kernel $\mathcal{K}_D \neq 0$ free of 2-torsion, which can be done using Nikulin's theory, see Section A.1.1. More details can be found in [45]. □

It is also relatively easy to enumerate all irreducible simple sextics of torus type (*i.e.*, $\mathbb{D}_6$-sextics) of weight $w \geqslant 8$: such sextics can be characterized by the existence of a quotient $\pi_1(\mathbb{P}^2 \smallsetminus D) \twoheadrightarrow \mathbb{D}[\mathbb{Z}_3 \oplus \mathbb{Z}_3]$. There are eight equisingular deformation families, one family for each of the following set of singularities (see [45]):

- $w = 8$: $8\mathbf{A}_2 \oplus k\mathbf{A}_1$, $\mathbf{A}_5 \oplus 6\mathbf{A}_2 \oplus k\mathbf{A}_1$, $0 \leqslant k \leqslant 1$, $2\mathbf{A}_5 \oplus 4\mathbf{A}_2$, $\mathbf{E}_6 \oplus 6\mathbf{A}_2$, $\mathbf{E}_6 \oplus \mathbf{A}_5 \oplus 4\mathbf{A}_2$;
- $w = 9$: $9\mathbf{A}_2$.

All these curves, as well as those listed in Theorem 7.40, exhibit a number of special geometric properties, such as the existence of stable (under deformations) *projective symmetries* and existence of the so-called *Z-splitting curves*, see [45, 47, 149]. In view of Theorem 7.45, it is natural to assign weight eight to the non-simple sextics with the sets of singularities $\mathbf{J}_{10} \oplus 4\mathbf{A}_2$ and $\mathbf{J}_{13} \oplus 3\mathbf{A}_2$.

The number of irreducible sextics of torus type of weight $w = 6, 7$ is rather large and their deformation classification has not been completed yet. The singularities of such curves are listed in [129]; the fundamental groups are discussed in Section 7.2.3.

7.2.2 Classification and fundamental groups

Let $D \subset \mathbb{P}^2$ be a plane sextic with a non-simple triple singular point P; in other words, P is a point of type $\mathbf{J}_{r,p}$, $r \geqslant 2$, $p \geqslant 0$, or $\mathbf{E}_{6r+\epsilon}$, $r \geqslant 2$, $\epsilon = 0, 1, 2$.

Throughout this section, we assume that *D has no linear components through P*.

Consider the blow-up $\mathbb{P}^2(P) \cong \Sigma_1$ and the proper transform $\tilde{C}$ of D. It has one improper fiber $\tilde{F}_{\mathrm{I}}$, at which $\tilde{C}$ has a triple point and intersects E with multiplicity three. Hence, a positive Nagata transformation converts $\tilde{C}$ to a proper trigonal curve $C \subset \Sigma_2$, which is called the *proper model* of the original sextic D. The inverse Nagata transformation is centered at a point $Q \in \Sigma_2 \smallsetminus (C \cup E)$; the fiber F_{I} through this point is called the *distinguished fiber*. Clearly, the point Q can freely be moved in the connected space $F_{\mathrm{I}} \smallsetminus (C \cup E)$, and this motion is followed by an equisingular deformation of D. Thus, the equisingular deformation classification of pairs (D, P) as above is reduced to the classification of pairs (C, F_{I}), and the latter is given by

- Theorem 7.1, if F_{I} is a nonsingular fiber, hence P is of type $\mathbf{J}_{10} = \mathbf{J}_{2,0}$,
- Theorem 7.5, if F_{I} is a stable singular fiber, hence P is of type $\mathbf{J}_{2,p}$, $p > 0$, $\mathbf{J}_{3,p}$, $p \geqslant 0$, $\mathbf{E}_{18}$, $\mathbf{E}_{19}$, or $\mathbf{E}_{20}$, and
- Theorem 7.6, if F_{I} is an unstable fiber, hence P is of type $\mathbf{E}_{12}$, $\mathbf{E}_{13}$, or $\mathbf{E}_{14}$.

In addition, there are two deformation families with non-simple proper models; their sets of singularities are $2\mathbf{J}_{10}$ and $\mathbf{J}_{4,0}$. Each of these sextics splits into three conics, respectively tangent at two common points or four-fold tangent at one common point.

Not trying to restate the classification theorems in terms of the singularities of D itself, we only mention one corollary.

Corollary 7.42. *The equisingular deformation type of an* irreducible *sextics D with a non-simple triple singular point is determined by the set of singularities of D.* $\triangleleft$

Lemma 7.43. *There is a canonical one-to-one correspondence between the set of torus structures on a plane sextic D with a non-simple triple singular point P and the set of torus structures on the proper model C of D.*

Proof. Choose homogeneous coordinates $[z_0 : z_1 : z_2]$ in $\mathbb{P}^2$ so that P has coordinates $[0 : 0 : 1]$ and the tangent to D at P is the line $\{z_0 = 0\}$. Denote $x := z_2/z_0$ and

$t := z_1/z_0$. Then (x,t) can be regarded as affine coordinates both in $\mathbb{P}^2$ and in Σ_2, and in these coordinates D and C have the same equation, which is of the form

$$x^3 + x^2 b_1(t) + x b_2(t) + b_3(t) = 0,$$

$\deg b_i \leqslant 2i$, $i = 1,2,3$. It is immediate that any torus structure on C, see (3.33), is also a torus structure on D. Conversely, given a torus structure $f_2^3 + f_3^2$ on D, Lemma 7.29 implies that P must be an inner singularity, and a simple local analysis shows that $f_2 = x + a_2(t)$ and $f_3 = x a_1(t) + a_3(t)$ with $\deg a_i \leqslant 2i$, $i = 1,2,3$. Hence, the pair (f_2, f_3) constitutes a torus structure on C, see (3.33). □

Remark 7.44. In the realm of sextics with non-simple singular points, we encounter a new phenomenon, which should be regarded as an 'outer degeneration' of an inner singularity, *i.e.*, a degeneration preserving the topology of the pair $(\mathbb{P}^2, Q_2 + Q_3)$. For example, assume that P is a node of Q_3 and $(Q_2 \circ Q_3)_P = 3$. Generically, the corresponding sextic D has P as a point of type $\mathbf{J}_{10}$: the distinguished fiber F_{I} is nonsingular for the proper model C of D. However, F_{I} may degenerate to a fiber of type $\tilde{\mathbf{A}}_0^*$ or $\tilde{\mathbf{A}}_1$ (*e.g.*, if C has the set of singularities $(3\mathbf{A}_2) \oplus \mathbf{A}_1$), and the type of P would change to $\mathbf{J}_{11}$ or $\mathbf{J}_{12}$, respectively.

According to Proposition 5.71, we have $\pi_1 := \pi_1(\mathbb{P}^2 \smallsetminus D) = \pi^{\mathrm{proj}}(C)$. Observe that, in particular, π_1 does not depend on the choice of the distinguished fiber F_{I}. The latter group is given by Theorems 7.7 and 7.8, and we have the following statement.

Theorem 7.45. *With the exception of the families*

- $\mathbf{J}_{10} \oplus 2\mathbf{A}_4$, $\mathbf{J}_{11} \oplus 2\mathbf{A}_4$, $\mathbf{J}_{15} \oplus \mathbf{A}_4$, *where* $\pi_1 \cong \mathbb{D}_{10} \times \mathbb{Z}_3$,
- $\mathbf{J}_{10} \oplus 4\mathbf{A}_2$, $\mathbf{J}_{13} \oplus 3\mathbf{A}_2$, *where* π_1 *is given by* (7.10), *and*
- *other sextics of torus type, where* $\pi_1 \cong \Gamma$,

the fundamental group $\pi_1 := \pi_1(\mathbb{P}^2 \smallsetminus D)$ *of an irreducible plane sextic* D *with a triple non-simple singular point is abelian.* ◁

As another consequence, in view of Lemma 7.43 and Theorem 6.11, we obtain a proof of Theorems 7.34 and 7.33 for sextics with a triple non-simple singular point.

7.2.3 A summary of further results

We conclude with a brief survey of other known results concerning the fundamental groups of plane sextics. The groups of the following irreducible sextics are known:

- all sextics with a non-simple triple (Section 7.2.2) or quadruple (Theorem B.13) point; the group of an irreducible sextic with a quintuple point is obviously abelian;
- all simple sextics with a triple singular point, see Chapter 8;

- all but one $\mathbb{D}_{2n}$-sextics, $n \geqslant 5$, see [51, 63, 73] and Theorem 7.46 below; the exception is the set of singularities $3\mathbf{A}_6 \oplus \mathbf{A}_1$. For the non-special counterparts with the sets of singularities $\mathbf{A}_9 \oplus 2\mathbf{A}_4$ and $4\mathbf{A}_4$ the groups are abelian, see [71].

There also are a great number of sporadic computations, for both irreducible and reducible curves; for references, see recent papers [7, 8, 10, 69, 71, 72, 165].

The known groups of special sextics are given by the following theorem.

Theorem 7.46 (see [51, 63, 73]). *Let D be a $\mathbb{D}_{2n}$-sextic, $n \geqslant 5$. Then the group $\pi_1 := \pi_1(\mathbb{P}^2 \smallsetminus D)$ is as follows (depending on the set of singularities $\mathbf{S}$ of D):*

- $\mathbf{S} = \mathbf{W}_{12} \oplus 2\mathbf{A}_4$: *one has* $\pi_1 \cong \langle u, v \mid u^4 = v^4, \{u, v\}_5 = 1, v^4 = (uv)^5 \rangle$;
- $\mathbf{S} = 4\mathbf{A}_4 \oplus 2\mathbf{A}_1$: *one has* $\pi_1''/\pi_1''' = \mathbb{Z}_2^4$ *and* $\pi_1''' = \mathbb{Z}_2$, *so that* $\operatorname{ord} \pi_1 = 960$;
- $\mathbf{S} = \mathbf{A}_9 \oplus 2\mathbf{A}_4 \oplus \mathbf{A}_2$: π_1'' *is the only perfect group of order* 720;
- *other* $\mathbb{D}_{10}$-*sextics: one has* $\pi_1 = \mathbb{Z}_3 \times \mathbb{D}_{10}$;
- $\mathbf{S} = 3\mathbf{A}_6$: *one has* $\pi_1 = \mathbb{Z}_3 \times \mathbb{D}_{14}$.

(We use $'$ for the commutant of a group.) In all cases $\pi_1/\pi_1' = \mathbb{Z}_6$ and $\pi_1'/\pi_1'' = \mathbb{Z}_n$. For the $\mathbb{D}_{14}$-sextic with $\mathbf{S} = 3\mathbf{A}_6 \oplus \mathbf{A}_1$, the group is unknown. ▷

The evidence at hand suggests the following conjecture, which replaces original Oka's conjecture [69] on the fundamental group.

Conjecture 7.47. Let $D \subset \mathbb{P}^2$ be a simple irreducible plane sextic not of torus type. Then the group $\pi_1(\mathbb{P}^2 \smallsetminus D)$ is finite.

There is one non-simple sextic not of torus type with infinite fundamental group; its set of singularities is $\mathbf{W}_{12} \oplus 2\mathbf{A}_4$, see Theorem B.13.

According to Lemma 7.30, the fundamental group of a sextic of torus type is always infinite. The groups of most irreducible sextics have been computed, see [48] for a 'map' of known results and [53] for references to a few recent updates. A feasible conjecture is that a non-maximizing irreducible sextic of torus type is determined by its set of singularities uniquely up to equisingular deformation and, possibly, complex conjugation. Assuming this conjecture, there are only six irreducible sextics of torus type whose groups are still unknown:

$$\begin{array}{ll} (\mathbf{A}_8 \oplus 3\mathbf{A}_2) \oplus \mathbf{A}_4 \oplus \mathbf{A}_1, & (\mathbf{A}_{11} \oplus 2\mathbf{A}_2) \oplus \mathbf{A}_4, \\ (\mathbf{A}_8 \oplus \mathbf{A}_5 \oplus \mathbf{A}_2) \oplus \mathbf{A}_4, & (\mathbf{A}_{14} \oplus \mathbf{A}_2) \oplus \mathbf{A}_3, \\ (\mathbf{A}_8 \oplus \mathbf{A}_5 \oplus \mathbf{A}_2) \oplus \mathbf{A}_2 \oplus \mathbf{A}_1, & (\mathbf{A}_{14} \oplus \mathbf{A}_2) \oplus \mathbf{A}_2 \oplus \mathbf{A}_1. \end{array}$$

In most cases, the group $\pi_1(\mathbb{P}^2 \smallsetminus D)$ is isomorphic to Γ, *i.e.*, the minimal group given by Lemma 7.30. Below is a brief list of the known larger groups.

For sextics of weight eight and nine, introduce the group

$$G_\infty := \langle \alpha, \bar{\alpha}, \beta, \gamma, \bar{\gamma} \mid \{\alpha, \beta\}_3 = \{\bar{\alpha}, \beta\}_3 = \{\gamma, \beta\}_3 = \{\bar{\gamma}, \beta\}_3 = \beta\gamma\alpha\beta\bar{\gamma}\bar{\alpha} = 1 \rangle.$$

Note that G_∞ has an involutive automorphism $\alpha \leftrightarrow \bar{\alpha}$, $\gamma \leftrightarrow \bar{\gamma}$. This automorphism has a geometric meaning: any sextic $D \subset \mathbb{P}^2$ of weight eight or nine is the double covering of the maximal trigonal curve $C \subset \Sigma_2$ with the set of singularities $4\mathbf{A}_2$ ramified at E and at another section L disjoint from E, see [49]; hence, all quotients below retain this symmetry. The fundamental groups $\pi_1 := \pi_1(\mathbb{P}^2 \smallsetminus D)$ are as follows:

- weight nine (the set of singularities $9\mathbf{A}_2$): $\pi_1 = G_3$ is the quotient of G_∞ by the relations

$$\{\bar{\gamma}, \alpha\} = \{\gamma, \bar{\alpha}\} = \{\gamma, \bar{\gamma}\} = [\beta, \alpha^{-1}\gamma^{-1}\bar{\alpha}\bar{\gamma}] = 1, \quad \bar{\gamma}^{-1}\alpha\bar{\gamma} = \gamma^{-1}\bar{\alpha}\gamma$$

  (for alternative presentations of this group, see [35] and [169]);
- the set of singularities $\mathbf{E}_6 \oplus \mathbf{A}_5 \oplus 4\mathbf{A}_2$:

$$\pi_1 = G_2 := G_\infty / \{\bar{\alpha}\gamma\alpha = \alpha\bar{\gamma}\bar{\alpha} = \gamma\alpha\bar{\gamma} = \bar{\gamma}\bar{\alpha}\gamma\}$$

  (for alternative presentations, see (8.47) with $\bar{s} = (3, 6, 3, 3, 6, 3)$);
- the set of singularities $\mathbf{A}_5 \oplus 6\mathbf{A}_2 \oplus \mathbf{A}_1$:

$$\pi_1 = G_1 := G_\infty / \{\{\alpha, \gamma\}_3 = \{\bar{\alpha}, \bar{\gamma}\}_3 = [\gamma, \bar{\gamma}] = 1,\ \gamma\alpha\bar{\gamma} = \bar{\gamma}\bar{\alpha}\gamma\};$$

- other sextics of weight eight:

$$\pi_1 = G_0 := G_\infty / \{\alpha = \bar{\alpha}, \gamma = \bar{\gamma}, \{\alpha, \gamma\}_3 = 1\}$$

  (for alternative presentations, see (8.47) with $\bar{s} = (3, 3, 3, 3, 3, 3)$ or [130]).

All perturbation epimorphisms $G_3 \twoheadrightarrow G_0$ and $G_2 \twoheadrightarrow G_1 \twoheadrightarrow G_0$, see Proposition 5.33, lift to the identity id: $G_\infty \to G_\infty$. I do not know whether $G_2 \twoheadrightarrow G_1$ is proper; the others are. The Alexander modules are as expected, *i.e.*, $\Lambda^3/(t^2 - t + 1)$ for G_3 and $\Lambda^2/(t^2 - t + 1)$ for the other three groups.

Most other sextics with excessive fundamental groups have a singular point of type $\mathbf{E}_6$, see Theorem 8.2. The two exceptions are $(6\mathbf{A}_2) \oplus \mathbf{A}_3 \oplus 2\mathbf{A}_1 \to (6\mathbf{A}_2) \oplus 4\mathbf{A}_1$, which degenerate to $(\mathbf{E}_6 \oplus 4\mathbf{A}_2) \oplus \mathbf{A}_3 \oplus \mathbf{A}_1$, see Argument 8.40. The perturbation epimorphisms are isomorphisms and the group is $\mathbb{B}_4/\sigma_2\sigma_1^2\sigma_2\sigma_3^2$.

Perturbing the singularities, see Theorem 7.28, and using Proposition 5.33, one can obtain a great deal of other sextics with controllable fundamental groups. To complete the computation and, in particular, to settle Conjecture 7.47, it remains to consider about three dozens of maximizing irreducible sextics with $\mathbf{A}$ type singularities only. One may hope that this can be done using the approach suggested in [8], reducing sextics to trigonal curves with a number of sections.

Together with Theorem 7.28, the following statement gives rise to a great deal of sextics with abelian groups, substantiating Conjecture 7.47.

Theorem 7.48. *Let D be a simple plane sextic with $\pi_1 := \pi_1(\mathbb{P}^2 \smallsetminus D) \cong \Gamma$. Then, for any perturbation D' of D, one has either $\pi_1(\mathbb{P}^2 \smallsetminus D') \cong \Gamma$ (if D' is of torus type) or $\pi_1(\mathbb{P}^2 \smallsetminus D') \cong \mathbb{Z}_6$ (if D' is not of torus type).*

Proof. According to Proposition 7.36 and Theorem 7.33, the original sextic D is of torus type and irreducible. If D' is also of torus type, the statement follows from Proposition 5.33, Lemma 7.30, and the fact that Γ is a Hopfian group.

Assume that D' is not of torus type. For the torus structure to be destroyed, at least one inner singular point P must undergo a perturbation that is not of torus type, see, *e.g.*, Theorems 7.34 and 7.37. Consider a Milnor ball $U \ni P$. Under the assumptions, the inclusion induces an epimorphism $\pi_1(U \smallsetminus D) \twoheadrightarrow \Gamma$. (For proof, one can perturb D to a Zariski sextic, for which this statement is known.) Furthermore, if P is of type $\mathbf{A}_{3k-1}$, $k \in \mathbb{N}$, this epimorphism is $\beta_i \mapsto \sigma_i \bmod \Delta^2$, $i = 1, 2$, where $\{\beta_1, \beta_2\}$ is a basis for the local fundamental group as in Theorem 7.23. Hence, the statement of the theorem follows from Theorems 7.22 and 7.23: any perturbation that is not of torus type either makes the group $\pi_1(U \smallsetminus D')$ abelian (for P of type $\mathbf{E}_6$) or introduces an extra relation $\{\beta_1, \beta_2\}_s = 1$ with g.c.d.$(s, 3) = 1$. □

7.3 Plane quintics

As yet another application, we outline simple combinatorial proofs of the following two theorems, originally found in [41, 43]. Stating the results in full generality, we consider quintics with $\mathbf{A}$ type singular points only; those with triple or quadruple points are treated in Appendix B, see Theorem B.12.

Theorem 7.49 (see [41]). *The equisingular deformation type of an irreducible plane quintic D is determined by its set of singularities. Realizable are the simple sets of singularities*

$$\mathbf{A}_{12},\ \mathbf{A}_8 \oplus \mathbf{A}_4,\ \mathbf{A}_6 \oplus 3\mathbf{A}_2,\ 3\mathbf{A}_4,$$
$$\mathbf{E}_8 \oplus \mathbf{A}_4,\ \mathbf{E}_7 \oplus 2\mathbf{A}_2,\ \mathbf{E}_6 \oplus \mathbf{A}_6,\ \mathbf{E}_6 \oplus \mathbf{A}_4 \oplus \mathbf{A}_2,$$

all perturbations thereof, and the non-simple sets $\mathbf{X}_{10}$, $\mathbf{X}_{11}$, $\mathbf{Y}^1_{1,1}$, $\mathbf{Z}_{11}$, $\mathbf{W}_{12}$.

Theorem 7.50 (see [43]). *The group $\pi_1 := \pi_1(\mathbb{P}^2 \smallsetminus D)$ of an irreducible plane quintic D is abelian unless D has one of the following two sets of singularities*:

- $\mathbf{A}_6 \oplus 3\mathbf{A}_2$: *$\pi_1$ is the infinite group given by* (7.51);
- $3\mathbf{A}_4$: *π_1 is the group of order* 320 *given by* (7.52).

Proof of Theorem 7.49 (an outline). Assume that D has a singular point P of type $\mathbf{A}_p$, $p \geqslant 3$. As in Section 7.2.2, we can represent D by its *proper model* C, which is a proper simple trigonal curve in Σ_2 with a distinguished fiber F_{I} of type I_{p-3}. According to Lemma 7.4, C degenerates to a maximal irreducible curve (and, if F_{I} is of type I_0, it can be merged with a singular fiber of this curve), and this degeneration is followed by a degeneration of quintics. Hence, it suffices to consider the curves with the skeletons shown in Figure 4.2 (a), (b), or (d); non-maximal dessins sharing the same set of singularities are shown to be equivalent using Lemma 4.51.

Now, assume that all singular points of D are nodes or cusps, and take for P one of these points, with a preference towards cusps. Then, the proper model of D is a trigonal curve $C \subset \Sigma_3$ with a distinguished singular fiber F_{I} of type $\mathrm{I}_{\geqslant 4}$ (if P is a cusp) or two distinguished fibers $F_{\mathrm{I}}, F_{\mathrm{II}}$ of types $\mathrm{I}_{\geqslant 2}$ (if P is a node). As in Section 8.3.2 below, see page 268, one can prove that any *irreducible* curve C with these properties degenerates to the maximal one with the skeleton shown in Figure 8.23 (with the grey insertions disregarded) and that perturbed dessins sharing the same set of singularities are equivalent. The latter maximal curve corresponds to a quintic with the set of singularities $\mathbf{A}_6 \oplus 3\mathbf{A}_2$; for the next proof, we observe that, in fact, the proper model of a quintic with nodes and cusps only degenerates to that of $\mathbf{A}_3 \oplus 4\mathbf{A}_2$. □

Proof of Theorem 7.50. Assume that D has a singular point P of type $\mathbf{A}_p$, $p \geqslant 3$, and consider its proper model (C, F_{I}), see the previous proof. The fiber F_{I} is as in Case 1 in Section 5.1.3 with $n = 1$ and $m = p - 3$. Choosing the reference edge e as explained in Section 5.1.3, we have the braid relations $\alpha_3\alpha_i\alpha_3^{-1} = \alpha_i \uparrow \sigma_1^{p-3}$, $i = 1, 2$ (hence $[\alpha_1\alpha_2, \alpha_3] = 1$), and the slope $\varkappa_{\mathrm{I}} = \alpha_3$, see (5.72), so that the relation at infinity becomes $(\alpha_1\alpha_2)^2\alpha_3 = 1$. This allows us to rewrite the above braid relations in the form $\sigma_1^{p+1} = \mathrm{id}$, or $\sigma_1^{m+4} = \mathrm{id}$, where m is the width of the region R_{I} containing F_{I}, and obtain the presentation

$$\pi_1 = \langle \alpha_1, \alpha_2, \alpha_3 \mid \sigma_1^{m+4} = \mathrm{id},\ (\alpha_1\alpha_2)^2\alpha_3 = 1,\ \ldots\rangle,$$

where dots stand for the braid relations resulting from all but one remaining regions.

The skeleton $\mathfrak{S} := \operatorname{Sk} C$ is one of those shown in Figure 4.2 (a), (b), or (d) and, up to isomorphism, there are five pairs $(\mathfrak{S}, F_{\mathrm{I}})$, which are to be considered one by one. Since π_1 is generated by α_1, α_2, it is abelian whenever R_{I} is adjacent to a monogon, as then we have $\alpha_2 = \alpha_3$. For the remaining three cases and their perturbations, the `GAP` input is found in the file "`misc/quintics.txt`". The two nonabelian groups thus obtained are those listed in the statement.

For the set of singularities $\mathbf{A}_6 \oplus 3\mathbf{A}_2$, Figure 4.2 (a), the relations are

$$\{\alpha_1, \alpha_2\}_7 = \{\alpha_2, \alpha_3\}_3 = \{\alpha_3, \rho^{-1}\alpha_1\rho\}_3 = (\alpha_1\alpha_2)^2\alpha_3 = 1. \tag{7.51}$$

The commutant of this group is perfect. Changing the basis to $u := \alpha_1\alpha_2$ and $v := (\alpha_1\alpha_2)^3\alpha_1$, we have the defining relations $u^7 = v^2$ and $(v^{-1}u^2)^2vu^2 = 1$. Hence, the quotient π_1/v^2 is Coxeter's group

$$(2, 3, 7) := \langle \alpha, \beta, \gamma \mid \alpha^2 = \beta^3 = \gamma^7 = \alpha\beta\gamma = 1\rangle,$$

which is infinite, see [39]. (Here, $u \mapsto \gamma^3$ and $v \mapsto \alpha$, so that $vu^2 \mapsto (\gamma\alpha)^{-1}$.)

For the set of singularities $3\mathbf{A}_4$, Figure 4.2 (d) with R_{I} a monogon, the relations are

$$\{\alpha_1, \alpha_2\}_5 = \{\alpha_2, \alpha_3\}_5 = \{\alpha_2, \rho^{-1}\alpha_1\rho\}_5 = (\alpha_1\alpha_2)^2\alpha_3 = 1. \tag{7.52}$$

Using $'$ for the commutant, we have $\pi_1'/\pi_1'' = \mathbb{Z}_2^4$ and $\pi_1'' = \mathcal{Z}(\pi_1) = \mathbb{Z}_2^2$, so that $\operatorname{ord}\pi_1 = 320$. All statements are easily obtained with `GAP`. Since the Alexander module $\mathrm{A}_D := \pi_1'/\pi_1''$ is annihilated by the cyclotomic polynomial $\tilde{\varphi}_5$ and the latter is irreducible over $\Bbbk_2$, we also have $\mathrm{A}_D = (\Lambda/\tilde{\varphi}_5) \otimes \Bbbk_2$.

An irreducible quintic with nodes and cusps only admits a degeneration to a quintic with the set of singularities $4\mathbf{A}_2 \oplus \mathbf{A}_3$ (see the proof of Theorem 7.49) whose group has already been shown to be abelian. □

Digression: geometric surjections

Let $D \subset \mathbb{P}^2$ be a plane curve, and assume that D can be included into a pencil over a base B, not necessarily rational, with connected (after resolving the base points) fibers. Then the projection $\mathbb{P}^2 \smallsetminus D \to B^\sharp$ gives rise to an epimorphism $\pi_1(\mathbb{P}^2 \smallsetminus D) \twoheadrightarrow \pi_1^{\mathrm{orb}}(B^\sharp)$, where $B^\sharp$ is B punctured at the image of D and the orbifold structure is given by the multiple fibers of the pencil. In [12], the group $\pi_1 := \pi_1(\mathbb{P}^2 \smallsetminus D)$ is said to possess a *geometric surjection* if one can find a pencil as above with $\pi_1^{\mathrm{orb}}(B^\sharp)$ nonabelian; then π_1 is also nonabelian.

Numerous examples exhibited in [12] show that many known nonabelian fundamental groups of plane curves do possess a geometric surjection. Thus, if D is a sextic of torus type, $D = \{f_2^3 + f_3^2 = 0\}$, the rational pencil generated by f_2^3 and f_3^2 gives rise to an epimorphism $\pi_1 \twoheadrightarrow \mathbb{Z}_2 * \mathbb{Z}_3 \cong \Gamma$, *cf.* Lemma 7.30. More generally, if $D = \{f_p^q + f_q^p = 0\}$ is a curve of (p,q)-*torus type*, g.c.d.$(p,q) = 1$, the pencil generated by f_p^q and f_q^p surjects π_1 onto $\mathbb{Z}_p * \mathbb{Z}_q$. According to [127], for a generic curve of (p,q)-torus type, this surjection is an isomorphism.

In this respect, the plane quintic D with the set of singularities $3\mathbf{A}_4$ is a remarkable exception. It is shown in [12] that the group $\pi_1 := \pi_1(\mathbb{P}^2 \smallsetminus D)$, see (7.52), does not surject to any nonabelian orbifold fundamental group of a punctured surface. Hence, π_1 does not possess a geometric surjection.

Chapter 8

Fundamental groups of plane sextics

In this longest chapter, we compute the fundamental groups of all irreducible simple plane sextics with a triple singular point. The groups of some reducible curves are also computed (when it is not too expensive).

8.1 Statements

The principal results of the chapter are Theorems 8.1 and 8.2. Their proof is outlined in Section 8.1.2, and the details occupy the rest of the chapter. The classification of irreducible maximizing sextics, their geometric representation, and references to the computation of the group are summarized in a number of tables, see Table 8.1, 8.3, 8.5, 8.6, 8.8, and 8.10. A few other tables list selected reducible curves.

The classification of maximizing sextics could as well be obtained from the results of [166] and [148], but the approach of these papers, based on Theorem 7.27, is not constructive. We employ the reduction to trigonal curves, which gives us explicit geometric models.

8.1.1 Principal results

Throughout the chapter, D stands for a simple plane sextic with a distinguished triple singular point P. We use the shortcut $\pi_1 := \pi_1(\mathbb{P}^2 \smallsetminus D)$.

Theorem 8.1. *With few exceptions, the fundamental group π_1 of an irreducible simple plane sextic $D \subset \mathbb{P}^2$ with a triple singular point and not of torus type is abelian. The exceptions are:*

- $\mathbf{E}_8 \oplus \mathbf{A}_4 \oplus \mathbf{A}_3 \oplus 2\mathbf{A}_2$, *Line* 1 *in Table* 8.1 *and Argument* 8.14 (*one curve*):

$$\pi_1 = \langle \alpha_1, \alpha_2 \mid (\alpha_1\alpha_2^{-1})^5\alpha_2^6 = [\alpha_1, \alpha_2^3] = \{\alpha_1, \alpha_2\}_5 = 1 \rangle$$

  is the central product $SL(2, \Bbbk_5) \odot \mathbb{Z}_{12}$;
- $\mathbf{E}_7 \oplus 2\mathbf{A}_4 \oplus 2\mathbf{A}_2$, *Line* 1 *in Table* 8.3 *and Argument* 8.23 (*one curve*):

$$\pi_1 = \langle \alpha_1, \alpha_2, \alpha_3 \mid (8.22),\ \{\alpha_1, \alpha_3\}_5 = \{\alpha_2, \alpha_3\}_5 = 1 \rangle \cong SL(2, \Bbbk_{19}) \times \mathbb{Z}_6;$$

- $2\mathbf{E}_6 \oplus \mathbf{A}_4 \oplus \mathbf{A}_3$, *Lines* 5, 6 *in Table* 8.5 *and Arguments* 8.38, 8.32 (*two curves*):

$$\pi_1 = \langle \alpha_1, \alpha_2, \alpha_3 \mid \rho^4 = (\alpha_1\alpha_2)^3,\ (8.39) \rangle \quad \textit{or} \quad \langle \alpha_1, \alpha_2 \mid (8.30),\ (8.33) \rangle,$$

  respectively. Both groups are semidirect products $SL(2, \Bbbk_5) \rtimes \mathbb{Z}_6$;

- $2\mathbf{D}_7 \oplus 2\mathbf{A}_2$, *Argument* 8.56 (*one family*) *and its perturbations* $\mathbf{D}_7 \oplus \mathbf{D}_4 \oplus 3\mathbf{A}_2$ *and* $2\mathbf{D}_4 \oplus 4\mathbf{A}_2$, *Argument* 8.70: *one has* $\pi_1 \cong SL(2, \Bbbk_3) \times \mathbb{Z}_2$.

Theorem 8.2. *With few exceptions, the fundamental group* π_1 *of an irreducible simple plane sextic* $D \subset \mathbb{P}^2$ *with a triple singular point and of torus type is isomorphic to* Γ. *The exceptions are*:

- $(3\mathbf{E}_6) \oplus \mathbf{A}_1$, *Line* 1 *in Table* 8.5 *and its perturbations* $(2\mathbf{E}_6 \oplus 2\mathbf{A}_2) \oplus 2\mathbf{A}_1$ *and* $(\mathbf{E}_6 \oplus 4\mathbf{A}_2) \oplus 3\mathbf{A}_1$: *one has* $\pi_1 = \mathbb{B}_4/\sigma_2\sigma_1^2\sigma_2\sigma_3^2$, *see Argument* 8.40;
- $(\mathbf{E}_6 \oplus 4\mathbf{A}_2) \oplus \mathbf{A}_3 \oplus \mathbf{A}_1$: *one has* $\pi_1 = \mathbb{B}_4/\sigma_2\sigma_1^2\sigma_2\sigma_3^2$, *see Argument* 8.40;
- $(2\mathbf{E}_6 \oplus \mathbf{A}_5) \oplus \mathbf{A}_2$, *Lines* 7 *and* 8 *in Table* 8.5: *see* (8.43) *and* (8.36), *respectively*;
- $(2\mathbf{E}_6 \oplus 2\mathbf{A}_2) \oplus \mathbf{A}_3$, *Line* 3 *in Table* 8.5: *see* (8.42);
- (*sextics of weight eight*) $\mathbf{E}_6 \oplus \mathbf{A}_5 \oplus 4\mathbf{A}_2$, *Line* 43 *in Table* 8.5 *and* $\mathbf{E}_6 \oplus 6\mathbf{A}_2$: *see* (8.47) *with* $\bar{s} = (3, 6, 3, 3, 6, 3)$ *and* $(3, 3, 3, 3, 3, 3)$, *respectively*;
- $(\mathbf{E}_6 \oplus \mathbf{A}_5 \oplus 2\mathbf{A}_2) \oplus \mathbf{A}_4$, *Lines* 13 *and* 45 *in Table* 8.6: *see* (8.37) *and* (8.47) *with* $\bar{s} = (6, 5, 3, 3, 5, 6)$, *respectively*;

Most results concerning maximizing sextics were published in [57], [54], [55], and [56]. The systematic extension to non-maximizing sextics is new. In particular, newly discovered are the three non-maximizing curves with finite nonabelian groups, the last item in Theorem 8.1.

Theorem 8.1 provides extra evidence for Conjecture 7.47.

8.1.2 Beginning of the proof

Let $D \subset \mathbb{P}^2$ be a simple plane sextic and let P be a distinguished triple point of D. Throughout the chapter we assume that *D has no linear components through P*.

Let $\mathbb{P}^2(P) = \Sigma_1$ be the plane blown up at P, and define the *trigonal model* of D as its proper transform $\tilde{C} \in \Sigma_1$. It is a trigonal curve with one or several improper fibers $\tilde{F}_{\mathrm{I}}, \tilde{F}_{\mathrm{II}}, \ldots$, and one obviously has

$$\pi_1 := \pi_1(\mathbb{P}^2 \smallsetminus D) = \pi^{\mathrm{proj}}(\tilde{C}).$$

In order to classify the pairs (D, P) and compute the fundamental groups, we consider the *proper model* C of (D, P), which is the minimal proper model of $\tilde{C}$. It is a proper trigonal curve $C \subset \Sigma_d$, $d = 3$ or 4, equipped with one or several *distinguished fibers* $F_{\mathrm{I}}, F_{\mathrm{II}}, \ldots$, corresponding to the improper fibers of $\tilde{C}$. Generically (on a Zariski open set dense in each equisingular stratum), the topological types of the distinguished fibers depend on the type of P only; with one exception, they are stable and of Kodaira type I. The following statements are obvious:

1. a sextic D is irreducible if and only if so is its proper model C;
2. the singular points of D other than P are in a one-to-one correspondence with the singular points of C other than those in the distinguished fibers.

Conventions and notation

For the rest of the chapter, we fix the notation introduced above:

- D is a simple plane sextic with a distinguished triple point P, satisfying a certain set of conditions $(*)$ stated at the beginning of each subsection;
- $\tilde{C} \subset \Sigma_1$ is the trigonal model of D, and $C \subset \Sigma_d$ is its proper model;
- both D and C are assumed generic in their equisingular deformation classes, in the sense explained in the corresponding classification statements;
- $\tilde{F}_{\mathrm{I}}, \tilde{F}_{\mathrm{II}}, \ldots$ and $F_{\mathrm{I}}, F_{\mathrm{II}}, \ldots$ are the distinguished fibers of $\tilde{C}$ and C, respectively; $\varkappa_{\mathrm{I}}, \varkappa_{\mathrm{II}}, \ldots$ are the slopes at these fibers;
- t is the number of triple singular points of C;
- $\mathfrak{S} = \operatorname{Sk} C$ and $R_{\mathrm{I}}, R_{\mathrm{II}}, \ldots$ or $v_{\mathrm{I}}, v_{\mathrm{II}}, \ldots$ are, respectively, the regions or vertices, whichever is applicable, containing/representing the distinguished singular fibers;
- occasionally, we also use the notation $\mathfrak{S} = \operatorname{Sk}(D, P) = \operatorname{Sk} D$.

Classification of maximizing sextics

On a case-by-case basis, we use Theorem 4.33 and conclude that, with one isotrivial exception, a sextic D is maximizing if and only if C is generic (in the sense described above), maximal, and has no unstable fibers other than the distinguished ones. For maximizing sextics, the correspondence $(D, P) \mapsto (C, F_{\mathrm{I}}, \ldots)$ is invertible (if P is of type $\mathbf{E}_7$, a certain additional structure should be fixed, see Proposition 8.18) and pairs (D, P) are classified by the skeletons $\mathfrak{S} := \operatorname{Sk} C$ equipped with certain extra decorations, mainly, regions R_i containing the distinguished fibers F_i of type I and/or singular vertices v_j representing distinguished fibers F_j of other (exceptional) types. We explain how such skeletons can be enumerated and list corresponding maximizing sextics in tables. (If P is of type $\mathbf{D}$, we omit the lists of reducible sextics, as they are too long. However, we still enumerate the skeletons and their decorations.)

Another extra decoration of $\mathfrak{S}$ is its type specification. If $C \subset \Sigma_d$, the number t of triple singular points of C is found from the identity $\deg \mathfrak{S} = 3d - 3t$. By definition ($C$ is minimal), for each distinguished region R_i and each distinguished vertex v_j one has $\operatorname{tp}_C(R_i) = \operatorname{tp}_0(R_i)$ and $\operatorname{tp}_C(v_j) = \operatorname{tp}_0(v_j)$, respectively. For each singular $\bullet$- or $\circ$-vertex w other than the distinguished ones, one must have $\operatorname{tp}_C(w) = \operatorname{tp}_0(w) + 6$, as we assume that C has no unstable singular fibers. Denote by δ the difference between t and the number of such singular vertices w. Then $\delta \geqslant 0$ and for exactly δ regions R of $\mathfrak{S}$ one has $\operatorname{tp}_C(R) = \operatorname{tp}_0(R) + 6$; these regions contain $\tilde{\mathbf{D}}$ type singular fibers. It is due to this extra choice that some skeletons are represented by several lines in the tables (see, *e.g.*, Lines 18–22 in Table 8.1).

Ultimately, with Observation 4.41 taken into account, the classification is reduced to the list of the 191 regular skeletons of degree twelve found in [20]. However, to make both the enumeration and the computation of the groups more reader friendly, we use a number of tricks trying to reduce most skeletons to those listed in Figures 4.2

and 4.3. Most notably, we use the concept of *insertion*, which is a fragment of a skeleton, depending on the type(s) of the distinguished fibers(s), with two 'hanging' edges, *cf.* Figure 8.1, left, on page 233. Removing this fragment from $\mathfrak{S}$ produces an auxiliary skeleton $\mathfrak{S}'$ of smaller degree, *cf.* Figure 8.1, right, and the original skeleton $\mathfrak{S}$ is obtained from $\mathfrak{S}'$ by inserting the fragment at the center of any of its edges.

Convention 8.3. Strictly speaking, the passage from $\mathfrak{S}$ to $\mathfrak{S}'$ is more complicated than just explained: one removes the stars of all vertices contained in the insertion and identifies the two $\circ$-vertices w', w'' that become monovalent or isolated, *cf.* Figure 8.1. In view of Convention 1.7, in the drawings this *looks* like merely erasing the insertion, the inverse procedure being inserting a fragment at the middle of an edge. We will use this simplified language when describing similar modifications in the sequel.

Another trick, used in Section 8.3.3, is collapsing a triangular region of a regular skeleton to a single trivalent $\bullet$-vertex. This operation reduces degree by three, and its inverse consists in inflating a trivalent $\bullet$-vertex to a triangle.

The fundamental groups

The fundamental groups of all maximizing sextics are computed using their skeletons and the strategy outlined in Section 5.2.3. The common reference edge e is usually chosen close to a distinguished fiber. We routinely omit one of the non-distinguished regions, see Lemma 5.59, and use just enough relations to show that the group is as expected. In few cases (usually, when the existence of a certain fragment in $\mathfrak{S}$ implies that π_1 is abelian), we analyze the group manually; otherwise, we use GAP.

We mainly deal with irreducible curves. In the reducible case, a presentation for π_1 can still be written down, using the same techniques, but the analysis is difficult as the abelianization $\pi_1/[\pi_1, \pi_1]$ is infinite and GAP fails. In some selected cases we do make an attempt to analyze these groups, usually by finding an appropriate central element and referring to Corollary A.30.

Non-maximizing sextics

Let C be a proper model, not necessarily maximal, generic in the sense explained above. Then C satisfies the hypotheses of Corollary 4.44 and admits a degeneration to a maximal trigonal curve C', see Corollary 4.47. We call this degeneration and the inverse perturbation $C' \to C$ *mild* if

- the degeneration is simple, *i.e.*, only non-exceptional fibers collide;
- the curve C' is also simple, *i.e.*, pairs of $\tilde{\mathbf{D}}$ type fibers do not collide;
- distinguished fiber(s) do not collide with any other fibers.

For the construction, we need to *isolate* each non-exceptional distinguished fiber F_i, *i.e.*, replace the dessin of C with an equivalent dessin $\mathfrak{D}$ with connected skeleton $\mathfrak{S}$

and such that F_i is the only $\times$-vertex in a region of $\mathfrak{S}$. Then, as in Proposition 4.45, we can collide all $\times$-vertices within each region of $\mathfrak{S}$ to a single $\times$-vertex. Unless $\mathfrak{S}$ has a region shared by two $\tilde{\mathbf{D}}$ type fibers, the resulting degeneration is mild.

If a sextic D admits a mild degeneration to a maximizing sextic D', the group π_1 of D can be computed together with that of D', by merely changing the braid indices of some of the regions in the input.

Observation 8.4. To obtain all maximal (with respect to the natural partial order) irreducible mild perturbations of a maximizing sextic with two or three components, it suffices to perturb, respectively, one or two singular fibers of one of the following types:

1. $\tilde{\mathbf{A}}_p$ with p odd: in the computation of π_1, the width $(p+1)$ of the corresponding region is replaced with its maximal odd divisor;
2. $\tilde{\mathbf{D}}_q$ with q even: in the computation, the $\tilde{\mathbf{D}}$ type relation is retained and the braid index of the corresponding region is changed from 0 to an odd integer s, $1 \leqslant s < q-4$. If $s = 1$, the group is abelian, see Theorem 7.24.

Note that some (pairs of) fibers do not produce irreducible curves when perturbed and hence should not be considered. In the context of the computation of the fundamental group, the simplest irreducibility criterion is the identity $\pi_1/[\pi_1, \pi_1] = \mathbb{Z}_6$, *i.e.*, the call `AbelianInvariants(g)` should return `[ 2, 3 ]`.

Similarly, given a sextic of torus type, its maximal perturbations that are *not* of torus type are obtained by replacing the braid index of the region containing an inner singularity, which is of type $\tilde{\mathbf{A}}_p$ with $(p+1) = 0 \bmod 3$, with its maximal divisor prime to 3.

For the vast majority of perturbed curves, no computation is needed as the groups are as expected (abelian or isomorphic to Γ) due to Lemmas 5.82 and 5.83. For the few remaining cases, the `GAP` input is found in the same files as the computation for the corresponding maximal curves.

Observation 8.5. If a region of $\mathfrak{S}$ is shared by two $\tilde{\mathbf{D}}$ type fibers, the degeneration constructed above produces a non-simple curve. Such collisions of $\tilde{\mathbf{D}}$ type fibers are considered below on a case-by-case basis. Since a singular fiber of type $\tilde{\mathbf{D}}_4$ is not detected by the j-invariant and can freely be moved to any region of the skeleton, we can always assume that both fibers are of types $\tilde{\mathbf{D}}_q$ with $q \geqslant 5$. Similarly, if a $\tilde{\mathbf{D}}$ type fiber F collides with a distinguished fiber F_i, we can assume that F is of type $\tilde{\mathbf{D}}_q$ with $q \geqslant 5$ and F_i is of Kodaira type I_p with $p \geqslant 1$.

8.2 A distinguished point of type E

If the distinguished point P is of type $\mathbf{E}$, the trigonal model $\tilde{C}$ has a single improper fiber $\tilde{F}_{\mathrm{I}}$ and, choosing the reference edge e close to this fiber, we can write down the braid relations about $\tilde{F}_{\rm I}$ and the relation at infinity (5.55) common for all curves with the given topological type of P. Furthermore, in this case the number of skeletons is relatively small and we classify reducible maximizing sextics as well, see Tables 8.2, 8.4, and 8.7, and analyze their fundamental groups, unless they are too large.

8.2.1 A point of type $\mathbf{E}_8$

Considered in this subsection are all sextics with a distinguished point P of type $\mathbf{E}_8$.

Proposition 8.6. *There is a natural bijection ϕ, invariant under equisingular deformations, between the sets of*:

- *simple plane sextics $D \subset \mathbb{P}^2$ with a distinguished type $\mathbf{E}_8$ point P, and*
- *proper simple trigonal curves $C \subset \Sigma_3$ with a distinguished type $\tilde{\mathbf{A}}_1^*$ fiber $F_{\rm I}$.*

With one exception, a sextic D is maximizing if and only if $C := \phi(D)$ is maximal and has no unstable fibers other than $F_{\rm I}$. The exception is the reducible sextic D with the set of singularities $\mathbf{E}_8 \oplus \mathbf{E}_7 \oplus \mathbf{D}_4$; in this case, $\phi(D)$ is an isotrivial curve with the set of singular fibers $\tilde{\mathbf{A}}_1^* + \tilde{\mathbf{E}}_7 + \tilde{\mathbf{D}}_4$.

Proof. The trigonal model $\tilde{C}$ has a single improper fiber $\tilde{F}_{\rm I}$, at which a cusp of $\tilde{C}$ is tangent to the exceptional section. Two positive Nagata transformations convert $\tilde{F}_{\rm I}$ to a type $\tilde{\mathbf{A}}_1^*$ singular fiber $F_{\rm I}$ of C.

If C is non-isotrivial then, using Theorem 4.33, one can see that D is maximizing if and only if C is maximal and $t_{\rm u} = 1$. (Since $F_{\rm I}$ is unstable, one always has $t_{\rm u} \geqslant 1$.) If C is isotrivial, then $j_C \equiv 1$ (as $j_C(F_{\rm I}) = 1$, see Table 3.1) and all singular fibers of C are of types $\tilde{\mathbf{A}}_1^*$, $\tilde{\mathbf{D}}_4$, or $\tilde{\mathbf{E}}_7$, corresponding, respectively, to the simple, double, or triple roots of the polynomial $g_2(t)$, see page 75. Since $\deg g_2 \leqslant 5$, the only such set of singularities with the total Milnor number 12 is $\tilde{\mathbf{E}}_7 + \tilde{\mathbf{D}}_4 + \tilde{\mathbf{A}}_1^*$. In view of the presence of a $\tilde{\mathbf{D}}_4$ type fiber, this set of singularities cannot be realized by a maximal non-isotrivial curve. □

If a proper model C as in Proposition 8.6 is isotrivial, it is obviously unique up to isomorphism and gives rise to a single class of sextics, see Line 17′ in Table 8.2. Thus, assume that C is not isotrivial and consider its skeleton $\mathfrak{S} := \operatorname{Sk} C$.

According to Proposition 8.6, the skeleton $\mathfrak{S}$ has a distinguished monovalent $\circ$-vertex $v_{\rm I}$ representing the distinguished fiber $F_{\rm I}$. Let e be the edge incident to $v_{\rm I}$, and let u be the other end of e. If $\operatorname{ind} u = 1$, then e is the only edge of $\mathfrak{S}$ and the skeleton is as shown in Figure 8.2 (k). If $\operatorname{ind} u = 2$, then $t \geqslant 1$ and, due to (4.32), the skeleton has at most one other $\bullet$-vertex, which must be trivalent. The only skeleton with these properties is shown in Figure 8.2 (j).

Assume that $\operatorname{ind} u = 3$ and consider the skeleton $\mathfrak{S}'$ obtained from $\mathfrak{S}$ by removing the *insertion* $[u, v_{\rm I}]$, see Figure 8.1 and Convention 8.3. (The exceptional case, when $\mathfrak{S}'$ has no $\bullet$-vertices and hence is not a valid skeleton, is shown in Figure 8.3 (g).)

Figure 8.1. Modification of a skeleton.

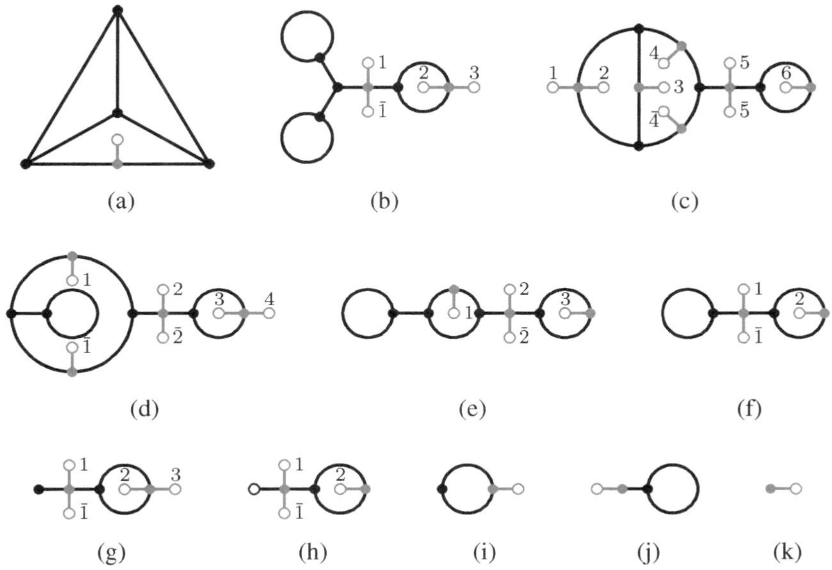

Figure 8.2. Type $\mathbf{E}_8$ singularity: Irreducible curves.

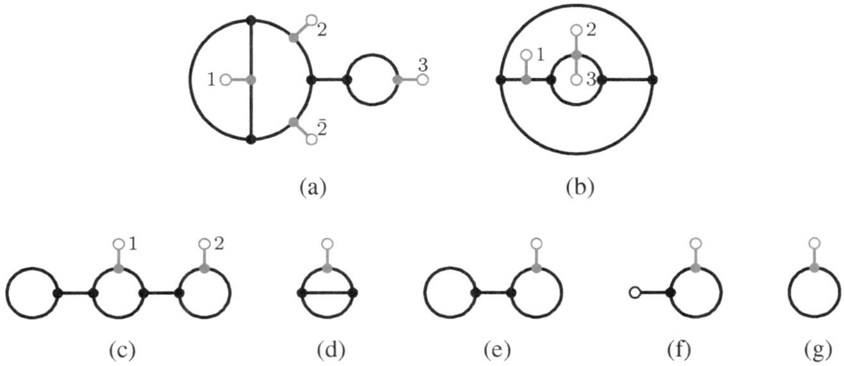

Figure 8.3. Type $\mathbf{E}_8$ singularity: Reducible curves.

For $\mathfrak{S}'$, the vertex count (4.32) takes the form

$$\#_\bullet + \#^2_\bullet + \#^1_\circ = 4 - 2t, \qquad t \geqslant \#^1_\bullet + \#^2_\bullet + \#^1_\circ; \tag{8.7}$$

hence $\mathfrak{S}'$ is one of the skeletons shown in Figures 4.2 and 4.3 and, up to symmetries of $\mathfrak{S}'$, the position of the insertion is one of those listed in Figure 8.2 (irreducible curves) and 8.3 (reducible curves); the insertion is shown in grey. The resulting sextics are listed in Tables 8.1 and 8.2.

Convention 8.8. Throughout the chapter, pairs of insertions that differ by an orientation reversing symmetry of the auxiliary skeleton $\mathfrak{S}'$ are marked with $n, \bar{n}$ for some integer n. Such pairs result in pairs of complex conjugate sextics.

Remark 8.9. Irreducible curves are detected using Theorem 5.92. One can easily see that any splitting marking of $\mathfrak{S}$ restricts to $\mathfrak{S}'$, whereas a marking μ of $\mathfrak{S}'$ extends to a splitting marking of $\mathfrak{S}$ if and only if in Figure 8.1, right, $\mu(e') = 1$ and $\mu(e'') = -1$.

The fundamental groups

Take $e = [u, v_{\mathrm{I}}]$ for the common reference edge. Then, combining (5.80) with $n = 2$ and the relation at infinity (5.55), we have

$$\rho^3 = \alpha_1\alpha_2^2, \qquad \alpha_3 = \alpha_2\alpha_1\alpha_2^{-1}, \qquad [\alpha_1, \alpha_2^3] = 1, \tag{8.10}$$

so that the group is generated by α_1, α_2 and α_2^3 is central. Eliminating α_3, one can rewrite these relations in the form

$$(\alpha_1\alpha_2^{-1})^5\alpha_2^6 = [\alpha_1, \alpha_2^3] = 1. \tag{8.11}$$

For the other relations, in most cases it suffices to consider the three regions closest to e, *viz.* $\mathrm{reg}_-(e \uparrow \zeta)$ for $\zeta = 1, \mathbf{x}^2, \mathbf{x}^2\mathbf{y}\mathbf{x}^2$. The functions "`size`" (for irreducible curves) and "`size2`" (for reducible curves) in Listing 8.4 implement this approach. (For reducible curves, we consider the quotient by α_2^3 and refer to Corollary A.30.) The group is stored in "g", and the functions return the size of its commutant. The parameters p, q, r are the braid indices of the corresponding regions, in the order listed.

Convention 8.12. Thus, in this simplistic approach we merely ignore the $\tilde{\mathbf{D}}$ type regions using 0 as a parameter. To emphasize the fact that a region is ignored, we use the notation - instead of 0 in the parameter lists.

Argument 8.13 (special fragments). Assume that $\mathfrak{S}$ has one of the three fragments shown in Figure 8.5, with a type $\tilde{\mathbf{A}}$ singular fiber inside the region R depicted as a loop. (More generally, one can assume that the braid index of R divides 1, 3, or 2, respectively.) Then the group π_1 is abelian. Indeed, in cases (a) and (b), the calls

Table 8.1. Irreducible maximizing sextics with a type $\mathbf{E}_8$ point[1]

#	Set of singularities	Figure	Count	π_1
* 1	$\mathbf{E}_8 \oplus \mathbf{A}_4 \oplus \mathbf{A}_3 \oplus 2\mathbf{A}_2$	8.2(a)	$(1,0)$	8.14
2	$\mathbf{E}_8 \oplus \mathbf{A}_{11}$	8.2(b)-1	$(0,1)$	8.13
3	[1] $\mathbf{E}_8 \oplus \mathbf{A}_9 \oplus \mathbf{A}_2$	8.2(b)-2	$(1,0)$	8.13
4	[2] $\mathbf{E}_8 \oplus \mathbf{A}_{10} \oplus \mathbf{A}_1$	8.2(b)-3	$(1,0)$	8.13
5	$\mathbf{E}_8 \oplus \mathbf{A}_7 \oplus 2\mathbf{A}_2$	8.2(c)-1	$(1,0)$	`(8,3,3)`[2]
6	$\mathbf{E}_8 \oplus \mathbf{A}_6 \oplus \mathbf{A}_3 \oplus \mathbf{A}_2$	8.2(c)-2	$(1,0)$	`(4,7,3)`[2]
7	[3] $\mathbf{E}_8 \oplus \mathbf{A}_5 \oplus \mathbf{A}_4 \oplus \mathbf{A}_2$	8.2(c)-3	$(1,0)$	`(5,3,6)`[2]
8	[4] $\mathbf{E}_8 \oplus \mathbf{A}_6 \oplus \mathbf{A}_4 \oplus \mathbf{A}_1$	8.2(c)-4	$(0,1)$	`(5,7,2)`[2]
9	[5] $\mathbf{E}_8 \oplus \mathbf{A}_8 \oplus \mathbf{A}_2 \oplus \mathbf{A}_1$	8.2(c)-5	$(0,1)$	8.13
10	$\mathbf{E}_8 \oplus \mathbf{A}_6 \oplus 2\mathbf{A}_2 \oplus \mathbf{A}_1$	8.2(c)-6	$(1,0)$	8.13
11	$\mathbf{E}_8 \oplus \mathbf{A}_6 \oplus \mathbf{A}_5$	8.2(d)-1	$(0,1)$	`(7,-,-)`[2]
12	$\mathbf{E}_8 \oplus \mathbf{A}_7 \oplus \mathbf{A}_4$	8.2(d)-2	$(0,1)$	8.13
13	[3] $\mathbf{E}_8 \oplus \mathbf{A}_5 \oplus \mathbf{A}_4 \oplus \mathbf{A}_2$	8.2(d)-3	$(1,0)$	8.13
14	[4] $\mathbf{E}_8 \oplus \mathbf{A}_6 \oplus \mathbf{A}_4 \oplus \mathbf{A}_1$	8.2(d)-4	$(1,0)$	8.13
15	$\mathbf{E}_8 \oplus \mathbf{A}_8 \oplus \mathbf{A}_3$	8.2(e)-1	$(1,0)$	`(4,9,9)`[2]
16	[2] $\mathbf{E}_8 \oplus \mathbf{A}_{10} \oplus \mathbf{A}_1$	8.2(e)-2	$(0,1)$	8.13
17	[5] $\mathbf{E}_8 \oplus \mathbf{A}_8 \oplus \mathbf{A}_2 \oplus \mathbf{A}_1$	8.2(e)-3	$(1,0)$	8.13
18	$\mathbf{E}_8 \oplus \mathbf{D}_{11}$	8.2(f)-1	$(1,0)$	8.13
19	$\mathbf{E}_8 \oplus \mathbf{D}_5 \oplus \mathbf{A}_6$	8.2(f)-1	$(0,1)$	8.13
20	$\mathbf{E}_8 \oplus \mathbf{D}_9 \oplus \mathbf{A}_2$	8.2(f)-2	$(1,0)$	8.13
21	$\mathbf{E}_8 \oplus \mathbf{D}_7 \oplus \mathbf{A}_4$	8.2(f)-2	$(1,0)$	`(-,5,5)`[2]
22	$\mathbf{E}_8 \oplus \mathbf{D}_5 \oplus \mathbf{A}_4 \oplus \mathbf{A}_2$	8.2(f)-2	$(1,0)$	8.13
23	$\mathbf{E}_8 \oplus \mathbf{E}_6 \oplus \mathbf{A}_5$	8.2(g)-1	$(0,1)$	8.13
24	$\mathbf{E}_8 \oplus \mathbf{E}_6 \oplus \mathbf{A}_3 \oplus \mathbf{A}_2$	8.2(g)-2	$(1,0)$	8.13
25	$\mathbf{E}_8 \oplus \mathbf{E}_6 \oplus \mathbf{A}_4 \oplus \mathbf{A}_1$	8.2(g)-3	$(1,0)$	8.13
26	$\mathbf{E}_8 \oplus \mathbf{E}_7 \oplus \mathbf{A}_4$	8.2(h)-1	$(0,1)$	8.13
27	$\mathbf{E}_8 \oplus \mathbf{E}_7 \oplus 2\mathbf{A}_2$	8.2(h)-2	$(1,0)$	8.13
28	$2\mathbf{E}_8 \oplus \mathbf{A}_2 \oplus \mathbf{A}_1$	8.2(i)	$(1,0)$	8.13
29	$2\mathbf{E}_8 \oplus \mathbf{A}_3$	8.2(j)	$(1,0)$	8.13
30	$\mathbf{E}_8 \oplus \mathbf{E}_6 \oplus \mathbf{D}_5$	8.2(k)	$(1,0)$	8.15

[1] There are 39 classes realizing 26 sets of singularities
[2] See "`size`" in Listing 8.4. All `GAP` input is found in "`misc/E8-ir.txt`"

Comments to all tables Marked with a * are curves with nonabelian groups. We list:

- line number,
- set of singularities; marked with equal prefixes are those shared by several curves,
- reference to the figure,
- count in the form (c_r, c_i), where c_r is the number of real curves and c_i is the number of pairs of complex conjugate curves,
- reference to the computation of the group $\pi_1 := \pi_1(\mathbb{P}^2 \smallsetminus D)$.

Table 8.2. Reducible maximizing sextics with a type $\mathbf{E}_8$ point[1]

#	Set of singularities	Figure	Count	π_1
$1'$	$\mathbf{E}_8 \oplus \mathbf{A}_5 \oplus 2\mathbf{A}_3$	8.3(a)-1	$(1,0)$	`(4,4,6)`[2]
$2'$	$\mathbf{E}_8 \oplus \mathbf{A}_7 \oplus \mathbf{A}_3 \oplus \mathbf{A}_1$	8.3(a)-2	$(0,1)$	`(8,4,2)`[2]
$3'$	$\mathbf{E}_8 \oplus \mathbf{A}_7 \oplus \mathbf{A}_2 \oplus 2\mathbf{A}_1$	8.3(a)-3	$(1,0)$	8.13
$4'$	$\mathbf{E}_8 \oplus \mathbf{A}_5 \oplus \mathbf{A}_4 \oplus 2\mathbf{A}_1$	8.3(b)-1	$(1,0)$	`(6,5,2)`[2]
$5'$	$\mathbf{E}_8 \oplus \mathbf{A}_5 \oplus \mathbf{A}_3 \oplus \mathbf{A}_2 \oplus \mathbf{A}_1$	8.3(b)-2	$(1,0)$	`(6,3,4)`[2]
$6'$	$\mathbf{E}_8 \oplus \mathbf{A}_4 \oplus 2\mathbf{A}_3 \oplus \mathbf{A}_1$	8.3(b)-3	$(1,0)$	`(4,5,4)`[2]
$7'$	[1] $\mathbf{E}_8 \oplus \mathbf{A}_9 \oplus \mathbf{A}_2$	8.3(c)-1	$(1,0)$	`(10,3,-)`[2]
$8'$	$\mathbf{E}_8 \oplus \mathbf{A}_9 \oplus 2\mathbf{A}_1$	8.3(c)-2	$(1,0)$	8.13
$9'$	$\mathbf{E}_8 \oplus \mathbf{D}_8 \oplus \mathbf{A}_2 \oplus \mathbf{A}_1$	8.3(d)	$(1,0)$	`(-,3,2)`[2]
$10'$	$\mathbf{E}_8 \oplus \mathbf{D}_7 \oplus \mathbf{A}_3 \oplus \mathbf{A}_1$	8.3(d)	$(1,0)$	`(4,-,2)`[2]
* $11'$	$\mathbf{E}_8 \oplus \mathbf{D}_6 \oplus \mathbf{A}_3 \oplus \mathbf{A}_2$	8.3(d)	$(1,0)$	8.16
$12'$	$\mathbf{E}_8 \oplus \mathbf{D}_{10} \oplus \mathbf{A}_1$	8.3(e)	$(1,0)$	8.13
$13'$	$\mathbf{E}_8 \oplus \mathbf{D}_6 \oplus \mathbf{A}_5$	8.3(e)	$(1,0)$	8.15
$14'$	$\mathbf{E}_8 \oplus \mathbf{D}_5 \oplus \mathbf{A}_5 \oplus \mathbf{A}_1$	8.3(e)	$(1,0)$	8.13
$15'$	$\mathbf{E}_8 \oplus \mathbf{E}_7 \oplus \mathbf{A}_3 \oplus \mathbf{A}_1$	8.3(f)	$(1,0)$	8.13
$16'$	$\mathbf{E}_8 \oplus \mathbf{D}_6 \oplus \mathbf{D}_5$	8.3(g)	$(1,0)$	8.13
$17'$	$\mathbf{E}_8 \oplus \mathbf{E}_7 \oplus \mathbf{D}_4$	isotrivial	$(1,0)$	8.17

[1] There are 18 classes realizing 17 sets of singularities

[2] See "`size2`" in Listing 8.4. All `GAP` input is found in "`misc/E8-r.txt`"

`size(-,-,1)` and `size(3,-,-)` return 1. (In case (a), we assume that the insertion is as shown in solid lines; otherwise one can use symmetry.) In particular, this fact implies that the curve is necessarily irreducible in these cases. In case (c), the call `size2(-,2,-)` returns 1.

In most other cases, functions "`size`" or "`size2`" with the parameters listed in the tables return 1. The few curves that need special attention are considered below, see "`misc/E8-ir.txt`" and "`misc/E8-r.txt`" for the `GAP` input.

Argument 8.14 ($\mathbf{E}_8 \oplus \mathbf{A}_4 \oplus \mathbf{A}_3 \oplus 2\mathbf{A}_2$, Line 1). Due to Lemma 5.59, in the presence of (8.10) the three regions used in Listing 8.4 do provide a complete presentation of the group. Creating the group *via* `size(5,4,3)`, one can use `GAP` to check that

- $\operatorname{ord} \pi_1 = 720$ and the commutant $H := [\pi_1, \pi_1] \cong SL(2, \Bbbk_5)$ is the only perfect group of order 120;
- the last two relations in the presentation can be omitted;
- the centralizer $C := \mathcal{Z}_{\pi_1}(H)$ equals $\mathbb{Z}_{12}$ and $\operatorname{ord}(C \cap H) = 2$; it follows that the projection $C \to \pi_1/H$ is onto.

Summarizing and replacing (8.10) with (8.11), one arrives at the presentation stated in Theorem 8.1 and shows that π_1 is the central product

$$H \odot C = SL(2, \Bbbk_5) \odot \mathbb{Z}_{12} .$$

```
# The computation of the fundamental groups of plane sextics with a type E8 singularity.
# We use the three type Ã regions closest to the insertion, viz. reg_(e ↑ ζ) for ζ = 1, x^2, and x^2yx^2.
    Their widths, in the order listed, are the parameters.

Reread("pi1sc.txt");    # (see Listing 5.3)
# Relations at infinity and regions to be used
fixed := [rho^3/(G.1*G.2^2), G.2*G.1*G.2^-1/G.3, G.1*G.2^3/(G.2^3*G.1)];
list := [[], x^2, x^2*y*x^2];
# This function handles irreducible curves (see "AA" and "DS" in Listing 5.3)
size := function(arg) return Size(DS(AA(list, arg))); end;
# For reducible curves, we consider the quotient by the central element α_2^3
size2 := function(arg) return Size(DS(AA(list, arg), [G.2^3])); end;
```

Listing 8.4. Computing π_1 for sextics with a type $\mathbf{E}_8$ point ("E8.txt").

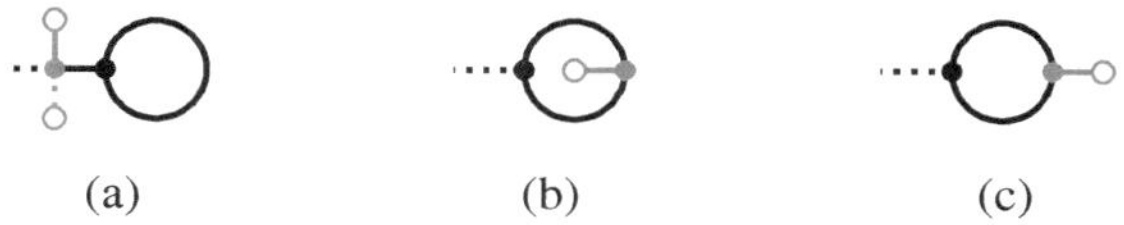

Figure 8.5. Fragments resulting in an abelian group.

Argument 8.15 ($\mathbf{E}_8 \oplus \mathbf{E}_6 \oplus \mathbf{D}_5$, Line 30 and $\mathbf{E}_8 \oplus \mathbf{D}_6 \oplus \mathbf{A}_5$, Line 13′). To show that π_1 is abelian, it suffices to add to (8.10) the braid relations arising from, respectively, the type $\mathbf{E}_6$ fiber over u or the monogonal region of $\mathfrak{S}$.

Argument 8.16 ($\mathbf{E}_8 \oplus \mathbf{D}_6 \oplus \mathbf{A}_3 \oplus \mathbf{A}_2$, Line 11′). The group π_1 is given by

$$\pi_1 = \langle \alpha_1, \alpha_2 \mid (\alpha_1\alpha_2^{-1})^5\alpha_2^6 = [\alpha_1, \alpha_2^3] = \{\alpha_1, \alpha_2\}_4 = 1 \rangle \cong SL(2, \Bbbk_5) \times \mathbb{Z}.$$

Indeed, due to Lemma 5.59, the relation arising from the region with the $\tilde{\mathbf{D}}$ type fiber can be omitted. Creating the quotient π_1/α_2^3 *via* `size2(4,3,-)`, one can see that the last relation can be omitted (hence the presentation stated) and that $H := [\pi_1, \pi_1]$ is isomorphic to $SL(2, \Bbbk_5)$. In view of the last relation, $(\alpha_1\alpha_2)^2$ is a central element; hence, so is $\beta := (\alpha_1\alpha_2)^2\alpha_2^{-3}$ and $\pi_1 = H \times \langle \beta \rangle$.

Argument 8.17 ($\mathbf{E}_8 \oplus \mathbf{E}_7 \oplus \mathbf{D}_4$, Line 17′). The curve C is isotrivial. Its monodromy group contains $(\sigma_1\sigma_2\sigma_1)^3$ and $(\sigma_1\sigma_2\sigma_1)^2$ (from the singular fibers of type $\tilde{\mathbf{E}}_7$ and $\tilde{\mathbf{D}}_4$, respectively, see page 169); hence it also contains $\sigma_1\sigma_2\sigma_1$, and π_1 is abelian.

Any non-maximizing simple sextic C with an $\mathbf{E}_8$ type singular point admits a mild degeneration to a maximizing one, and the fundamental group of C is abelian. In most cases, the existence of a degeneration is given by Corollary 4.47. In the exceptional case of the degree 3 skeleton $\mathfrak{S}$ shown in Figure 8.3 (g), the curve may have two $\tilde{\mathbf{D}}_5$ type singular fibers in the same bigonal region of $\mathfrak{S}$, see Observation 8.5. This region has a cutting sequence of width one and, applying a bold modification, one can place the two fibers to two distinct regions of the new skeleton; hence, the curve degenerates to $\mathbf{E}_8 \oplus \mathbf{D}_6 \oplus \mathbf{D}_5$, Line $16'$.

8.2.2 A point of type $\mathbf{E}_7$

Here, we consider simple plane sextics $D \subset \mathbb{P}^2$ satisfying the following conditions:

(∗) D has a distinguished singular point P of type $\mathbf{E}_7$,
D has no linear components through P, and
D has no singular points of type $\mathbf{E}_8$ or $\mathbf{E}_6$.

(All sextics with a singular point of type $\mathbf{E}_8$ were considered in Section 8.2.1, and those with a point of type $\mathbf{E}_6$ will be treated in Section 8.2.3 below.)

Proposition 8.18. *There is a natural bijection ϕ, invariant under equisingular deformations, between Zariski open and dense, in each equisingular stratum, subsets of the sets of:*

- *simple plane sextics $D \subset \mathbb{P}^2$ with a distinguished type $\mathbf{E}_7$ point P and without linear components through P, and*
- *proper simple trigonal curves $C \subset \Sigma_3$ with a distinguished type $\tilde{\mathbf{A}}_1$ fiber F_{I} and a distinguished branch of C at the node at F_{I}.*

A sextic D is maximizing if and only if $C := \phi(D)$ is maximal and stable.

Proof. The only improper singular fiber $\tilde{F}_{\mathrm{I}}$ of the trigonal model $\tilde{C}$ has a node of $\tilde{C}$ with one of the branches tangent to the exceptional section and, typically, a pair of positive Nagata transformations converts $\tilde{F}_{\mathrm{I}}$ to a type $\tilde{\mathbf{A}}_1$ singular fiber F_{I} of C. The blow up centers of the inverse transformations select a branch of C at the node in F_{I}, see Case 4 with $m = n = 2$ in Section 5.1.3. A new phenomenon is the possibility of an extra degeneration: the smooth branch of D at P may have P as a point of inflection; then the other branch at the node is tangent to $\tilde{F}_{\mathrm{I}}$ and F_{I} is of type $\tilde{\mathbf{A}}_2$ (Case 3 with $m = 3$ in Section 5.2.2). Such a fiber can be perturbed to one of type $\tilde{\mathbf{A}}_1$, and the perturbation is followed by a perturbation of sextics, as the multiplicities of the blow-up centers of the negative Nagata transformations remain constant. □

Fix a maximizing sextic D satisfying conditions 8.2.2(∗) and let $\mathfrak{S} := \operatorname{Sk} C$. The bigonal region R_{I} of $\mathfrak{S}$ containing F_{I} is called the *insertion*. If the two •-vertices of R_{I} coincide, R_{I} is the outer region shown in Figure 8.7 (e). Otherwise, one can remove R_{I}, *cf.* Convention 8.3, to obtain a new skeleton $\mathfrak{S}'$ with at most four •-vertices. (The exceptional case when $\mathfrak{S}'$ has no •-vertices is shown in Figure 8.8 (c).) The vertex count (8.7) applies to $\mathfrak{S}'$ and hence $\mathfrak{S}$ is one of the skeletons in Figure 8.6–8.8, where the possible positions of the insertion are shown in grey. (The meaning of the ∘-vertices in Figure 8.7 is explained further in this section.) The assumption that D has no singular points of types $\mathbf{E}_8$ or $\mathbf{E}_6$ excludes skeletons with singular •-vertices.

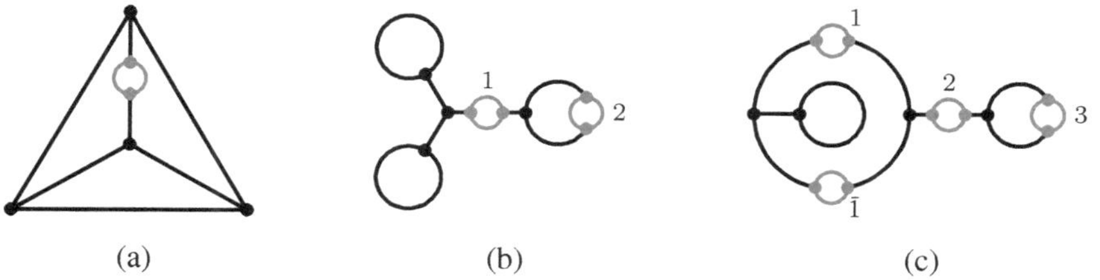

Figure 8.6. Type $\mathbf{E}_7$ singularity: Irreducible curves.

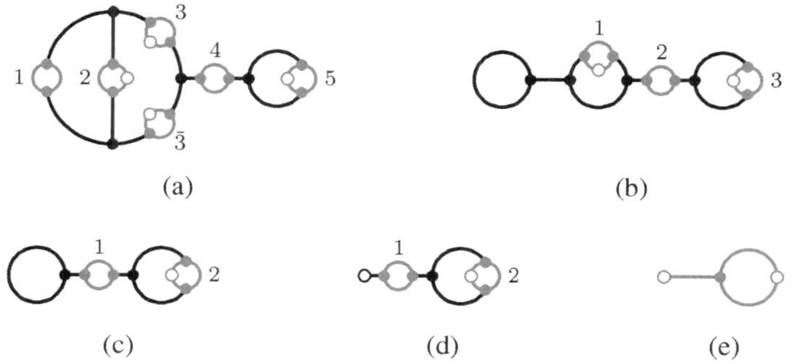

Figure 8.7. Type $\mathbf{E}_7$ singularity: Two component curves.

Let e_1, e_2 be the two edges constituting $\partial_- R_{\mathrm{I}}$. The distinguished fiber F_{I} of C is as in Case 4 in Section 5.1.3, with $m = n = 2$, and the second blow-up center distinguishes a branch $\mathfrak{b}_j$ of C at its singular point in F_{I}. The index $j = 1$ or 2 depends on the choice of a reference edge in $\partial_- R_{\mathrm{I}}$, and we number the two edges so that the distinguished branch is $\mathfrak{b}_1$ with respect to e_1 (and hence it is $\mathfrak{b}_2$ with respect to e_2). Thus, a sextic D is determined by a skeleton $\mathfrak{S}$, a distinguished bigonal region R_{I} of $\mathfrak{S}$, and a distinguished edge $e_1 \in \partial_- R_{\mathrm{I}}$. In terms of the auxiliary skeleton $\mathfrak{S}'$, one needs to specify an edge to hold the insertion and its coorientation. One can easily

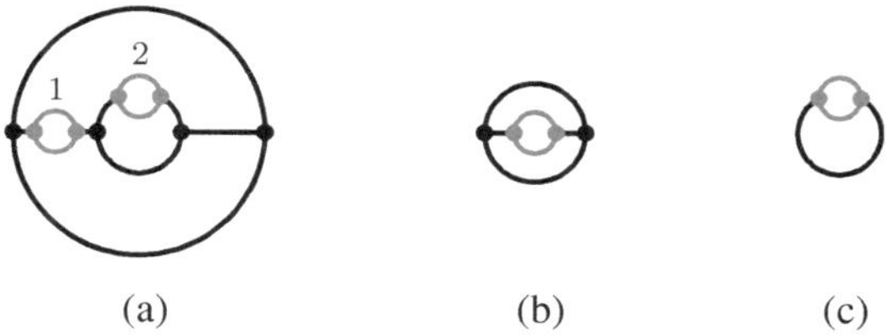

Figure 8.8. Type $\mathbf{E}_7$ singularity: Three component curves.

Table 8.3. Irreducible maximizing sextics with a type $\mathbf{E}_7$ point[1]

#	Set of singularities	Figure	Count	π_1
* 1	$\mathbf{E}_7 \oplus 2\mathbf{A}_4 \oplus 2\mathbf{A}_2$	8.6(a)	$(1,0)$	8.23
2	$\mathbf{E}_7 \oplus \mathbf{A}_{12}$	8.6(b)-1	$(0,1)$	`(13,-,1)`[2]
3	$\mathbf{E}_7 \oplus \mathbf{A}_{10} \oplus \mathbf{A}_2$	8.6(b)-2	$(2,0)$	`(11,3)(3,11)`[2]
4	$\mathbf{E}_7 \oplus 2\mathbf{A}_6$	8.6(c)-1	$(0,1)$	`(7,7,-,1)`[2]
5	$\mathbf{E}_7 \oplus \mathbf{A}_8 \oplus \mathbf{A}_4$	8.6(c)-2	$(0,1)$	`(9,-,1)`[2]
6	$\mathbf{E}_7 \oplus \mathbf{A}_6 \oplus \mathbf{A}_4 \oplus \mathbf{A}_2$	8.6(c)-3	$(2,0)$	`(7,3)(3,7)`[2]

[1] There are 11 classes realizing 6 sets of singularities

[2] See "`size`" in Listing 8.9. All `GAP` input is found in "`misc/E7-ir.txt`"

see that, geometrically, the sextics corresponding to the two opposite coorientations differ by a quadratic birational transformation of the plane.

Remark 8.19. In the oriented surface $S \cong S^2$, an orientation and a coorientation of an edge of $\mathfrak{S}'$ determine each other. However, the situation changes if we disregard the orientation of S^2. If D and $\bar{D}$ is a pair of complex conjugate sextics, the corresponding skeletons $\mathfrak{S}$, $\bar{\mathfrak{S}}$ and $\mathfrak{S}'$, $\bar{\mathfrak{S}}'$ differ by an orientation *reversing* automorphism φ of S, and it is the *coorientation* of the insertion that is preserved by φ. For example, in Figure 8.6 (b), the two coorientations of insertion 1 result in a pair of non-real complex conjugate sextics, whereas the two coorientations of insertion 2 produce two distinct real sextics. It is this reason why we speak about a coorientation of an insertion rather than about its orientation.

Remark 8.20. The irreducible components of the curves C and D are determined using Theorem 5.92. The splitting markings of $\mathfrak{S}$ are in a one-to-one correspondence with those of $\mathfrak{S}'$. The degrees of the components of D are discussed below; they can be determined in terms of the fundamental group.

The classification is summarized in Tables 8.3 and 8.4 (see comments to Table 8.1); in Table 8.4, we also list the commutants of the group π_1. The count $(2,0)$, Lines 3, 6, and $1'$, signifies pairs of real families sharing the same combinatorial configuration of singularities/components that differ by the coorientation of the insertion. (In the pairs with the sets of singularities $\mathbf{E}_7 \oplus \mathbf{A}_5 \oplus 2\mathbf{A}_3 \oplus \mathbf{A}_1$ and $\mathbf{E}_7 \oplus \mathbf{D}_8 \oplus \mathbf{A}_3 \oplus \mathbf{A}_1$, the two sextics are listed separately as they differ by the fundamental groups.) Observe that, in Table 8.4, each sextic splitting into a quintic and a line ($D_5 + D_1$) shares the set of singularities with one splitting into a quartic and a conic ($D_4 + D_2$), and *vice versa*. Such pairs also differ by the coorientation of the insertion. In Figure 8.7, in the corresponding insertions, we mark the $\circ$-end of the edge e_1 corresponding to the splitting $D_5 + D_1$.

Table 8.4. Reducible maximizing sextics with a type $\mathbf{E}_7$ point.[1]

#	Set of singularities	Figure	Count	π_1	$[\pi_1, \pi_1]$
		The splitting $D_3 + D_3$			
* 1′	$\mathbf{E}_7 \oplus \mathbf{A}_7 \oplus \mathbf{A}_3 \oplus \mathbf{A}_2$	8.7(a)-1	$(2,0)$	(8,4,3)(4,8,3)[2]	Q_8
2′	$\mathbf{E}_7 \oplus \mathbf{A}_9 \oplus \mathbf{A}_2 \oplus \mathbf{A}_1$	8.7(a)-4	$(0,1)$	(10,-,1)[2]	0
* 3′	$\mathbf{E}_7 \oplus \mathbf{A}_{11} \oplus \mathbf{A}_1$	8.7(b)-2	$(0,1)$	(12,-,1,-,2)[2]	$\mathbb{Z}_3$
4′	$\mathbf{E}_7 \oplus \mathbf{D}_{12}$	8.7(c)-1	$(1,0)$	8.24	0
5′	$\mathbf{E}_7 \oplus \mathbf{D}_5 \oplus \mathbf{A}_7$	8.7(c)-1	$(0,1)$	(8,-,1)[2]	0
* 6′	$2\mathbf{E}_7 \oplus \mathbf{A}_5$	8.7(d)-1	$(0,1)$	(6,-,1)[2]	$\mathbb{Z}_3$
		The splitting $D_5 + D_1$			
7′	$\mathbf{E}_7 \oplus \mathbf{A}_5 \oplus \mathbf{A}_4 \oplus \mathbf{A}_3$	8.7(a)-2	$(1,0)$	(5,4)[2]	0
8′	$\mathbf{E}_7 \oplus \mathbf{A}_7 \oplus \mathbf{A}_4 \oplus \mathbf{A}_1$	8.7(a)-3	$(0,1)$	(5,8,-,1)[2]	0
* 9′	$\mathbf{E}_7 \oplus \mathbf{A}_7 \oplus 2\mathbf{A}_2 \oplus \mathbf{A}_1$	8.7(a)-5	$(1,0)$	(3,8,-,3)[2]	$SL(2,\Bbbk_5)$
10′	$\mathbf{E}_7 \oplus \mathbf{A}_9 \oplus \mathbf{A}_3$	8.7(b)-1	$(1,0)$	(4,-,-,1)[2]	0
11′	$\mathbf{E}_7 \oplus \mathbf{A}_9 \oplus \mathbf{A}_2 \oplus \mathbf{A}_1$	8.7(b)-3	$(1,0)$	(3,-,-,2)[2]	0
12′	$\mathbf{E}_7 \oplus \mathbf{D}_{10} \oplus \mathbf{A}_2$	8.7(c)-2	$(1,0)$	(3,-,-,1)[2]	0
13′	$\mathbf{E}_7 \oplus \mathbf{D}_7 \oplus \mathbf{A}_5$	8.7(c)-2	$(1,0)$	8.24	0
14′	$\mathbf{E}_7 \oplus \mathbf{D}_5 \oplus \mathbf{A}_5 \oplus \mathbf{A}_2$	8.7(c)-2	$(1,0)$	(3,6)[2]	0
* 15′	$2\mathbf{E}_7 \oplus \mathbf{A}_3 \oplus \mathbf{A}_2$	8.7(d)-2	$(1,0)$	(3,4)[2]	$SL(2,\Bbbk_5)$
16′	$2\mathbf{E}_7 \oplus \mathbf{D}_5$	8.7(e)	$(1,0)$	8.24	0
		The splitting $D_4 + D_2$			
* 17′	$\mathbf{E}_7 \oplus \mathbf{A}_5 \oplus \mathbf{A}_4 \oplus \mathbf{A}_3$	8.7(a)-2	$(1,0)$	(4,5,6)[3]	$SL(2,\Bbbk_5)$
18′	$\mathbf{E}_7 \oplus \mathbf{A}_7 \oplus \mathbf{A}_4 \oplus \mathbf{A}_1$	8.7(a)-3	$(0,1)$	(8,5,2)[3]	0
* 19′	$\mathbf{E}_7 \oplus \mathbf{A}_7 \oplus 2\mathbf{A}_2 \oplus \mathbf{A}_1$	8.7(a)-5	$(1,0)$	8.25	∞
20′	$\mathbf{E}_7 \oplus \mathbf{A}_9 \oplus \mathbf{A}_3$	8.7(b)-1	$(1,0)$	(-,4,-,1)[3]	0
21′	$\mathbf{E}_7 \oplus \mathbf{A}_9 \oplus \mathbf{A}_2 \oplus \mathbf{A}_1$	8.7(b)-3	$(1,0)$	(-,3,-,2)[3]	0
22′	$\mathbf{E}_7 \oplus \mathbf{D}_{10} \oplus \mathbf{A}_2$	8.7(c)-2	$(1,0)$	8.24	0
23′	$\mathbf{E}_7 \oplus \mathbf{D}_7 \oplus \mathbf{A}_5$	8.7(c)-2	$(1,0)$	8.24	0
* 24′	$\mathbf{E}_7 \oplus \mathbf{D}_5 \oplus \mathbf{A}_5 \oplus \mathbf{A}_2$	8.7(c)-2	$(1,0)$	(6,3)[3]	$\mathbb{Z}_3$
* 25′	$2\mathbf{E}_7 \oplus \mathbf{A}_3 \oplus \mathbf{A}_2$	8.7(d)-2	$(1,0)$	8.25	∞
* 26′	$2\mathbf{E}_7 \oplus \mathbf{D}_5$	8.7(e)	$(1,0)$	8.24	$\mathbb{Z}_3$
		The splitting $D_3 + D_2 + D_1$			
* 27′	$\mathbf{E}_7 \oplus 2\mathbf{A}_5 \oplus 2\mathbf{A}_1$	8.8(a)-1	$(1,0)$	8.25	∞
* 28′	${}^1\mathbf{E}_7 \oplus \mathbf{A}_5 \oplus 2\mathbf{A}_3 \oplus \mathbf{A}_1$	8.8(a)-2	$(1,0)$	8.25	∞
* 29′	${}^1\mathbf{E}_7 \oplus \mathbf{A}_5 \oplus 2\mathbf{A}_3 \oplus \mathbf{A}_1$	8.8(a)-2	$(1,0)$	(6,4,4)[4]	$SL(2,\Bbbk_5)$
* 30′	${}^2\mathbf{E}_7 \oplus \mathbf{D}_8 \oplus \mathbf{A}_3 \oplus \mathbf{A}_1$	8.8(b)	$(1,0)$	8.24, 8.25	∞
31′	${}^2\mathbf{E}_7 \oplus \mathbf{D}_8 \oplus \mathbf{A}_3 \oplus \mathbf{A}_1$	8.8(b)	$(1,0)$	8.24	0
* 32′	$\mathbf{E}_7 \oplus \mathbf{D}_6 \oplus 2\mathbf{A}_3$	8.8(b)	$(1,0)$	(4,4)[4]	Q_8
33′	$\mathbf{E}_7 \oplus 2\mathbf{D}_6$	8.8(c)	$(1,0)$	8.24	0

[1] There are 39 classes realizing 30 combinatorial types

[2] See "size2" in Listing 8.9. All GAP input is found in "misc/E7-r.txt"

[3,4] See, respectively, "size3", "group3" and "size4", "group4" in Listing 8.9

```
# The computation of the fundamental groups of plane sextics with a type E7 singularity.
# We use the type Ã regions reg_(e ↑ ζ), where ζ = x, x^2, xyx^2, and either x(yx^2)^2, (yx^2)^2 or
    (xy)^2x^2. Their widths (in the order listed) are the parameters.

Reread("pi1sc.txt");    # (see Listing 5.3)
# Relations at infinity
fixed := [G.3*rho^2/G.1, G.1*G.2/(G.2*G.1), G.1^2*G.2*G.3/(G.3*G.1^2*G.2)];
# Regions and the central element to be used
list := [x, x^2, x*y*x^2, x*(y*x^2)^2, (y*x^2)^2];   kappa := G.1^2*G.2;
# For irreducible curves (see "AA" and "DS" in Listing 5.3)
size := function(arg) return Size(DS(AA(list, arg))); end;
# For the splittings D5 + D1 and D3 + D3, consider the quotient π1/ϰ, ϰ := α1^2α2
size2 := function(arg) return Size(DS(AA(list, arg), [kappa])); end;
# For the splitting D4 + D2, consider the quotient π1/(α1α3)^s, where s = wd reg_(e ↑ x^2)
# Also change the last chain to (xy)^2x^2
list3 := [x, x^2, x*y*x^2, (x*y)^2*x^2];
AAc := function(w) return Union(AA(list3, w), [(G.1*G.3)^w[2]]); end;
group3 := function(arg) return DS(AAc(arg)); end;
size3 := function(arg) return Size(CallFuncList(group3, arg)); end;
# Finally, for D3 + D2 + D1, use both central elements ϰ and (α1α3)^s
group4 := function(arg) return DS(AAc(arg), [kappa]); end;
size4 := function(arg) return Size(CallFuncList(group4, arg)); end;
```

Listing 8.9. Computing π_1 for sextics with a type $\mathbf{E}_7$ point ("E7.txt").

Remark 8.21. The set of singularities $2\mathbf{E}_7 \oplus \mathbf{A}_5$, Line 6′ can be realized by a real sextic, but the two type $\mathbf{E}_7$ points of such a sextic are necessarily complex conjugate.

The fundamental groups

Taking $e := e_1$ for the common reference edge and combining relations (5.75) with $m = n = 2$ and the relation at infinity (5.55), we have

$$[\alpha_1, \alpha_2] = [\alpha_1^2\alpha_2, \alpha_3] = 1, \qquad \alpha_3\rho^2 = \alpha_1. \tag{8.22}$$

Hence, $\varkappa_{\mathrm{I}} = \alpha_1^2\alpha_2 = \rho^3$ is a central element. The abelianization of this group is generated by the images $\bar{\alpha}_i$ of α_i, $i = 1, 2, 3$, subject to the relation $\bar{\alpha}_1 + 2\bar{\alpha}_2 + 3\bar{\alpha}_3 = 0$ and, possibly, some relations of the form $\bar{\alpha}_i = \bar{\alpha}_j$ arising from the braid relations in π_1. Hence, the components of D and their degrees are easily found as follows:

- if $\bar{\alpha}_1 = \bar{\alpha}_2 = \bar{\alpha}_3$, the curve is irreducible;
- if no other relation is present, the curve splits into a cubic, a conic, and a line;
- if $\bar{\alpha}_1 = \bar{\alpha}_2$, the curve splits into two cubics ($D_3 + D_3$);
- if $\bar{\alpha}_2 = \bar{\alpha}_3$, the curve splits into a quintic and a line ($D_5 + D_1$);
- if $\bar{\alpha}_1 = \bar{\alpha}_3$, the curve splits into a quartic and a conic ($D_4 + D_2$).

Argument 8.23 ($\mathbf{E}_7 \oplus 2\mathbf{A}_4 \oplus 2\mathbf{A}_2$, Line 1). Creating the group *via* `size(5,5,3)`, one can see that

- $\operatorname{ord} \pi_1 = 41040$ and the commutant $H := [\pi_1, \pi_1] \cong SL(2, \Bbbk_{19})$ is the only perfect group of order 6840;
- the last relation in the presentation can be omitted; the first two relations, when combined with (8.22), constitute the presentation in Theorem 8.1;
- the centralizer $C := \mathcal{Z}_{\pi_1}(H)$ equals $\mathbb{Z}_3 \times \mathbb{Z}_2^2$ and $\operatorname{ord}(C \cap H) = 2$.

From the last statement one concludes that there is a central subgroup $C_1 \cong \mathbb{Z}_6$ such that $C_1 \cap H = 0$. Then $\pi_1 = H \times C_1 \cong SL(2, \Bbbk_{19}) \times \mathbb{Z}_6$.

Argument 8.24 (a $\tilde{\mathbf{D}}$ type region). If the edge $e = e_1$ belongs to the boundary of a $\tilde{\mathbf{D}}$ type region, the group π_1 is abelian. Indeed, one of the braid relations about the $\tilde{\mathbf{D}}$ type region $\operatorname{reg}_-(e \uparrow \mathbf{x})$ is $[\alpha_2\alpha_3, \alpha_1] = 1$; in the presence of (8.22), this implies $[\alpha_3, \alpha_1] = [\alpha_3, \alpha_2] = 1$.

If it is e_2 that belongs to the boundary of a $\tilde{\mathbf{D}}$ type region, then one of the relations is $[\alpha_2, \rho] = 1$, which implies that α_2 is a central element. We use this element instead of $(\alpha_1\alpha_3)^s$ for the sets of singularities $\mathbf{E}_7 \oplus \mathbf{D}_7 \oplus \mathbf{A}_5$, Line 23′, $2\mathbf{E}_7 \oplus \mathbf{D}_5$, Line 26′, and $\mathbf{E}_7 \oplus \mathbf{D}_8 \oplus \mathbf{A}_3 \oplus \mathbf{A}_1$, Line 30′.

Argument 8.25 (infinite commutants). Five maximizing sextics have groups with infinite commutants:

- $\mathbf{E}_7 \oplus \mathbf{A}_7 \oplus 2\mathbf{A}_2 \oplus \mathbf{A}_1$, Line 19′ and $2\mathbf{E}_7 \oplus \mathbf{A}_3 \oplus \mathbf{A}_2$, Line 25′: $\pi_1'/\pi_1'' = \mathbb{Z}_3^2$ and $\pi_1''/\pi_1''' = \mathbb{Z}^2$ (we temporarily use $'$ for the commutant);
- $\mathbf{E}_7 \oplus 2\mathbf{A}_5 \oplus 2\mathbf{A}_1$, Line 27′: $\pi_1'/\pi_1'' = \mathbb{Z}^4$;
- $\mathbf{E}_7 \oplus \mathbf{A}_5 \oplus 2\mathbf{A}_3 \oplus \mathbf{A}_1$, Line 28′: $\pi_1'/\pi_1'' = \mathbb{Z}^2 \times \mathbb{Z}_3$;
- $\mathbf{E}_7 \oplus \mathbf{D}_8 \oplus \mathbf{A}_3 \oplus \mathbf{A}_1$, Line 30′: $\pi_1'/\pi_1'' = \mathbb{Z}^2$.

For the last curve, the pair of independent central elements used in the computation is α_1^2 and α_2, see Argument 8.24.

Non-maximizing sextics

Consider a trigonal model C as in Proposition 8.18. If its skeleton $\mathfrak{S}$ is regular, the fiber F_{I} can be isolated using Lemma 4.52. Otherwise, $\mathfrak{S}$ has a unique singular $\circ$-vertex and thus is one of the skeletons shown in Figure 7.2 (a)–(e); for each of these skeletons and each region R_{I} of width $\operatorname{wd} R_{\mathrm{I}} \geqslant 3$, one can find a cutting sequence of width two and apply Lemma 4.51 directly.

Thus, with one exception discussed below, a sextic satisfying conditions 8.2.2(∗) admits a mild degeneration and its group is computed as explained in Section 8.1.2, see page 230. The groups of all non-maximizing irreducible sextics are abelian.

The exception is the skeleton $\mathfrak{S}$ shown in Figure 8.8 (c), where the trigonal model may have two $\tilde{\mathbf{D}}_5$ type fibers inside the same region R of $\mathfrak{S}$. If e_1 is in ∂R, the group is abelian due to Argument 8.24. Otherwise, α_2 is a central element, see Argument 8.24, and, using (8.22) and the other braid relations from the $\tilde{\mathbf{D}}_5$ type fibers, we conclude that $\alpha_1 = \alpha_2$; hence, the group is also abelian.

8.2.3 A point of type $\mathbf{E}_6$

Here, we deal with simple plane sextics $D \subset \mathbb{P}^2$ satisfying the following conditions:

$(*)$ D has a distinguished singular point P of type $\mathbf{E}_6$ and
D has no singular points of type $\mathbf{E}_8$.

(All sextics with a singular point of type $\mathbf{E}_8$ were considered in Section 8.2.1.) The proof of the next statement is similar to that of Proposition 8.18.

Proposition 8.26. *There is a natural bijection ϕ, invariant under equisingular deformations, between Zariski open and dense, in each equisingular stratum, subsets of the following two sets*:

- *simple plane sextics D with a distinguished type $\mathbf{E}_6$ point P, and*
- *proper simple trigonal curves $C \subset \Sigma_4$ with a distinguished type $\tilde{\mathbf{A}}_5$ fiber F_{I}.*

A sextic D is maximizing if and only if $C := \phi(D)$ is maximal and stable. ◁

Consider a maximizing sextic D satisfying conditions 8.2.3$(*)$ and let $\mathfrak{S} := \operatorname{Sk} C$. The distinguished region R_{I} containing F_{I} is a hexagon, and we consider separately several cases, summarizing the results in Tables 8.5 (more than one $\mathbf{E}$ type singular point), 8.6 (irreducible sextics), and 8.7 (reducible sextics).

The sextic D has more than one point of type E In this case, $\mathfrak{S}$ has a monovalent •- or ○-vertex. Fix such a vertex w, let e' be the edge incident to w, and let u be the other •-end, necessarily trivalent, of this edge. (More precisely, if w is a •-vertex, then e' is a pair of edges sharing the ○-end.) The fragment $[u, w]$ is called an *insertion*; removing it from the skeleton, see Convention 8.3, we obtain another skeleton $\mathfrak{S}'$, which must be one of those shown in Figures 4.2 and 4.3. Conversely, $\mathfrak{S}$ is obtained from $\mathfrak{S}'$ by attaching an edge (pair of edges) $[u, w]$ at one of the bivalent ○-vertices, *provided that the result has a hexagonal region.* Trying the possible combinations, we arrive at Figure 8.10 and Table 8.5. Note that insertion 2 in Figure 8.10 (f) produces a skeleton $\mathfrak{S}$ with two distinct hexagonal regions, which results in two distinct sextics (Lines 7 and 8 in Table 8.5) sharing the same set of singularities $(2\mathbf{E}_6 \oplus \mathbf{A}_5) \oplus \mathbf{A}_2$.

Hexagon with a loop Assume that the hexagon R_{I} has a monogonal region inside, see Figure 8.11, left, where R_{I} is shaded. The fragment shown in the figure is called an *insertion*; removing this fragment from $\mathfrak{S}$, *cf.* Convention 8.3, produces a new skeleton $\mathfrak{S}'$ satisfying the vertex count (8.7); in particular, $\mathfrak{S}'$ has at most four •-vertices. To avoid repetition, we refer to Figures 8.2 and 8.3, where the possible positions of the insertion are indicated by a fragment of the form •—○, cooriented as shown in Figure 8.11, right. (Note that the widths of the regions of $\mathfrak{S}$ differ from what they appear in the figures.) Under this convention, irreducible curves are detected as explained in Remark 8.9. Disregarding the skeletons with singular •- or ○-vertices (these curves were considered in the previous paragraph), we arrive at the curves listed in Lines 13–36 in Table 8.6 and Lines 6′–24′ in Table 8.7.

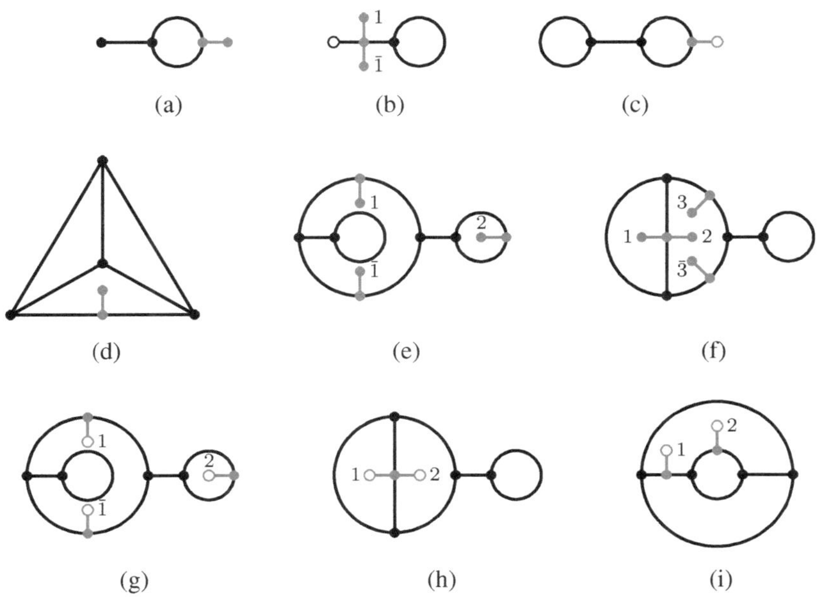

Figure 8.10. Type $\mathbf{E}_6$ singularity: Two or more type **E** points.

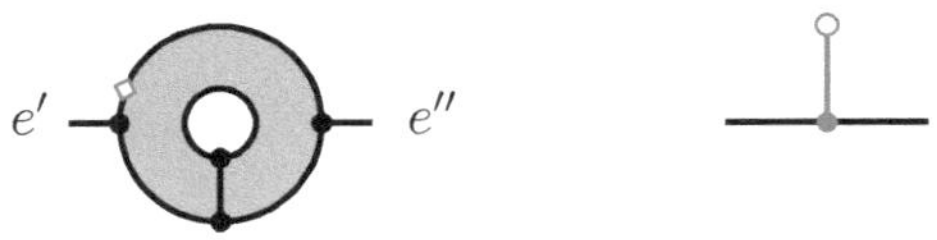

Figure 8.11. Hexagon with a loop (compared to Figure 8.1).

The pairs of curves in Lines 31, 32, Lines 35, 36, and Lines 20′, 21′ differ by the type specifications: the skeleton has two monogonal regions, containing one type $\tilde{\mathbf{A}}_0^*$ and one type $\tilde{\mathbf{D}}_5$ singular fiber of the curve. The latter may be either inside of one of the 'free' monogons of the auxiliary skeleton $\mathfrak{S}'$ or inside the insertion. By convention, the latter case corresponds to Lines 32, 36, and 21′.

Hexagon with a double loop Now, assume that the distinguished hexagon R_{I} looks like the outer region in Figure 8.12, left. Each of the remaining fragments A, B of $\mathfrak{S}$ has an odd number of •-vertices, and the total number of remaining •-vertices is at most four. Hence, one can assume that A has one trivalent •-vertex and B has at most three •-vertices. Then A is a single loop and the graph can be redrawn as shown in Figure 8.12, right, where R_{I} is represented by the shaded area. In other words, $\mathfrak{S}$ is obtained from an auxiliary skeleton $\mathfrak{S}'$ of degree 3 or 6 and with a loop (a monogonal region) by replacing a loop with the fragment shown in Figure 8.12, right.

Table 8.5. Maximizing sextics with two or more type $\mathbf{E}$ points.[1]

#	Set of singularities	Figure	Count	π_1
** 1	$(3\mathbf{E}_6) \oplus \mathbf{A}_1$	8.10(a)	$(1,0)$	8.40
2	$2\mathbf{E}_6 \oplus \mathbf{E}_7$	8.10(b)-1	$(1,0)$	8.40
** 3	$(2\mathbf{E}_6 \oplus 2\mathbf{A}_2) \oplus \mathbf{A}_3$	8.10(d)	$(1,0)$	(8.42)
4	$2\mathbf{E}_6 \oplus \mathbf{A}_7$	8.10(e)-1, 8.2 (g)–1	$(0,1)$	8.28
* 5	$^{1}\,2\mathbf{E}_6 \oplus \mathbf{A}_4 \oplus \mathbf{A}_3$	8.10(e)-2, 8.13 (f)	$(1,0)$	8.38
* 6	$^{1}\,2\mathbf{E}_6 \oplus \mathbf{A}_4 \oplus \mathbf{A}_3$	8.10(f)-1, 8.2 (g)–3	$(1,0)$	8.32
** 7	$^{2}\,(2\mathbf{E}_6 \oplus \mathbf{A}_5) \oplus \mathbf{A}_2$	8.10(f)-2	$(1,0)$	(8.43)
** 8	$^{2}\,(2\mathbf{E}_6 \oplus \mathbf{A}_5) \oplus \mathbf{A}_2$	8.10(f)-2, 8.2 (g)–2	$(1,0)$	(8.36)
9	$2\mathbf{E}_6 \oplus \mathbf{A}_6 \oplus \mathbf{A}_1$	8.10(f)-3	$(0,1)$	8.40
10	$\mathbf{E}_6 \oplus \mathbf{E}_7 \oplus \mathbf{A}_6$	8.10(g)-1, 8.2 (h)–1	$(0,1)$	8.28
11	$^{3}\,\mathbf{E}_6 \oplus \mathbf{E}_7 \oplus \mathbf{A}_4 \oplus \mathbf{A}_2$	8.10(g)-2, 8.13 (g)	$(1,0)$	8.38
12	$^{3}\,\mathbf{E}_6 \oplus \mathbf{E}_7 \oplus \mathbf{A}_4 \oplus \mathbf{A}_2$	8.10(h)-2, 8.2 (h)–2	$(1,0)$	8.28
$1'$	$\mathbf{E}_6 \oplus \mathbf{E}_7 \oplus \mathbf{D}_6$	8.10(c)	$(1,0)$	8.44
$2'$	$\mathbf{E}_6 \oplus \mathbf{E}_7 \oplus \mathbf{D}_5 \oplus \mathbf{A}_1$	8.10(c)	$(1,0)$	8.44
$3'$	$\mathbf{E}_6 \oplus \mathbf{E}_7 \oplus 2\mathbf{A}_3$	8.10(h)-1, 8.3 (f)	$(1,0)$	8.28
$4'$	$\mathbf{E}_6 \oplus \mathbf{E}_7 \oplus \mathbf{A}_4 \oplus 2\mathbf{A}_1$	8.10(i)-1	$(1,0)$	8.44
$5'$	$\mathbf{E}_6 \oplus \mathbf{E}_7 \oplus \mathbf{A}_3 \oplus \mathbf{A}_2 \oplus \mathbf{A}_1$	8.10(i)-2	$(1,0)$	8.44

[1] There are 20 classes realizing 14 sets of singularities

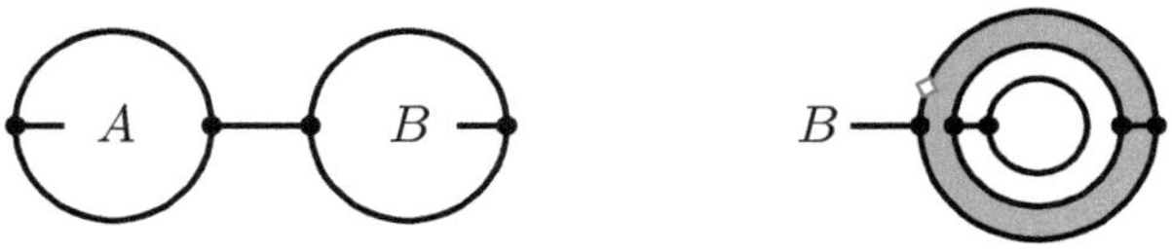

Figure 8.12. Hexagon with a double loop.

The seven possibilities are listed in Figure 8.13; the resulting sextics are Lines 37–42 in Table 8.6. (For the classification, we omit the last two skeletons as they have singular vertices; such skeletons were considered above as sextics with two type $\mathbf{E}$ singular points.) Using Theorem 5.92, one can show that the existence of a fragment as in Figure 8.12, right implies that the curve is irreducible.

The skeleton in Figure 8.13 (e) has a symmetry interchanging its two monogons and two pentagons. For this reason, unlike previous paragraph, this skeleton admits only two essentially different type specifications, resulting in the sextics with the sets of singularities $\mathbf{E}_6 \oplus \mathbf{D}_9 \oplus \mathbf{A}_4$, Line 41 and $\mathbf{E}_6 \oplus \mathbf{D}_5 \oplus 2\mathbf{A}_4$, Line 42.

Genuine hexagon Finally, assume that R_{I} is a *genuine hexagon, i.e.*, all six vertices in the boundary ∂R_{I} are pairwise distinct. Exclude singular $\bullet$- and $\circ$-vertices (all skeletons with such vertices were treated above) and consider regular skeletons only. Thus, the boundary ∂R_{I} can be regarded as the equator in the sphere S^2, and $\mathfrak{S}$ is

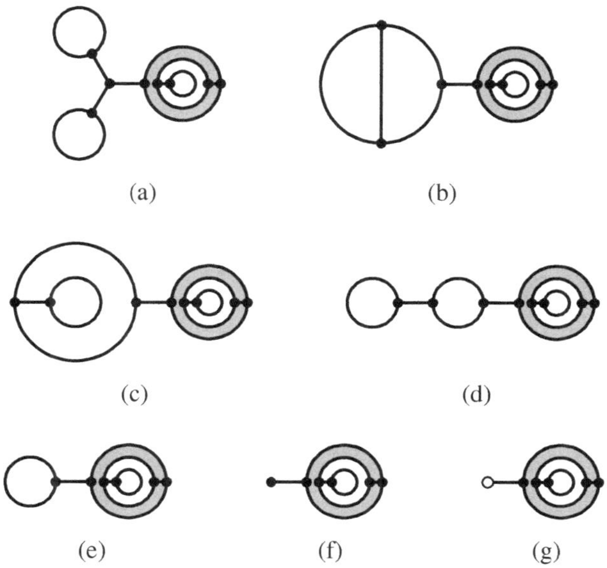

Figure 8.13. Hexagon with a double loop: Skeletons.

obtained from ∂R_{I} by completing it to a regular skeleton by inserting two or four $\bullet$-vertices and connecting edges into one of the hemispheres. All possibilities are shown in Figures 8.14 (irreducible curves) and 8.15 (reducible curves); the resulting sextics are listed in Lines 43–46 and 25′–43′. (The marked edges in the figures are used below in the computation of the fundamental groups.)

The lists In addition to the common notation, see comments to Table 8.1, for sextics with a type $\mathbf{E}_6$ singular point we use the following conventions. For sextics of torus type, the inner singularities are grouped in the parentheses; an exception is the set of singularities $\mathbf{E}_6 \oplus \mathbf{A}_5 \oplus 4\mathbf{A}_2$, Line 43, which admits four distinct torus structures. Besides, the lines containing sextics with nonabelian fundamental group are prefixed

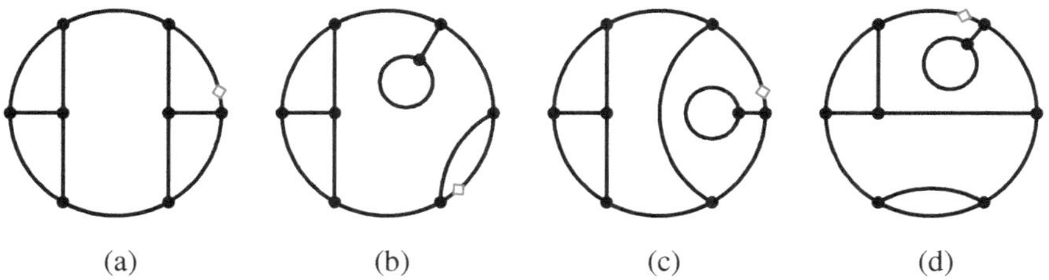

Figure 8.14. Genuine hexagon: Irreducible curves.

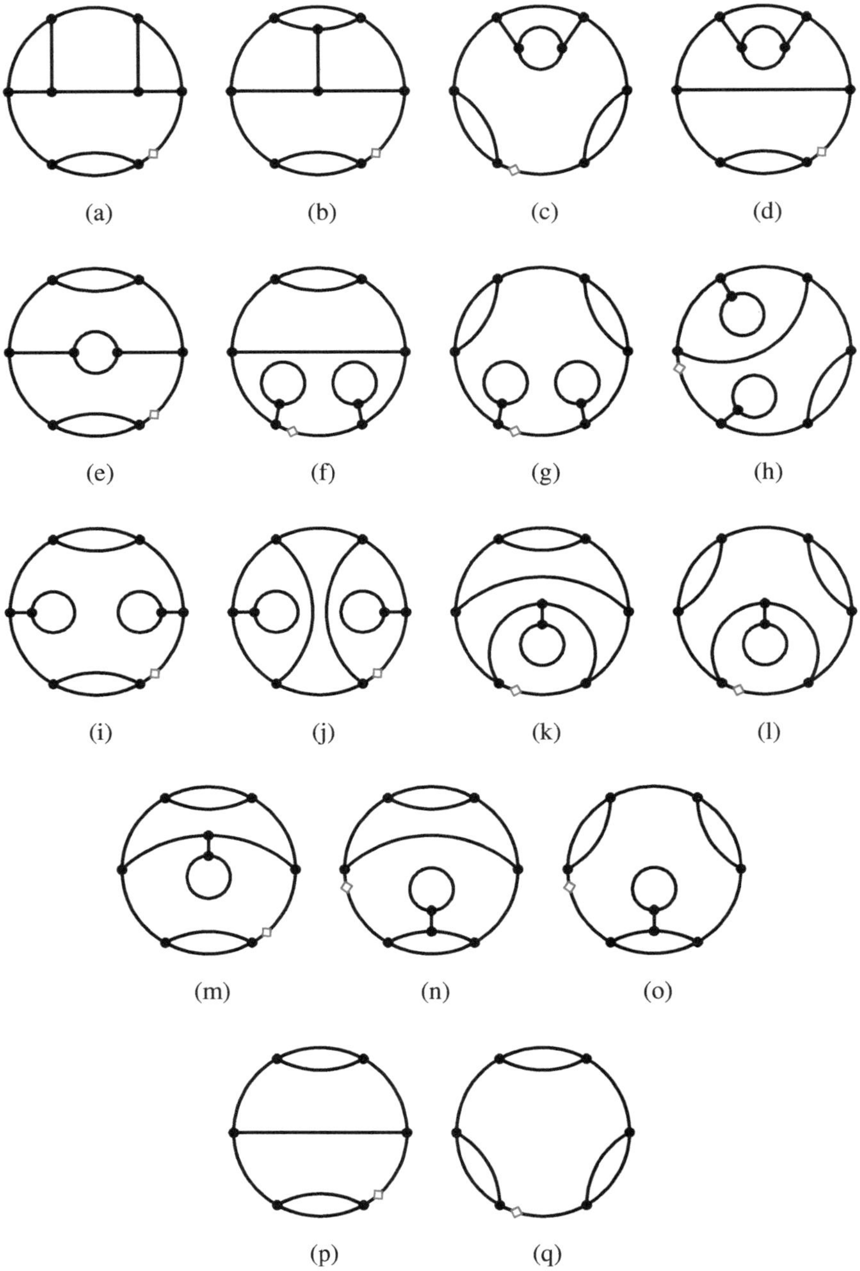

Figure 8.15. Genuine hexagon: Reducible curves.

Table 8.6. Irreducible maximizing sextics with a type $\mathbf{E}_6$ point[1]

#	Set of singularities	Figure	Count	π_1
** 13	[4] $(\mathbf{E}_6 \oplus \mathbf{A}_5 \oplus 2\mathbf{A}_2) \oplus \mathbf{A}_4$	8.2(a)	$(1,0)$	(8.37)
14	$\mathbf{E}_6 \oplus \mathbf{A}_{13}$	8.2(b)-1	$(0,1)$	8.28
* 15	$(\mathbf{E}_6 \oplus \mathbf{A}_{11}) \oplus \mathbf{A}_2$	8.2(b)-2	$(1,0)$	5.83
16	[5] $\mathbf{E}_6 \oplus \mathbf{A}_{10} \oplus \mathbf{A}_3$	8.2(b)-3	$(1,0)$	8.28
17	[6] $\mathbf{E}_6 \oplus \mathbf{A}_7 \oplus \mathbf{A}_4 \oplus \mathbf{A}_2$	8.2(c)-1	$(1,0)$	8.28
* 18	$(\mathbf{E}_6 \oplus \mathbf{A}_8 \oplus \mathbf{A}_2) \oplus \mathbf{A}_3$	8.2(c)-2	$(1,0)$	8.35
19	[7] $\mathbf{E}_6 \oplus \mathbf{A}_5 \oplus 2\mathbf{A}_4$	8.2(c)-3	$(1,0)$	8.28
20	[8] $\mathbf{E}_6 \oplus \mathbf{A}_8 \oplus \mathbf{A}_4 \oplus \mathbf{A}_1$	8.2(c)-4	$(0,1)$	8.28
21	[9] $\mathbf{E}_6 \oplus \mathbf{A}_{10} \oplus \mathbf{A}_2 \oplus \mathbf{A}_1$	8.2(c)-5	$(0,1)$	8.28
* 22	[10] $(\mathbf{E}_6 \oplus \mathbf{A}_8 \oplus \mathbf{A}_2) \oplus \mathbf{A}_2 \oplus \mathbf{A}_1$	8.2(c)-6	$(1,0)$	8.35
23	$\mathbf{E}_6 \oplus \mathbf{A}_7 \oplus \mathbf{A}_6$	8.2(d)-1	$(0,1)$	8.28
24	[11] $\mathbf{E}_6 \oplus \mathbf{A}_9 \oplus \mathbf{A}_4$	8.2(d)-2	$(0,1)$	8.28
25	[6] $\mathbf{E}_6 \oplus \mathbf{A}_7 \oplus \mathbf{A}_4 \oplus \mathbf{A}_2$	8.2(d)-3	$(1,0)$	8.28
26	$\mathbf{E}_6 \oplus \mathbf{A}_6 \oplus \mathbf{A}_4 \oplus \mathbf{A}_3$	8.2(d)-4	$(1,0)$	8.28
27	[5] $\mathbf{E}_6 \oplus \mathbf{A}_{10} \oplus \mathbf{A}_3$	8.2(e)-1	$(1,0)$	8.28
28	$\mathbf{E}_6 \oplus \mathbf{A}_{12} \oplus \mathbf{A}_1$	8.2(e)-2	$(0,1)$	8.28
29	[9] $\mathbf{E}_6 \oplus \mathbf{A}_{10} \oplus \mathbf{A}_2 \oplus \mathbf{A}_1$	8.2(e)-3	$(1,0)$	8.28
30	$\mathbf{E}_6 \oplus \mathbf{D}_{13}$	8.2(f)-1	$(1,0)$	8.28
31	[12] $\mathbf{E}_6 \oplus \mathbf{D}_5 \oplus \mathbf{A}_8$	8.2(f)-1	$(0,1)$	8.28
32	[12] $\mathbf{E}_6 \oplus \mathbf{D}_5 \oplus \mathbf{A}_8$	8.2(f)-1	$(1,0)$	8.31
33	$\mathbf{E}_6 \oplus \mathbf{D}_{11} \oplus \mathbf{A}_2$	8.2(f)-2	$(1,0)$	8.28
34	$\mathbf{E}_6 \oplus \mathbf{D}_7 \oplus \mathbf{A}_6$	8.2(f)-2	$(1,0)$	8.28
35	[13] $\mathbf{E}_6 \oplus \mathbf{D}_5 \oplus \mathbf{A}_6 \oplus \mathbf{A}_2$	8.2(f)-2	$(1,0)$	8.28
36	[13] $\mathbf{E}_6 \oplus \mathbf{D}_5 \oplus \mathbf{A}_6 \oplus \mathbf{A}_2$	8.2(f)-2	$(1,0)$	8.31
37	[11] $\mathbf{E}_6 \oplus \mathbf{A}_9 \oplus \mathbf{A}_4$	8.13(a)	$(1,0)$	8.38
38	[14] $\mathbf{E}_6 \oplus \mathbf{A}_6 \oplus \mathbf{A}_4 \oplus \mathbf{A}_2 \oplus \mathbf{A}_1$	8.13(b)	$(1,0)$	8.38
39	[7] $\mathbf{E}_6 \oplus \mathbf{A}_5 \oplus 2\mathbf{A}_4$	8.13(c)	$(1,0)$	8.38
40	[8] $\mathbf{E}_6 \oplus \mathbf{A}_8 \oplus \mathbf{A}_4 \oplus \mathbf{A}_1$	8.13(d)	$(1,0)$	8.38
41	$\mathbf{E}_6 \oplus \mathbf{D}_9 \oplus \mathbf{A}_4$	8.13(e)	$(1,0)$	8.38
42	$\mathbf{E}_6 \oplus \mathbf{D}_5 \oplus 2\mathbf{A}_4$	8.13(e)	$(1,0)$	8.38
** 43	$\mathbf{E}_6 \oplus \mathbf{A}_5 \oplus 4\mathbf{A}_2$	8.14(a)	$(1,0)$	8.45
* 44	[10] $(\mathbf{E}_6 \oplus \mathbf{A}_8 \oplus \mathbf{A}_2) \oplus \mathbf{A}_2 \oplus \mathbf{A}_1$	8.14(b)	$(0,1)$	8.45
** 45	[4] $(\mathbf{E}_6 \oplus \mathbf{A}_5 \oplus 2\mathbf{A}_2) \oplus \mathbf{A}_4$	8.14(c)	$(1,0)$	8.45
46	[14] $\mathbf{E}_6 \oplus \mathbf{A}_6 \oplus \mathbf{A}_4 \oplus \mathbf{A}_2 \oplus \mathbf{A}_1$	8.14(d)	$(0,1)$	8.45

[1] There are 49 classes realizing 23 sets of singularities

Table 8.7. Reducible maximizing sextics with a type $\mathbf{E}_6$ point[1]

#		Set of singularities	Figure	Count	π_1
** 6′	15	$(\mathbf{E}_6 \oplus 2\mathbf{A}_5) \oplus \mathbf{A}_3$	8.3(a)-1	$(1,0)$	8.35
7′	16	$\mathbf{E}_6 \oplus \mathbf{A}_7 \oplus \mathbf{A}_5 \oplus \mathbf{A}_1$	8.3(a)-2	$(0,1)$	8.28
* 8′	17	$\mathbf{E}_6 \oplus \mathbf{A}_7 \oplus \mathbf{A}_3 \oplus \mathbf{A}_2 \oplus \mathbf{A}_1$	8.3(a)-3	$(1,0)$	8.34(1)
9′	18	$\mathbf{E}_6 \oplus \mathbf{A}_6 \oplus \mathbf{A}_5 \oplus 2\mathbf{A}_1$	8.3(b)-1	$(1,0)$	8.28
* 10′	19	$\mathbf{E}_6 \oplus \mathbf{A}_5 \oplus \mathbf{A}_4 \oplus \mathbf{A}_3 \oplus \mathbf{A}_1$	8.3(b)-2	$(1,0)$	8.34(2)
11′		$\mathbf{E}_6 \oplus \mathbf{A}_6 \oplus 2\mathbf{A}_3 \oplus \mathbf{A}_1$	8.3(b)-3	$(1,0)$	8.28
12′	11	$\mathbf{E}_6 \oplus \mathbf{A}_9 \oplus \mathbf{A}_4$	8.3(c)-1	$(1,0)$	8.28
13′	20	$\mathbf{E}_6 \oplus \mathbf{A}_9 \oplus \mathbf{A}_3 \oplus \mathbf{A}_1$	8.3(c)-2	$(1,0)$	8.28
14′		$\mathbf{E}_6 \oplus \mathbf{D}_9 \oplus \mathbf{A}_3 \oplus \mathbf{A}_1$	8.3(d)	$(1,0)$	8.28
15′		$\mathbf{E}_6 \oplus \mathbf{D}_8 \oplus \mathbf{A}_4 \oplus \mathbf{A}_1$	8.3(d)	$(1,0)$	8.28
16′		$\mathbf{E}_6 \oplus \mathbf{D}_6 \oplus \mathbf{A}_4 \oplus \mathbf{A}_3$	8.3(d)	$(1,0)$	8.28
17′		$\mathbf{E}_6 \oplus \mathbf{D}_5 \oplus \mathbf{A}_4 \oplus \mathbf{A}_3 \oplus \mathbf{A}_1$	8.3(d)	$(1,0)$	8.31
18′		$\mathbf{E}_6 \oplus \mathbf{D}_{10} \oplus \mathbf{A}_3$	8.3(e)	$(1,0)$	8.28
19′		$\mathbf{E}_6 \oplus \mathbf{D}_8 \oplus \mathbf{A}_5$	8.3(e)	$(1,0)$	8.28
* 20′	21	$\mathbf{E}_6 \oplus \mathbf{D}_5 \oplus \mathbf{A}_5 \oplus \mathbf{A}_3$	8.3(e)	$(1,0)$	8.34(3)
* 21′	21	$\mathbf{E}_6 \oplus \mathbf{D}_5 \oplus \mathbf{A}_5 \oplus \mathbf{A}_3$	8.3(e)	$(1,0)$	8.31
22′		$\mathbf{E}_6 \oplus \mathbf{D}_7 \oplus \mathbf{D}_6$	8.3(g)	$(1,0)$	8.28
23′		$\mathbf{E}_6 \oplus \mathbf{D}_7 \oplus \mathbf{D}_5 \oplus \mathbf{A}_1$	8.3(g)	$(1,0)$	8.31
24′		$\mathbf{E}_6 \oplus \mathbf{D}_6 \oplus \mathbf{D}_5 \oplus \mathbf{A}_2$	8.3(g)	$(1,0)$	8.31
** 25′		$(\mathbf{E}_6 \oplus \mathbf{A}_5 \oplus 2\mathbf{A}_2) \oplus \mathbf{A}_3 \oplus \mathbf{A}_1$	8.15(a)	$(1,0)$	8.45
? 26′		$\mathbf{E}_6 \oplus \mathbf{A}_4 \oplus 2\mathbf{A}_3 \oplus \mathbf{A}_2 \oplus \mathbf{A}_1$	8.15(b)	$(1,0)$	8.45
* 27′		$\mathbf{E}_6 \oplus \mathbf{A}_7 \oplus \mathbf{A}_3 \oplus 3\mathbf{A}_1$	8.15(c)	$(1,0)$	8.50
* 28′		$\mathbf{E}_6 \oplus \mathbf{A}_5 \oplus 2\mathbf{A}_3 \oplus 2\mathbf{A}_1$	8.15(d)	$(1,0)$	8.50
** 29′		$(\mathbf{E}_6 \oplus 2\mathbf{A}_5) \oplus 3\mathbf{A}_1$	8.15(e)	$(1,0)$	8.50
30′	20	$\mathbf{E}_6 \oplus \mathbf{A}_9 \oplus \mathbf{A}_3 \oplus \mathbf{A}_1$	8.15(f)	$(1,0)$	8.49
31′	22	$\mathbf{E}_6 \oplus \mathbf{A}_{11} \oplus 2\mathbf{A}_1$	8.15(g)	$(1,0)$	8.49
32′	16	$\mathbf{E}_6 \oplus \mathbf{A}_7 \oplus \mathbf{A}_5 \oplus \mathbf{A}_1$	8.15(h)	$(0,1)$	8.49
** 33′	22	$(\mathbf{E}_6 \oplus \mathbf{A}_{11}) \oplus 2\mathbf{A}_1$	8.15(i)	$(1,0)$	8.49
** 34′	15	$(\mathbf{E}_6 \oplus 2\mathbf{A}_5) \oplus \mathbf{A}_3$	8.15(j)	$(1,0)$	8.49
* 35′	19	$\mathbf{E}_6 \oplus \mathbf{A}_5 \oplus \mathbf{A}_4 \oplus \mathbf{A}_3 \oplus \mathbf{A}_1$	8.15(k)	$(1,0)$	8.48(1)
36′	18	$\mathbf{E}_6 \oplus \mathbf{A}_6 \oplus \mathbf{A}_5 \oplus 2\mathbf{A}_1$	8.15(l)	$(1,0)$	8.48
37′		$\mathbf{E}_6 \oplus \mathbf{A}_7 \oplus \mathbf{A}_4 \oplus 2\mathbf{A}_1$	8.15(m)	$(1,0)$	8.48
* 38′	17	$\mathbf{E}_6 \oplus \mathbf{A}_7 \oplus \mathbf{A}_3 \oplus \mathbf{A}_2 \oplus \mathbf{A}_1$	8.15(n)	$(1,0)$	8.48(2)
* 39′		$\mathbf{E}_6 \oplus \mathbf{A}_9 \oplus \mathbf{A}_2 \oplus 2\mathbf{A}_1$	8.15(o)	$(1,0)$	8.48(3)
40′		$\mathbf{E}_6 \oplus \mathbf{D}_8 \oplus \mathbf{A}_3 \oplus 2\mathbf{A}_1$	8.15(p)	$(1,0)$	8.50
* 41′		$\mathbf{E}_6 \oplus \mathbf{D}_6 \oplus 2\mathbf{A}_3 \oplus \mathbf{A}_1$	8.15(p)	$(1,0)$	8.50
* 42′		$\mathbf{E}_6 \oplus \mathbf{D}_{10} \oplus 3\mathbf{A}_1$	8.15(q)	$(1,0)$	8.50
43′		$\mathbf{E}_6 \oplus \mathbf{D}_6 \oplus \mathbf{A}_5 \oplus 2\mathbf{A}_1$	8.15(q)	$(1,0)$	8.50

[1] There are 40 classes realizing 31 sets of singularities

with one of the following symbols:

* * the group π_1 is not abelian;
* $^{?}$ the group π_1 is not known to be abelian;
* ** for curves of torus type, $\pi_1 \neq \Gamma$.

For the sets of singularities $2\mathbf{E}_6 \oplus \mathbf{A}_4 \oplus \mathbf{A}_3$, Lines 5 and 6, and $(2\mathbf{E}_6 \oplus \mathbf{A}_5) \oplus \mathbf{A}_2$, Lines 7 and 8, one may ask if the two curves remain non-equivalent if a permutation of the two type $\mathbf{E}_6$ points is allowed. For $(2\mathbf{E}_6 \oplus \mathbf{A}_5) \oplus \mathbf{A}_2$, there still are two distinct deformation families, see [44] or [130]. For $2\mathbf{E}_6 \oplus \mathbf{A}_4 \oplus \mathbf{A}_3$, the two curves become equivalent, see [148].

Note also that the sets of singularities $2\mathbf{E}_6 \oplus \mathbf{A}_7$, Line 4 and $2\mathbf{E}_6 \oplus \mathbf{A}_6 \oplus \mathbf{A}_1$, Line 9 with $(c_r, c_i) = (0, 1)$ can be realized by real curves, see [148], but with respect to this real structure the two type $\mathbf{E}_6$ points must be complex conjugate.

The fundamental groups

The distinguished fiber $\tilde{F}_{\mathrm{I}}$ is as in Case 2 with $m = 2n = 6$, see Section 5.2.2, and, choosing the reference edge e in $\partial_- R_{\mathrm{I}}$ and combining (5.73) with the relation at infinity (5.55), we have

$$\rho^4 = (\alpha_1 \alpha_2)^3. \tag{8.27}$$

This relation implies that $[\alpha_3, (\alpha_1\alpha_2)^3] = 1$. We consider separately a number of cases, according to the shape of the distinguished region R_{I}.

Argument 8.28 (hexagon with a loop). Choosing the reference edge as shown in Figure 8.11, left, from the inner monogonal region (assuming the latter of type $\tilde{\mathbf{A}}$) we have the relation

$$\alpha_3 = (\alpha_1\alpha_2)^{-2}\alpha_1(\alpha_1\alpha_2)^2. \tag{8.29}$$

Then (8.27) implies that $\varkappa_{\mathrm{I}} = (\alpha_1\alpha_2)^3$ is a central element and, after eliminating α_3, the relation becomes

$$(\alpha_1\alpha_2\alpha_1)^3 = \alpha_2\alpha_1\alpha_2. \tag{8.30}$$

The abelianization of the group defined by (8.27) and (8.29) is $\mathbb{Z}$; hence, each curve in question either is irreducible or splits into a quintic and a line.

For most irreducible curves, it suffices to consider the braid relations from the $\tilde{\mathbf{A}}$ type regions reg_$(e \uparrow \zeta)$, where $\zeta = \mathbf{x}$, $\mathbf{x}^2$, $\mathbf{xyx}^2$, and $\mathbf{x}(\mathbf{yx}^2)^2$, see functions "`group11`" and "`size11`" in Listing 8.16 and `GAP` input in "`misc/E6-2.txt`" and "`misc/E6-ir.txt`". For reducible curves, we use the chains $\zeta = \mathbf{x}$, $\mathbf{x}^2$, $\mathbf{xyx}^2$, and $(\mathbf{xy})^2\mathbf{x}^2$ and analyze the quotient $\pi_1/\varkappa_{\mathrm{I}}$, see functions "`group21`" and "`size21`" and `GAP` input in "`misc/E6-r.txt`".

A few special cases are discussed below.

Argument 8.31 ($\tilde{\mathbf{D}}$ type fiber inside the insertion). If the monogonal region inside the insertion is of type $\tilde{\mathbf{D}}$, relation (8.29) is not present. For the two irreducible curves

```
# The computation of the fundamental groups of plane sextics with a type E6 singularity.
Reread("pi1sc.txt");     # (see Listing 5.3)
# Relations at infinity and the slope
fixed := [rho^4/(G.1*G.2)^3];  kappa := (G.1*G.2)^3;
# Hexagon with a loop, irreducible curves (see Argument 8.28)
loop := [(G.1*G.2)^-2*G.1*(G.1*G.2)^2/G.3];  # relation from the inner loop
c1l := [x, x^2, x*y*x^2, x*(y*x^2)^2];         # chains to be used
group1l := function(arg) return DS(AA(c1l, arg), loop); end;
size1l := function(arg) return Size(CallFuncList(group1l, arg)); end;
# Hexagon with a loop, reducible curves (see Argument 8.28)
loop2 := [loop[1], kappa];                      # add a central element
c2l := [x, x^2, x*y*x^2, (x*y)^2*x^2];          # chains to be used
group2l := function(arg) return DS(AA(c2l, arg), loop2); end;
size2l := function(arg) return Size(CallFuncList(group2l, arg)); end;
# Other sextics with two type E points (see Argument 8.40)
ce := [x, y*x^2];                               # chains to be used
group1e := function(arg) return DS(AA(ce, arg), E6(x^2*y)); end;
# Genuine hexagon (see Argument 8.45)
ch := List([0..5], i -> (y*x)^i*x);            # chains to be used
group0h := function(arg) return DS(AA(ch, arg)); end;
# One monogonal region inside B \ R_I (see Argument 8.48)
loop1h := s -> [RA((y*x^2)^3, 1), (G.2*G.3)^(s[1]/2)];
group1h := function(arg) return DS(AA(ch, arg), loop1h(arg)); end;
# Two monogonal regions inside B \ R_I (see Argument 8.49)
loop2h := s -> [RA((y*x^2)^2, 1), (G.2*G.3)^(s[1]/2)];
group2h := function(arg) return DS(AA(ch, arg), loop2h(arg)); end;
```

Listing 8.16. Computing π_1 for sextics with a type $\mathbf{E}_6$ point ("E6.txt").

(Lines 32 and 36), the braid relations arising from the other regions of the skeleton suffice to show that the groups are abelian. For the reducible curves (Lines 17′, 21′, 23′, and 24′), a partial relation from the $\tilde{\mathbf{D}}$ type region is

$$[(\alpha_1\alpha_2)^{-1}\alpha_2(\alpha_1\alpha_2), \rho] = 1.$$

In view of (8.27), this still implies that $\varkappa_{\mathrm{I}} = \rho^4 = (\alpha_1\alpha_2)^3$ is a central element, and we can consider the quotient $\pi_1/\varkappa_{\mathrm{I}}$. Three of the four group are abelian, and the case $\mathbf{E}_6 \oplus \mathbf{D}_5 \oplus \mathbf{A}_5 \oplus \mathbf{A}_3$, Line 21′ is discussed in Argument 8.34, see Item 3.

Argument 8.32 ($2\mathbf{E}_6 \oplus \mathbf{A}_4 \oplus \mathbf{A}_3$, Line 6). Creating the group *via* `size1l(5,4)`, one can see that

- $\operatorname{ord}\pi_1 = 720$ and the commutant $H := [\pi_1, \pi_1] \cong SL(2, \Bbbk_5)$ is the only perfect group of order 120;
- each generator has order 6; hence, the abelianization epimorphism splits;
- the centralizer $C := \mathcal{Z}_{\pi_1}(H)$ equals $\mathbb{Z}_6$ and one has $\operatorname{ord}(C \cap H) = 2$; hence, π_1 is a semidirect but not direct product $H \rtimes \mathbb{Z}_6$.

After eliminating α_3, the braid relations take the form

$$\{\alpha_2, (\alpha_1\alpha_2)\alpha_1(\alpha_1\alpha_2)^{-1}\}_5 = \{\alpha_1, \alpha_2\alpha_1\alpha_2^{-1}\}_4 = 1 \tag{8.33}$$

and one obtains the presentation stated in Theorem 8.1.

Argument 8.34 (hexagon with a loop, nonabelian groups). In the range $6'$–$24'$, there are four reducible sextics not of torus type and with nonabelian fundamental group:

1. $\mathbf{E}_6 \oplus \mathbf{A}_7 \oplus \mathbf{A}_3 \oplus \mathbf{A}_2 \oplus \mathbf{A}_1$, Line $8'$: one has $\pi_1 = SL(2, \Bbbk_7) \times \mathbb{Z}$;
2. $\mathbf{E}_6 \oplus \mathbf{A}_5 \oplus \mathbf{A}_4 \oplus \mathbf{A}_3 \oplus \mathbf{A}_1$, Line $10'$: one has $\pi_1 = SL(2, \Bbbk_5) \rtimes \mathbb{Z}$;
3. $\mathbf{E}_6 \oplus \mathbf{D}_5 \oplus \mathbf{A}_5 \oplus \mathbf{A}_3$, Lines $20'$ and $21'$: one has $\pi_1 = (\mathbb{Z}_3^2 \rtimes \mathbb{Z}_3) \rtimes \mathbb{Z}$.

The groups are computed as explained in Argument 8.28. In Items 1, 2, and 3, the centralizer of $[\pi_1, \pi_1]$ projects to a subgroup of index 1, 2, and 4, respectively, in the abelianization. In the former case, it follows that the product is direct.

The group in Item 1 was first computed in [10].

Argument 8.35 (hexagon with a loop, sextics of torus type). For the two sets of singularities $(\mathbf{E}_6 \oplus \mathbf{A}_8 \oplus \mathbf{A}_2) \oplus \mathbf{A}_3$, Line 18 and $(\mathbf{E}_6 \oplus \mathbf{A}_8 \oplus \mathbf{A}_2) \oplus \mathbf{A}_2 \oplus \mathbf{A}_1$, Line 22, the call `simplify(g.1/g.2)` in Listing 5.3 returns a presentation with two generators and no relations. Hence, $[\pi_1, \pi_1] \cong \mathbb{F}_2$ and $\pi_1 \cong \Gamma$, see Lemma 7.31. For the reducible curve $(\mathbf{E}_6 \oplus 2\mathbf{A}_5) \oplus \mathbf{A}_3$, Line $6'$, the same approach applies to the irreducible perturbations of torus type.

For $(2\mathbf{E}_6 \oplus \mathbf{A}_5) \oplus \mathbf{A}_2$, Line 8, the group π_1 is

$$\langle \alpha_1, \alpha_2 \mid (8.30),\ \{\alpha_2, (\alpha_1\alpha_2)\alpha_1(\alpha_1\alpha_2)^{-1}\}_3 = \{\alpha_1, \alpha_2\alpha_1\alpha_2^{-1}\}_6 = 1\rangle, \tag{8.36}$$

and one has $\pi_1/\alpha_1^3 = \mathbb{Z}_4^2 \rtimes \mathbb{Z}_3$, whereas $\Gamma/(\mathbb{X}\mathbb{Y})^3 \cong \mathbb{A}_4$. Alternative presentations of this group are found in [44, 70]. The groups π_1 given by (8.36) or (8.43) below are the only known examples of the fundamental groups of irreducible sextics of torus type whose Alexander module has integral torsion: the abelianization of $[\pi_1, \pi_1]$ equals $\mathbb{Z}^2 \times \mathbb{Z}_2^2$, see [70].

For $(\mathbf{E}_6 \oplus \mathbf{A}_5 \oplus 2\mathbf{A}_2) \oplus \mathbf{A}_4$, Line 13, the group π_1 is

$$\begin{aligned}\langle \alpha_1, \alpha_2 \mid (8.30),\ \{\alpha_2, (\alpha_1\alpha_2)\alpha_1(\alpha_1\alpha_2)^{-1}\}_5 &= \{\alpha_1, \alpha_2\alpha_1\alpha_2^{-1}\}_6 \\ &= \{\alpha_2, (\alpha_1\alpha_2^2)\alpha_1(\alpha_1\alpha_2^2)^{-1}\}_3 = 1\rangle;\end{aligned} \tag{8.37}$$

the quotient π_1/α_1^5 is a perfect group of order 7680, whereas $\Gamma/(\mathbb{X}\mathbb{Y})^5 \cong \mathbb{A}_5$.

Argument 8.38 (hexagon with a double loop, Lines 5, 11, and 37–42). Choosing the reference edge e as shown in Figure 8.12, from the two inner regions in the figure we have the relations

$$\{\alpha_1\alpha_2\alpha_1^{-1}, \alpha_3\}_5 = 1, \qquad \alpha_1\alpha_2\alpha_1\alpha_2^{-1}\alpha_1^{-1} = \alpha_3^{-1}\alpha_1\alpha_2\alpha_1^{-1}\alpha_3. \tag{8.39}$$

Since all generators are conjugate to each other, all curves in question are irreducible. Using `GAP`, see "`misc/E6-ir.txt`", we can see that

- the fundamental group π_1 defined by (8.27) and (8.39) is $SL(2, \Bbbk_5) \rtimes \mathbb{Z}_6$, *cf.* Argument 8.32;
- in this group π_1, relation $\{\alpha_2, \alpha_3\}_4 = 1$ holds;
- if the outer region adjacent to $R_{\rm I}$ in Figure 8.12 is of type $\tilde{\mathbf{D}}$ or has braid index $s \neq 0 \bmod 4$, adding the corresponding relations to π_1 makes this group abelian.

These arguments apply as well to the sets of singularities $2\mathbf{E}_6 \oplus \mathbf{A}_4 \oplus \mathbf{A}_3$, Line 5 and $\mathbf{E}_6 \oplus \mathbf{E}_7 \oplus \mathbf{A}_4 \oplus \mathbf{A}_2$, Line 11, as their skeletons can be redrawn as shown in Figures 8.13 (f) and (g), respectively, and the former is the only one resulting in a nonabelian group π_1, which is defined by relations (8.27) and (8.39).

Argument 8.40 (two or more type **E** points). In all cases not considered above, the skeleton has a singular •- or ∘-vertex w in the boundary of the distinguished hexagon $R_{\rm I}$. If w is a •-vertex, choose the reference edge $e := e' \uparrow \mathbf{yx}$, where e' is the edge incident to w. Then the braid relations about w are

$$\rho\alpha_2\rho^{-1} = \alpha_2^{-1}\alpha_1\alpha_2 = \rho^{-1}\alpha_3\rho \tag{8.41}$$

and π_1 is computed *via* "group1e" in Listing 8.16, see "misc/E6-2.txt".

For the curve $(3\mathbf{E}_6) \oplus \mathbf{A}_1$, Line 1, the only extra relation is $[\alpha_2, \alpha_3] = 1$. In its presence, (8.41) simplifies to $\alpha_1\alpha_2\alpha_1 = \alpha_2\alpha_1\alpha_2$ and $\alpha_1\alpha_3\alpha_1 = \alpha_3\alpha_1\alpha_3$ and, due to (8.27), we have an isomorphism $\pi_1 = \mathbb{B}_4/\sigma_2\sigma_1^2\sigma_2\sigma_3^2$ given by $\alpha_1 \mapsto \sigma_2$, $\alpha_2 \mapsto \sigma_1$, $\alpha_3 \mapsto \sigma_3$. (This group was first computed in [130].)

Worth mentioning is the set of singularities $(2\mathbf{E}_6 \oplus 2\mathbf{A}_2) \oplus 2\mathbf{A}_1$, which is obtained by a maximal perturbation of the singular fiber over w. Using Theorem 7.22, one can see that no new relations are introduced and the perturbation epimorphism is an isomorphism. Since the same argument applies to the other type $\tilde{\mathbf{E}}_6$ fiber, their common perturbation to $(\mathbf{E}_6 \oplus 4\mathbf{A}_2) \oplus 3\mathbf{A}_1$ does not produce any new relations and results in the same fundamental group.

For the curve $(2\mathbf{E}_6 \oplus 2\mathbf{A}_2) \oplus \mathbf{A}_3$, Line 3, the presentation obtained is

$$\pi_1 = \langle \alpha_1, \alpha_2, \alpha_3 \mid (3.27),\ (8.41),\ \{\alpha_2, \alpha_3\}_4 = \{\alpha_1, \alpha_3\}_3 = 1 \rangle, \tag{8.42}$$

and $\pi_1/\alpha_1^2 = SL(2, \Bbbk_3)$ is a group of order 48, whereas $\Gamma/(\mathbb{X}\mathbb{Y})^2 \cong \mathbb{S}_3$. A maximal perturbation of the fiber over w produces a plane sextic with the set of singularities $(\mathbf{E}_6 \oplus 4\mathbf{A}_2) \oplus \mathbf{A}_3 \oplus \mathbf{A}_1$, and Theorem 7.22 establishes an epimorphism $\mathbb{B}_4 \twoheadrightarrow \pi_1$ given by $\sigma_1 \mapsto \alpha_1$, $\sigma_2 \mapsto \alpha_2\alpha_3\alpha_2^{-1}$, $\sigma_3 \mapsto \alpha_2$. One can check that the other braid relations hold in $\mathbb{B}_4$ (*e.g.*, one can use "IsBraidID" in Listing 2.5); as above, the relation at infinity (8.27) turns into $\sigma_2\sigma_1^2\sigma_2\sigma_3^2 = 1$.

Finally, for the curve $(2\mathbf{E}_6 \oplus \mathbf{A}_5) \oplus \mathbf{A}_2$, Line 7 we have

$$\pi_1 = \langle \alpha_1, \alpha_2, \alpha_3 \mid (3.27),\ (8.41),\ \{\alpha_2, \alpha_3\}_3 = \{\alpha_1, \alpha_3\}_6 = 1 \rangle. \tag{8.43}$$

The properties of this group resemble those of (8.36), but I do not know if the two groups are isomorphic. For alternative presentations, see [44, 70].

Argument 8.44 (a point of type $\mathbf{E}_7$). If w as above is a $\circ$-vertex, then ρ^3 is a central element; hence, so is $\rho\varkappa_{\mathrm{I}}^{-1}$ and one has $[\rho, \alpha_1\alpha_2] = [\rho, \alpha_3] = [\alpha_3, \alpha_1\alpha_2] = 1$. Letting $e = e' \uparrow \mathbf{x}$, where e' is the edge incident to w, from the braid relations about w we have $\alpha_3 = \alpha_2$. Then $[\alpha_3, \alpha_1] = 1$ and the group is abelian.

For further references, notice that we only used the fact that the distinguished hexagon R_{I} contains a singular $\circ$-vertex in its boundary.

Argument 8.45 (genuine hexagon). If the distinguished hexagon R_{I} is genuine, the group π_1 has the relations

$$\{\alpha_2 \uparrow \sigma_1^i, \alpha_3\}_{s_i} = 1, \quad i = 0, \ldots, 5, \tag{8.46}$$

where s_i, $i = 0, \ldots, 5$, is the braid index of the region $R_i := \mathrm{reg}_+(e \uparrow (\mathbf{yx})^i)$. In most cases, (8.27) and (8.46) define the group, see Lemma 5.59:

$$\pi_1 = \langle \alpha_1, \alpha_2, \alpha_3 \mid \rho^4 = (\alpha_1\alpha_2)^3,\ \{\alpha_2 \uparrow \sigma_1^i, \alpha_3\}_{s_i} = 1,\ i = 0, \ldots, 5 \rangle \tag{8.47}$$

for a certain vector $\bar{s} := (s_0, \ldots, s_5)$ as above. In particular, (8.47) presents π_1 for all four irreducible curves (Lines 43–46), see function "`qroup0h`" in Listing 8.16 and `GAP` input in "`misc/E6-ir.txt`"; the reference edge e used is marked in the figures.

The set of singularities $\mathbf{E}_6 \oplus \mathbf{A}_5 \oplus 4\mathbf{A}_2$, Line 43 and its perturbation $\mathbf{E}_6 \oplus 6\mathbf{A}_2$ are of weight eight and their fundamental groups are 'large', see [49] for more details; the latter group is 'minimal' for a sextic of weight eight, see G_0 in Section 7.2.3. The quotients π_1/α_1^2 are $(\mathbb{Z}_3 \times \mathbb{A}_4) \rtimes \mathbb{Z}_2$ and $\mathbb{Z}_3^2 \rtimes \mathbb{Z}_2$, respectively, and hence the perturbation epimorphism is proper. (This question was left unresolved in [49].) This curve has two perturbations of weight six with $\pi_1 \neq \Gamma$, *viz.* $(\mathbf{E}_6 \oplus 4\mathbf{A}_2) \oplus \mathbf{A}_3 \oplus \mathbf{A}_1$ and $(\mathbf{E}_6 \oplus 4\mathbf{A}_2) \oplus 3\mathbf{A}_1$. They were considered in Argument 8.40.

For the set of singularities $(\mathbf{E}_6 \oplus \mathbf{A}_5 \oplus 2\mathbf{A}_2) \oplus \mathbf{A}_4$, Line 45, the quotient π_1/α_1^5 is a perfect group of order 7680, *cf.* (8.37). I do not know if π_1 is isomorphic to (8.37).

The other two groups are as expected, *i.e.*, Γ and $\mathbb{Z}_6$.

This approach applies to most reducible curves as well. The first two classes do not seem to have any features that would facilitate the study of their fundamental groups. The curve $(\mathbf{E}_6 \oplus \mathbf{A}_5 \oplus 2\mathbf{A}_2) \oplus \mathbf{A}_3 \oplus \mathbf{A}_1$, Line 25′ is of torus type, and its group has infinite commutant. The curve $\mathbf{E}_6 \oplus \mathbf{A}_4 \oplus 2\mathbf{A}_3 \oplus \mathbf{A}_2 \oplus \mathbf{A}_1$, Line 26′ is not of torus type. The commutant $[\pi_1, \pi_1]$ is perfect (one can manually compute the Alexander module and see that it is trivial) and appears infinite, but at the moment I do not even know whether it is nontrivial. The other curves are discussed below.

Argument 8.48 (one monogon inside $S \smallsetminus R_{\mathrm{I}}$). Let $\mathfrak{S}$ be one of the skeletons shown in Figures 8.15 (k)–(o), *i.e.*, Lines 35′–39′ in Table 8.7. Choose the reference edge e as shown in the figures, so that the only monogonal region of $\mathfrak{S}$ is $\mathrm{reg}_-(e \uparrow (\mathbf{yx}^2)^3)$. Then we have a relation $\alpha_1 = (\alpha_2\alpha_3)^{-1}\alpha_3(\alpha_2\alpha_3)$ (which formally follows from the others); hence, π_1 is generated by α_2 and α_3 and, in view of (8.47), $(\alpha_2\alpha_3)^{s_0/2}$ is a central element. (Note that s_0 is even in all cases.) The quotient group $\pi_1/(\alpha_2\alpha_3)^{s_0/2}$ is computed using the function "group1h" in Listing 8.16.

The result of the computation is as follows:

1. $\mathbf{E}_6 \oplus \mathbf{A}_5 \oplus \mathbf{A}_4 \oplus \mathbf{A}_3 \oplus \mathbf{A}_1$, Line 35′: one has $\pi_1 = SL(2, \Bbbk_5) \rtimes \mathbb{Z}$, *cf.* 8.34 (2);
2. $\mathbf{E}_6 \oplus \mathbf{A}_7 \oplus \mathbf{A}_3 \oplus \mathbf{A}_2 \oplus \mathbf{A}_1$, Line 38′: one has $\pi_1 = SL(2, \Bbbk_7) \times \mathbb{Z}$, *cf.* 8.34 (1);
3. $\mathbf{E}_6 \oplus \mathbf{A}_9 \oplus \mathbf{A}_2 \oplus 2\mathbf{A}_1$, Line 39′: the commutant is a perfect group, which appears infinite, but I do not know a proof. The commutants of the quotients π_1/α_3^2 and π_1/α_3^3 are of orders 60 and 51840, respectively.

The group in Item 2 was first computed in [10].

Argument 8.49 (two monogons inside $S \smallsetminus R_{\mathrm{I}}$). Let $\mathfrak{S}$ be one of the skeletons shown in Figures 8.15 (f)–(j), *i.e.*, Lines 30′–34′ in Table 8.7. Choose the reference edge e as shown in the figures, so that one of the monogonal regions is $\operatorname{reg}_-(e \uparrow (\mathbf{y}\mathbf{x}^2)^2)$. Then we have a relation $\alpha_1 = \alpha_3^{-1}\alpha_2\alpha_3$, *which should be added to* (8.47). Hence, π_1 is generated by α_2 and α_3 and, in view of (8.47), $(\alpha_2\alpha_3)^{s_0/2}$ is a central element. (Note that s_0 is even in all cases.) The group $\pi_1/(\alpha_2\alpha_3)^{s_0/2}$ is computed using the function "group2h" in Listing 8.16.

The first three groups are abelian, whereas the curves $(\mathbf{E}_6 \oplus \mathbf{A}_{11}) \oplus 2\mathbf{A}_1$, Line 33′ and $(\mathbf{E}_6 \oplus 2\mathbf{A}_5) \oplus \mathbf{A}_3$, Line 34′ are of torus type, so that their fundamental groups have infinite commutants. (For proof, one can argue that these curves admit irreducible perturbations of torus type.)

Argument 8.50 (splitting into three components). Each of the remaining seven sextics splits into a quartic and two lines. Their groups have presentation (8.47).

If the proper model C has a $\tilde{\mathbf{D}}$ type fiber (Figures 8.15 (p), (q) and Lines 40′–43′), choose the reference edge e so that $\operatorname{reg}_+ e$ is the $\tilde{\mathbf{D}}$ type region. Then, in the notation of Argument 8.45, we have relations

$$\alpha_1^{-1}\alpha_j\alpha_1 = \alpha_j \uparrow \sigma_2^{s_0+2}, \quad j = 2, 3.$$

As a consequence, $[\alpha_1, \alpha_2\alpha_3] = 1$, hence $[\alpha_1, \rho] = [\alpha_2\alpha_3, \rho] = 1$, and the element $\delta := \alpha_1(\alpha_2\alpha_3)^{1+s_0/2}$ is central. Since also $[\alpha_1\alpha_2, \rho^4] = 1$, see (8.27), from the commutativity relations above one has $[\alpha_3, \rho^4] = [\alpha_2, \rho^4] = 1$ and the element $\rho^4 = \varkappa_{\mathrm{I}}$ is also central. Thus, applying Corollary A.30 twice, we can use GAP to analyze the quotient $\pi_1/\langle \delta, \varkappa_{\mathrm{I}} \rangle$. For two curves, *viz.* $\mathbf{E}_6 \oplus \mathbf{D}_8 \oplus \mathbf{A}_3 \oplus 2\mathbf{A}_1$, Line 40′ and $\mathbf{E}_6 \oplus \mathbf{D}_6 \oplus \mathbf{A}_5 \oplus 2\mathbf{A}_1$, Line 43′, the groups are abelian.

The groups of the remaining five curves have infinite commutants. This statement can be proved by computing their Alexander modules, see [57].

Curves admitting a mild degeneration

Consider a proper model C as in Proposition 8.26. With Lemma 4.51 in mind, we need to show that, if $\operatorname{wd} R_{\mathrm{I}} > 6$, then ∂R_{I} contains a cutting sequence of width six.

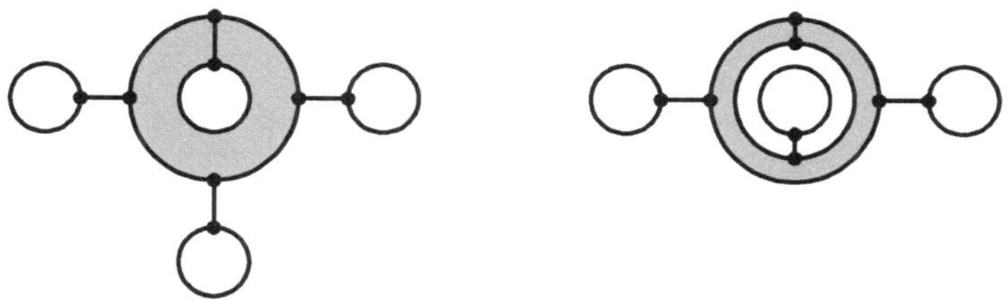

(a) $(13, 7, 1, 1, 1, 1)$ (b) $(9, 7, 5, 1, 1, 1)$

Figure 8.17. The two exceptional heptagons.

Unfortunately, this statement does not hold for all pairs $(\mathfrak{S}, R_{\mathrm{I}})$, see, *e.g.*, Figures 8.17 and 8.19 below, and we have to engage into a case-by-case analysis.

Regular skeleton of degree twelve In this case, $\mathfrak{S}$ is one of the skeletons found in [20] and, considering pairs $(\mathfrak{S}, R_{\mathrm{I}})$ one by one, we conclude that, with the exception of the two heptagons shaded in Figure 8.17, a cutting sequence does exist. In each of the exceptional cases, there is a sequence satisfying condition 2, but not 1 in Lemma 4.49. Then, using Lemma 4.51, we can cut R_{I} into a larger region (of width 19 or 13, respectively) and reduce these cases to those already considered.

Regular skeleton of degree nine or six If $\mathfrak{S}$ has an edge e not in the boundary of R_{I}, then, inserting one or two monogonal regions, each connected to e by an edge (*cf.* Convention 8.3), we can formally reduce this case to that of eight $\bullet$-vertices. The six skeletons with all edges in the boundary of a single region are shown in Figure 8.18 and Figure 4.2 (b), (e); each if these skeletons has a cutting sequence of width six.

Skeleton $\mathfrak{S}$ has singular $\bullet$- or $\circ$-vertices If none of these vertices is in ∂R_{I}, then, 'inflating' each singular vertex to a monogon, we can formally reduce this case to one of those already considered. If the boundary of R_{I} contains a singular $\circ$-vertex, the fundamental group π_1 is abelian due to Argument 8.44 (and we leave the question of the existence of a mild degeneration open). One can enumerate all skeletons with a monovalent $\bullet$-vertex in the boundary of a distinguished region of width at least seven: they are all obtained by inserting a fragment $\bullet\!\!-\!\!\bullet$ at the center of an appropriate edge of one of the skeletons shown in Figures 4.2 (b)–(f) or 4.3 (b), (c). In all cases except those shown in Figure 8.19, the skeleton has a cutting sequence of width six.

The fundamental groups The groups of perturbed curves are computed in the files "`misc/E6-ir.txt`" and "`misc/E6-r.txt`" together with their maximizing degenerations. With few exceptions, mentioned in the corresponding arguments, these groups are as expected, *i.e.*, Γ or $\mathbb{Z}_6$.

For the three sets of singularities $(2\mathbf{E}_6 \oplus 2\mathbf{A}_2) \oplus 2\mathbf{A}_1$, $(\mathbf{E}_6 \oplus 4\mathbf{A}_2) \oplus \mathbf{A}_3 \oplus \mathbf{A}_1$, and $(\mathbf{E}_6 \oplus 4\mathbf{A}_2) \oplus 3\mathbf{A}_1$, the groups π_1 were computed using the non-simple perturbation

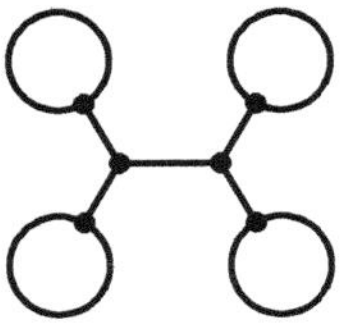

(a) $(14, 1, 1, 1, 1)$

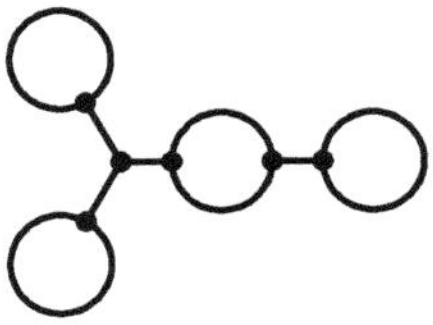

(b) $(13, 2, 1, 1, 1)$

(c) $(12, 2, 2, 1, 1)$

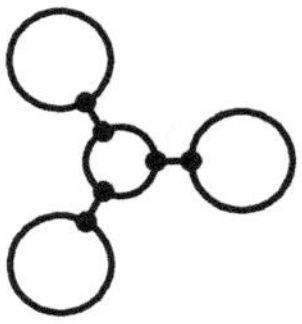

(d) $(12, 3, 1, 1, 1)$

Figure 8.18. The four skeletons with six •-vertices.

(a) $3\mathbf{E}_6$

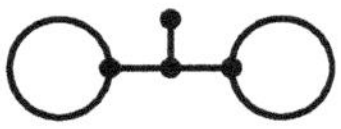

(b) $2\mathbf{E}_6 \oplus \mathbf{D}_5 \oplus \mathbf{A}_1$

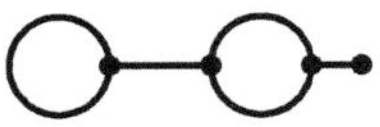

(c) $2\mathbf{E}_6 \oplus \mathbf{D}_6$

Figure 8.19. Sextics admitting no mild degenerations.

$\mathbf{E}_6 \to 2\mathbf{A}_2 \oplus \mathbf{A}_1$; thus, we need to show that each of these sets of singularities is realized by a unique deformation family of sextics of torus type, so that simple perturbation would not produce a different curve with possibly different fundamental group. For the set $(\mathbf{E}_6 \oplus 4\mathbf{A}_2) \oplus 3\mathbf{A}_1$, the uniqueness is given by [3]. As shown above, any sextic realizing one of the first two sets of singularities admits a mild degeneration. According to Tables 8.5–8.7, these degenerations can only be $(2\mathbf{E}_6 \oplus 2\mathbf{A}_2) \oplus \mathbf{A}_3$, Line 3 and $(\mathbf{E}_6 \oplus \mathbf{A}_5 \oplus 2\mathbf{A}_2) \oplus \mathbf{A}_3 \oplus \mathbf{A}_1$, Line 25′, respectively; in each case, the inverse perturbation is unique up to equivalence due to Proposition 4.48.

Exceptional cases

Skeleton $\mathfrak{S}$ as in Figure 8.19 Skeletons (b) and (c) are equivalent: they are related by a modification as in Lemma 4.51 satisfying condition 2, but not 1. In these two cases, the curve has a $\tilde{\mathbf{D}}$ type fiber and, choosing the correct model, one can assume that this fiber is not in R_{I} and that the set of singularities of the corresponding sextic is as indicated in the figures, in case (a) the sextic being *not* of torus type.

All three fundamental groups π_1 are abelian. For proof, it suffices to consider the braid relations arising from a singular •-vertex in the boundary ∂R_{I} and the mono- or bivalent ×-vertex inside R_{I}. From the point of view of the braid monodromy, the corresponding pair of singular fibers can be regarded as a perturbation of a fiber of type $\mathbf{E}_7$ or $\mathbf{E}_8$, *cf.* Figures 7.2 (f) and 7.1 (f), respectively, and the statement follows from Theorems 7.20 and 7.18. In fact, using Theorem 7.28, one can easily show that the corresponding sextics do degenerate to those with the sets of singularities $2\mathbf{E}_6 \oplus \mathbf{E}_7$, $\mathbf{E}_6 \oplus \mathbf{E}_8 \oplus \mathbf{D}_5$, and $\mathbf{E}_6 \oplus \mathbf{E}_7 \oplus \mathbf{D}_6$, respectively, whose groups have already been found to be abelian.

Collision of two $\tilde{\mathbf{D}}$ type fibers The skeleton $\mathfrak{S}$ is as shown in Figure 8.3 (g), with the insertion as in Figure 8.11, and we can assume that the two fibers are either $2\tilde{\mathbf{D}}_5$ in the bigonal region of $\mathfrak{S}$ or $\tilde{\mathbf{D}}_6 + \tilde{\mathbf{D}}_5$ in its triangular region, see Observation 8.5. In both cases, applying Lemma 4.51 twice, we can find an equivalent dessin with the two $\tilde{\mathbf{D}}$ type fibers in two distinct regions of the new skeleton; hence, the curve admits a mild degeneration to $\mathbf{E}_6 \oplus \mathbf{D}_6 \oplus \mathbf{D}_5 \oplus \mathbf{A}_2$, Line 24′ or $\mathbf{E}_6 \oplus \mathbf{D}_7 \oplus \mathbf{D}_5 \oplus \mathbf{A}_1$, Line 23′, respectively.

8.3 A distinguished point of type D

The trigonal model of a sextic D with a distinguished singular point P of type $\mathbf{D}$ has two or three improper fibers, one of them being of type $\tilde{\mathbf{A}}_1$; hence, unlike Section 8.2, we do not have a 'common' relation at infinity. To reduce the number of cases, we confine ourselves to sextics without $\mathbf{E}$ type singular points (and, hence, proper models with regular skeletons), as such points were considered in Section 8.2. The only case that may be missing, *viz.* that of a point of type $\mathbf{E}_7$ with a linear component, can be disregarded as we are mainly interested in irreducible sextics.

8.3.1 A point of type $\mathbf{D}_p$, $p \geqslant 6$

We consider simple plane sextics $D \subset \mathbb{P}^2$ satisfying the following conditions:

($*$) D has a distinguished singular point P of type $\mathbf{D}_p$, $p \geqslant 6$,
D has no linear components through P, and
D has no singular points of type $\mathbf{E}$.

To make the following statement more regular, we use Kodaira's notation for the types of singular fibers.

Proposition 8.51. *There is a natural bijection ϕ, invariant under equisingular deformations, between Zariski open and dense, in each equisingular stratum, subsets of the sets of:*

- *simple plane sextics $D \subset \mathbb{P}^2$ with a distinguished type $\mathbf{D}_p$, $p \geqslant 6$, point P and without linear components through P, and*
- *proper simple trigonal curves $C \subset \Sigma_3$ with a distinguished fiber F_{I} of type I_2 and a distinguished fiber F_{II} of type I_{p-6}.*

Table 8.8. Irreducible maximizing sextics with a type $\mathbf{D}_p$, $p \geqslant 7$, point[1]

#	Set of singularities	Figure	Count	π_1
1	$\mathbf{D}_{11} \oplus \mathbf{A}_4 \oplus 2\mathbf{A}_2$	8.6(a)	$(1,0)$	8.55
2	$\mathbf{D}_9 \oplus 2\mathbf{A}_4 \oplus \mathbf{A}_2$	8.6(a)	$(1,0)$	
3	$\mathbf{D}_{19}$	8.6(b)-1	$(1,0)$	8.55
4	[1] $\mathbf{D}_7 \oplus \mathbf{A}_{12}$	8.6(b)-1	$(1,0)$	
5	[1] $\mathbf{D}_7 \oplus \mathbf{A}_{12}$	8.6(b)-1	$(0,1)$	
6	$\mathbf{D}_{17} \oplus \mathbf{A}_2$	8.6(b)-2	$(1,0)$	8.55
7	$\mathbf{D}_9 \oplus \mathbf{A}_{10}$	8.6(b)-2	$(1,0)$	8.55
8	$\mathbf{D}_7 \oplus \mathbf{A}_{10} \oplus \mathbf{A}_2$	8.6(b)-2	$(0,1)$	
9	$\mathbf{D}_{13} \oplus \mathbf{A}_6$	8.6(c)-1	$(0,1)$	8.55
10	$\mathbf{D}_7 \oplus 2\mathbf{A}_6$	8.6(c)-1	$(0,1)$	
11	$\mathbf{D}_{15} \oplus \mathbf{A}_4$	8.6(c)-2	$(1,0)$	8.55
12	$\mathbf{D}_{11} \oplus \mathbf{A}_8$	8.6(c)-2	$(1,0)$	
13	[2] $\mathbf{D}_7 \oplus \mathbf{A}_8 \oplus \mathbf{A}_4$	8.6(c)-2	$(1,0)$	
14	[2] $\mathbf{D}_7 \oplus \mathbf{A}_8 \oplus \mathbf{A}_4$	8.6(c)-2	$(1,0)$	
15	$\mathbf{D}_{13} \oplus \mathbf{A}_4 \oplus \mathbf{A}_2$	8.6(c)-3	$(1,0)$	8.55
16	$\mathbf{D}_{11} \oplus \mathbf{A}_6 \oplus \mathbf{A}_2$	8.6(c)-3	$(1,0)$	
17	$\mathbf{D}_9 \oplus \mathbf{A}_6 \oplus \mathbf{A}_4$	8.6(c)-3	$(1,0)$	8.55
18	$\mathbf{D}_7 \oplus \mathbf{A}_6 \oplus \mathbf{A}_4 \oplus \mathbf{A}_2$	8.6(c)-3	$(1,0)$	

[1] There are 22 classes realizing 16 sets of singularities

A sextic D is maximizing if and only if $C := \phi(D)$ is maximal and stable and the fiber F_{II} is singular, i.e., $p \geqslant 7$. $\triangleleft$

Consider a maximizing sextic D satisfying conditions 8.3.1$(*)$. Then $\mathfrak{S} := \operatorname{Sk} C$ is a regular skeleton with two distinguished regions: a bigon R_{I} and a $(p-6)$-gon R_{II}. Considering the bigon R_{I} as an *insertion*, we conclude that $\mathfrak{S}$ is one of the skeletons shown in Figure 8.6 (irreducible curves), Figure 8.7 (a)–(c) (two component curves), or Figure 8.8 (a), (b) (three component curves). The irreducible components of C are given by Remark 8.20. Unlike the case of a point P of type $\mathbf{E}_7$, the insertion has no distinguished coorientation. The two distinguished fibers are canonically ordered (even if $p = 8$), as the corresponding improper fibers have distinct topological types.

The complete list of pairs (D, P) satisfying conditions 8.3.1$(*)$ and with sextic D maximizing and irreducible is given by Table 8.8. Multiple entries corresponding to a single figure are due to the fact that one of the regions of the skeletons should be chosen for R_{II} and, in Figures 8.7 (c) and 8.8 (b), another region should be chosen to contain a $\tilde{\mathbf{D}}$ type fiber. In particular, the pairs of curves in Lines 4, 5 and those in Lines 13, 14 differ by the choice of one of the two monogonal regions of the skeleton in Figures 8.6 (b) and (c), respectively.

```
# The computation of the fundamental groups of plane sextics with a type Dp, p ≥ 6, singularity.
Reread("pi1sc.txt");    # (see Listing 5.3)
# Paths to be used
p := i -> (x*y)^i*x^2; list := List([-1..8], p); list[1] := x;
# Relations (8.52), (8.53), and (8.54); p is the path from e to eII and w = wd RII.
inf := function(p, w)
 fixed := Braid(S.1^w, G, [G.3^-1*G.1*G.3, G.3^-1*G.2*G.3]);
 fixed := bm(p, [fixed[1]/G.1, fixed[2]/G.2, G.3]);
 fixed := [G.3*G.1*G.2/rho, fixed[1], fixed[2], rho^2*G.3/fixed[3]];
end;
# Use "inf(p,w);" and then "size(n1,n2,...);"
# Parameters are the braid indices of reg_(e ↑ ζ) for ζ = x and ζ = (xy)^i x^2, i = 0, 1, ....
size := function(arg) return Size(DS(AA(list, arg))); end;
```

Listing 8.20. Computing π_1 for sextics with a type $\mathbf{D}_p$ point ("Dp.txt").

The fundamental groups

To compute the fundamental group π_1, we choose a common reference edge $e = e_{\mathrm{I}}$ in $\partial_- R_{\mathrm{I}}$ and an additional reference edge e_{II}, with a canonical basis $\{\beta_1, \beta_2, \beta_3\}$, in $\partial_- R_{\mathrm{II}}$. The fiber $\tilde{F}_{\mathrm{I}}$ is as in Case 2 with $m = 2n = 2$, see Section 5.2.2, and we have $\varkappa_{\mathrm{I}} = \alpha_1\alpha_2$ and

$$[\alpha_1\alpha_2, \alpha_3] = 1. \tag{8.52}$$

The fiber $\tilde{F}_{\mathrm{II}}$ is as in Case 1 with $m = p - 6$ and $n = 1$, and we have $\varkappa_{\mathrm{II}} = \beta_3$ and

$$\beta_3\beta_i\beta_3^{-1} = \beta_i \uparrow \sigma_1^{p-6}, \quad i = 1, 2. \tag{8.53}$$

In view of (8.52), the relation at infinity (5.55) cancels to

$$(\alpha_1\alpha_2)^2\alpha_3^3 = \beta_3. \tag{8.54}$$

Argument 8.55 (adjacent regions R_{I}, R_{II}). If the distinguished regions R_{I} and R_{II} are adjacent to each other, the group π_1 is abelian. Indeed, replacing, if necessary, $\mathfrak{S}$ with its mirror image, we can select the reference edges so that $e_{\mathrm{II}} = e \uparrow \mathbf{x}$. Then $\beta_1 = \alpha_2$, $\beta_2 = \alpha_3$, and $\beta_3 = (\beta_1\beta_2)\beta_3(\beta_1\beta_2)^{-1} = \alpha_1$, see (8.53), and the relation at infinity (8.54) simplifies to $\alpha_2\alpha_1\alpha_2\alpha_3^3 = 1$. Hence, α_3 commutes with $\alpha_2\alpha_1\alpha_2$ and, in view of (8.52), this implies $[\alpha_3, \alpha_2] = 1$. On the other hand, $\alpha_1 = \alpha_2^{-1}\alpha_3^{-3}\alpha_2^{-1}$ belongs to the abelian subgroup generated by α_2 and α_3.

This observation applies as well to a number of reducible curves; these curves are listed in Table 8.9.

The other groups are computed using functions "inf" and "size" in Listing 8.20: the former initializes relations (8.52)–(8.54), and the latter adds the braid relations from the other regions and returns the size of the commutant $[\pi_1, \pi_1]$. The GAP input is found in "misc/Dp-ir.txt". All groups obtained are abelian.

Table 8.9. Some reducible sextics with abelian fundamental groups (see Argument 8.55).

Set of singularities	Figure
The splitting $D_3 + D_3$	
$\mathbf{D}_{14} \oplus \mathbf{A}_3 \oplus \mathbf{A}_2$	8.7(a)-1
$\mathbf{D}_{10} \oplus \mathbf{A}_7 \oplus \mathbf{A}_2$	8.7(a)-1
$\mathbf{D}_{16} \oplus \mathbf{A}_2 \oplus \mathbf{A}_1$	8.7(a)-4
$\mathbf{D}_{18} \oplus \mathbf{A}_1$	8.7(b)-2
$\mathbf{D}_{14} \oplus \mathbf{D}_5$	8.7(c)-1
The splitting $D_4 + D_2$	
$\mathbf{D}_{11} \oplus \mathbf{A}_5 \oplus \mathbf{A}_3$	8.7(a)-2
$\mathbf{D}_{11} \oplus \mathbf{A}_7 \oplus \mathbf{A}_1$	8.7(a)-3
$\mathbf{D}_9 \oplus \mathbf{A}_7 \oplus \mathbf{A}_2 \oplus \mathbf{A}_1$	8.7(a)-5
$\mathbf{D}_{10} \oplus \mathbf{A}_9$	8.7(b)-1
$\mathbf{D}_9 \oplus \mathbf{A}_9 \oplus \mathbf{A}_1$	8.7(b)-3
$\mathbf{D}_9 \oplus \mathbf{D}_{10}$	8.7(c)-2
$\mathbf{D}_9 \oplus \mathbf{D}_5 \oplus \mathbf{A}_5$	8.7(c)-2

Set of singularities	Figure
The splitting $D_5 + D_1$	
$\mathbf{D}_{10} \oplus \mathbf{A}_5 \oplus \mathbf{A}_4$	8.7(a)-2
$\mathbf{D}_{14} \oplus \mathbf{A}_4 \oplus \mathbf{A}_1$	8.7(a)-3
$\mathbf{D}_{14} \oplus 2\mathbf{A}_2 \oplus \mathbf{A}_1$	8.7(a)-5
$\mathbf{D}_{16} \oplus \mathbf{A}_3$	8.7(b)-1
$\mathbf{D}_{16} \oplus \mathbf{A}_2 \oplus \mathbf{A}_1$	8.7(b)-3
$\mathbf{D}_{12} \oplus \mathbf{D}_7$	8.7(c)-2
$\mathbf{D}_{12} \oplus \mathbf{D}_5 \oplus \mathbf{A}_2$	8.7(c)-2
The splitting $D_3 + D_2 + D_1$	
$\mathbf{D}_{12} \oplus \mathbf{A}_5 \oplus 2\mathbf{A}_1$	8.8(a)-1
$\mathbf{D}_{12} \oplus 2\mathbf{A}_3 \oplus \mathbf{A}_1$	8.8(a)-2
$\mathbf{D}_{10} \oplus \mathbf{A}_5 \oplus \mathbf{A}_3 \oplus \mathbf{A}_1$	8.8(a)-2
$\mathbf{D}_{10} \oplus \mathbf{D}_8 \oplus \mathbf{A}_1$	8.8(b)
$\mathbf{D}_{10} \oplus \mathbf{D}_6 \oplus \mathbf{A}_3$	8.8(b)

Non-maximizing irreducible curves

Consider a proper model C as in Proposition 8.51. The distinguished fiber F_{I} can be isolated using Lemma 4.52. Since we did not make any assumption about the width of the other distinguished region R_{II}, we will not try to isolate F_{II}; in other words, we do not insist that the degeneration of C should be mild and the fiber F_{II} is allowed to collide with another type $\tilde{\mathbf{A}}$ singular fiber of C. In terms of the corresponding sextics, this means a degeneration $\mathbf{D}_p \to \mathbf{D}_{p'}$, $p' > p$, of the distinguished singular point P.

Thus, in addition to the perturbations of $\tilde{\mathbf{A}}$ and $\tilde{\mathbf{D}}$ type fibers, see Cases 1 and 2 in Observation 8.4, we should also consider the following perturbations, *cf.* Case 2:

3. fiber F_{II} of type I_{p-6} with p even: relation (8.53) is retained and the braid index of R_{II} is changed from 0 to an odd integer s, $1 \leqslant s < p-6$. If $s = 1$, the group is abelian, *cf.* Theorem 7.24.

(Recall that we are interested in maximal irreducible perturbations only.) All other details are found in the file "`misc/Dp-r.txt`".

Argument 8.56 (the set of singularities $2\mathbf{D}_7 \oplus 2\mathbf{A}_2$). As a result of the computation, we discover two irreducible perturbations with nonabelian fundamental groups, *viz.* $\mathbf{D}_7 \oplus \mathbf{D}_{10} \oplus \mathbf{A}_2 \to 2\mathbf{D}_7 \oplus 2\mathbf{A}_2$ and $2\mathbf{D}_7 \oplus \mathbf{A}_5 \to 2\mathbf{D}_7 \oplus 2\mathbf{A}_2$. In both cases, the skeleton $\mathfrak{S}$ is as shown in Figure 8.7 (c), insertion 2, and $\pi_1 \cong SL(2, \Bbbk_3) \times \mathbb{Z}_2$. Due to Lemma 4.51, the dessins of the perturbed curves are equivalent; hence, the two curves belong to a single deformation family.

(a) A bigon

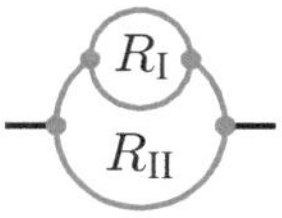

(b) A bibigon

Figure 8.21. Bigonal and bibigonal insertions.

Since, in the former case, we only use the braid indices of the regions, the further perturbation $\mathbf{D}_7 \oplus \mathbf{D}_4 \oplus 3\mathbf{A}_2$ has the same group, see also Argument 8.70 below.

Collision of F_{II} and a $\tilde{\mathbf{D}}$ type fiber Another new phenomenon is the case where the fiber F_{II} shares a region of the skeleton with a $\tilde{\mathbf{D}} = \mathrm{I}^*$ type fiber F of C. This pair (F_{II}, F) cannot collide and we do not get a degeneration of C to a maximal curve. For C to have a $\tilde{\mathbf{D}}$ type fiber, its skeleton must be one of those shown in Figures 8.7 (c) or 8.8 (b), (c). (Note that the latter cannot appear as the skeleton of a maximal proper model, as the curve would have only one $\tilde{\mathbf{A}}$ type fiber.) Taking into account Argument 8.55 and Observation 8.5, we are left with only one case, where $\mathfrak{S}$ is as in Figure 8.8 (b) and the offending fibers F_{II} and F are represented by a pair of monovalent $\times$-vertices in the outer region of $\mathfrak{S}$. Using Lemma 4.51, one can show that the corresponding dessin is equivalent to one with the skeleton as in Figure 8.7 (c), insertion 1; hence, C admits a mild degeneration to a maximal curve.

8.3.2 A point of type $\mathbf{D}_5$

Here, we consider simple plane sextics $D \subset \mathbb{P}^2$ satisfying the following conditions:

($*$) D has a distinguished singular point P of type $\mathbf{D}_5$,
D has no linear components through P, and
D has no singular points of type $\mathbf{E}$.

Proposition 8.57. *There is a natural bijection ϕ, invariant under equisingular deformations, between Zariski open and dense, in each equisingular stratum, subsets of the sets of:*

- *simple plane sextics $D \subset \mathbb{P}^2$ with a distinguished type $\mathbf{D}_5$ point P and without linear components through P, and*
- *proper simple trigonal curves $C \subset \Sigma_4$ with a distinguished fiber F_{I} of type $\tilde{\mathbf{A}}_1$ and a distinguished fiber F_{II} of type $\tilde{\mathbf{A}}_3$.*

A sextic D is maximizing if and only if $C := \phi(D)$ is maximal and stable. ◁

Thus, if C is a maximal trigonal model as in Proposition 8.57, then $\mathfrak{S} := \operatorname{Sk} C$ is a regular skeleton with two distinguished regions: a bigon R_{I} and a tetragon R_{II}.

A bibigonal insertion If the distinguished regions $R_{\rm I}$, $R_{\rm II}$ are adjacent to each other, together they form a fragment shown in Figure 8.21 (b), which we call a *bibigon*. Declare this fragment an *insertion* and remove it from $\mathfrak{S}$, see Convention 8.3, to obtain a regular skeleton $\mathfrak{S}'$ satisfying the vertex count (8.7). In terms of $\mathfrak{S}'$, the components of D are given by Remark 8.20. Hence, assuming D irreducible, $\mathfrak{S}$ must be one of the skeletons shown in Figure 8.6, where the possible positions of the insertion are shown in grey. Since a bibigonal insertion has a clearly defined distinguished coorientation, each position shown in the figures results in two skeletons $\mathfrak{S}$, which may or may not be isomorphic. Note also that the widths of the regions of $\mathfrak{S}$ differ from what they appear in Figure 8.6; they depend on the coorientation.

Table 8.10. Irreducible maximizing sextics with a type $\mathbf{D}_5$ point[1]

#	Set of singularities	Figure	Count	π_1
1	$\mathbf{D}_5 \oplus \mathbf{A}_6 \oplus \mathbf{A}_4 \oplus 2\mathbf{A}_2$	8.6(a)	$(1,0)$	8.62
2	$\mathbf{D}_5 \oplus \mathbf{A}_{14}$	8.6(b)-1	$(0,1)$	8.62
3	$\mathbf{D}_5 \oplus \mathbf{A}_{12} \oplus \mathbf{A}_2$	8.6(b)-2	$(1,0)$	8.62
4	[1] $\mathbf{D}_5 \oplus \mathbf{A}_{10} \oplus \mathbf{A}_4$	8.6(b)-2	$(1,0)$	8.62
5	$\mathbf{D}_5 \oplus \mathbf{A}_8 \oplus \mathbf{A}_6$	8.6(c)-1	$(0,1)$	8.62
6	[1] $\mathbf{D}_5 \oplus \mathbf{A}_{10} \oplus \mathbf{A}_4$	8.6(c)-2	$(0,1)$	8.62
7	[2] $\mathbf{D}_5 \oplus \mathbf{A}_8 \oplus \mathbf{A}_4 \oplus \mathbf{A}_2$	8.6(c)-3	$(1,0)$	8.62
8	[3] $\mathbf{D}_5 \oplus \mathbf{A}_6 \oplus 2\mathbf{A}_4$	8.6(c)-3	$(1,0)$	8.62
9	[2] $\mathbf{D}_5 \oplus \mathbf{A}_8 \oplus \mathbf{A}_4 \oplus \mathbf{A}_2$	8.23-1	$(0,1)$	8.65
10	[3] $\mathbf{D}_5 \oplus \mathbf{A}_6 \oplus 2\mathbf{A}_4$	8.23-2	$(1,0)$	8.65
11	$\mathbf{D}_5 \oplus \mathbf{A}_{10} \oplus 2\mathbf{A}_2$	8.23-3	$(1,0)$	8.65
12	$\mathbf{D}_5 \oplus (\mathbf{A}_8 \oplus 3\mathbf{A}_2)$	8.23-4	$(1,0)$	8.66

[1] There are 16 classes realizing 9 sets of singularities

The irreducible maximizing sextics whose skeleton has a bibigonal insertion are listed in Lines 1–8 in Table 8.10. Reducible sextics are obtained in a similar way from Figure 8.7 (a)–(c) (two component curves, see Table 8.11) and Figure 8.8 (three component curves, not listed).

A genuine tetragon If $R_{\rm II}$ is not adjacent to $R_{\rm I}$, we refer to it as a *genuine tetragon*. In this case, declaring $R_{\rm I}$ an *insertion* and removing it from $\mathfrak{S}$, see Convention 8.3, we obtain a regular skeleton $\mathfrak{S}'$ with six or four $\bullet$-vertices and a distinguished tetragonal region $R'_{\rm II}$. Such skeletons can easily be listed, *cf.* the case of genuine hexagon in Section 8.2.3; they are shown in Figure 8.22, where $R'_{\rm II}$ is the outer region. The original skeleton $\mathfrak{S}$ is obtained from $\mathfrak{S}'$ by inserting a bigon $R_{\rm I}$ at the center of any edge that is *not in the boundary of* $R'_{\rm II}$. The resulting curve is irreducible if and only if $\mathfrak{S}'$ does not admit a splitting marking. The only skeleton with this property is the one in Figure 8.22 (h); an alternative drawing of this skeleton and the possible positions of the bigonal insertion $R_{\rm I}$ are shown in Figure 8.23, and the corresponding sextics are listed in Lines 9–12 in Table 8.10.

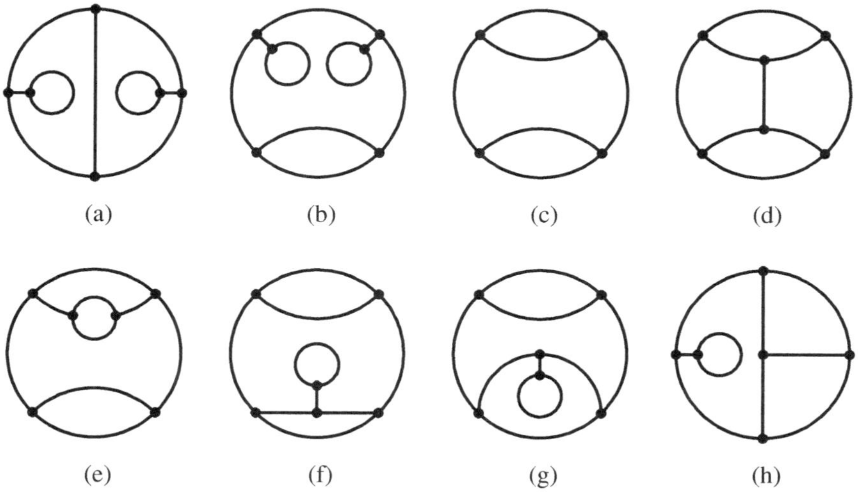

Figure 8.22. Genuine tetragon: All curves.

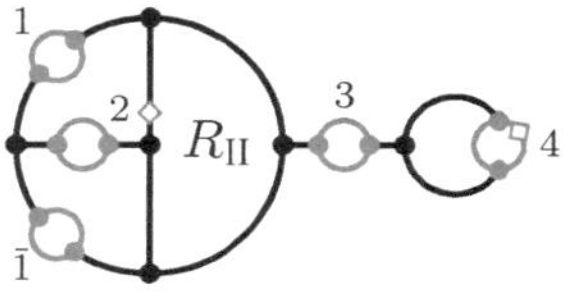

Figure 8.23. Genuine tetragon: Irreducible curves, see Figure 8.22 (h).

The fundamental groups

Choose a common reference edge $e := e_{II}$ in $\partial_- R_{II}$ and an extra reference edge e_I, with a canonical basis $\{\beta_1, \beta_2, \beta_3\}$, in $\partial_- R_I$. The fiber $\tilde{F}_I$ is as in Case 2 with $m = 2n = 2$, see Section 5.2.2, and we have $\varkappa_I = \beta_1\beta_2$ and

$$[\beta_1\beta_2, \beta_3] = 1, \quad \text{hence} \quad [\rho, \beta_3] = 1. \tag{8.58}$$

The fiber $\tilde{F}_{II}$ is as in Case 2 with $m = 2n = 4$, and we have $\varkappa_{II} = (\alpha_1\alpha_2)^2$ and

$$[(\alpha_1\alpha_2)^2, \alpha_3] = 1. \tag{8.59}$$

The relation at infinity (5.55) takes the form

$$\rho^4 = (\alpha_1\alpha_2)^2(\beta_1\beta_2). \tag{8.60}$$

Assume that the skeleton $\mathfrak{S}$ has a bibigonal insertion. Then the reference edges can be chosen so that $e_I = e \uparrow \mathbf{x}$ and hence $\beta_1 = \alpha_2$, $\beta_2 = \alpha_3$, and $\beta_3 = \alpha_1$ in view

Table 8.11. Some reducible maximizing sextics with a type $\mathbf{D}_5$ point![1]

#	Set of singularities	Figure	Count	π_1
	The splitting $C_3 + C_3$ $(G/[G,G] = \mathbb{Z}_4 \oplus \mathbb{Z}_3)$			
* 1′	$\mathbf{D}_5 \oplus \mathbf{A}_7 \oplus \mathbf{A}_5 \oplus \mathbf{A}_2$	8.7(a)-1	$(1,0)$	8.63(1)
2′	$\mathbf{D}_5 \oplus \mathbf{A}_9 \oplus \mathbf{A}_3 \oplus \mathbf{A}_2$	8.7(a)-1	$(1,0)$	8.63
3′	$\mathbf{D}_5 \oplus \mathbf{A}_{11} \oplus \mathbf{A}_2 \oplus \mathbf{A}_1$	8.7(a)-4	$(0,1)$	8.63
4′	$\mathbf{D}_5 \oplus \mathbf{A}_{13} \oplus \mathbf{A}_1$	8.7(b)-2	$(0,1)$	8.63
5′	$\mathbf{D}_5 \oplus \mathbf{D}_{14}$	8.7(c)-1	$(1,0)$	8.63
* 6′	$2\mathbf{D}_5 \oplus \mathbf{A}_9$	8.7(c)-1	$(0,1)$	8.63(2)
	The splitting $C_4 + C_2$ $(G/[G,G] = \mathbb{Z}_2 \oplus \mathbb{Z}_2)$			
* 7′	$\mathbf{D}_5 \oplus 2\mathbf{A}_5 \oplus \mathbf{A}_4$	8.7(a)-2	$(1,0)$	8.63(3)
8′	$\mathbf{D}_5 \oplus \mathbf{A}_9 \oplus \mathbf{A}_4 \oplus \mathbf{A}_1$	8.7(a)-3	$(0,1)$	8.63
* 9′	$\mathbf{D}_5 \oplus \mathbf{A}_9 \oplus 2\mathbf{A}_2 \oplus \mathbf{A}_1$	8.7(a)-5	$(1,0)$	8.63(4)
* 10′	$\mathbf{D}_5 \oplus \mathbf{A}_{11} \oplus \mathbf{A}_3$	8.7(b)-1	$(1,0)$	8.63(5)
* 11′	$\mathbf{D}_5 \oplus \mathbf{A}_{11} \oplus \mathbf{A}_2 \oplus \mathbf{A}_1$	8.7(b)-3	$(1,0)$	8.63(6)
12′	$\mathbf{D}_5 \oplus \mathbf{D}_{12} \oplus \mathbf{A}_2$	8.7(c)-2	$(1,0)$	8.63
* 13′	$\mathbf{D}_5 \oplus \mathbf{D}_7 \oplus \mathbf{A}_7$	8.7(c)-2	$(1,0)$	8.63(5)
* 14′	$2\mathbf{D}_5 \oplus \mathbf{A}_7 \oplus \mathbf{A}_2$	8.7(c)-2	$(1,0)$	8.64
	The splitting $C_5 + C_1$ $(G/[G,G] = \mathbb{Z}_8)$			
15′	$\mathbf{D}_5 \oplus \mathbf{A}_6 \oplus \mathbf{A}_5 \oplus \mathbf{A}_3$	8.7(a)-2	$(1,0)$	8.63
16′	$\mathbf{D}_5 \oplus \mathbf{A}_7 \oplus \mathbf{A}_6 \oplus \mathbf{A}_1$	8.7(a)-3	$(0,1)$	8.63
* 17′	$\mathbf{D}_5 \oplus \mathbf{A}_7 \oplus \mathbf{A}_4 \oplus \mathbf{A}_2 \oplus \mathbf{A}_1$	8.7(a)-5	$(1,0)$	8.63(8)
18′	$\mathbf{D}_5 \oplus \mathbf{A}_9 \oplus \mathbf{A}_5$	8.7(b)-1	$(1,0)$	8.63
19′	$\mathbf{D}_5 \oplus \mathbf{A}_9 \oplus \mathbf{A}_4 \oplus \mathbf{A}_1$	8.7(b)-3	$(1,0)$	8.63
20′	$\mathbf{D}_5 \oplus \mathbf{D}_{10} \oplus \mathbf{A}_4$	8.7(c)-2	$(1,0)$	8.63
21′	$\mathbf{D}_5 \oplus \mathbf{D}_9 \oplus \mathbf{A}_5$	8.7(c)-2	$(1,0)$	8.63
22′	$2\mathbf{D}_5 \oplus \mathbf{A}_5 \oplus \mathbf{A}_4$	8.7(c)-2	$(1,0)$	8.63

[1] Marked with a * are the curves with nonabelian fundamental groups

of (8.58). It follows that $[\rho, \alpha_1] = [(\alpha_1\alpha_2)^2, \alpha_1] = 1$, see (8.58) and (8.60). Then also $[(\alpha_1\alpha_2)^2, \alpha_2] = 1$ and, in view of (8.59), $(\alpha_1\alpha_2)^2$ is a central element. We will consider the quotient group $G := \pi_1/(\alpha_1\alpha_2)^2$, reducing relations (8.58)–(8.60) to

$$(\alpha_1\alpha_2)^2 = 1, \quad \rho^4 = \alpha_2\alpha_3. \tag{8.61}$$

(Without this reduction, the GAP coset enumeration algorithm fails for some curves.)

Argument 8.62 (irreducible curves with a bibigonal insertion). Using the function "size" in Listing 8.24, we show that, for all irreducible maximizing sextics with a bibigonal insertion, $G = \mathbb{Z}_2$. Hence, π_1 is a central extension of a cyclic group and thus is abelian, see Lemma A.31. For the set of singularities $\mathbf{D}_5 \oplus \mathbf{A}_{10} \oplus \mathbf{A}_4$, Line 4, we should also take into account the braid relations from the monogonal region reg_$(e \uparrow \mathbf{x}^2\mathbf{y}\mathbf{x}(\mathbf{y}\mathbf{x}^2)^3)$. See "misc/D5-ir.txt" for the GAP input.

```
# The computation of the fundamental groups of plane sextics with a type D5 singularity.
Reread("pi1sc.txt");    # (see Listing 5.3)
# Skeletons with a bibigonal insertion
lbb := [x^2,x^2*y*x^2,x^2*y*x*y*x^2,x^2*y*(x*y)^2*x^2,x^2*y*x*(y*x^2)^2];
bb := [(G.1*G.2)^2, G.2*G.3/rho^4];        # fixed relations (8.61)
group := function(arg) return DS(AA(lbb, arg), bb); end;
size := function(arg) return Size(CallFuncList(group, arg)); end;
# Genuine tetragon: use "inf(p);" and then "size2(s0,s1,s2,s3);" where p is a path from
    e = eII to eI and si is the braid index of reg_(e ↑ (yx)^i x), cf. "group*h" in Listing 8.16
lbg := List([0..3], i -> (y*x)^i*x); bg := [];
inf := function(p)    # initialize relations (8.58)–(8.60)
 bg := bm(p, [G.1*G.2*G.3/(G.3*G.1*G.2), G.1*G.2]);
 bg := [bg[1], (G.1*G.2)^2*bg[2]/rho^4, (G.1*G.2)^2*G.3/(G.3*(G.1*G.2)^2)];
end;
group2 := function(arg) return DS(AA(lbg, arg), bg); end;
size2 := function(arg) return Size(CallFuncList(group2, arg)); end;
```

Listing 8.24. Computing π_1 for sextics with a type $\mathbf{D}_5$ point ("D5.txt").

Argument 8.63 (reducible curves with a bibigonal insertion). The functions "size" and "group" in Listing 8.24 apply as well to all two component curves with a bibigonal insertion in the skeleton, see Figure 8.7 (a)–(c) and Table 8.11. This time, the central element $(\alpha_1\alpha_2)^2$ has infinite order in the abelianization and, hence, $[\pi_1, \pi_1] = [G, G]$, see Corollary A.30. Nonabelian are the following groups:

1. $\mathbf{D}_5 \oplus \mathbf{A}_7 \oplus \mathbf{A}_5 \oplus \mathbf{A}_2$, Line $1'$: $[\pi_1, \pi_1] = SL(2, \Bbbk_7)$;
2. $2\mathbf{D}_5 \oplus \mathbf{A}_9$, Line $6'$: $[\pi_1, \pi_1] = \mathbb{Z}_5$;
3. $\mathbf{D}_5 \oplus 2\mathbf{A}_5 \oplus \mathbf{A}_4$, Line $7'$: $[\pi_1, \pi_1]$ is one of the five perfect groups of order 7680;
4. $\mathbf{D}_5 \oplus \mathbf{A}_9 \oplus 2\mathbf{A}_2 \oplus \mathbf{A}_1$, Line $9'$: $[\pi_1, \pi_1] = \mathbb{Z}_3$;
5. $\mathbf{D}_5 \oplus \mathbf{A}_{11} \oplus \mathbf{A}_3$, Line $10'$, and $\mathbf{D}_5 \oplus \mathbf{D}_7 \oplus \mathbf{A}_7$, Line $13'$: $[\pi_1, \pi_1] = \mathbb{Z}_4$;
6. $\mathbf{D}_5 \oplus \mathbf{A}_{11} \oplus \mathbf{A}_2 \oplus \mathbf{A}_1$, Line $11'$: $[\pi_1, \pi_1] = \mathbb{Z}$ (the call `simplify(g.2/g.3)`, see Listing 5.3, returns a presentation with one generator and no relations);
7. $2\mathbf{D}_5 \oplus \mathbf{A}_7 \oplus \mathbf{A}_2$, Line $14'$: see Argument 8.64;
8. $\mathbf{D}_5 \oplus \mathbf{A}_7 \oplus \mathbf{A}_4 \oplus \mathbf{A}_2 \oplus \mathbf{A}_1$, Line $17'$: $[\pi_1, \pi_1] = SL(2, \Bbbk_{25})$.

All irreducible perturbations of these curves have abelian fundamental groups.

The same approach is used to compute the fundamental groups of the irreducible perturbations of the three component curves, see Figure 8.8. All groups are abelian. The GAP input is found in the file "misc/D5-r.txt".

Argument 8.64 ($2\mathbf{D}_5 \oplus \mathbf{A}_7 \oplus \mathbf{A}_2$, Line $14'$). One has $\pi_1'/\pi_1'' = \mathbb{Z} \oplus \mathbb{Z}_3$ and $\pi_1'' = Q_8$, where we temporarily use $'$ for the commutant of a group. The former statement is straightforward. For the latter, the function `simplify(g.2/g.3)`, see Listing 5.3, returns a presentation with two generators κ_3, κ_7 and four relations, of which the first,

$\kappa_3^{-2}\kappa_7^{-1}\kappa_3^3\kappa_7\kappa_3^{-1} = 1$, implies that κ_3^3 is a central element. This element has infinite order in π_1'/π_1'' and, in view of Corollary A.30, we have $\pi_1'' = (\pi_1'/\kappa_3^3)' = Q_8$.

Argument 8.65 (genuine tetragon). If R_{II} is a genuine tetragon, Figure 8.22, we use an approach similar to that of Argument 8.45. The common relations (8.58)–(8.60) depend on the position of the bigonal insertion R_{I}; they are initialized by the function "`inf`" in Listing 8.24. In addition, the group π_1 has relations

$$\{\alpha_2 \uparrow \sigma_1^i, \alpha_3\}_{s_i} = 1, \quad i = 0, \ldots, 3,$$

where s_i, $i = 0, \ldots, 3$, is the braid index of the region $R_i := \mathrm{reg}_+(e \uparrow (\mathbf{yx})^i)$; these numbers are passed as parameters to "`group2`" or "`size2`". If the skeleton $\mathfrak{S}'$ has at most one region strictly inside the complement of R_{II}, these relations define π_1, see Lemma 5.59. The following curves need special attention:

- Figures 8.22 (a) and (b): we need to add manually the braid relations given by one of the monogonal regions of $\mathfrak{S}'$; we choose the monogon (which may become a triangle in $\mathfrak{S}$) closest to R_{I};
- Figure 8.22 (d), with R_{I} at the center of the inner edge: some of the irreducible perturbations of this curve are of torus type. The corresponding dessins can be modified so that their skeletons are as in Figure 8.23; hence, we do not need to consider these perturbations explicitly;
- Figure 8.22 (e): if the fiber inside the inner bigonal region is perturbed so that the braid index becomes odd, the new dessin can be modified so that its skeleton is as in Figure 8.22 (b); hence, we do not need to consider these perturbations;
- insertion 4 in Figure 8.23: this curve is discussed in details in Argument 8.66.

For details, see "`misc/D5-ir.txt`" and "`misc/D5-r.txt`". All irreducible curves that are not of torus type have abelian fundamental groups.

Argument 8.66 ($\mathbf{D}_5 \oplus (\mathbf{A}_8 \oplus 3\mathbf{A}_2)$, Line 12). The curve is of torus type and its group is isomorphic to Γ. For proof, we choose the edges e_{I}, e_{II} as shown in Figure 8.23 and take e_{I} for the common reference edge, thus computing π_1 in the basis $\{\beta_1, \beta_2, \beta_3\}$. Then $e_{\mathrm{II}} = e_{\mathrm{I}} \uparrow (\mathbf{xy})^3\mathbf{x}^2\mathbf{y}$. The other regions used are those to the right from $e_{\mathrm{I}} \uparrow \mathbf{x}$, $e_{\mathrm{I}} \uparrow \mathbf{x}^2$, $e_{\mathrm{II}} \uparrow \mathbf{x}$, and $e_{\mathrm{II}} \uparrow \mathbf{x}^2$. Crucial is the fact that we use (8.59) to simplify the braid relations arising from the triangle $\mathrm{reg}_-(e_{\mathrm{I}} \uparrow \mathbf{x}^2)$; it becomes $\{\beta_2, \beta_3\}_3 = 1$. Then the call `simplify(g.2/g.3)`, see Listing 5.3, succeeds to return a presentation with two generators and no relations; hence $\pi_1 \cong \Gamma$, see Lemma 7.31.

Non-maximizing irreducible curves

Consider a proper model C as in Proposition 8.57. Using Lemma 4.52, we can assume that the fiber F_{I} is already isolated, and we need to isolate the other fiber F_{II}.

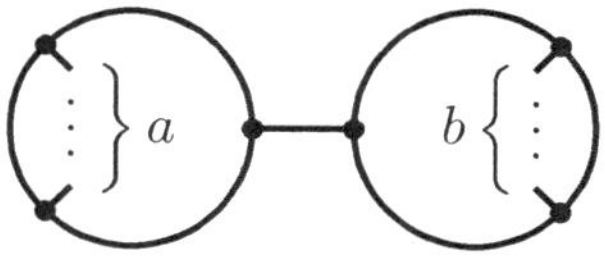

Figure 8.25. An n-gonal region, $4 < n = 4 + a + b \leqslant 8$.

If $R_{\rm I}$ and $R_{\rm II}$ are adjacent, then, contracting the bigon $R_{\rm I}$ to a single ∘-vertex u, see Convention 8.3, and applying Lemma 4.52 to the resulting auxiliary skeleton, we isolate the other fiber $F_{\rm II}$ and obtain a skeleton with a bibigonal insertion.

If $R_{\rm I}$ and $R_{\rm II}$ are not adjacent, the fiber $F_{\rm II}$ can still be isolated and, hence, C admits a mild degeneration to a maximal curve. For proof, we refer to the list found in [20]. To reduce the number of skeletons to be considered, we make a 'general' observation concerning the case $5 \leqslant \operatorname{wd} R_{\rm II} \leqslant 8$; then it remains to find a splitting sequence of width four in about a dozen of skeletons with a region $R_{\rm II}$ of width $\operatorname{wd} R_{\rm II} \geqslant 9$ and a bigonal region $R_{\rm I}$ not adjacent to $R_{\rm II}$.

Thus, assume that $5 \leqslant \operatorname{wd} R_{\rm II} \leqslant 8$. If all •-vertices in $\partial R_{\rm II}$ are pairwise distinct, then any sequence of consecutive edges in $\partial R_{\rm II}$ satisfies condition 1 in Lemma 4.49. Furthermore, since $\mathfrak{S}$ is a regular skeleton and the total number of •-vertices is at most eight, not all regions adjacent to $R_{\rm II}$ coincide. (Each •-vertex in $\partial R_{\rm II}$ is adjacent to another •-vertex, and these ten vertices cannot be pairwise distinct.) Hence, there is a sequence satisfying condition 2 as well.

If $R_{\rm II}$ is not a genuine polygon, at most two pairs of •-vertices in $\partial R_{\rm II}$ may coincide and $R_{\rm II}$ looks like the outer region in Figure 8.25. Unless $\operatorname{wd} R_{\rm II} = 5$, this region has a cutting sequence of width four: one can find such a sequence with the first and last edges in the two distinct circles in the figure. If $\operatorname{wd} R_{\rm II} = 5$, there is a sequence of edges satisfying condition 2, but not 1 in Lemma 4.49, and the region can be cut into a larger one, thus reducing this case to one of those already considered.

Collision of two $\tilde{\mathbf{D}}$ type fibers The skeleton $\mathfrak{S}$ is as shown in Figure 8.8 (c), with a bibigonal insertion as in Figure 8.21 (b), and the two $\tilde{\mathbf{D}}$ type fibers are either $2\tilde{\mathbf{D}}_5$ in the bigonal region of $\mathfrak{S}$ or $\tilde{\mathbf{D}}_7 + \tilde{\mathbf{D}}_5$ in its tetragonal region, see Observation 8.5. (A curve with two $\tilde{\mathbf{D}}_6$ type fibers in the tetragonal region cannot be irreducible.) Furthermore, for the curve to be irreducible, the other region must have braid index one. In the former case $2\tilde{\mathbf{D}}_5$, the call `size(1)` returns 1; in the latter case $\tilde{\mathbf{D}}_7 + \tilde{\mathbf{D}}_5$, adding to `group(0,1)` the braid relation from the $\tilde{\mathbf{D}}_5$ fiber, we also obtain an abelian group, see "`misc/D5-r.txt`".

8.3.3 A point of type $\mathbf{D}_4$

Finally, consider simple plane sextics $D \subset \mathbb{P}^2$ satisfying the following conditions:

(∗) D has a distinguished singular point P of type $\mathbf{D}_4$,
D has no linear components through P, and

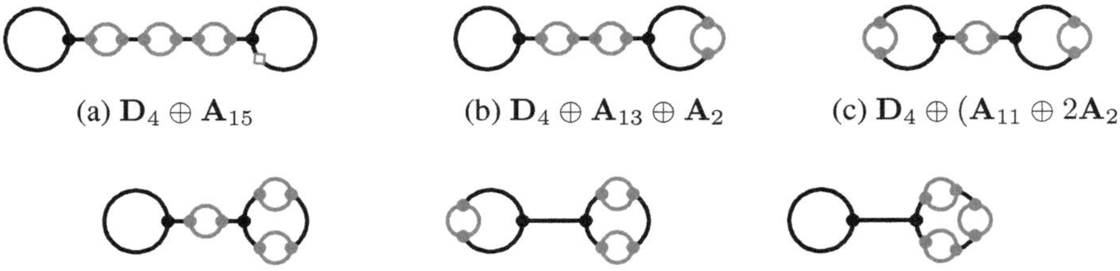

(a) $\mathbf{D}_4 \oplus \mathbf{A}_{15}$ (b) $\mathbf{D}_4 \oplus \mathbf{A}_{13} \oplus \mathbf{A}_2$ (c) $\mathbf{D}_4 \oplus (\mathbf{A}_{11} \oplus 2\mathbf{A}_2)$

(d) $\mathbf{D}_4 \oplus \mathbf{A}_{11} \oplus \mathbf{A}_4$ (e) $\mathbf{D}_4 \oplus \mathbf{A}_9 \oplus \mathbf{A}_4 \oplus \mathbf{A}_2$ (f) $\mathbf{D}_4 \oplus \mathbf{A}_9 \oplus \mathbf{A}_6$

Figure 8.26. Type $\mathbf{D}_4$ singularity: Two component curves.

(a) $\mathbf{D}_4 \oplus 2\mathbf{A}_7 \oplus \mathbf{A}_1$ (b) $\mathbf{D}_4 \oplus \mathbf{A}_7 \oplus \mathbf{A}_5 \oplus \mathbf{A}_3$ (c) $\mathbf{D}_4 \oplus (3\mathbf{A}_5)$ (d) $\mathbf{D}_4 \oplus \mathbf{D}_{10} \oplus \mathbf{A}_5$

Figure 8.27. Type $\mathbf{D}_4$ singularity: Three component curves.

D has no singular points of type $\mathbf{E}$.

Proposition 8.67. *There is a natural bijection ϕ, invariant under equisingular deformations, between Zariski open and dense, in each equisingular stratum, subsets of the sets of*:

- *simple plane sextics $D \subset \mathbb{P}^2$ with a distinguished type $\mathbf{D}_4$ point P and without linear components through P, and*
- *proper simple trigonal curves $C \subset \Sigma_4$ with an unordered triple $(F_{\mathrm{I}}, F_{\mathrm{II}}, F_{\mathrm{III}})$ of distinguished fibers of type $\tilde{\mathbf{A}}_1$.*

A sextic D is maximizing if and only if $C := \phi(D)$ is maximal and stable. ◁

In view of Corollary 7.38, all maximizing sextics with a type $\mathbf{D}_4$ singular point are reducible; their skeletons and sets of singularities are listed in Figures 8.26 and 8.27, where the distinguished bigonal regions are shown in grey. (Unlike most other figures in this chapter, all three bigonal regions are present in each skeleton.)

The fundamental groups

For each $i = \mathrm{I}, \mathrm{II}, \mathrm{III}$, pick a reference edge e_i in the boundary $\partial_- R_i$ and a canonical basis $\{\beta_1^i, \beta_2^i, \beta_3^i\}$ over e_i. Each distinguished fiber is as in Case 2 with $m = 2n = 2$, see Section 5.2.2, and we have $\varkappa_i = \beta_1^i \beta_2^i$. Furthermore, since all three products $\beta_1^i \beta_2^i \beta_3^i$ are equal to ρ, the braid relations can be written in the form

$$[\rho, \beta_3^{\mathrm{I}}] = [\rho, \beta_3^{\mathrm{II}}] = [\rho, \beta_3^{\mathrm{III}}] = 1 \tag{8.68}$$

```
# The computation of the fundamental groups of plane sextics with a type D4 singularity.
# Parameters: n1 and n2 are the braid indices of reg e and reg(e ↑ x), respectively, and li is the
    exponent l in the chain y(xy)^l from e to ei in the boundary of the outer region, i = 1, 2, 3.
```

```
Reread("pi1sc.txt");    # (see Listing 5.3)
group := function(n1, n2, l1, l2, l3)
 b := List([l1, l2, l3], i -> bm(y*(x*y)^i, [G.3])[1]);
 b := Union(List(b, i -> rho*i/(i*rho)), [rho*b[3]*b[2]*b[1]]);
 return DS(AA([[], x], [n1, n2]), b);
end;
size := function(arg) return Size(CallFuncList(group, arg)); end;
```

Listing 8.28. Computing π_1 for sextics with a type $\mathbf{D}_4$ point ("D4.txt").

and the relation at infinity (5.55) simplifies to

$$\rho\beta_3^{\mathrm{III}}\beta_3^{\mathrm{II}}\beta_3^{\mathrm{I}} = 1. \tag{8.69}$$

The order of the factors in this relation depends on the choice of chains connecting edges e_i to the common reference edge e, see Section 5.1.3. We use these relations to compute the groups of all irreducible perturbations of the maximal curves.

Two component curves, Figure 8.26 The common reference edge e is chosen in the boundary of the outer region R in the figures, see Figure 8.26 (a), and the edges e_i and chains ζ_i from e to e_i, $i = \mathrm{I}, \mathrm{II}, \mathrm{III}$ are also in ∂R, so that each ζ_i has the form $\mathbf{y}(\mathbf{xy})^l$. We use relations (8.68), (8.69) and the braid relations from $R = \mathrm{reg}_- e$ and the region $\mathrm{reg}_+ e$, see "group" and "size" in Listing 8.28. In all but one cases, "size" returns 1 and the group is abelian, see "misc/D4-r.txt".

The exception is Figure 8.26 (c) with the braid index of the outer region R set to 3. This curve is of torus type, and the call `simplify(g.1/g.2)` returns a presentation with two generators and no relations. Hence, $\pi_1 \cong \Gamma$, see Lemma 7.31.

Three component curves, Figure 8.27 The same approach as above applies to the curve shown in Figure 8.27 (d), with the edge e as in the figure. All but one maximal irreducible perturbations have abelian fundamental groups.

Argument 8.70 ($\mathbf{D}_4 \oplus \mathbf{D}_7 \oplus 3\mathbf{A}_2$). The perturbation $\mathbf{D}_4 \oplus \mathbf{D}_7 \oplus 3\mathbf{A}_2$ of the set of singularities $\mathbf{D}_4 \oplus \mathbf{D}_{10} \oplus \mathbf{A}_5$, Figure 8.27 (d), has nonabelian fundamental group. Taking for P the point of type $\mathbf{D}_7$, we see that this curve is covered by Argument 8.56. Since the computation only uses the braid index of the $\tilde{\mathbf{D}}$ type region, the further perturbation $2\mathbf{D}_4 \oplus 4\mathbf{A}_2$ has the same group.

If $\mathfrak{S}$ is the skeleton in Figure 8.27 (c) and the perturbed sextic is irreducible and of torus type, then at least one region of $\mathfrak{S}$ contains two $\times$-vertices of index three and, using Lemma 4.51, we can modify $\mathfrak{S}$ so that two distinguished fibers share the same region. Hence, due to Lemma 8.71 below and the results of Section 8.3.2, we have $\pi_1 \cong \Gamma$. (The same argument could be applied to Figure 8.26 (c).)

In all other cases, $\mathfrak{S}$ is as in Figure 8.27 (a)–(c) with at least one region of braid index one. Using Lemma 4.51, we can modify $\mathfrak{S}$ along a simple path in this region to obtain one of the skeletons shown in Figure 8.26.

Non-maximal irreducible curves

Consider an irreducible proper model C as in Proposition 8.67. Using Lemma 4.52, we can assume that the fiber F_{I} is isolated. Below we show that either

- the two other distinguished fibers can also be isolated, and hence D degenerates to a maximizing sextic satisfying conditions 8.3.3($*$), or
- D degenerates to a sextic with a distinguished point of type $\mathbf{D}_5$, or
- the group π_1 is abelian.

Due to the results of Section 8.3.2 and the computation for the maximal curves, in each case the group π_1 is as expected, *i.e.*, Γ or $\mathbb{Z}_6$. Two exceptions (perturbations of the sextics described in Argument 8.56) are discussed in Argument 8.70.

Lemma 8.71. *If two distinguished fibers share a region of $\mathfrak{S}$, the sextic D admits a degeneration to a sextic with a distinguished singular point of type $\mathbf{D}_5$; the latter is irreducible if and only if so is D.*

Proof. Degenerate the proper model C by merging the two type $\tilde{\mathbf{A}}_1$ fibers to a single fiber of type $\tilde{\mathbf{A}}_3$. This deformation is followed by a deformation of trigonal models $\tilde{C}$ (two simple Nagata transformations degenerate to a two-fold one) and, hence, by a deformation of sextics. □

Lemma 8.72. *Assume that the skeleton $\mathfrak{S}$ has an edge shared by two distinguished regions, not necessarily distinct. Then $\rho \in \pi_1$ is a central element.*

Proof. Let R', R'' be the two regions. The reference edges e', e'' in their boundaries can be chosen so that $e'' = e' \uparrow \mathbf{x}$. Then $\beta_3'' = \rho^{-1}\beta_1'\rho$ and $[\rho, \beta_3'] = [\rho, \beta_1'] = 1$, see (8.60). Since ρ, β_1', and β_3' generate the group, the statement follows. □

Corollary 8.73. *Under the hypotheses of Lemma* 8.72*, the braid relations arising from a singular fiber of type I_p^* are equivalent to the braid relations arising from a fiber of type I_p.* ◁

Lemma 8.74. *Under the hypotheses of Lemma* 8.72*, assume that the skeleton $\mathfrak{S}$ has a distinguished region of width* 3 *or* 4 *or a non-distinguished region of width* 1 *or* 2*. Then the group π_1 is abelian.*

Proof. Choose a common reference edge e in the boundary $\partial_- R$ of a region R as in the statement. The braid index of R is 1 or 2, and we have relations $\alpha_1 = \alpha_2$ or $[\alpha_1, \alpha_2] = 1$. (Due to Corollary 8.73, we do not need to distinguish between $\tilde{\mathbf{D}}$ and $\tilde{\mathbf{A}}$ type regions.) Since α_1, α_2, and ρ generate the group, the statement follows. □

(a) $(6, 4, 4, 2, 2)$ (b) $(5, 5, 3, 3, 2)$

Figure 8.29. Two special skeletons.

Skeleton without adjacent distinguished regions From now on, we assume that all distinguished regions $R_{\rm I}$, $R_{\rm II}$, $R_{\rm III}$ are pairwise distinct, see Lemma 8.71. Denote by $2 \leqslant m_{\rm I} \leqslant m_{\rm II} \leqslant m_{\rm III}$ their widths. In the regular $(3, 2)$-skeleton $\mathfrak{S}$, two regions share an edge if and only if they share a vertex. Hence, the inequality

$$m_{\rm I} + m_{\rm II} + m_{\rm III} > \#_{\bullet}(\mathfrak{S}) \tag{8.75}$$

is sufficient for $\mathfrak{S}$ to satisfy the hypotheses of Lemma 8.72.

If $\#_{\bullet}(\mathfrak{S}) \leqslant 6$, the only triple $\bar{m} := (m_{\rm I}, m_{\rm II}, m_{\rm III})$ violating (8.75) is $(2, 2, 2)$ with $\#_{\bullet}(\mathfrak{S}) = 6$; in this case, all distinguished fibers are isolated and $\mathfrak{S}$ is the skeleton shown in Figure 8.27 (d). If $\#_{\bullet}(\mathfrak{S}) = 8$, we have $\bar{m} = (2, 2, 2)$ (all distinguished fibers are isolated) or $\bar{m} = (2, 2, 3)$, $(2, 3, 3)$, or $(2, 2, 4)$. In the last case, $\mathfrak{S}$ is obtained by inserting two bigonal regions, not in the boundary of a distinguished tetragon $R_{\rm III}$, to the skeleton shown in Figure 4.2 (f), and $R_{\rm III}$ has a cutting sequence of width two. If $\bar{m} = (2, 2, 3)$ or $(2, 3, 3)$, then $\mathfrak{S}$ can be obtained by inserting two or one bigonal regions and 'inflating' to triangles, respectively, one or two original •-vertices from one of the skeletons shown in Figure 4.3 (a), (b); each new triangle obtained by the 'inflation' has a cutting sequence of width two.

Skeletons satisfying the hypotheses of Lemma 8.72 Let $n_1, n_2, \ldots$ be the widths of the non-distinguished regions of $\mathfrak{S}$. In view of Lemma 8.74, we can assume that $m_{\rm I} = 2$, $m_i = 2$ or $m_i \geqslant 5$ for $i = {\rm II}, {\rm III}$, and $n_j \geqslant 3$ for all j.

Consider the case $m_{\rm II} = 2$. Then $\mathfrak{S}$ is obtained by inserting two bigonal regions into a regular skeleton $\mathfrak{S}'$ of degree 6 or 3, see Figures 4.2 and 4.3 (a), (b), with the extra condition (imposed by the inequalities $n_j \geqslant 3$) that there must be an insertion in the boundary of each mono- or bigonal region of $\mathfrak{S}'$. The latter condition eliminates Figures 4.2 (b) and (e) and determines the position of the insertions almost uniquely in all other cases. With one exception, see Figure 8.29 (a), any region $R_{\rm III}$ of width $\operatorname{wd} R_{\rm III} \geqslant 5$ (after the insertion) has a cutting sequence of width two.

In the exceptional case, if the curve C is to be irreducible, one of the tetragons in the figure must have braid index one, and Lemma 8.74 implies that $\pi_1 \cong \mathbb{Z}_6$.

If $m_{\rm III} \geqslant m_{\rm II} \geqslant 5$, we show that the fiber $F_{\rm II}$ can be isolated and reduce this case to the one just considered. We have $2 \deg \mathfrak{S} = m_{\rm I} + m_{\rm II} + m_{\rm III} + \sum_j n_j > 12$.

If $\deg \mathfrak{S} = 9$, then $m_{\rm II} = m_{\rm III} = 5$ and $n_1 = n_2 = 3$. The only skeleton with these properties is shown in Figure 8.29 (b), and each of the two pentagonal regions (which are, in fact, related by a symmetry) has a cutting sequence of width two such that the corresponding modification preserves the insertions.

If $\deg \mathfrak{S} = 12$, we address the tables found in [20]. We are interested in the skeletons with no monogons, exactly one bigon, and at least two regions of width at least five. There are three skeletons with these properties, $(\underline{7}, \underline{5}, 4, 3, 3, \underline{2})$, $(\underline{6}, \underline{6}, 4, 3, 3, \underline{2})$, and $(\underline{6}, \underline{5}, 4, 4, 3, \underline{2})$. The distinguished regions (underlined) are uniquely determined by the assumption $m_{\mathrm{III}} \geqslant m_{\mathrm{II}} \geqslant 5$; they do have cutting sequences of width two.

Chapter 9

The transcendental lattice

Consider a compact Jacobian elliptic surface J with a distinguished section S and let $\mathbf{S}'_J \subset H_2(J)$ be the sublattice spanned by the classes of S and the components of the singular fibers of J. (We assume that J has at least one such fiber.) If J is generic in its equisingular deformation class, then the primitive hull $\tilde{\mathbf{S}}'_J := (\mathbf{S}'_J \otimes \mathbb{Q}) \cap H_2(J)$ coincides with the Néron–Severi lattice $NS(J)$. Define the *transcendental lattice* $\mathbf{T}_J$ of J as the orthogonal complement $(\mathbf{S}'_J)^\perp$. The computation of this lattice in terms of the homological invariant of J is the principal goal of this chapter.

9.1 Extremal elliptic surfaces without exceptional fibers

Throughout this chapter, we consider a compact relatively minimal Jacobian elliptic fibration $p\colon J \to B$ with a distinguished section S and at least one singular fiber. We fix the notation $\mathbf{S}' := \mathbf{S}'_J$, $\tilde{\mathbf{S}}' := \tilde{\mathbf{S}}'_J$, and $\mathbf{T} := \mathbf{T}_J$ for the three lattices introduced above. (To be consistent with the previous chapters, we reserve $\mathbf{S} := \mathbf{S}_J$ for the set of singularities of the ramification curve of J, *i.q.* the sublattice spanned by the components of the singular fibers that are disjoint from S.) Let, further, $\tilde{F}_*$ be the union of all singular fibers $\tilde{F}_i$ of J, and let $J^\sharp := J \smallsetminus (S \cup \tilde{F}_*)$. The groups $H_2(J)$ and $H_2(J^\sharp)$ are regarded as lattices *via* the intersection index pairing.

9.1.1 The tripod calculus

Assume that J is an extremal elliptic surface without exceptional singular fibers and consider the skeleton $\mathfrak{S}$ of the ramification curve of J. Under the assumptions, $\mathfrak{S}$ is a regular skeleton embedded into the set $B^\sharp$ of regular values of p.

The homological invariant $\mathcal{M}_J$ of J restricts to a local system over $\mathfrak{S}$ and, due to Proposition 5.69 and Theorem 5.62, the Γ-valued reduction of its monodromy $\mathfrak{h} := \mathfrak{h}_J$ is given by $\gamma \mapsto \operatorname{val}\gamma \in \Gamma$. To describe the $\tilde{\Gamma}$-valued monodromy, consider a fiber $\tilde{F}_e$ (or punctured fiber $\tilde{F}_e^\circ := \tilde{F}_e \smallsetminus S$) over each edge e of $\mathfrak{S}$ and choose a canonical identification $H_1(\tilde{F}_e) = \mathcal{H}$ as explained prior to Proposition 5.69; this identification is defined up to sign. With respect to these identifications, the monodromy along a chain $(e, \mathbf{x})$ is $\pm\mathbb{X}$, and we assume that, *at each trivalent $\bullet$-vertex of $\mathfrak{S}$, the identifications are chosen in a coherent way, so that always* $\mathfrak{h}(e, \mathbf{x}) = -\mathbb{X}$. Such a coherent choice of identifications is called a *reference set*. At each $\circ$-vertex, we have $\mathfrak{h}(e, \mathbf{y}) = \pm\mathbb{Y}$, and then necessarily $\mathfrak{h}(e \uparrow \mathbf{y}, \mathbf{y}) = \mp\mathbb{Y} = (\pm\mathbb{Y})^{-1}$. Hence, the pair of edges incident to a $\circ$-vertex can be ordered, (e^-, e^+), so that $\mathfrak{h}(e^\epsilon, \mathbf{y}) = \epsilon\mathbb{Y}$, $\epsilon = \pm$.

Alternatively, since all $\circ$-vertices are bivalent, we can regard $\mathfrak{S}$ as the bipartite subdivision of an ordinary ribbon graph $\mathfrak{S}'$, *cf.* Convention 1.7, and this auxiliary graph $\mathfrak{S}'$ is directed: we direct each edge of $\mathfrak{S}'$ from e^- to e^+. For this reason, we refer to the ordering (e^-, e^+) of the edges of the original graph $\mathfrak{S}$ as an *orientation* of $\mathfrak{S}$; this should not be confused with the canonical orientation of a bipartite graph. An orientation constructed as above, using a reference set $H_1(\tilde{F}_e) = \mathcal{H}$, $e \in \operatorname{Edg} \mathfrak{S}$, is said to be *compatible* with the homological invariant $\mathfrak{h}$.

Conversely, given an oriented regular skeleton $\mathfrak{S}$, we can recover the homological invariant $\mathfrak{h}\colon \pi_1(\mathfrak{S}) \to \tilde{\Gamma}$ and, more generally, the monodromy $\mathfrak{h}(\gamma)\colon \mathcal{H} \to \mathcal{H}$ along a chain γ in $\mathfrak{S}$: for a one letter chain, let $\mathfrak{h}(e, \mathbf{x}) = -\mathbb{X}$ and $\mathfrak{h}(e^\epsilon, \mathbf{y}) = \epsilon\mathbb{Y}$, $\epsilon = \pm$, and extend this definition to all chains by the multiplicativity. Note that $\mathfrak{h}(e^\epsilon, \mathbf{y}^{-1}) = \epsilon\mathbb{Y}$. Note also that $\mathfrak{h}(\gamma) = \pm \operatorname{val} \gamma$, where the sign depends on the orientation of $\mathfrak{S}$.

The orientation of $\mathfrak{S}$ depends on the choice of a reference set $H_1(\tilde{F}_e) = \mathcal{H}$, which can be composed with $-\operatorname{id}$ simultaneously over all edges incident to any fixed set of $\bullet$-vertices of $\mathfrak{S}$. Hence, in terms of the auxiliary graph $\mathfrak{S}'$, the orientation is defined up to a *flip* at a subset $V \subset \operatorname{Vtx} \mathfrak{S}'$, which consists in reversing the direction of each edge that has *exactly one* end incident to a vertex in V.

Theorem 9.1. *There is a canonical bijection between the set of isomorphism classes of extremal elliptic surfaces without exceptional fibers and the set of isomorphism and flip equivalence classes of directed finite* 3*-regular ribbon graphs.*

Proof. The map sending an elliptic surface to a directed ribbon graph is constructed above. For the converse, observe that the j-invariant j_J of an extremal elliptic surface J must have extremal branching behavior, see Theorem 3.67, and, on the other hand, the bipartite subdivision $\mathfrak{S}$ of a ribbon graph $\mathfrak{S}'$ as in the statement gives rise to a unique function $j\colon B \cong \operatorname{Supp} \mathfrak{S} \to \mathbb{P}^1_{\mathbb{R}}$ with this property, see Proposition 4.38. In the absence of exceptional singular fibers, $\mathfrak{S}$ is a strict deformation retract of $B^\sharp = j^{-1}(\mathbb{P}^1 \smallsetminus \infty)$ and the homological invariant $\mathfrak{h}_J\colon \pi_1(B^\sharp) \to \tilde{\Gamma}$ is determined by its restriction to $\mathfrak{S}$, which is given by the orientation of $\mathfrak{S}'$. In view of Theorem 3.52, the pair $(j_J, \mathfrak{h}_J)$ determines a unique Jacobian elliptic surface J. □

Consider a skeleton $\mathfrak{S}$ and let $\mathcal{H} \otimes \mathfrak{S} := \bigoplus_e \mathcal{H} \otimes e$, where $\mathcal{H} \otimes e$ is a copy of $\mathcal{H}$, $e \in \operatorname{Edg} \mathfrak{S}$, and the summation runs over all edges of $\mathfrak{S}$. The projection $\mathcal{H} \otimes \mathfrak{S} \to \mathcal{H} \otimes e$ is denoted by $h \mapsto h_e$. For $u \in \mathcal{H}$ and $e \in \operatorname{Edg} \mathfrak{S}$, let $u \otimes e$ be the element of $\mathcal{H} \otimes \mathfrak{S}$ whose only nontrivial projection is $(u \otimes e)_e = u$. Convert $\mathcal{H} \otimes \mathfrak{S}$ to a rational lattice by letting

$$h^2 := \frac{1}{3} \sum_{e \in \operatorname{Edg} \mathfrak{S}} h_e \cdot (h_{e \uparrow \mathbf{x}} \uparrow \mathbb{X}^{-1}), \quad h \in \mathcal{H} \otimes \mathfrak{S}, \tag{9.2}$$

where $\cdot$ stands for the symplectic product on $\mathcal{H}$. (Recall that a $\mathbb{Q}$-valued symmetric bilinear form φ is determined by its squares, as $2\varphi(x, y) = (x + y)^2 - x^2 - y^2$.)

Similarly, let $\mathcal{H}^* := \operatorname{Hom}(\mathcal{H}, \mathbb{Z})$ with the adjoint left action of $\tilde{\Gamma}$, and define the dual group $\mathcal{H}^* \otimes \mathfrak{S} := \bigoplus_e \mathcal{H}^* \otimes e$. It is indeed dual to $\mathcal{H} \otimes \mathfrak{S}$ *via* the canonical nonsingular pairing $h \otimes h^* \mapsto \sum_e \langle h_e, h_e^* \rangle$, $h \in \mathcal{H} \otimes \mathfrak{S}$, $h^* \in \mathcal{H}^* \otimes \mathfrak{S}$.

Definition 9.3. Given an ordered regular skeleton $\mathfrak{S}$, define $\mathcal{H}_{\mathfrak{S}}$ as the sublattice of $\mathcal{H} \otimes \mathfrak{S}$ consisting of the elements h satisfying the conditions

1. $h_e - (h_{e\uparrow \mathbf{x}} \uparrow \mathbb{X}^{-1}) + (h_{e\uparrow \mathbf{x}^2} \uparrow \mathbb{X}^{-2}) = 0$ for each edge $e \in \operatorname{Edg} \mathfrak{S}$;
2. $h_{e^+} - (h_{e^-} \uparrow \mathbb{Y}) = 0$ for each ordered $\circ$-vertex $(e^-, e^+) \in \operatorname{Vtx}_\circ \mathfrak{S}$.

Similarly, let $\mathcal{H}^*_{\mathfrak{S}}$ be the quotient of $\mathcal{H}^* \otimes \mathfrak{S}$ by all relations of the form

3. $u \otimes e - (\mathbb{X}^{-1} \downarrow u) \otimes (e \uparrow \mathbf{x}) + (\mathbb{X}^{-2} \downarrow u) \otimes (e \uparrow \mathbf{x}^2)$ for all $u \in \mathcal{H}^*$, $e \in \operatorname{Edg} \mathfrak{S}$;
4. $u \otimes e^+ - (\mathbb{Y} \downarrow u) \otimes e^-$ for all $u \in \mathcal{H}^*$ and $(e^-, e^+) \in \operatorname{Vtx}_\circ \mathfrak{S}$.

Remark 9.4. In 9.3 (1), the relations arising from the three edges incident to a single $\bullet$-vertex are equivalent to each other. Hence, it suffices to pick one of the edges and consider one relation at each such vertex.

It is immediate that $\mathcal{H}_{\mathfrak{S}}$ annihilates the subgroup defined by (3) and (4). Hence, the duality above restricts to a well defined pairing $\mathcal{H}_{\mathfrak{S}} \otimes \mathcal{H}^*_{\mathfrak{S}} \to \mathbb{Z}$. Note that, in general, $\operatorname{Hom}(\mathcal{H}_{\mathfrak{S}}, \mathbb{Z}) \neq \mathcal{H}^*_{\mathfrak{S}}$, as the latter group may have torsion.

The next two theorems are proved in Section 9.1.2.

Theorem 9.5. *Let J be an extremal elliptic surface without exceptional singular fibers, and let $\mathfrak{S}$ be the skeleton of J, equipped with an orientation compatible with the homological invariant $\mathfrak{h}_J$. Then there are canonical (up to sign) isomorphisms $\mathcal{H}_{\mathfrak{S}} = H_2(J^\sharp)$ and $\mathcal{H}^*_{\mathfrak{S}} = H^2(J^\sharp)$; the former is an isomorphism of lattices, and the latter takes $\mathcal{H}_{\mathfrak{S}} \otimes \mathcal{H}^*_{\mathfrak{S}} \to \mathbb{Z}$ to the Kronecker pairing $H_2(J^\sharp) \otimes H^2(J^\sharp) \to \mathbb{Z}$.*

Theorem 9.6. *Under the hypotheses of Theorem 9.5, there are canonical (up to sign) isomorphisms $\mathbf{T}_J = \mathcal{H}_{\mathfrak{S}} / \ker$ and $MW(J) = \operatorname{Tors} \mathcal{H}^*_{\mathfrak{S}}$.*

Corollary 9.7 (of Theorems 9.1 and 9.6). *For any oriented regular skeleton $\mathfrak{S}$, the group $\mathcal{H}_{\mathfrak{S}}$ is an even integral positive semidefinite lattice.* $\triangleleft$

9.1.2 Proofs and further observations

We start with a few topological observations, that hold for any compact Jacobian elliptic surface J, not necessarily extremal or without exceptional singular fibers.

Lemma 9.8. *One has* $(\operatorname{ord} \operatorname{Tors} MW(J))^2 = |\det \mathbf{S}_J| / |\det \mathbf{T}_J|$.

Proof. According to Theorem 3.60, $\mathcal{K} := \operatorname{Tors} MW(J)$ is the kernel of the finite index extension $\tilde{\mathbf{S}}'_J \supset \mathbf{S}'_J$; hence, one has

$$(\operatorname{ord} \mathcal{K})^2 = \operatorname{ord}(\operatorname{discr} \tilde{\mathbf{S}}'_J)/\operatorname{ord}(\operatorname{discr} \mathbf{S}'_J) = |\det \tilde{\mathbf{S}}'_J|/|\det \mathbf{S}'_J|,$$

see Theorem A.4. On the other hand, due to Theorem A.5, the discriminant forms $\operatorname{discr} \tilde{\mathbf{S}}'_J$ and $\operatorname{discr} \mathbf{T}_J$ are anti-isometric and $|\det \tilde{\mathbf{S}}'_J| = |\det \mathbf{T}_J|$. It remains to observe that $\mathbf{S}'_J = \mathbf{S}_J \oplus (\mathbb{Z}[S] + \mathbb{Z}[\tilde{F}])$, where $\tilde{F}$ is a fiber, and the parenthesized summand is unimodular; hence, $|\det \tilde{\mathbf{S}}'_J| = |\det \mathbf{S}_J|$. □

A more precise description of the relation between $\operatorname{discr} \mathbf{S}_J$ and $\operatorname{discr} \mathbf{T}_J$ (in a slightly more general setting) is given by Theorem 10.47 below.

Lemma 9.9. *There is a canonical isomorphism of lattices* $\mathbf{T}_J = H_2(J^\sharp)/\ker$.

Proof. By Poincaré–Lefschetz duality, $H_2(J^\sharp) = H^2(J, S \cup \tilde{F}_*)$, and from the exact sequence of pair we have an exact sequence

$$H^1(J) \xrightarrow{j^1} H^1(S \cup \tilde{F}_*) \xrightarrow{\delta} H^2(J, S \cup \tilde{F}_*) \xrightarrow{i^*} H^2(J) \xrightarrow{j^2} H^2(S \cup \tilde{F}_*), \quad (9.10)$$

where i^* is Poincaré dual to the inclusion homomorphism $i_*\colon H_2(J^\sharp) \to H_2(J)$ and j^* are induced by the inclusion $j\colon S \cup \tilde{F}_* \hookrightarrow J$. Since the 1- and 2-dimensional homology groups of both $S \cup \tilde{F}_*$ and J are all torsion free, j^2 is the adjoint of the homomorphism $j_*\colon H_2(S \cup \tilde{F}_*) \to H_2(J)$; hence, $\operatorname{Ker} j^2$ can be identified with $\mathbf{T}_J$ and $\mathbf{T}_J = H_2(J^\sharp)/\operatorname{Ker} i_*$. Since i_* is a lattice morphism, $\operatorname{Ker} i_* \subset \ker H_2(J^\sharp)$. On the other hand, $\mathbf{S}' = \mathbf{S} \oplus (\mathbb{Z}[E] \oplus \mathbb{Z}[\tilde{F}_\infty])$ is a nondegenerate lattice; hence, so is $\mathbf{T}_J$ and we also have the opposite inclusion $\operatorname{Ker} i_* \supset \ker H_2(J^\sharp)$. □

Consider disjoint closed tubular neighborhoods U_i of the singular fibers $\tilde{F}_i$ of J, denote $U_i^\circ := U_i \smallsetminus S$, and let $U_* := \bigcup_i U_i$ and $U_*^\circ := \bigcup_i U_i^\circ$. Each boundary ∂U_i° is fibered over a circle with the fiber punctured torus $\tilde{F}^\circ$, and from the Wang exact sequence we have

$$H_2(\partial U_i) = \operatorname{Ker}[(\mathfrak{m}_i - 1)\colon H_1(\tilde{F}^\circ) \to H_1(\tilde{F}^\circ)], \quad (9.11)$$

where $\mathfrak{m}_i$ is the local monodromy about $\tilde{F}_i$, see Table 3.2. It follows that $H_2(\partial U_*^\circ) = \bigoplus_i H_2(\partial U_i^\circ)$ is a free abelian group and its rank equals the number of singular fibers of Kodaira type I_p, $p > 0$.

Lemma 9.12. *If J has at least one singular fiber, the inclusion $\partial U_*^\circ \hookrightarrow J^\sharp$ induces an isomorphism* $H_2(\partial U_*^\circ) = \ker H_2(J^\sharp)$.

Proof. Under the assumptions, the inclusion homomorphism $H^1(J) \to H^1(S)$ is an isomorphism, see Theorem 3.57, and, in view of (9.10) and Lemma 9.9, the boundary homomorphism $\delta\colon H^1(S \cup \tilde{F}_*, S) \to H^2(J, S \cup \tilde{F}_*)$ is a monomorphism and its

Figure 9.1. Shifting a skeleton.

image equals $\ker H_2(J^\sharp)$. Now, δ is Poincaré dual to $\partial\colon H_3(U_*^\circ, \partial U_*^\circ) \to H_2(J^\sharp)$, which factors through the isomorphism $\partial\colon H_3(U_*^\circ, \partial U_*^\circ) \to H_2(\partial U_*^\circ)$ given by the exact sequence of pair (as $H^i(U_*^\circ) = H^i(\tilde{F}_*^\circ) = 0$ for $i \geqslant 2$). □

Lemma 9.13. *There is a canonical isomorphism* $\operatorname{Tors} \mathit{MW}(J) = \operatorname{Tors} H^2(J^\sharp)$.

Proof. By Poincaré–Lefschetz duality, $H^2(J^\sharp) = H_2(J, S \cup \tilde{F}_*)$, and from the exact sequence of pair we have an exact sequence

$$0 \to H_2(J)/\mathbf{S}' \to H^2(J^\sharp) \to H_1(S \cup \tilde{F}_*).$$

Since the group $H_1(S \cup \tilde{F}_*)$ is torsion free and $\mathit{NS}(J)$ is a torsion free primitive subgroup of $H_2(J)$, the statement of the lemma follows from Theorem 3.60. □

Corollary 9.14 (of Lemmas 9.9 and 9.13). *Both* $\mathbf{T}_J$ *and* $\operatorname{Tors} \mathit{MW}(J)$ *are invariant under bi-oriented homeomorphisms of Jacobian elliptic surfaces.* ◁

Proof of Theorems 9.5 *and* 9.6. Theorem 9.6 follows immediately from Theorem 9.5 and Lemmas 9.9 and 9.13.

The additive isomorphisms in Theorem 9.5, as well as the Kronecker pairing, are given by the Leray–Serre spectral sequence. Indeed, $J^\sharp$ is fibered over $B^\sharp$ with a typical fiber $\tilde{F}^\circ$ homeomorphic to a punctured torus, and we have a spectral sequence $H_p(B^\sharp; H_q(\tilde{F}^\circ)) \Rightarrow H_{p+q}(J^\sharp)$. Since both the base and the fiber have homotopy type of CW-complexes of dimension 1, the sequence degenerates and

$$H_2(J^\sharp) = H_1(B^\sharp; H_1(\tilde{F}^\circ)) = H_1(B^\sharp; \mathcal{M}_J).$$

Under the assumptions, $\mathfrak{S}$ is a strict deformation retract of $B^\sharp$ and the latter group can be replaced with $H_1(\mathfrak{S}, \mathcal{M}_J)$. It remains to observe that the definition of $\mathcal{H}_\mathfrak{S}$ is merely the cellular computation of the homology: $\mathcal{H} \otimes \mathfrak{S}$ is the group of 1-chains (representing 2-chains in $J^\sharp$), and relations 9.3 (1) and (2) state that the boundary over each, respectively, •- or ○vertex should vanish. The computation of the cohomology group is literally the same.

Crucial is the description of the intersection index form. Consider a 2-chain $h = \sum h_e \otimes e$. Geometrically, it can be realized as the union of cylinders $l_e \times e$, where $l_e \subset \tilde{F}_e^\circ$ is a circle realizing $h_e \in H_1(\tilde{F}_e)$. To compute the self-intersection h^2, shift $\mathfrak{S}$ in $B^\sharp$ and consider a new copy $\mathfrak{S}'$ transversal to $\mathfrak{S}$ (grey lines in Figure 9.1). The shift can be chosen so that, next to each $\bullet$-vertex v, *one* edge e_v incident to v (depending on the shift) intersects its image $e_v \uparrow \mathbf{x}$, see Figure 9.1, left; this intersection point contributes $h_{e_v} \cdot (h_{e_v \uparrow \mathbf{x}} \uparrow \mathbb{X}^{-1})$ to h^2. (We apply $-\mathbb{X}^{-1}$ to the second factor to bring the two classes to the same fiber over e_v.) The other points of intersection of $\mathfrak{S}$ and $\mathfrak{S}'$ are as in Figure 9.1, right, and such points do not contribute to h^2, as the intersection form in the fibers is symplectic and all squares are trivial. Thus, we obtain an alternative expression

$$h^2 = \sum_{v \in \operatorname{Vtx}_\bullet \mathfrak{S}} h_{e_v} \cdot (h_{e_v \uparrow \mathbf{x}} \uparrow \mathbb{X}^{-1}), \tag{9.15}$$

the summation running over (any) *one* representative of each $\bullet$-vertex v of $\mathfrak{S}$. This expression could as well be taken for the definition of the quadratic form on $\mathcal{H} \otimes \mathfrak{S}$. In general, (9.15) differs from (9.2), and neither has a geometric meaning, as intersection of chains is not well defined. However, their restrictions to $\mathcal{H}_{\mathfrak{S}}$ coincide. Indeed, if h is a 2-*cycle*, it satisfies 9.3 (1) and, since $\mathbb{X}$ is a symplectomorphism of $\mathcal{H}$, we have $h_e \cdot (h_{e \uparrow \mathbf{x}} \uparrow \mathbb{X}^{-1}) = (h_{e \uparrow \mathbf{x}} \uparrow \mathbb{X}^{-1}) \cdot (h_{e \uparrow \mathbf{x}^2} \uparrow \mathbb{X}^{-2})$ for any edge e of $\mathfrak{S}$; averaging over all three edges incident to each $\bullet$-vertex converts (9.15) to (9.2). □

Remark 9.16. In practice, instead of using formal expressions (9.2) or (9.15), it is often much easier to compute the (self-)intersections of 2-chains directly, by a shift of the skeleton, referring to the spirit of Theorem 9.5 rather than to its statement.

Consider an oriented regular skeleton $\mathfrak{S}$. Given a chain $\gamma := (e, w)$ in $\mathfrak{S}$, where $w = w_1 \cdots w_r$ is a *reduced* word in $\{\mathbb{X}^{\pm 1}, \mathbb{Y}\}$, see Definition 2.14, let $e_0 := e$ and $e_i = e_{i-1} \uparrow w_i$, $i = 1, \ldots, r$. For an element $h \in \mathcal{H}$, define

$$h \otimes \gamma := \sum_{i=0}^{r} (-1)^i (h \uparrow \mathfrak{h}(e, w_1 \cdots w_i)) \otimes e_i \in \mathcal{H} \otimes \mathfrak{S}. \tag{9.17}$$

The assumption that w should be reduced ensures that, homologically, the boundaries of the 1-chains $(-1)^i[e_i]$ cancel each other and the sum $\sum_i (-1)^i [e_i]$ is a 1-cycle except, possibly, at the very first vertex u and very last vertex v. Hence, $h \otimes \gamma$ satisfies all relations 9.3 (1), (2) except over u and v. If γ is closed and h is a $\mathfrak{h}(\gamma)$-invariant vector, the *fundamental cycle*

$$[h \otimes \gamma] := h \otimes \gamma - h \otimes e = \sum_{i=1}^{r} (-1)^i (h \uparrow \mathfrak{h}(e, w_1 \cdots w_i)) \otimes e_i \tag{9.18}$$

is an element of $\mathcal{H}_{\mathfrak{S}}$.

As an example, assume that e belongs to an n-gonal region R of $\mathfrak{S}$ and let $\gamma := (e, (\mathbb{X}\mathbb{Y})^{-n})$ be the boundary of R. We have $\mathfrak{h}(\gamma) = \pm(\mathbb{X}\mathbb{Y})^{-n}$. In terms of the extremal elliptic surface J defined by $\mathfrak{S}$, the sign $+$ or $-$ corresponds to a singular fiber of type $\mathrm{I}_n = \tilde{\mathbf{A}}_{n-1}$ ($\tilde{\mathbf{A}}_0^*$ if $n = 1$) or $\mathrm{I}_n^* = \tilde{\mathbf{D}}_{n+4}$, respectively, inside R; according to this sign, we will call R a *region of type* $\tilde{\mathbf{A}}$ *or* $\tilde{\mathbf{D}}$, *cf.* Remark 5.77. The monodromy $\mathfrak{h}(\gamma)$ has a nontrivial invariant vector, *viz.* $\mathbf{b} \in \mathcal{H}$, if and only if R is of type $\tilde{\mathbf{A}}$, and in this case the fundamental cycle $[\mathbf{b} \otimes \partial R] := [\mathbf{b} \otimes \gamma]$ belongs to $\ker \mathcal{H}_{\mathfrak{S}}$. (The easiest way to see this is to shift γ inside R, making it disjoint from any other 2-chain.) Note that $[\mathbf{b} \otimes \partial R]$ does not depend on the choice of an edge $e \in R$. In fact, $[\mathbf{b} \otimes \partial R]$ is a generator of $H_2(\partial U_i^\circ) \cong \mathbb{Z}$, where U_i is a tubular neighborhood of the singular fiber $\tilde{F}_i$ inside R, see (9.11), and, in view of Theorem 9.5 and Lemma 9.12, we have the following combinatorial description of the kernel $\ker \mathcal{H}_{\mathfrak{S}}$.

Proposition 9.19. *The kernel* $\ker \mathcal{H}_{\mathfrak{S}}$ *is freely generated by the elements* $[\mathbf{b} \otimes \partial R]$, *where* R *runs over all type* $\tilde{\mathbf{A}}$ *regions of* $\mathfrak{S}$. ◁

9.2 Generalizations and examples

9.2.1 A computation *via* the homological invariant

We start with generalizing Theorems 9.5 and 9.6 to arbitrary Jacobian elliptic surfaces. Instead of trying to develop a formal calculus based on the skeleton of J, we assume that the homological invariant $\mathfrak{h} := \mathfrak{h}_J$ of J is known (for example, from the skeleton or the dessin of the ramification curve and appropriate type specification) and compute $\mathbf{T}_J$ and Tors $MW(J)$ in terms of $\mathfrak{h}$. This approach was motivated by the original proof of Theorem 9.31; a similar approach is developed in [5].

We keep the notation introduced at the beginning of Section 9.1.1. As usual, $B^\sharp$ is the set of regular values of p. The punctured surface $B^\sharp$ has a strict deformation retract $\mathfrak{G}$ homeomorphic to a wedge of circles. Orient each circle and regard $\mathfrak{G}$ as a directed ribbon graph with a single vertex v. Let $\bar{\mathfrak{G}}$ be the bipartite subdivision of $\mathfrak{G}$.

As an exception, take for the reference the fiber $\tilde{F}_v$ over v. Fix an identification $H_1(\tilde{F}_v^\circ) = \mathcal{H}$ and let $\mathfrak{h}(e) \in \tilde{\Gamma}$ be the monodromy along an edge e of $\mathfrak{G}$. Since each edge e defines an ordered pair (e^-, e^+) of edges of $\bar{\mathfrak{G}}$, we can also speak about the monodromies $\mathfrak{h}(e^-) := \mathfrak{h}(e)$ and $\mathfrak{h}(e^+) := \mathfrak{h}(e)^{-1}$. As in Section 9.1.1, define the groups

$$\mathcal{H} \otimes \mathfrak{G} := \bigoplus_{e \in \operatorname{Edg} \mathfrak{G}} \mathcal{H} \otimes e \hookrightarrow \mathcal{H} \otimes \bar{\mathfrak{G}} := \bigoplus_{e \in \operatorname{Edg} \bar{\mathfrak{G}}} \mathcal{H} \otimes e; \qquad (9.20)$$

the inclusion is given by $u \otimes e \mapsto u \otimes e^- - (u \uparrow \mathfrak{h}(e)) \otimes e^+$, $u \in \mathcal{H}$, $e \in \operatorname{Edg} \mathfrak{G}$. The cohomology versions are defined *via*

$$\mathcal{H}^* \otimes \bar{\mathfrak{G}} := \bigoplus_{e \in \operatorname{Edg} \bar{\mathfrak{G}}} \mathcal{H}^* \otimes e \twoheadrightarrow \mathcal{H}^* \otimes \mathfrak{G} := \bigoplus_{e \in \operatorname{Edg} \mathfrak{G}} \mathcal{H}^* \otimes e, \qquad (9.21)$$

the projection being $u \otimes e^- \mapsto u \otimes e$, $u \otimes e^+ \mapsto (-\mathfrak{h}(e) \downarrow u) \otimes e$.

Definition 9.22. Define $\mathcal{H}_{\bar{\mathfrak{G}}} = \mathcal{H}_{\mathfrak{G}}$ as the subgroup of $\mathcal{H} \otimes \bar{\mathfrak{G}}$ formed by the elements h satisfying the conditions

1. $\sum_e h_e = 0$, the summation running over all edges $e \in \operatorname{Edg} \bar{\mathfrak{G}}$;
2. $h_e \uparrow \mathfrak{h}(e) + h_{e\uparrow\mathbf{y}} = 0$ for each edge $e \in \operatorname{Edg} \bar{\mathfrak{G}}$.

Similarly, let $\mathcal{H}^*_{\bar{\mathfrak{G}}} = \mathcal{H}^*_{\mathfrak{G}}$ be the quotient of $\mathcal{H}^* \otimes \bar{\mathfrak{G}}$ by all relations of the form

3. $\sum_e u \otimes e$, $u \in \mathcal{H}^*$ and the summation runs over all edges $e \in \operatorname{Edg} \bar{\mathfrak{G}}$;
4. $(\mathfrak{h}(e) \downarrow u) \otimes e + u \otimes (e \uparrow \mathbf{y})$ for all $u \in \mathcal{H}^*$ and $e \in \operatorname{Edg} \bar{\mathfrak{G}}$.

There is a canonical pairing $\mathcal{H}_{\mathfrak{G}} \otimes \mathcal{H}^*_{\mathfrak{G}} \to \mathbb{Z}$.

Remark 9.23. One can see that $\mathcal{H} \otimes \mathfrak{G} \subset \mathcal{H} \otimes \bar{\mathfrak{G}}$ is the subgroup of the elements h satisfying 9.22 (2), whereas $\mathcal{H}^* \otimes \mathfrak{G}$ is the quotient of $\mathcal{H}^* \otimes \bar{\mathfrak{G}}$ by the relations 9.22 (4). Hence, $\mathcal{H}_{\mathfrak{G}}$ can be defined as the subgroup of $\mathcal{H} \otimes \mathfrak{G}$ subject to the only condition

1. $\sum_e h_e \uparrow (\mathfrak{h}(e) - 1) = 0$, the summation running over all edges $e \in \operatorname{Edg} \mathfrak{G}$,

and $\mathcal{H}^*_{\mathfrak{G}}$ can be defined as the quotient of $\mathcal{H}^* \otimes \mathfrak{G}$ by the relations

3. $\sum_e (\mathfrak{h}(e) - 1) \downarrow u \otimes e$, $u \in \mathcal{H}^*$ and the summation runs over all edges $e \in \operatorname{Edg} \mathfrak{G}$.

The advantage of the group $\mathcal{H} \otimes \bar{\mathfrak{G}}$ is in the definition that follows, as *a priori* the order of the edge ends of $\mathfrak{G}$ at v is unknown.

Convert $\mathcal{H} \otimes \bar{\mathfrak{G}}$ (hence also $\mathcal{H} \otimes \mathfrak{G}$ and $\mathcal{H}_{\mathfrak{G}}$) to a rational lattice by letting

$$h^2 := -\sum_{i=1}^{d-1} \frac{d-i}{d} \sum_{e \in \operatorname{Edg} \bar{\mathfrak{G}}} h_e \cdot h_{e\uparrow\mathbf{x}^i}, \quad h \in \mathcal{H} \otimes \bar{\mathfrak{G}}, \tag{9.24}$$

where d is the degree of the only vertex v of $\bar{\mathfrak{G}}$ and $\cdot$ is the symplectic product in $\mathcal{H}$.

Theorem 9.25. *There are canonical isomorphisms $\mathcal{H}_{\mathfrak{G}} = H_2(J^\sharp)$, $\mathcal{H}^*_{\mathfrak{G}} = H^2(J^\sharp)$; the former is an isomorphism of lattices, and the latter takes $\mathcal{H}_{\mathfrak{G}} \otimes \mathcal{H}^*_{\mathfrak{G}} \to \mathbb{Z}$ to the Kronecker pairing $H_2(J^\sharp) \otimes H^2(J^\sharp) \to \mathbb{Z}$.*

Corollary 9.26 (of Theorem 9.25 and Lemmas 9.9 and 9.13). *There are canonical isomorphisms* $\mathbf{T}_J = \mathcal{H}_{\mathfrak{G}} / \ker$ *and* $\operatorname{Tors} MW(J) = \operatorname{Tors} \mathcal{H}^*_{\mathfrak{G}}$. ◁

Proof of Theorem 9.25. As in Theorem 9.5, the additive isomorphisms are obtained by computing the (co-)homology $H_1(\mathfrak{G}; \mathcal{M}_J)$ and $H^1(\mathfrak{G}; \mathcal{M}^*_J)$; for example, one can use the Mayer–Vietoris exact sequence.

Pick an edge e_1 of $\bar{\mathfrak{G}}$ and number the other edges consecutively according to their cyclic order at v, so that $e_i = e_1 \uparrow \mathbf{x}^{i-1}$, $i = 2, \dots, d$. To compute the self-intersection of a 2-cycle $h := \sum_i u_i \otimes e_i$, 'spread out' the vertex v and shift the graph $\bar{\mathfrak{G}}$ to make it transversal to itself, see Figure 9.2. To keep h a cycle, the elements $s_i \in \mathcal{H}$ over the newly inserted edges must be $s_i = \sum_{j=1}^{i} u_j$, $i = 1, \dots, d-1$, and the contribution

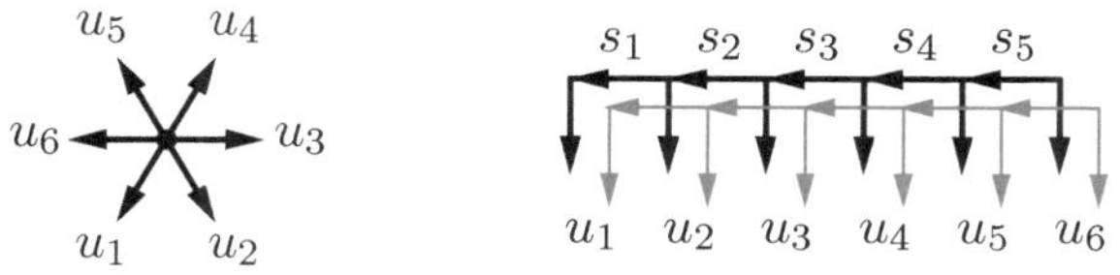

Figure 9.2. Spreading out and shifting a vertex ($d = 6$).

to h^2 of the intersection points shown in Figure 9.2, right, is $-\sum_{i=1}^{d-1} s_i \cdot u_{i+1}$. Hence,

$$h^2 = - \sum_{1 \leqslant j < i \leqslant d} u_j \cdot u_i. \tag{9.27}$$

(As in the proof of Theorem 9.5, the other intersection points can be ignored, as they contribute summands of the form $u_i^2 = 0$.) If h *is* a cycle, we must obtain the same result taking for e_1 any other edge of $\bar{\mathfrak{G}}$. (In fact, it suffices that $\sum_i u_i = 0$, see condition 9.22 (1).) Averaging over all edges, we arrive at (9.24). □

The case of the rational base $B \cong \mathbb{P}^1$

If the base B is rational, we can take for $\mathfrak{G}$ a 'thickening', in the obvious sense, of a system of lassoes representing a geometric basis $\{\gamma_1, \ldots, \gamma_r\}$ for the group $\pi_1(B^\sharp, v)$. Then, the edges of $\mathfrak{G}$ and $\bar{\mathfrak{G}}$ are *linearly* ordered, $e_1, \ldots, e_r$ and $e_1^-, e_1^+, \ldots, e_r^-, e_r^+$, respectively, and the elements $\mathfrak{h}_i := \mathfrak{h}(e_i)$ are the monodromies $\mathfrak{h}(\gamma_i)$ about all but one singular fibers $\tilde{F}_i$ of J. The linear order of the edges of $\bar{\mathfrak{G}}$ induces their cyclic order at v.

If $\mathfrak{G}$ is as above, we can define $\mathcal{H}_{\mathfrak{G}}$ as a subgroup of $\mathcal{H} \otimes \mathfrak{G}$, see Remark 9.23, and use (9.27) rather than (9.24) in the definition of the product. Expression (9.27) should be applied to the sequence $u_1, -u_1 \uparrow \mathfrak{h}_1, \ldots, u_r, -u_r \uparrow \mathfrak{h}_r$, *cf.* the definition of inclusion (9.20); hence, in view of 9.22 (1), it takes the form

$$h^2 := \sum_{i=1}^{r} u_i \cdot (u_i \uparrow \mathfrak{h}_i) - \sum_{1 \leqslant i < j \leqslant r} u_i \uparrow (\mathfrak{h}_i - 1) \cdot u_j \uparrow (\mathfrak{h}_j - 1), \tag{9.28}$$

where $h = \sum_{i=1}^{r} u_i \otimes e_i$. Although the quadratic forms (9.24) and (9.28) differ on $\mathcal{H} \otimes \mathfrak{G}$, their restrictions to $\mathcal{H}_{\mathfrak{G}}$ are equal.

The first term in (9.28) can be rewritten in the form $\sum_i u_i \cdot u_i \uparrow (\mathfrak{h}_i - 1)$, and we have the following simple statement.

Proposition 9.29. *The kernel* $\ker(\mathcal{H} \otimes \mathfrak{G})$ *contains the subgroup* $\mathcal{H}_{\mathfrak{G}}^0$ *generated by the elements of the form* $u_i \otimes e_i$*, where* $u_i \in \mathcal{H}$ *is an invariant vector of* $\mathfrak{h}_i$. ◁

9.2.2 An example

As an example, we use simple pseudo-trees to construct large collections of extremal elliptic surfaces sharing the same combinatorial sets of singular fibers and compute some of their topological invariants.

Theorem 9.30. *For each integer $k \geqslant 1$, there are exactly $T(k)$ pairwise distinct extremal elliptic surfaces $J \to \mathbb{P}^1$ with the combinatorial sets of singular fibers*

1. $(k+1)\tilde{\mathbf{A}}_0^* + \tilde{\mathbf{D}}_{5k+3}$, *if k is odd, or*
2. $(k+1)\tilde{\mathbf{A}}_0^* + \tilde{\mathbf{A}}_{5k-2}$, *if k is even.*

Besides, there are exactly $C(k-1)$ pairwise distinct extremal elliptic surfaces $J \to \mathbb{P}^1$ with the combinatorial sets of singular fibers

3. $k\tilde{\mathbf{A}}_0^* + \tilde{\mathbf{D}}_5 + \tilde{\mathbf{A}}_{5k-2}$, *if k is odd, or*
4. $k\tilde{\mathbf{A}}_0^* + \tilde{\mathbf{D}}_5 + \tilde{\mathbf{D}}_{5k+3}$, *if k is even.*

Proof. All surfaces are given by Theorem 9.1, where we take for $\mathfrak{S}$ a simple pseudo-tree with $2k$ $\bullet$-vertices. For Items 1 and 2, orient each loop so that the corresponding monogonal region is of type $\tilde{\mathbf{A}}$. (The orientations of the other edges are irrelevant as all such orientations are flip equivalent.) For Items 3 and 4, the orientation of one of the loops is reversed to form a $\tilde{\mathbf{D}}$ type monogon. This distinguished loop defines a marking of $\mathfrak{T}$ and the count is $C(k-1)$ rather than $T(k)$, see Lemma 1.66. This construction does produce all surfaces as in the statement due to Corollary 1.64. □

For the next statement, define $f_l := 3(\mathbf{v}_1 + \cdots + \mathbf{v}_s) + \mathbf{v}_{s+1} + \cdots + \mathbf{v}_l \in \mathbf{B}_l^+$, where $l = 2s - \epsilon$, $\epsilon = 0, 1$, and introduce the rational lattice $\mathbf{B}_l'$ isomorphic to $\mathbf{B}_l^+$ as a group and with the modified product $u \otimes v \mapsto u \cdot v + \frac{1}{4}(u \cdot f_l)(v \cdot f_l)$; here and in Item 3 of Theorem 9.31, $\cdot$ stands for the original product on $\mathbf{B}_l^+$.

Theorem 9.31. *Let $J \to \mathbb{P}^1$ be an elliptic surface as in Theorem 9.30. Then the transcendental lattice $\mathbf{T}_J$ depends on the set of singular fibers only; it is as follows (in the order of appearance in Theorem 9.30):*

1. $\mathbf{D}_{k-1}^+$ (*see Convention* 1.11);
2. $f_{k-1}^\perp \subset \mathbf{B}_{k-1}^+$;
3. $\{u \in \mathbf{B}_{k-1}' \mid f_{k-1} \cdot u = 0 \bmod 4\} \subset \mathbf{B}_{k-1}'$;
4. $\mathbf{D}_{k-1}^+ \oplus \mathbf{D}_1^+$ (*see Convention* 1.11).

Furthermore, one has $MW(J) = \mathbb{Z}_2$ if $k = 1$, $MW(J) = \mathbb{Z}_3$ if the singular fibers are $3\tilde{\mathbf{A}}_0^* + \tilde{\mathbf{A}}_8$ (Item 2 with $k = 2$), and $MW(J) = 0$ otherwise.

In spite of the somewhat disappointing outcome of Theorem 9.31, in Chapter 10 we show that the surfaces in question do differ topologically, see Corollary 10.67.

Proof. The statement on $MW(J)$ is given by Lemma 9.8; alternatively, it follows from Addendum 9.36 below and Corollary 6.13.

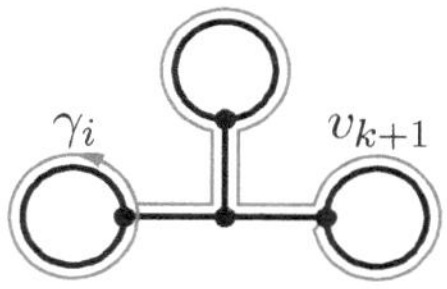

Figure 9.3. A chain $\gamma_i := (\bar{e}, (\mathbf{xy})^{n_i}\mathbf{xyx}^{-2}(\mathbf{xy})^{-n_i})$.

Consider the admissible tree $\mathfrak{T}$ used in the construction of J. For Items 1 and 2, mark $\mathfrak{T}$ arbitrary; for Items 3 and 4, mark it so that the $\tilde{\mathbf{D}}$ type monogon of $\operatorname{Sk}\mathfrak{T}$ is attached to the *last* leaf v_{k+1}. Let $(m_1, \dots, m_k)$ be the signature of $(\mathfrak{T}, v_1)$, and let $m_{ij} := \sum_{s=i}^{j-1} m_s$, $1 \leqslant i \leqslant j \leqslant k+1$, be the corresponding circular distances; denote $n_i = m_{i,k+1}$. In the bipartite subdivision $\bar{\mathfrak{T}}$ of $\mathfrak{T}$, there is a unique edge $\bar{e}$ incident to v_{k+1}, and the chains $\gamma_i := (\bar{e}, (\mathbf{xy})^{n_i}\mathbf{xyx}^{-2}(\mathbf{xy})^{-n_i})$, $i = 1, \dots, k+1$, form a geometric basis for the group $\pi_1(\operatorname{Sk}\mathfrak{T}, \bar{e}) = \pi_1(B^\sharp, v_{k+1})$, see Figure 9.3. Realize this basis by a system of lassoes in $B^\sharp$ and consider the corresponding graph $\mathfrak{G}$ with linearly ordered edges $e_1, \dots, e_{k+1}$, see Section 9.2.1, page 283. In the notation introduced on page 283, we have $\mathfrak{h}_i = L^{n_i} D L^{-n_i}$ for $i = 1, \dots, k$ and $\mathfrak{h}_{k+1} = \pm D$, where $L := \mathbb{X}\mathbb{Y} \in \tilde{\Gamma}$ is the matrix given by (2.3) and $D := \mathbb{X}L^{-1}\mathbb{X}^{-1}$.

Assume that J is as in Item 1 or 2, so that $\mathfrak{h}_{k+1} = D$, and consider the vectors

$$u_i := \mathbf{b} \uparrow L^{-n_i} = \mathbf{b}, \quad w_i := (\mathbf{a} - \mathbf{b}) \uparrow L^{-n_i} = \mathbf{a} - (n_i + 1)\mathbf{b}. \tag{9.32}$$

Each pair $\{u_i, w_i\}$ forms a basis for $\mathcal{H}$; hence, the lattice $\mathcal{H} \otimes \mathfrak{G}$ is freely generated by $u_i \otimes e_i$, $w_i \otimes e_i$, $i = 1, \dots, k+1$. Furthermore, we have $u_i \uparrow (\mathfrak{h}_i - 1) = w_i$ and $w_i \uparrow (\mathfrak{h}_i - 1) = 0$. Thus, the quotient $\mathcal{H}'_{\mathfrak{G}} := (\mathcal{H} \otimes \mathfrak{G})/\mathcal{H}^0_{\mathfrak{G}}$ is generated by $u_i \otimes e_i$ and (9.28) becomes

$$\Big(\sum_i c_i u_i \otimes e_i\Big)^2 = -\sum_i c_i^2 - \sum_{i<j} m_{ij} c_i c_j. \tag{9.33}$$

We are interested in the values of this expression only on the elements h satisfying relations 9.23 (1), which take the form

$$\sum_i c_i = \sum_i n_i c_i = 0; \tag{9.34}$$

hence, we can add to (9.33) the quadratic form $-(\sum_i c_i)^2 + (\sum_i c_i)(\sum_i n_i c_i)$, and then $(u_i \otimes e_i)^2 = n_i - 2$ and $(u_i \otimes e_i) \cdot (u_j \otimes e_j) = n_j - 1$ for $i < j$.

The subgroup of the elements of $\mathcal{H}'_{\mathfrak{G}}$ satisfying the first relation in (9.34) has basis $\mathbf{q}_i := (u_i \otimes e_i) - (u_{i+1} \otimes e_{i+1})$, $i = 1, \dots, k$, and in this basis it turns into the associated lattice $\mathbf{Q}_{\mathfrak{T}}$, see Section 1.3.3. Furthermore, the second relation in (9.34) turns into $\xi_{\mathfrak{T}}(h) = 0$, where $\xi_{\mathfrak{T}}$ is the characteristic functional of $\mathfrak{T}$. Hence, we have an isomorphism

$$\mathbf{T}_J = \operatorname{Ker} \xi_{\mathfrak{T}}/\ker. \tag{9.35}$$

The structure of this lattice is given by Lemma 1.71 and Propositions 1.79 and 1.81.

Now, assume that J is as in Items 3 or 4, so that $\mathfrak{h}_{k+1} = -D$ has no invariant vectors, and let $u_i, w_i \in \mathcal{H}$ be as in (9.32). This time, $\mathcal{H}^0_{\mathfrak{G}}$ is generated by $w_i \otimes e_i$, $i = 1, \ldots, k$, and we consider the subgroup $\mathcal{H}''_{\mathfrak{G}} \subset (\mathcal{H}_{\mathfrak{G}}/\mathcal{H}^0_{\mathfrak{G}}) \otimes \mathbb{Q}$ generated over $\mathbb{Z}$ by the cycles

$$\mathbf{r}_i := u_i \otimes e_i - \frac{n_i}{2} u_{k+1} \otimes e_{k+1} + \frac{n_i + 2}{4} w_{k+1} \otimes e_{k+1}, \quad i = 1, \ldots, k.$$

Then $\mathcal{H}_{\mathfrak{G}}/\mathcal{H}^0_{\mathfrak{G}}$ is the subgroup of $\mathcal{H}''_{\mathfrak{G}}$ defined by the condition $\psi(h) = 0 \bmod 4$, where $\psi := \sum_i (n_i - 2)\mathbf{r}^*_i$. In the basis $\mathbf{q}_i := \mathbf{r}_i - \mathbf{r}_{i+1}$, $i = 1, \ldots, k-1$, $\mathbf{q}_k := \mathbf{r}_k$, one has $\psi = \xi_{\mathfrak{T}} - 2\mathbf{q}^*_k$ and $\mathcal{H}''_{\mathfrak{G}}$ turns into the associated lattice $\mathbf{Q}_{\mathfrak{T}}$ with the modified bilinear form $u \otimes v \mapsto u \cdot v + \frac{1}{4}\psi(u)\psi(v)$.

Contract $\mathfrak{T}$ towards its last leaf v_{k+1}, establishing an isomorphism $\mathbf{Q}_{\mathfrak{T}} \cong \tilde{\mathbf{B}}^+_{k-1}$, see Lemma 1.71. According to Propositions 1.79 and 1.81, ψ contracts to $\bar{\xi}_{k-1}$ if k is odd (Item 3) or $\bar{\psi} := \bar{\xi}_{k-1} - 2\mathbf{w}^*$ if k is even (Item 4), see Lemma 1.78; the correction term $-2\mathbf{w}^*$ is given by Lemma 1.72. In the former case, $\operatorname{Ker} \bar{\xi}_{k-1} = \ker \tilde{\mathbf{B}}^+_{k-1}$ and we obtain the description of $\mathbf{T}_J$ as in the statement. In the latter case, we have an orthogonal decomposition $\mathbf{T}_J = \operatorname{Ker} \bar{\psi} \oplus \mathbb{Z}(2\mathbf{w})$, and $\operatorname{Ker} \bar{\psi}$ is generated by the vectors $\mathbf{v}_s - \mathbf{v}_{s+1} + \mathbf{w}$, $\mathbf{v}_1 + \mathbf{v}_2 + 3\mathbf{w}$, $\mathbf{v}_i - \mathbf{v}_{i+1}$, $i = 1, \ldots, s-1, s+1, \ldots, k-2$. It is immediate that $\operatorname{Ker} \bar{\psi} \cong \mathbf{D}^+_{k-1}$. (If $s = 1$ and $k = 2$, the kernel $\operatorname{Ker} \bar{\psi} \cong \mathbf{D}^+_1$ is generated by a single vector $2\mathbf{v}_1 + 3\mathbf{w}$.) □

Proof of Lemma 1.77. The lemma follows from (9.35). □

We conclude this section with a computation of the fundamental groups of the ramification curves of the surfaces given by Theorem 9.30.

Addendum 9.36. *Let C be the ramification curve of a Jacobian elliptic surface J as in Theorem* 9.30. *Then the group $\pi_1 := \pi^{\mathrm{proj}}(C)$ depends on the set of singular fibers of J only and is as follows*:

- $2\tilde{\mathbf{A}}^*_0 + \tilde{\mathbf{D}}_8$ (*Item* 1 *with* $k = 1$): $\pi_1 = \mathbb{Z} \times \mathbb{Z}_2$;
- $3\tilde{\mathbf{A}}^*_0 + \tilde{\mathbf{A}}_8$ (*Item* 2 *with* $k = 2$): $\pi_1 = \Gamma$;
- $\tilde{\mathbf{A}}^*_0 + \tilde{\mathbf{D}}_5 + \tilde{\mathbf{A}}_3$ (*Item* 3 *with* $k = 1$): $\pi_1 = (\Lambda/(t^2 - 1)) \rtimes \mathbb{Z}_2$, *see Argument* 7.13.

In all other cases, π_1 is cyclic.

Proof. The three exceptional groups are computed in Section 7.1.1, see Lines 5′, 2, and 6′, respectively, in Table 7.1. The other groups are computed similarly. Let $\mathfrak{G} := \operatorname{Sk} \mathfrak{T}$ and take for reference the edge $e := \bar{e} \uparrow \mathbf{x}$, where $\bar{e}$ is the edge of $\bar{\mathfrak{T}}$ incident to v_{k+1}. The chains $\gamma_i := (e, (\mathbf{yx})^{n_i}\mathbf{yx}^{-1}(\mathbf{yx})^{-n_i})$, $i = 1, \ldots, k+1$

(*cf.* Figure 9.3) form a basis for $\pi_1(\mathfrak{S}, e)$, and the braid monodromy is found using Theorem 5.62 and 5.68: one has $\mathfrak{m}_i := \mathfrak{m}(\gamma_i) = \sigma_1^{-n_i}\sigma_2\sigma_1^{n_i}$ for $i = 1, \dots, k$ and $\mathfrak{m}_{k+1} := \mathfrak{m}(\gamma_{k+1}) = \sigma_1\Delta^{2\epsilon}$, where $\epsilon = 0$ in Items 1, 2 and $\epsilon = 1$ in Items 3, 4.

The braid relation $\alpha_3 \uparrow \mathfrak{m}_i = \alpha_3$, $i = 1, \dots, k$, implies $\alpha_2 \uparrow \sigma_1^{n_i} = \alpha_3$; hence, $\alpha_2 \uparrow \sigma_1^{n_i} = \alpha_2 \uparrow \sigma_1^{n_j}$ for $1 \leqslant i < j \leqslant k$. Since σ_1 preserves α_3 and ρ, these relations can be written in the form $\sigma_1^{n_i} = \sigma_1^{n_j}$, or $\sigma_1^{n_i - n_j} = \text{id}$. If $k \geqslant 3$, the differences $n_i - n_j$ are coprime, see Corollary 1.60, and we also have $\sigma_1 = \text{id}$. Then $\alpha_1 = \alpha_2$, the original relation $\alpha_2 \uparrow \sigma_1^{n_i} = \alpha_3$ implies $\alpha_1 = \alpha_3$, and the group is cyclic. In the remaining case of Item 4 with $k = 2$, the group π_1 is abelian due to Lemma 5.82; since the curve is irreducible, π_1 is cyclic. □

Chapter 10

Monodromy factorizations

The concept of *monodromy factorization* as an algebraic means for the topological study of singular fibrations over a disk was probably introduced by Hurwitz [89]. In this chapter, we make a few steps towards the problem of Hurwitz equivalence of simple factorizations in the modular group Γ, see Section 10.2, and consider a few geometric applications, mainly deformation *vs.* topological classification of elliptic surfaces, real trigonal curves, and real Lefschetz fibrations, see Section 10.3. In Section 10.1.1, we discuss the general problem and cite a few significant results concerning other groups.

10.1 Hurwitz equivalence

Hurwitz equivalence of monodromy factorizations is an algebraic way to describe singular fibrations over a disk. We start with stating the general problem, explaining its relation to topology, and citing a few known results and examples, mainly related to the free group $\mathbb{F}_n$ and symmetric group $\mathbb{S}_n$.

10.1.1 Statement of the problem

Let G be a group. A G-valued (*monodromy*) *factorization* of *length* r is an ordered sequence $\bar{\mathfrak{m}} := (\mathfrak{m}_1, \ldots, \mathfrak{m}_r)$ of elements of G of length r. The entries $\mathfrak{m}_i$ are called the *factors* of $\bar{\mathfrak{m}}$, and the product $\mathfrak{m}_\infty := \mathfrak{m}_1 \cdots \mathfrak{m}_r$ is the *monodromy at infinity*. The factorization $\bar{\mathfrak{m}}$ itself is often referred to as a *factorization of* $\mathfrak{m}_\infty$ and written in the form $\mathfrak{m}_\infty = \mathfrak{m}_1 \odot \cdots \odot \mathfrak{m}_r$ or $\mathfrak{m}_\infty = \bigodot_{i=1}^r \mathfrak{m}_i$ (left to right product).

Definition 10.1. A *Hurwitz move* σ_i (respectively, *inverse Hurwitz move* σ_i^{-1}) of a factorization $\bar{\mathfrak{m}}$ is the modification $\bar{\mathfrak{m}} \mapsto \bar{\mathfrak{m}} \uparrow \sigma_i^{\pm 1}$ defined *via*

$$\begin{aligned} \sigma_i &\colon (\ldots, \mathfrak{m}_i, \mathfrak{m}_{i+1}, \ldots) \mapsto (\ldots, \mathfrak{m}_i \mathfrak{m}_{i+1} \mathfrak{m}_i^{-1}, \mathfrak{m}_i, \ldots), \\ \sigma_i^{-1} &\colon (\ldots, \mathfrak{m}_i, \mathfrak{m}_{i+1}, \ldots) \mapsto (\ldots, \mathfrak{m}_{i+1}, \mathfrak{m}_{i+1}^{-1} \mathfrak{m}_i \mathfrak{m}_{i+1}, \ldots), \end{aligned}$$

$i = 1, \ldots, \operatorname{length}(\bar{\mathfrak{m}}) - 1$. The *global conjugation* of $\bar{\mathfrak{m}}$ by an element $g \in G$ is the modification

$$\bar{\mathfrak{m}} \mapsto g^{-1} \bar{\mathfrak{m}} g := \bar{\mathfrak{m}} \uparrow g := (g^{-1} \mathfrak{m}_1 g, \ldots, g^{-1} \mathfrak{m}_r g).$$

Two factorizations $\bar{\mathfrak{m}}'$, $\bar{\mathfrak{m}}''$ are said to be *strongly* (*weakly*) *Hurwitz equivalent* if they can be related by a sequence of Hurwitz moves (respectively, a sequence of

Hurwitz moves and/or global conjugation). In this chapter, we routinely abbreviate strong/weak Hurwitz equivalence to just strong/weak equivalence.

It is straightforward from the definition that the monodromy at infinity $\mathfrak{m}_\infty$ is a strong Hurwitz invariant of a factorization $\bar{\mathfrak{m}}$, whereas its conjugacy class $[\![\mathfrak{m}_\infty]\!]$ is a weak Hurwitz invariant. (In general, we have $(g^{-1}\bar{\mathfrak{m}}g)_\infty = g^{-1}\mathfrak{m}_\infty g$; hence, the global conjugation by g preserves $\mathfrak{m}_\infty$ if and only if g commutes with $\mathfrak{m}_\infty$.)

Another obvious invariant is the multiset $\mathcal{P}(\bar{\mathfrak{m}}) := \{\mathcal{P}_i := [\![\mathfrak{m}_i]\!] \mid i = 1, \dots, r\}$ of the conjugacy classes of the factors. (Here, by a multiset we mean a sequence of objects, not necessarily distinct, regarded up to permutation.) Usually, this multiset is assumed fixed in advance; sometimes it is called the *passport* of $\bar{\mathfrak{m}}$.

The passport and the monodromy at infinity are the two most important invariants (their geometric meaning is explained below); they are usually assumed fixed and a typical problem in the subject questions the number of classes of factorizations of a given element $g \in G$ and with a given passport $\mathcal{P}$.

Problem 10.2. Characterize the elements $g \in G$ admitting a factorization of a given length r and with a given passport $\mathcal{P} = (\mathcal{P}_1, \dots, \mathcal{P}_r)$.

Problem 10.3. Is a factorization of a given element $g \in G$ and with a given passport $\mathcal{P}$ unique up to strong/weak equivalence? If not, how many classes are there? What are the invariants distinguishing these classes?

Problem 10.4. If two factorizations $\bar{\mathfrak{m}}'$, $\bar{\mathfrak{m}}''$ are weakly equivalent and share the same monodromy at infinity, do they need to be strongly equivalent?

In some important cases (see, *e.g.*, Theorems 10.7, 10.10, 10.18, 10.30 below in this chapter), the answer to Problems 10.3 and 10.4 is in the affirmative. On the other hand, numerous examples show that, even in very simple settings, the answer may be in the negative (see, *e.g.*, Example 10.25). In general, the situation seems very far from its complete understanding and we only as little as scratch the surface.

Remark 10.5. If G is finite, the answer to Problem 10.2 is given by the Frobenius type formula counting the number $N(\mathcal{P})$ of solutions to the equation $g_1 \cdots g_n = 1$, $g_i \in \mathcal{P}_i$, $i = 1, \dots, n$, for a given passport $\mathcal{P}$ of length n, see [4]:

$$N(\mathcal{P}) = \frac{1}{\operatorname{ord} G} \prod_{i=1}^{n} \operatorname{Card} \mathcal{P}_i \sum_{\chi} \frac{\chi(\mathcal{P}_1) \cdots \chi(\mathcal{P}_n)}{\chi(1)^{n-2}},$$

the summation running over all irreducible characters of G. Taking $n = r + 1$ and $\mathcal{P}_n = [\![g^{-1}]\!]$ in the notation of Problem 10.2, we conclude that g admits a factorization if and only if

$$\sum_{\chi} \chi(\mathcal{P}_1) \cdots \chi(\mathcal{P}_r) \chi(g^{-1}) \chi(1)^{1-r} > 0. \tag{10.6}$$

Problems 10.2 and 10.4, even in the case of a finite group G, are more involved, see, *e.g.*, Section 10.1.3 below.

For non-uniqueness statements, we need additional invariants that would distinguish factorizations of the same element $g \in G$. One obvious invariant, common to all groups, is the *monodromy group* $\operatorname{Im}\bar{\mathfrak{m}} \subset G$, which is the subgroup of G generated by all factors $\mathfrak{m}_i$ of $\bar{\mathfrak{m}}$. (The subgroup $\operatorname{Im}\bar{\mathfrak{m}}$ is indeed the image of $\bar{\mathfrak{m}}$ regarded as a homomorphism $\mathbb{F}_r \to G$, see the geometric interpretation below.) The subgroup $\operatorname{Im}\bar{\mathfrak{m}} \subset G$ is a strong Hurwitz invariant, whereas its conjugacy class $[\![\operatorname{Im}\bar{\mathfrak{m}}]\!]$ is a weak invariant; both statements follow immediately from Definition 10.1.

A number of other invariants can be produced for various specific groups G. Thus, if $G = \mathbb{B}_n$ (or, more generally, $G \subset \operatorname{Aut}\mathbb{F}_n$), one can consider the *fundamental group* $\pi_1(\bar{\mathfrak{m}}) := \mathbb{F}_n / \operatorname{Im}\bar{\mathfrak{m}}$, *cf.* Section 5.1.2, and the derived invariants such as the Alexander module or maximal dihedral quotient. (Often, it is $\pi_1(\bar{\mathfrak{m}})$ that is the ultimate goal of computing the monodromy, *cf.* Chapters 6, 7, and 8.) If $G = \tilde{\Gamma}$, one can consider the transcendental lattice $\mathbf{T}(\bar{\mathfrak{m}})$, see Section 9.2.1 for the geometric insight and Section 10.2.3 below for the precise definitions; this invariant is relatively new and not very well studied. Both can be generalized: the fundamental group can obviously be constructed starting from any fixed representation $G \to \operatorname{Aut} A$ (where A is a group), and the transcendental lattice, from a representation $G \to \operatorname{Sp}(\mathcal{U})$, see Section 10.2.3.

The geometric interpretation

A G-valued factorization $\bar{\mathfrak{m}} = (\mathfrak{m}_1, \ldots, \mathfrak{m}_r)$ can be regarded as a homomorphism $\mathbb{F}_r = \langle \gamma_1, \ldots, \gamma_r \rangle \to G$, $\gamma_i \mapsto \mathfrak{m}_i$, $i = 1, \ldots, r$. In this interpretation, Hurwitz moves generate the canonical action of the braid group $\mathbb{B}_r$ on the free group $\langle \gamma_1, \ldots, \gamma_r \rangle$, see (2.27) and Figure 2.4, and the global conjugation represents the adjoint action of G on itself. Geometrically, a homomorphism as above arises as the monodromy of a locally trivial fibration $X^\sharp \to B^\sharp \cong \bar{D}_r^\circ$ over a punctured disk; then G is the appropriately defined mapping class group of the fiber over a fixed point $b \in \partial B^\sharp$ and $\langle \gamma_1, \ldots, \gamma_r \rangle$ is a geometric basis for the fundamental group $\pi_1(B^\sharp, b)$. In this set-up, Hurwitz moves can be interpreted either as basis changes or as automorphisms of $B^\sharp$ fixed on the boundary, see Theorem 2.28, and the topological classification of fibrations reduces to the purely algebraic classification of G-valued factorizations up to weak Hurwitz equivalence. The best known examples are

- ramified coverings (the fiber is a finite set and $G = \mathbb{S}_n$, see Section 10.1.3 below and [89, 124, 19, 139]);
- algebraic or, more generally, pseudoholomorphic and Hurwitz curves in $\mathbb{C}^2$ (the fiber is a punctured plane, $G = \mathbb{B}_n$, and $\bar{\mathfrak{m}}$ is the braid monodromy of the curve, see Section 5.1.1 and [167, 161, 34, 33, 121, 120, 98, 99, 9, 10, 32, 131, 132]);
- (real) elliptic surfaces or, more generally, (real) elliptic Lefschetz fibrations (the fiber is an elliptic curve/topological torus, $G = \tilde{\Gamma}$, and $\bar{\mathfrak{m}}$ is the homological

invariant/monodromy of the fibration, see Sections 3.2.2 and 3.3.3 and [95, 151, 121, 25, 10, 54, 60, 131, 132, 142]).

The last two subjects are quite popular and the reference lists are far from complete: I tried to cite the founding papers and a few recent results/surveys only. For the free group $\mathbb{F}_n$, an alternative interpretation is discussed in Section 10.1.2.

As the examples suggest, usually $X^\sharp \to B^\sharp$ is the restriction of a singular fibration $p\colon X \to B \cong \bar{D}^2$ over the whole disk $\bar{D}^2$, the punctures corresponding to the singular fibers. Then the conjugacy class $\mathcal{P}_i := [\![\mathfrak{m}_i]\!]$ represents the topological type of the fiber F_i over the i-th puncture, *cf.* Table 3.2 or the computation of the braid monodromy and slopes in Section 5.2.2. From this point of view, Problems 10.2 and 10.3 question the existence and uniqueness, up to automorphism fixed on the boundary, of an extension of a given fibration over the circle $S^1 = \partial\bar{D}^2$ (the monodromy at infinity $g = \mathfrak{m}[\partial\bar{D}^2] \in G$) to a fibration over $\bar{D}^2$ with given number and topological types of singular fibers. An important special case is that of $g = 1$: if $\mathfrak{m}[\partial\bar{D}^2] = 1$, the fibration over $\bar{D}^2$ can be regarded as one over the sphere $S^2 \cong \bar{D}^2/\partial\bar{D}^2$.

In most cases (in fact, whenever $\mathfrak{m}[\partial\bar{D}^2] \neq 1$), singular fibers are unavoidable. However, one can require that such fibers should be as simple as possible (*e.g.*, double ramification points or Kodaira type I_1 singular fibers of trigonal curves or elliptic surfaces). This requirement leads to the notion of *simple* monodromy factorization, whose precise definition depends on a particular problem; usually, one merely restricts the passports to sequences of copies of one distinguished conjugacy class $\mathcal{P}_0$. Below, simple factorizations are defined separately for each group G considered.

10.1.2 $\mathbb{F}_n$-valued factorizations

If G is the free group $\mathbb{F}_n := \langle\alpha_1, \ldots, \alpha_n\rangle$, a factorization $\bar{\mathfrak{m}}$ is called *simple* if each factor $\mathfrak{m}_i$ is conjugate to one of the generators α_j. It is clear that the length of a simple factorization of an element $g \in \mathbb{F}_n$ is necessarily equal to $\deg g$ and its passport consists of r_j copies of $[\![\alpha_j]\!]$, $j = 1, \ldots, n$, where r_j is the exponent sum of α_j in a word representing g. One has $\deg g = \sum_j r_j$. In particular, a necessary condition for the existence of a simple factorization is $r_j \geqslant 0$, $j = 1, \ldots, n$.

One of the first and most important uniqueness statements, which lays a basis for the geometric applications of monodromy factorizations, is Artin's Theorem 2.28: it can be interpreted as the uniqueness, up to strong Hurwitz equivalence, of an $\mathbb{F}_n$-valued simple factorization of the element $\rho := \alpha_1 \cdots \alpha_n$. Artin's proof cited in Section 2.2.1 extends almost literally to the following slightly more general statement.

Theorem 10.7 (essentially Artin). *Let $G = \mathbb{F}_n = \langle\alpha_1, \ldots, \alpha_n\rangle$ be a free group. Then any element $g \in G$ represented by a positive word in $\{\alpha_1, \ldots, \alpha_n\}$ admits a unique, up to strong Hurwitz equivalence, simple factorization.* $\triangleleft$

The existence of simple factorizations is given by the following theorem.

Theorem 10.8 (Blank [24]). *Let $G = \mathbb{F}_n = \langle \alpha_1, \dots, \alpha_n \rangle$ be a free group. An element $g \in G$ represented by a word w in the alphabet $\{\alpha_j^{\pm 1}\}$, $j = 1, \dots, n$, admits a simple factorization if and only if* $\deg g \geqslant 0$ *and, after removing from w an appropriate collection of* $\deg g$ positive *generators, the new word represents the identity $1 \in G$.*
▷ [24, 137]

Simple factorizations of elements $g \in \mathbb{F}_n$ have another geometric interpretation. Let $\bar{D}^2$ be the closed unit disk, $S^1 = \partial \bar{D}^2$, and let B be another copy of $\bar{D}^2$. An immersion $\varphi\colon S^1 \to B$ is called *normal* if the image $\varphi(S^1)$ lies in the interior of B and has finitely many points of self-intersection, which are all simple nodes. We are interested in *extensions* of φ to orientation preserving immersions $\bar{\varphi}\colon \bar{D}^2 \to B$ of the whole disk. Two such extensions are said to be *equivalent* if they are related by an auto-diffeomorphism of $\bar{D}^2$.

Consider a normal immersion $\varphi\colon S^1 \to B$ and pick a point b_i in each connected component of the complement $B \smallsetminus \varphi(S^1)$ other than its outer component (*i.e.*, the one containing ∂B). Let $B^\sharp := B \smallsetminus \{b_1, \dots\}$ and consider the group $G := \pi_1(B^\sharp, b)$, where $b \in \partial B$ is a basepoint. Then the image $\varphi(S^1)$, with its positive orientation, defines a certain conjugacy class $[\![\varphi(S^1)]\!]$ in G. Given an extension $\bar{\varphi}\colon \bar{D}^2 \to B$ of φ, let $\bar{D}^\circ_r$ be the disk $\bar{D}^2$ punctured at all preimages $\bar{\varphi}^{-1}(b_i)$, and let $\{\gamma_1, \dots, \gamma_r\}$ be a geometric basis for $\pi_1(\bar{D}^\circ_r, d)$, $d \in \partial \bar{D}^2$. Then $\gamma_1 \cdots \gamma_r = [S^1]$, and the images $\bar{\varphi}_*(\gamma_j)$, extended towards b along a common arc connecting $\bar{\varphi}(d)$ and b, give rise to a simple factorization of a representative of $[\![\varphi(S^1)]\!]$. This factorization is well defined up to weak Hurwitz equivalence, essentially due to Theorem 2.28 and to the assumption that $\bar{\varphi}$ is an immersion.

The following theorem asserts that this construction is invertible.

Theorem 10.9 (Blank [24]). *Given a normal immersion $\varphi\colon S^1 \to B$, there is a canonical bijection between the set of equivalence classes of extensions $\bar{\varphi}\colon \bar{D}^2 \to B$ as above and that of weak equivalence classes of simple* (*with respect to any geometric basis*) *factorizations of $[\![\varphi(S^1)]\!] \subset \pi_1(B^\sharp, b)$.* ▷ [24, 137]

The first example of a normal immersion admitting two non-equivalent extensions is attributed to J. Milnor (oral tradition). Other examples are constructed in [24].

10.1.3 $\mathbb{S}_n$-valued factorizations

From the point of view of Hurwitz equivalence, the best studied group is probably the symmetric group $\mathbb{S}_n$. According to Hurwitz [89], the weak/strong equivalence classes of $\mathbb{S}_n$-valued factorizations are in a canonical one-to-one correspondence with the homeomorphism classes of ramified coverings $p\colon X \to \bar{D}^2$, the factors corresponding to the critical values of the projection. (For strong equivalence, we should fix a point $b \in \partial \bar{D}^2$ and an identification $p^{-1}(b) = \{1, \dots, n\}$.) An $\mathbb{S}_n$-valued factorization is called *simple* if all factors are transpositions. Simple factorizations classify *simple* ramified coverings, where each critical point has ramification index 2 and all critical values are pairwise distinct. Such coverings are generic: any other ramified covering can be perturbed to a simple one.

An $\mathbb{S}_n$-valued factorization $\bar{\mathfrak{m}}$ is called *transitive* if its monodromy group $\operatorname{Im}\bar{\mathfrak{m}}$ is transitive. Such factorizations correspond to ramified coverings $X \to \bar{D}^2$ with connected covering space X.

The following theorem is classical. Since it plays crucial rôle in the passage from dessins back to trigonal curves, see Section 4.1, we reproduce its proof, which is independent of the rest of the book.

Theorem 10.10 (essentially Hurwitz; see [124, 19]). *Let $g \in \mathbb{S}_n$, $n \geqslant 2$, be a product of c disjoint cycles (including those of length one). Then g admits a transitive simple $\mathbb{S}_n$-valued factorization $\bar{\mathfrak{m}}$ of length r if and only if $r \geqslant n+c-2$ and $r = n+c \bmod 2$. Such a factorization is unique up to strong Hurwitz equivalence.*

Proof. A factorization as in the statement gives rise to a simple ramified covering $X \to \bar{D}^2$, where X is a connected orientable surface with c boundary components, hence $\chi(X) \leqslant 2 - c$ and $\chi(X) = c \bmod 2$. Thus, the restriction on the length r (the number of ramification points) follows from the Riemann–Hurwitz formula. For the existence, observe that a cycle of length l does split into a product of $(l-1)$ transpositions: up to conjugation,

$$(1\,2\,\ldots\,l)^{-1} = (1\,2) \odot (2\,3) \odot \cdots \odot (l-1, l).$$

Representing each cycle in this form, we obtain a factorization $\bar{\mathfrak{m}}'$ of length $n-c$ with c disjoint orbits, one orbit for each cycle. To 'join' these orbits, pick a set $i_1, \ldots, i_c$ of representatives and add to $\bar{\mathfrak{m}}'$ the $2(c-1)$ elements

$$(i_1\, i_2), (i_1\, i_2), (i_2\, i_3), (i_2\, i_3), \ldots, (i_{c-1}\, i_c), (i_{c-1}\, i_c).$$

The result is a factorization of g of length $n+c-2$. To obtain a factorization of length $r = n + c - 2 + 2k$, $k \geqslant 0$, add $2k$ copies of the transposition $(1\,2)$.

For the uniqueness, we need the following lemma.

Lemma 10.11. *If $\bar{\mathfrak{m}}$ is a transitive simple $\mathbb{S}_n$-valued factorization, then, up to strong Hurwitz equivalence, one can assume that $\mathfrak{m}_1 = (i\, j)$ for any given pair (i, j), $1 \leqslant i < j \leqslant n$.*

Proof. Since i and j are in the same orbit of the monodromy group, there is a sequence $i = i_0, i_1, \ldots, i_k = j$ such that, for each $s = 1, \ldots, k$, the transposition $(i_{s-1}\, i_s)$ is a certain factor $\mathfrak{m}_{l_s}$ of $\bar{\mathfrak{m}}$. By eliminating unnecessary cycles, we can assume that all i_s are pairwise distinct. We assert that, by a number of Hurwitz moves, the length of this sequence can be reduced to one. To this end, move $\mathfrak{m}_{l_1}$ towards $\mathfrak{m}_{l_2}$ by the direct moves $\sigma_{l_1}, \ldots$, if $l_1 < l_2$, or inverse moves $\sigma_{l_1-1}^{-1}, \ldots$, if $l_1 > l_2$. All factors $\mathfrak{m}_{l_s}$ with

```
# For the proof of Hurwitz theorem (Theorem 10.10)
# Just check that the S4-valued monodromy factorizations ((1 2), (2 3), (1 2), (1 2)) and
    ((1 2), (2 3), (2 3), (2 3)) are strongly Hurwitz equivalent.
# The equivalence is given by "mu" = βσ3βσ3^-1 ∈ B4, where β = σ2σ1^2σ2

Reread("braid.txt");     # (see Listing 2.5)
F := FreeGroup(4);;      # The free group F4 to act on
B := FreeGroup(3);;      # The braid group B4 (we do not need the relations)
gens := GeneratorsOfGroup(F);;
beta := B.2*B.1^2*B.2; mu := beta*B.3*beta*B.3^-1;  # The braids β and μ
free := Braid(mu, F, gens);   # The action on F4 (see "Braid" in Listing 2.5)
old := [(1,2), (2,3), (1,2), (1,2)];  # Original monodromy factorization
new := List(free, x -> MappedWord(x, gens, old));  # New factorization
```

Listing 10.1. Comparing two factorizations of $(3\,2\,1)$ ("Hurwitz.txt").

$s > 2$ commute with $\mathfrak{m}_{l_1}$; hence, they are not affected by these moves (except for the order). As a result, $\mathfrak{m}_{l_1}$ and $\mathfrak{m}_{l_2}$ become neighbors in $\bar{\mathfrak{m}}$, *i.e.*, $l_2 = l_1 \pm 1$. After one extra move σ_l, where $l = \min\{l_1, l_2\}$, the transposition $(i_0\, i_2)$ becomes a factor of $\bar{\mathfrak{m}}$, and the element i_1 can be removed from the above sequence.

Applying the reduction procedure $(k-1)$ times, we arrive at an equivalent factorization which contains $(i\,j)$ as a certain factor $\mathfrak{m}_l$. By the sequence $\sigma_{l-1}^{-1}, \sigma_{l-2}^{-1}, \ldots$ of inverse Hurwitz moves this factor can be moved to the first place. □

Now, we prove the uniqueness by induction in n and r. The case $n = 2$, $r \geqslant 1$ is obvious (any factorization is just a sequence of copies of $(1\,2)$), and we can assume that $n \geqslant 3$.

First, consider the case $c > 1$. Pick a pair (i, j) of representatives of two distinct cycles of g. According to Lemma 10.11, we can assume that $\mathfrak{m}_1 = (i\,j)$. Remove $\mathfrak{m}_1$ and denote by $\bar{\mathfrak{m}}'$ the new factorization of length $r-1$. Clearly, $\mathfrak{m}'_\infty = g' := (i\,j)g$, and we assert that $\bar{\mathfrak{m}}'$ is transitive. Indeed, since $\bar{\mathfrak{m}}$ is transitive, $\bar{\mathfrak{m}}'$ may have at worst two distinct orbits, those of i and j. However, i and j are within one cycle of g'; hence, the two orbits coincide. By the induction hypothesis, $\bar{\mathfrak{m}}'$ is unique in its strong Hurwitz equivalence class; hence, so is $\bar{\mathfrak{m}}$.

Now, consider the case where $c = 1$, *i.e.*, g is a single cycle of length n. Up to conjugation, $g = (1\,2\,\ldots\,n)^{-1}$. As in the previous case, using Lemma 10.11, we can assume that $\mathfrak{m}_1 = (1\,2)$, remove this element, and consider the new factorization $\bar{\mathfrak{m}}'$ of $g' := (2\,\ldots\,n)^{-1}$. This time, either $\bar{\mathfrak{m}}'$ is transitive (and then necessarily $r \geqslant n+1$) or the monodromy group of $\bar{\mathfrak{m}}'$ has two orbits, $\{1\}$ and $\{2, \ldots, n\}$. By the induction hypothesis, in each case $\bar{\mathfrak{m}}'$ is unique up to strong Hurwitz equivalence. (In the case of two orbits, the induction hypothesis is applied to $\bar{\mathfrak{m}}'$ regarded as an $\mathbb{S}_{n-1}$-valued factorization of $g' \in \mathbb{S}_{n-1}$.) Thus, it remains to show that, if $r \geqslant n+1$, the two factorizations of g' become equivalent after prepending the transposition $(1\,2)$. In view of the uniqueness already proved, the two factorizations of g' can be assumed standard, *e.g.*, those described in the proof of the existence. Then the factorizations of g to be compared are

$$(1\,2) \odot (2\,3) \odot (1\,2) \odot (1\,2) \cdots \quad \text{and} \quad (1\,2) \odot (2\,3) \odot (2\,3) \odot (2\,3) \cdots,$$

with the omitted parts equal, *i.e.*, the problem reduces to comparing two particular transitive factorizations of length four of the cycle $(3\,2\,1) \in \mathbb{S}_3$. The former is taken to the latter by the sequence $\mu := \beta\sigma_3\beta\sigma_3^{-1}$, where $\beta = \sigma_2\sigma_1^2\sigma_2$. Indeed, one can check (see, *e.g.*, Listing 10.1) that, on the free group $\langle \alpha_1, \alpha_2, \alpha_3, \alpha_4 \rangle$, the braid μ acts *via* $\alpha_i \mapsto \gamma_i\alpha_i\gamma_i^{-1}$, $i = 1, \ldots, 4$, where

$$\gamma_1 = \gamma_2 = \alpha_1\alpha_2(\alpha_3\alpha_4)\alpha_2^{-1}\alpha_1^{-1} \mapsto 1 \in \mathbb{S}_3, \quad \gamma_3 = \gamma_4 = \alpha_1\alpha_2 \mapsto (3\,2\,1) \in \mathbb{S}_3,$$

and one has $(3\,2\,1) \cdot (1\,2) \cdot (3\,2\,1)^{-1} = (2\,3)$. □

Corollary 10.12. *Let X be an oriented surface and let $\varphi\colon \partial X \to S^1 = \partial\bar{D}^2$ be an orientation preserving covering. Assume, further, that either $X \cong \bar{D}^2$ or $\deg\varphi > 1$. Then φ extends to a simple ramified covering $\bar{\varphi}\colon X \to \bar{D}^2$; this extension is unique up to isotopy fixed on the boundary.*

Proof. The uniqueness of the extension is a consequence of the uniqueness part of Theorem 10.10. The existence also follows from Theorem 10.10 (assuming that $n := \deg\varphi \geqslant 2$): in geometric terms the latter states that, for any integer $g \geqslant 0$, there exists a surface X' of genus g and a simple ramified covering $\bar{\varphi}'\colon X' \to \bar{D}^2$ such that $\partial X' = \partial X$ and $\bar{\varphi}'|_{\partial X'} = \varphi$. (By the Riemann–Hurwitz formula, one has $r - 2g = n + c - 2$ in the notation of Theorem 10.10.) According to the classification of compact surfaces, if g is the genus of the original surface X, the identity map $\partial X = \partial X'$ extends to a homeomorphism $\psi\colon X \to X'$, and one can take $\bar{\varphi} = \bar{\varphi}' \circ \psi$. □

Corollary 10.13. *Under the hypotheses of Corollary* 10.12, *φ extends to a ramified covering $\bar{\varphi}\colon X \to \bar{D}^2$; this extension is unique up to homotopy in the class of ramified coverings fixed on the boundary.*

Proof. The existence is given by Corollary 10.12. For the uniqueness, observe that, by a small perturbation (homotopy), any ramified covering can be converted to a simple one, and any two simple coverings are isotopic due to Corollary 10.12. □

For completeness, we mention two other examples of uniqueness statements. The first one, Theorem 10.14, is a restatement of Proposition 4.48 in the algebraic terms, and the second one, Theorem 10.15, deals with coverings having generic branching behavior; it is the algebraic basis for Seiler's Theorem 3.56.

Theorem 10.14. *Let $g \in \mathbb{S}_n$, $n \geqslant 2$, be a product of c disjoint cycles (including those of length one). Then there is a unique, up to weak Hurwitz equivalence, $\mathbb{S}_n$-valued factorization of g of length c with one factor a cycle of length n and the other $(c-1)$ factors transpositions.* ◁

Theorem 10.15 (Seiler [144, 145]). *For each pair of integers $d > 0$ and $g \geqslant 0$, there is a unique, up to weak Hurwitz equivalence, transitive factorization $\bar{\mathfrak{m}}$ of $1 \in \mathbb{S}_{12d}$ of length $10d + 2g$ with $\mathfrak{m}_1$ a product of $6d$ cycles of length two, $\mathfrak{m}_2$ a product of $4d$ cycles of length three, and the other factors $\mathfrak{m}_i$ transpositions.* ▷ [144, 145, 75]

The proof of Theorem 10.15 found in [75] is purely algebraic, in the spirit of the proof of Theorem 10.10. Alternatively, one can prove this theorem geometrically, using equivalence of dessins. Indeed, due to Theorem 10.10, each embedded bipartite ribbon graph without monogonal regions of positive genus extends to a dessin with all $\times$-vertices of index one, and such an extension is unique up to equivalence. Hence, one can ignore the $\times$-vertices and incident edges, and it suffices to show that, by a sequence of moves shown in Figure 4.1 (a), any $(3, 2)$-regular embedded bipartite ribbon graph can be brought to a certain standard form.

Non-equivalent factorizations of $1 \in \mathbb{S}_n$ (see [138, 139])

If the factorizations are not assumed simple, the uniqueness may be lost. Probably, the simplest example is due to Protopopov, who gave a complete classification of the equivalence classes of factorizations of $1 \in \mathbb{S}_n$ with all factors cycles of length three.

To define extra invariants, recall that the weak equivalence classes of factorizations of $1 \in \mathbb{S}_n$ are in a one-to-one correspondence with the homeomorphism classes of ramified coverings $X \to S^2$ of degree n.

Let $\varphi\colon X \to S^2$ be a ramified covering of degree n. The only Spin-structure θ on the sphere S^2 restricts to a Spin-structure on the set $S^\sharp$ of regular values of φ, and the restriction lifts to a Spin-structure $\tilde{\theta}^\sharp$ on the pull-back $X^\sharp := \varphi^{-1}(S^\sharp)$. Let l be a small loop surrounding a critical value s. The restriction $\theta|_l$ is null-cobordant and, if all critical points over s are of odd ramification index, so is the lift to $\varphi^{-1}(l)$. Thus, if all critical points of φ are of odd ramification index, $\tilde{\theta}^\sharp$ extends to a unique Spin-structure $\tilde{\theta}$ on X. The Arf-invariant $\operatorname{Arf}\tilde{\theta} \in \mathbb{Z}_2$ is obviously a topological invariant of φ; hence, it is also a weak Hurwitz equivalence invariant of the factorization $\bar{\mathfrak{m}}$ of $1 \in \mathbb{S}_n$ corresponding to φ. Once again, $\operatorname{Arf}\bar{\mathfrak{m}}$ is well defined whenever each factor of $\bar{\mathfrak{m}}$ is a product of cycles of odd length.

For another invariant, consider an $\mathbb{S}_n$-valued, $n = 3$ or 4, factorization $\bar{\mathfrak{m}}$ with all factors conjugate to, respectively, $(1\,2\,3)$ or $(1\,2\,3)(4)$. Since cycles of length three are even, we have $\operatorname{Im}\bar{\mathfrak{m}} \subset \mathbb{A}_n$, and in $\mathbb{A}_n$ the conjugacy class of the factors of $\bar{\mathfrak{m}}$ splits into two, say, $\mathcal{C}_1$ and $\mathcal{C}_2$. Define the *signature* of $\bar{\mathfrak{m}}$ as

$$\sigma(\bar{\mathfrak{m}}) := \bigl|\operatorname{Card}\{i \mid \mathfrak{m}_i \in \mathcal{C}_1\} - \operatorname{Card}\{i \mid \mathfrak{m}_i \in \mathcal{C}_2\}\bigr|.$$

Theorem 10.16 (Protopopov [138, 139]). *Up to weak equivalence, a transitive $\mathbb{S}_n$-valued factorization $1 = \mathfrak{m}_1 \odot \cdots \odot \mathfrak{m}_r$ with each factor $\mathfrak{m}_i \overset{c}{\sim} (1\,2\,3)$ is determined by the signature σ, if $n = 3$, the pair $(\sigma, \operatorname{Arf})$, if $n = 4$, or the* Arf*-invariant* Arf, *if $n \geqslant 5$. The invariants $(n, r, \sigma, \operatorname{Arf})$ are subject to the restrictions $2 \leqslant n - 1 \leqslant r$,*

$0 \leqslant \sigma \leqslant r$, $\sigma = 3r \bmod 6$, *and* $\mathrm{Arf} = 0$ *if* $r = n - 1$. *Any collection* $(n, r, \sigma, \mathrm{Arf})$ *satisfying these restrictions is realized.* $\rhd$

10.2 Factorizations in Γ

For a group $G = \Gamma$, $\tilde{\Gamma}$, or $\mathbb{B}_3$, a *simple* G-valued factorization is a factorization with all factors conjugate to a positive Dehn twist (for $G = \Gamma$ or $\tilde{\Gamma}$) or an Artin generator σ_1 (for $G = \mathbb{B}_3$ or, more generally, $G = \mathbb{B}_n$). Since $\tilde{\Gamma} \twoheadrightarrow \Gamma$ and $\mathbb{B}_3 \twoheadrightarrow \Gamma$ are central extensions and each Dehn twist in Γ lifts to a unique Dehn twist in $\tilde{\Gamma}$ (respectively, a unique Artin generator in $\mathbb{B}_3$), the problems of the classification of simple factorizations in all three groups are equivalent. The lift of the monodromy at infinity is uniquely determined by the requirement $\operatorname{dg} \mathfrak{m}_\infty = r \bmod 12$ for $\tilde{\Gamma}$ or $\operatorname{dg} \mathfrak{m}_\infty = r$ for $\mathbb{B}_3$, where r is the length of $\bar{\mathfrak{m}}$. In particular, the length of a simple factorization of a braid $\mathfrak{m}_\infty \in \mathbb{B}_3$ equals $\operatorname{dg} \mathfrak{m}_\infty$; hence, it does not need to be specified.

To shorten the statements, we reserve the term *r-factorization* for a simple Γ-valued factorization of length r. In view of the equivalence between the three groups, for an r-factorization $\bar{\mathfrak{m}}$ we can freely speak about its fundamental group $\pi_1(\bar{\mathfrak{m}})$ (lifting $\bar{\mathfrak{m}}$ to $\mathbb{B}_3$) and its transcendental lattice $\mathbf{T}(\bar{\mathfrak{m}})$ (lifting $\bar{\mathfrak{m}}$ to $\tilde{\Gamma}$, see Section 10.2.3 below).

Throughout this section, we fix the notation $L := \mathbb{X}\mathbb{Y}$ and $R := (\mathbb{Y}\mathbb{X})^{-1} = \mathbb{X}^2\mathbb{Y}$, see (2.3), regarding these matrices as elements of Γ or $\tilde{\Gamma}$, depending on the context. Recall that $R = t_{\mathbf{a}}$ is a positive Dehn twist and $L \overset{c}{\sim} R^{-1}$.

The existence and partial uniqueness of simple factorizations in $\mathbb{B}_3$ are given by the following three theorems.

Theorem 10.17 (Orevkov [133]). *Let* $\beta_+\Delta^m$ *be the Garside normal form of a braid* $\beta \in \mathbb{B}_3$. *Then* β *admits a simple factorization if and only if either* $m \geqslant 0$ *or* $m < 0$ *and* β_+ *becomes equal to* Δ^{-m} *after removing an appropriate collection of* $\operatorname{dg} \beta$ *letters from its representation by a positive word in* $\{\sigma_1, \sigma_2\}$. $\rhd$

Theorem 10.18 (Moishezon, Livné [121]). *A central element* $\Delta^{2n} \in \mathbb{B}_3$, $n \in \mathbb{N}$, *has a unique, up to strong Hurwitz equivalence, simple monodromy factorization.* $\rhd$

In terms of Γ, Theorem 10.18 states that $1 \in \Gamma$ admits a unique $6n$-factorization for each $n \in \mathbb{N}$. This theorem is the algebraic version of Theorem 3.89 classifying elliptic Lefschetz fibrations; it was recently generalized by Orevkov.

Theorem 10.19 (Orevkov [132]). *Any element* $\beta \in \mathbb{B}_3^+$ *has a unique, up to strong Hurwitz equivalence, simple monodromy factorization.* $\rhd$

10.2.1 Exponential examples

We begin with an example showing that, in a group as simple (or as complicated?) as Γ, the Hurwitz equivalence problem is much harder than it may seem.

Theorem 10.20. *For each integer $r \geqslant 2$, the parabolic element L^{5r-6} admits at least $\tilde{T}(r-1)$ strong equivalence classes of r-factorizations, which also share the same fundamental group and transcendental lattice. These r-factorizations form $T(r-1)$ weak equivalence classes.*

Explicitly, the r-factorizations in Theorem 10.20 are

$$L^{5r-6} = \bigodot_{i=1}^{r} (L^{n_i - s}\mathbb{X})L^{-1}(L^{n_i - s}\mathbb{X})^{-1},$$

where $(m_1, \ldots, m_{r-1})$ is the signature of a marked admissible tree with $2(r-1)$ vertices (*cf.* Theorem 1.73), $n_i := \sum_{j=1}^{r-1} m_i$, $i = 1, \ldots, r$ (so that $n_r = 0$), and s is an additional shift in the range $0 \leqslant s < n_{r-1}$. These r-factorizations (weak equivalence classes) differ by their monodromy groups (respectively, conjugacy classes of the monodromy groups).

Since there is an inclusion $\mathbb{B}_3 \hookrightarrow \mathbb{B}_n$, $n \geqslant 3$, and the *strong* equivalence of two factorizations does not depend on the ambient group, we have a corollary.

Corollary 10.21. *In any braid group $\mathbb{B}_n$, $n \geqslant 3$, and for any integer $r \geqslant 2$, the element $\sigma_2^{6-5r}(\sigma_1\sigma_2)^{3r-3}$ admits at least $\tilde{T}(r-1)$ distinct strong Hurwitz equivalence classes of simple factorizations of length r.* ◁

Proof of Theorem 10.20. Pick and admissible tree $\mathfrak{T}$ and consider a properly marked simple pseudo-tree $(\mathfrak{S}, e)$, $\mathfrak{S} := \operatorname{Sk}\mathfrak{T}$. Let $\gamma_1, \ldots, \gamma_r$ be a geometric basis for the group $\pi_1(\mathfrak{S}, e) = \pi_1(\bar{D}_r^{\circ})$. (Due to Theorem 2.28, the particular choice of the generators is irrelevant; for example, one can mark $\mathfrak{T}$ so that e is between v_{r-1} and v_r and take for the basis elements $\gamma_i := (e, (\mathbf{xy})^{n_i - s}\mathbf{xyx}^{-2}\cdot(\mathbf{xy})^{-n_i+s})$, *cf.* Figure 9.3.) The r-factorization in question is $\mathfrak{m}_i := \operatorname{val}\gamma_i$, $i = 1, \ldots, r$. By construction, the product $\gamma_1 \cdots \gamma_r$ is homotopic to the boundary $(\mathbf{xy})^{5r-6}$ of the outer region of $\mathfrak{S}$; hence, $\mathfrak{m}_\infty = L^{5r-6}$. The r-factorizations obtained from distinct properly marked pseudo-trees $(\mathfrak{S}, e)$ are not equivalent, as their monodromy groups $\operatorname{Im}\bar{\mathfrak{m}} = \operatorname{stab} e \subset \Gamma$ differ. Disregarding the marking, we obtain conjugate (by a power of L) r-factorizations, and their weak equivalence classes are enumerated by simple pseudo-trees: they are distinguished by the conjugacy classes $[\![\operatorname{Im}\bar{\mathfrak{m}}]\!] = \operatorname{Stab}\mathfrak{S}$.

Clearly, any r-factorizations arising from a simple pseudo-tree is merely the Γ-valued reduction of the homological invariant of an appropriate elliptic surface given by Theorem 9.30 (with $k = r - 1$). Hence, all these r-factorizations share the same fundamental group (cyclic for $r \geqslant 4$), see Addendum 9.36, and transcendental lattice, see Theorem 9.31, Items 1 and 2, or Example 10.54 below. □

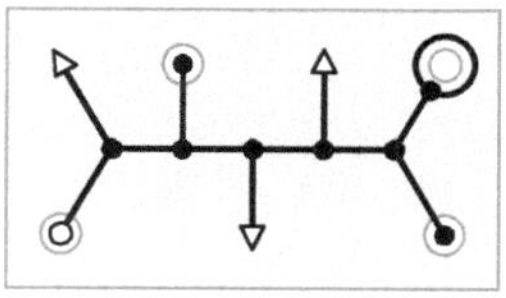

Figure 10.2. A pseudo-tree $\mathfrak{S}$ and punctured disk $B^\sharp$.

A generalization

As a generalization, consider a pseudo-tree $\mathfrak{S} := \operatorname{Sk}(\mathfrak{T}, \ell)$ defined by an admissible tree $\mathfrak{T}$ with $2k$ vertices and vertex function ℓ. Denote by $\#_*$ the number of leaves of $\mathfrak{T}$ on which ℓ takes value $* \in \{0, \bullet, \circ, \vartriangle\}$.

Consider the embedding $\mathfrak{S} \hookrightarrow \operatorname{Supp}\mathfrak{S}$ and denote by B a regular neighborhood of the union of the compact part $\mathfrak{S}^c$ and the disks patching the monogonal regions of $\mathfrak{S}$. Let $B^\sharp$ be the disk B punctured at the center of each monogonal region and at each monovalent vertex of $\mathfrak{S}$, see Figure 10.2. There is an epimorphism $\varphi\colon \pi_1(B^\sharp) \twoheadrightarrow \pi_1^{\mathrm{orb}}(\mathfrak{S})$. The conjugacy class $\mathfrak{m}_\infty(\mathfrak{S}) := \operatorname{val}\varphi[\partial B]$ in Γ is called the *monodromy at infinity* of $\mathfrak{S}$.

Pick a basepoint $b \in \partial B$ and a geometric basis $\{\gamma_1, \dots, \gamma_r\}$ for $\pi_1(B^\sharp, b)$ (the particular choice of a basis is irrelevant due to Theorem 2.28) and define a Γ-valued factorization $\bar{\mathfrak{m}}$ of $\mathfrak{m}_\infty(\mathfrak{S})$ *via* $\mathfrak{m}_i := \operatorname{val}\varphi(\gamma_i) \in \Gamma$, $i = 1, \dots, r$. Its passport consists of $\#_\bullet$ copies of $[\![\mathbb{X}]\!]$, $\#_\circ$ copies of $[\![\mathbb{Y}]\!]$, and $\#_0 = k + 1 - \#_\bullet - \#_\circ - \#_\vartriangle$ copies of $[\![R]\!]$. In particular, $r = k + 1 - \#_\vartriangle$ and $\bar{\mathfrak{m}}$ is simple if and only if $\#_\bullet = \#_\circ = 0$.

If $\#_\vartriangle = 0$, the skeleton is finite and $\bar{\mathfrak{m}}$ is the homological invariant of a certain extremal elliptic surface, *cf.* Sections 9.2.2 and 10.3.1. In this case, one has $\mathfrak{m}_\infty \in [\![L^m]\!]$, where $m = 5k - 1 - \#_\bullet - 2\#_\circ$.

In general, $\mathfrak{m}_\infty = \mathfrak{m}_\infty(\mathfrak{S})$ can be found as follows. Start at a $\vartriangle$-vertex of $\mathfrak{S}$ and denote it by t_1. Given a $\vartriangle$-vertex t_i, let e_i be the edge of $\mathfrak{S}$ incident to t_i and denote by s_i the smallest positive integer such that the edge $e_i \uparrow \mathbf{y}(\mathbf{xy})^{s_i}$ is incident to a $\vartriangle$-vertex; denote the latter $\vartriangle$-vertex by t_{i+1} and stop the process when $t_{i+1} = t_1$. We obtain a sequence of integers $(s_1, \dots, s_{\#_\vartriangle})$; intuitively, s_i is the vertex distance, with only $\bullet$-vertices counted, in the boundary ∂B between two consecutive $\vartriangle$-vertices t_i and t_{i+1}. For example, in Figure 10.2, starting from the upper left corner, one has $(s_1, s_2, s_3) = (4, 9, 6)$; for the pseudo-tree shown in Figure 10.5 below, $(s_1, s_2, s_3) = (2, 5, 6)$. Note that $\sum_i s_i = 5k - 1 - \#_\bullet - 2\#_\circ - 2\#_\vartriangle$ is the 'length' of the boundary ∂B and, in view of the restriction on the $\vartriangle$-values of ℓ, one has $s_i \geqslant 2$ for all i. The following statement is straightforward.

Proposition 10.22. *Let $\mathfrak{S} := \operatorname{Sk}(\mathfrak{T}, \ell)$ be a pseudo-tree with $\#_\vartriangle > 0$. Then $\mathfrak{m}_\infty(\mathfrak{S}) \overset{c}{\sim} \prod_{i=1}^{\#_\vartriangle} RL^{s_i - 2}$ (left to right product). Two pseudo-trees have conjugate monodromy at infinity if and only if their sequences (s'_i), (s''_j) differ by a cyclic permutation.* $\lhd$

Table 10.1. Irreducible maximizing sextics with a type $\mathbf{E}$ singular point.

$^*\mathbf{E}_8 \oplus \mathbf{A}_{10} \oplus \mathbf{A}_1$	$\mathbf{E}_6 \oplus \mathbf{D}_5 \oplus \mathbf{A}_8$	$(\mathbf{E}_6 \oplus \mathbf{A}_8 \oplus \mathbf{A}_2) \oplus \mathbf{A}_2 \oplus \mathbf{A}_1$
$\mathbf{E}_8 \oplus \mathbf{A}_8 \oplus \mathbf{A}_2 \oplus \mathbf{A}_1$	$^*\mathbf{E}_6 \oplus \mathbf{D}_5 \oplus \mathbf{A}_6 \oplus \mathbf{A}_2$	$^*\mathbf{E}_6 \oplus \mathbf{A}_7 \oplus \mathbf{A}_4 \oplus \mathbf{A}_2$
$^*\mathbf{E}_8 \oplus \mathbf{A}_6 \oplus \mathbf{A}_4 \oplus \mathbf{A}_1$	$^*\mathbf{E}_6 \oplus \mathbf{A}_{10} \oplus \mathbf{A}_3$	$^*\mathbf{E}_6 \oplus \mathbf{A}_6 \oplus \mathbf{A}_4 \oplus \mathbf{A}_2 \oplus \mathbf{A}_1$
$\mathbf{E}_8 \oplus \mathbf{A}_5 \oplus \mathbf{A}_4 \oplus \mathbf{A}_2$	$^*\mathbf{E}_6 \oplus \mathbf{A}_{10} \oplus \mathbf{A}_2 \oplus \mathbf{A}_1$	$\mathbf{E}_6 \oplus \mathbf{A}_5 \oplus 2\mathbf{A}_4$
$(2\mathbf{E}_6 \oplus \mathbf{A}_5) \oplus \mathbf{A}_2$	$\mathbf{E}_6 \oplus \mathbf{A}_9 \oplus \mathbf{A}_4$	$(\mathbf{E}_6 \oplus \mathbf{A}_5 \oplus 2\mathbf{A}_2) \oplus \mathbf{A}_4$
$2\mathbf{E}_6 \oplus \mathbf{A}_4 \oplus \mathbf{A}_3$	$\mathbf{E}_6 \oplus \mathbf{A}_8 \oplus \mathbf{A}_4 \oplus \mathbf{A}_1$	

Other examples

For completeness, we cite a few examples concerning monodromy factorizations (not necessarily simple) in the other braid groups $\mathbb{B}_n$.

Example 10.23 (Auroux, Kulikov, Shevchishin [101]). In the group $\mathbb{B}_4$, there are two nonequivalent simple factorizations

$$\begin{aligned}(bacb)^3a^{-3}c^{-3} &= a\uparrow(c^{-2}b)^{-1} \odot b \odot b\uparrow(ac^3)^{-1} \odot a\uparrow(ac^5b^{-1})^{-1} \odot c \odot c \\ &= b \odot b\uparrow(ac)^{-1} \odot b\uparrow(ac)^{-1} \odot b\uparrow(ac)^{-2} \odot b\uparrow(ac)^{-2} \odot b\uparrow(ac)^{-3}.\end{aligned}$$

Here, we shorten $\{a,b,c\} := \{\sigma_1,\sigma_2,\sigma_3\}$ and use $\uparrow$ for the adjoint action of $\mathbb{B}_4$ on itself. The identities are easily checked, *e.g.*, using `"IsBraidID"` in Listing 2.5. The two factorizations differ by their $\mathbb{S}_4$-valued reductions: the former is transitive (*i.e.*, the corresponding Hurwitz curve is irreducible), and the latter is not (the curve splits into two degree 2 components).

Example 10.24. A number of examples of nonequivalent non-simple factorizations arise from maximizing plane sextics with a distinguished type $\mathbf{E}$ singular point: the Γ-valued reduction of the braid monodromy of the proper model C of such a sextic can be regarded as a factorization of the (negative) monodromy about the distinguished singular fiber F_{I}. Listed in Table 10.1 are all sets of singularities realized by pairs D_1, D_2 of *irreducible* maximizing sextics whose proper models have essentially distinct skeletons (*i.e.*, we disregard pairs of complex conjugate curves), *cf.* Tables 8.1, 8.5, and 8.6. The corresponding factorizations differ by their monodromy groups. Since the curves are irreducible, all factorizations have transitive $\mathbb{S}_3$-valued reduction.

Pairs marked with a * are distinguished by the transcendental lattice, see Section 10.2.3 below, as explained in Example 10.52.

The non-equivalence of the factorizations arising from the two reducible sextics with the set of singularities $\mathbf{E}_6 \oplus \mathbf{A}_7 \oplus \mathbf{A}_3 \oplus \mathbf{A}_2 \oplus \mathbf{A}_1$, Figures 8.3 (a)–3 and 8.15 (n), was established in [8] by computing the Hurwitz orbits of the reductions modulo 32.

Example 10.25 (Moishezon [119]). This example is the most remarkable one: in $\mathbb{B}_{54}$, there are infinitely many equivalence classes of factorizations of the Garside element Δ^2 with the conjugacy classes of the factors as follows: 378 copies of $[\![\sigma_1^3]\!]$, 756 copies of $[\![\sigma_1^2]\!]$, and 216 copies of $[\![\sigma_1]\!]$. To my knowledge, even in $\mathbb{B}_3$, and even for simple factorizations, the finiteness of the number of classes of r-factorizations of a given element is still an open problem.

10.2.2 2-factorizations (see [64])

It appears that the only nontrivial case where the problem of Hurwitz equivalence of r-factorizations is solved completely is that of $r = 2$. Even this apparently simple case has interesting geometric applications; they are discussed in Sections 10.3.2 and 10.3.3.

For a positive word A in $\{L, R\}$ define the *transpose* A^{t} as the word obtained from A by interchanging $L \leftrightarrow R$ and reversing the order of the letters. Obviously, $(A^{\mathrm{t}})^{\mathrm{t}} = A$ and transposed words represent transposed matrices in Γ or $\tilde{\Gamma}$. We have the identity

$$g^{-1} = \mathbb{Y}^{-1} g^{\mathrm{t}} \mathbb{Y}, \quad g \in \tilde{\Gamma}. \tag{10.26}$$

(Both maps are anti-homomorphisms and they coincide on the generators L, R.)

Theorem 10.27 (see [64]). *An element $g \in \Gamma$ admits a 2-factorization $\bar{\mathfrak{m}}$ if and only if either*

1. $g \overset{c}{\sim} \mathbb{X} = R \odot L^{-1}$ *(and then* $\operatorname{Im} \bar{\mathfrak{m}} = \Gamma$*), or*
2. $g \overset{c}{\sim} R^2 = R \odot R$ *(and then* $\operatorname{Im} \bar{\mathfrak{m}} \cong \mathbb{Z}$*), or*
3. $g \overset{c}{\sim} L^2 A L^2 A^{\mathrm{t}}$ *for a positive word A in $\{L, R\}$ (and then* $\operatorname{Im} \bar{\mathfrak{m}} \cong \mathbb{F}_2$*).*

The 2-factorizations as in Item 1, 2, and 3 of Theorem 10.27 are called *elliptic*, *parabolic*, and *hyperbolic*, respectively. In the hyperbolic case, the weak equivalence class of $\bar{\mathfrak{m}}$ is determined by the representation as in the statement and is

$$L^2 A L^2 A^{\mathrm{t}} = \mathbb{X} L^{-1} \mathbb{X}^{-1} \odot (\mathbb{Y} A \mathbb{X}) L^{-1} (\mathbb{Y} A \mathbb{X})^{-1}; \tag{10.28}$$

this expression is verified using (10.26). Note also that L^4 ($A = \varnothing$ in 10.27 (3)) is a special case of Theorem 10.20 with $r = 2$.

Remark 10.29. As a curious fact, note that elements admitting a factorization into products of *positive* Dehn twists tend to be 'negative': in the general case 10.27 (3), the count of copies of L exceeds that of R by four. Another example is Theorem 10.20, where a 'very negative' element L^{5r-6} decomposes into a product of a large number of positive Dehn twists. At present, I do not know how general this phenomenon is.

Theorem 10.30 (see [64]). *An element $g \overset{c}{\sim} L^4$ has a unique weak equivalence class of 2-factorizations, which splits into 2 strong equivalence classes:*

$$L^4 = R \odot (R^{-1}L^2) R (R^{-1}L^2)^{-1} = LRL^{-1} \odot (LR^{-1}L^2) R (LR^{-1}L^2)^{-1}$$

(cf. Theorem 10.20 with $r = 2$). For all other elements, each weak equivalence class of 2-factorizations constitutes a single strong equivalence class.

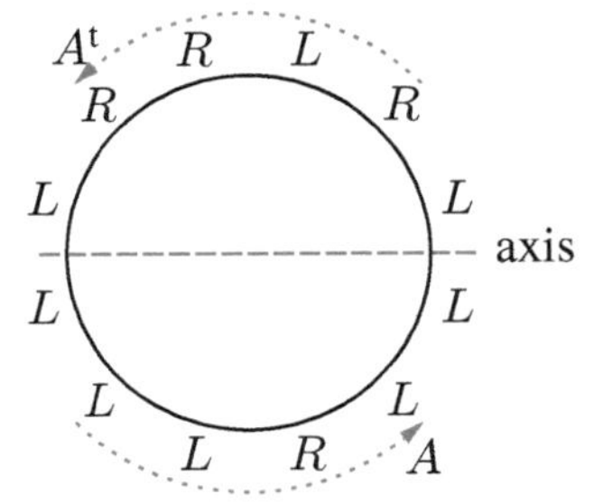

Figure 10.3. Cyclic diagram associated to $L^2AL^2A^t$ and its para-symmetry.

For the next statement, we need extra notation. For an integer $m \geqslant 0$, let

$$V_m := L^2(LR)^m L^2(LR)^m \overset{c}{\sim} L^2(RL)^m L^2(RL)^m. \tag{10.31}$$

For an odd integer $n = 2k+1 \geqslant 1$, consider the word $w_{1/n} := l(lr)^k l(lr)^k$ in the alphabet $\{l, r\}$. Denote by $w[i]$, $i \geqslant 0$, the i-th letter of a word w, the indexing starting from 0. Pick an odd integer $1 \leqslant m < n$ prime to n and let w_q, $q := m/n$, be the word in $\{l, r\}$ of length $2n$ defined by $w_q[i] = w_{1/n}[mi \bmod 2n]$, $i = 0, \ldots, 2n-1$. Given a positive word B in $\{L, R\}$, denote by $w_q\{B\}$ be the word obtained from w_q by inserting a copy of B between $w_q[2i]$ and $w_q[2i+1]$ and a copy of B^t between $w_q[2i+1]$ and $w_q[2i+2]$, $i = 0, \ldots, n-1$. Finally, let $W_q(B)$ be the word obtained from $w_q\{B\}$ by the substitution $l \mapsto L^2$, $r \mapsto R^2$.

Theorem 10.32 (see [64]). *An element $g \in \Gamma$ admits at most two distinct strong equivalence classes of 2-factorizations. The number of classes is two if and only if either $g \overset{c}{\sim} V_m$, $m \geqslant 0$, or $g \overset{c}{\sim} W_q(B)$, where $0 < q < 1$ is a rational number with odd numerator and denominator and B is a positive word in $\{L, R\}$, possibly empty.*

We outline the proofs of Theorems 10.27, 10.30, and 10.32 further in this section.

Recall that the conjugacy classes of non-elliptic elements of Γ are enumerated by positive cyclic words in $\{L, R\}$, see (2.16), and we represent the class of an element $g \in \Gamma$ by its *cyclic diagram* $\mathcal{D} := \mathcal{D}_g$, placing the letters equidistantly in the oriented unit circle. A *symmetry* of a cyclic diagram $\mathcal{D}$ is a symmetry of the underlying circle taking letters to letters. Each symmetry is either a rotation (by a rational angle) or reflection; in the latter case, each point of intersection of the axis and the circle either coincides with a letter or bisects the arc connecting a pair of consecutive letters. A symmetry of $\mathcal{D}$ is called *literal* if it preserves the letters, taking L to L and R to R.

Lemma 10.33 (obvious). *A non-elliptic element $g \in \Gamma$ is a power h^n, $h \in \Gamma$, $n \in \mathbb{N}$ if and only if the cyclic diagram $\mathcal{D}_g$ admits a literal rotation symmetry of order n.* ◁

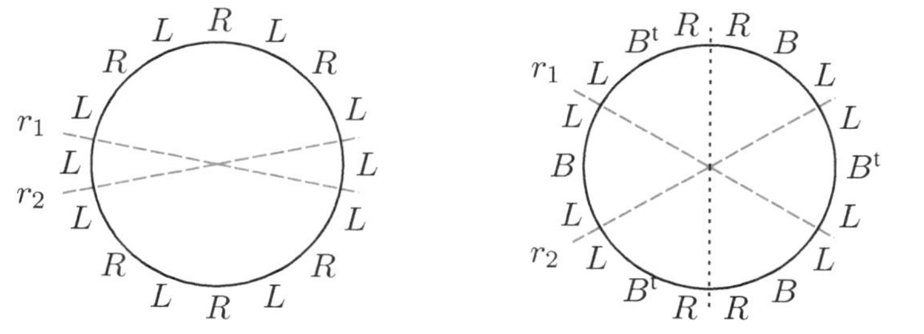

(a) V_m (two common anchors) (b) $W_q(B)$ (no common anchors)

Figure 10.4. Cyclic diagrams with two para-symmetries.

Definition 10.34. A *para-symmetry* of a cyclic diagram $\mathcal{D}$ is a reflection symmetry r of $\mathcal{D}$ with the following properties:

- the axis of r does not pass through a letter of $\mathcal{D}$;
- the four letters immediately adjacent to the axis (called *anchors*) are all L;
- the type of each other letter of $\mathcal{D}$ is reversed by r.

An example of a para-symmetry is shown in Figure 10.3. It is clear from the figure that para-symmetries of $\mathcal{D}_g$, $g \in \Gamma$, are in a one-to-one correspondence with pairs of representations $g \overset{c}{\sim} L^2AL^2A^{\mathrm{t}} \overset{c}{\sim} L^2A^{\mathrm{t}}L^2A$.

Lemma 10.35. *A non-empty cyclic diagram $\mathcal{D}$ may have at most two distinct para-symmetries. The number of para-symmetries is two if and only if $\mathcal{D}$ represents a word of the form V_m or $W_q(B)$ as in Theorem* 10.32.

Proof. Assume that $\mathcal{D}$ has two para-symmetries r_1, r_2. They generate a finite dihedral group $\mathbb{D}_{2n}$. Let $s := r_1r_2$ be the generator of the cyclic subgroup $\mathbb{Z}_n \subset \mathbb{D}_{2n}$; it is a rotation through angle 2θ, where θ is the angle between the two axes.

If r_1 and r_2 have a common anchor, the $\mathbb{D}_{2n}$ action on $\mathcal{D}$ is obviously transitive and $\mathcal{D}$ represents V_m, where $n = 2m + 1$, see Figure 10.4 (a). In this case, $\mathcal{D}$ has only four pairs of consecutive copies of L; hence, it has no other para-symmetries.

If there are no common anchors, Figure 10.4 (b), consider the orbits of the $\mathbb{Z}_n$-action on $\mathcal{D}$. Call an orbit *special* or *ordinary* if it, respectively, does or does not contain an anchor. Each ordinary orbit is 'constant', *i.e.*, either L^n or R^n. For a special orbit, start with an anchor a of r_1 and observe that s preserves the letter $a \uparrow s^i$ unless $i = 0 \bmod n$ or $i = k := [n/2] \bmod n$. In the latter case, $a \uparrow s^k$ is an anchor for r_1 if n is even and $a \uparrow s^k r_1$ is an anchor for r_2 if n is odd. Since $a \uparrow s^k$ is an R, we conclude that $n = 2k + 1$ must be odd, there are four special orbits, and each special orbit is of the form LR^kL^k (*in the orbit cyclic order*), where the first and the $(k + 1)$-st letters are anchors for r_1 and r_2, respectively.

If $\mathcal{D}$ has another para-symmetry r, together with r_1 and r_2 it generates a dihedral group $\mathbb{D}_{2n'} \supset \mathbb{D}_{2n}$. Since $\mathbb{Z}_n \subset \mathbb{D}_{2n'}$ is a normal subgroup, r must take s-orbits to s-orbits, reversing their orbit order. If $n \neq 3$, a special orbit is taken to a special one, with one of the two anchors preserved and the other letters reversed. If $n = 3$, a special orbit LRL can be taken to L^3, with both copies of L preserved. In both cases, r shares an anchor with r_1 or r_2, which contradicts to the first part of the proof.

It remains to show that $g \overset{c}{\sim} W_q(B)$. We have $\theta = \pi m/n$ for an integer m prime to n. Choosing for θ the minimal angle and replacing it, if necessary, with $\pi - \theta$, we can assume that m is also odd and $0 < m < n$. The union of the special orbits is determined by θ: if $m = 1$, then $g \overset{c}{\sim} W_{1/n}(\varnothing)$ (with the ordinary orbits disregarded), see Figure 10.4 (b); otherwise, each orbit is 'stretched' m times and 'wrapped' back around the circle, so that $g \overset{c}{\sim} W_q(\varnothing)$. In the union of the special orbits, adjacent to each semiaxis of a reflection contained in $\mathbb{D}_{2n}$ is a pair of equal letters, L^2 or R^2, and these pairs are encoded, respectively, by l and r in the definition of $w_{1/n}$. These pairs divide the circle into $2n$ arcs occupied by ordinary orbits and, taking into account the full $\mathbb{D}_{2n}$ action, one can see that the union of these ordinary orbits is $B, B^{\mathrm{t}}, \ldots, B, B^{\mathrm{t}}$, where B is the portion of the union in one of the arcs, see Figure 10.4 (b). □

Corollary 10.36. *Assume that $g \in \Gamma$ is not conjugate to L^4 and the cyclic diagram $\mathcal{D} := \mathcal{D}_g$ has a para-symmetry r. Then the group G of literal symmetries of $\mathcal{D}$ is at most $\mathbb{Z}_2 \times \mathbb{Z}_2$. More precisely, $G \neq 0$ only in one of the following mutually exclusive cases (where A, B are positive words in $\{L, R\}$ and q is as in Theorem* 10.32):

1. *$g \overset{c}{\sim} L^2AL^2A^{\mathrm{t}}$ with $A^{\mathrm{t}} = A$: the para-symmetry r is unique and $G = \mathbb{Z}_2$ is generated by a rotation commuting with r;*
2. *$g \overset{c}{\sim} L^2AL^2A^{\mathrm{t}}$ with A palindromic: the para-symmetry r is unique and $G = \mathbb{Z}_2$ is generated by a reflection commuting with r;*
3. *$g \overset{c}{\sim} W_q(B)$ with $B \neq \varnothing$ palindromic: $G = \mathbb{Z}_2$ is generated by a reflection interchanging the two para-symmetries of $\mathcal{D}$;*
4. *$g \overset{c}{\sim} V_m$, $m > 0$, or $g \overset{c}{\sim} W_q(\varnothing)$: $G \cong \mathbb{Z}_2 \times \mathbb{Z}_2$ is generated by two reflections, both interchanging the two para-symmetries of $\mathcal{D}$.*

Proof. It suffices to apply Lemma 10.35, observing that, for any $s \in G$, the product $s^{-1} \circ r \circ s$ is also a para-symmetry. As an extra restriction, since g is not conjugate to L^4, a para-symmetry cannot be literal and the axes of two para-symmetries cannot be orthogonal. Details are left to the reader, see [64]. □

Corollary 10.37. *Unless $g \overset{c}{\sim} L^4$, distinct para-symmetries of the cyclic diagram $\mathcal{D}_g$ are not isomorphic, i.e., not related by a literal* rotation *symmetry.* ◁

Theorem 10.38 (a refinement of Theorem 10.27 (3), see [64]). *The strong equivalence classes of hyperbolic 2-factorizations $\bar{\mathfrak{m}}$ with given $\mathfrak{m}_\infty = g \in \Gamma$ are in a one-to-one correspondence with the para-symmetries of the cyclic diagram $\mathcal{D}_g$.*

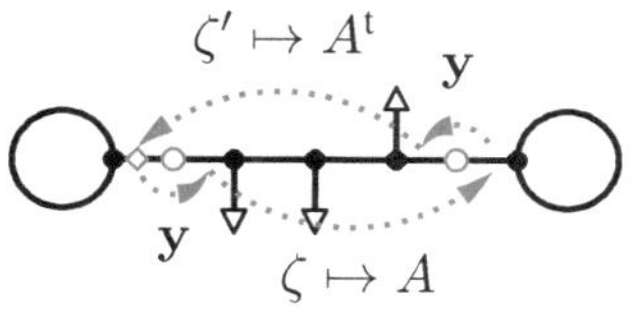

(a) The words A and A^{t}

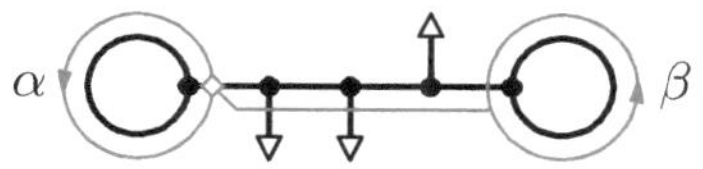

(b) A basis for $\pi_1(\mathfrak{S})$

Figure 10.5. A regular pseudo-tree $\mathfrak{S}$ with two loops.

The following simple lemma is crucial for the proofs: it is this lemma that does not generalize literally to larger free groups (*cf.* [15]) and thus prevents us from extending the approach to r-factorizations with $r \geqslant 3$. (We can still use pseudo-trees to construct examples, but not to study r-factorizations systematically.)

Lemma 10.39. *Let $\mathbb{F}_2 = \langle \alpha, \beta \rangle$, and let α', β' be a basis for $\mathbb{F}_2$ with each generator conjugate to α or β. Then the pair (α', β') is weakly Hurwitz equivalent to (α, β).*

Proof. After a global conjugation and, if necessary, Hurwitz move σ_1, we can assume that $\alpha' = \alpha$ and $\beta' = \gamma^{-1}\beta\gamma$ for some word γ in $\{\alpha^{\pm 1}, \beta^{\pm 1}\}$ that does *not* start in $\alpha^{\pm 1}$ and does *not* end in $\beta^{\pm 1}$. Then, any reduced word in $\{\alpha'^{\pm 1}, \beta'^{\pm 1}\}$ admits no cancellations. On the other hand, there is a word equal to β; hence, $\beta' = \beta$ □

Proof of Theorems 10.27 *and* 10.38. Consider a 2-factorization $\bar{\mathfrak{m}}$ of $g \in \Gamma$.

If $\operatorname{Im} \bar{\mathfrak{m}} = \Gamma$, the primitive eigenvectors u_1, u_2 of $\mathfrak{m}_1, \mathfrak{m}_2$ must span $\mathcal{H}$. (Indeed, Dehn twists acting by transvections, the image $\operatorname{Im} \bar{\mathfrak{m}}$ must preserve the subgroup $\mathbb{Z}u_1 + \mathbb{Z}u_2$; on the other hand, Γ acts transitively on the set of primitive vectors of $\mathcal{H}$, see Proposition 2.6.) Thus, the pair (u_1, u_2) is conjugate to $(\mathbf{a}, \pm\mathbf{b})$ and $\bar{\mathfrak{m}}$ is conjugate to $\mathbb{X} = R \odot L^{-1}$.

Assume that $\operatorname{Im} \bar{\mathfrak{m}}$ is a proper subgroup and consider its skeleton $\mathfrak{S} := \operatorname{Sk}(\operatorname{Im} \bar{\mathfrak{m}})$. According to Theorem 1.63, $\mathfrak{S}$ is a regular pseudo-tree with at most two monogonal regions (loops). If $\mathfrak{S}$ has one loop, $\operatorname{Im} \bar{\mathfrak{m}}$ is cyclic and hence $\bar{\mathfrak{m}}$ is conjugate to $R^2 = R \odot R$.

Assume that $\mathfrak{S}$ has two loops, so that $\operatorname{Im} \bar{\mathfrak{m}} = \pi_1(\mathfrak{S}) \cong \mathbb{F}_2$. Due to the restriction on the $\vartriangle$-values of the vertex function ℓ, see page 31 or Remark 1.53, a regular pseudo-tree with two loops looks as shown in Figure 10.5: the two loops are connected by a segment, to which a number of Farey branches are attached, some pointing upwards and some, downwards. This sequence of branches can be encoded by the matrix $\operatorname{val} \zeta \in \Gamma$, where ζ is the chain shown in Figure 10.5 (a): each branch pointing upwards results in a copy of $\mathbf{x}^2\mathbf{y} \mapsto R$, and each branch pointing downwards results in a copy of $\mathbf{xy} \mapsto L$. It is clear from the figure that the loop $\zeta\mathbf{y}\zeta'\mathbf{y}$ is contractible; hence, the same sequence read in the opposite direction produces $\operatorname{val} \zeta' = \mathbb{Y}A^{-1}\mathbb{Y} = A^{t}$.

After a global conjugation, we can assume that the reference edge e is chosen as in Figure 10.5 (b). Let α, β be the geometric generators for $\pi_1(\mathfrak{S}, e)$ shown in the figure. Then, due to Lemma 10.39, the original 2-factorization is weakly equivalent to $\alpha \odot \beta = \alpha\beta = L^2AL^2A^{\mathrm{t}}$, which is precisely (10.28). In particular, $\bar{\mathfrak{m}}$ gives rise to a para-symmetry of $\mathcal{D}_g$. Conversely, each para-symmetry of $\mathcal{D}_g$ results in 2-factorization (10.28) or its appropriate conjugate.

For the strong equivalence (Theorem 10.38), we need to fix a representative of the conjugacy class $[\![g]\!]$. Let $g = \alpha\beta$ and assume that $g \neq L^4$. Due to Corollary 10.37, distinct para-symmetries of $\mathcal{D}_g$ correspond to representations $g \overset{c}{\sim} L^2A_1L^2A_1^{\mathrm{t}} \overset{c}{\sim} L^2A_2L^2A_2^{\mathrm{t}}$ with $A_1 \neq A_2, A_2^{\mathrm{t}}$, hence non-isomorphic skeletons, hence not weakly equivalent 2-factorizations. One para-symmetry gives rise to two 2-factorizations, obtained by choosing the reference edge next to one of the two monogons of $\mathfrak{S}$. They are conjugate by L^2A, and their monodromies at infinity are equal if and only if L^2A commutes with g, *i.e.*, $g = (L^2A)^2$, *cf.* Corollary 10.36 (1). In this case, the *marked* skeletons are isomorphic and the 2-factorizations are strongly equivalent.

The special case $g \overset{c}{\sim} L^4$ is essentially covered by Theorem 10.20. □

Proof of Theorems 10.30 *and* 10.32. Recall that two elements of Γ commute if and only if they are powers of a common element, see, *e.g.*, Proposition 2.15. Recall also that, for a 2-factorization $\bar{\mathfrak{m}}$, the conjugation by $\mathfrak{m}_\infty$ is the Hurwitz move σ_1^{-2}.

The centralizer of $g = \mathbb{X}$ or R^2 is generated by $h := \mathbb{X}$ or R, respectively, and it is immediate that, in these cases, the conjugate $h^{-1}\mathfrak{m}h$ of the 2-factorization $\mathfrak{m}$ given by Theorem 10.27 (1), (2) is strongly equivalent to $\bar{\mathfrak{m}}$.

The parabolic element $g \overset{c}{\sim} L^4$ is covered by Theorem 10.20.

Assume g hyperbolic. Due to Lemma 10.33 and Corollary 10.36, $\mathcal{Z}_\Gamma(g) \neq \langle g \rangle$ if and only if g is as in 10.36 (1). In this case, $\mathcal{Z}_\Gamma(g) = \langle h \rangle$, where $h := L^2A = L^2\mathbb{Y}A^{-1}\mathbb{Y}$, and, for the 2-factorization $\bar{\mathfrak{m}}$ given by (10.28), one has the identity $h\bar{\mathfrak{m}}h^{-1} = \bar{\mathfrak{m}} \uparrow \sigma_1$.

Finally, Theorem 10.32 follows from Theorem 10.38 and Lemma 10.35. □

Real elements and 2-factorizations

A *real structure* on Γ is an involutive element of $PGL(2, \mathbb{Z}) \smallsetminus \Gamma$. A real structure τ defines an involutive anti-automorphism $\hat{\tau}\colon \Gamma \to \Gamma$, $g \mapsto \tau g^{-1}\tau$. Following [143], call an element $g \in \Gamma$ *real*, or τ-*real*, if $\hat{\tau}(g) = g$. If g is τ-real, then $g^{\pm 1}\tau$ and $\tau g^{\pm 1}$ are also real structures. Hence, real elements can equivalently be defined as those admitting a decomposition into product of two real structures. Geometrically, a real element appears as the monodromy at infinity of a real Lefschetz fibration over the closed unit disk $\bar{D}^2 \subset \mathbb{C}$ with the standard real structure $z \mapsto \bar{z}$.

Real elements of Γ are studied in [143]. It is shown that all elliptic and parabolic elements are real and a hyperbolic element $g \in \Gamma$ is real if and only if its cutting period-cycle is *odd bipalindromic*, *i.e.*, is the union of two palindromic pieces of odd length. This criterion can be restated in terms of cyclic diagrams.

Proposition 10.40 (Salepci [143]). *A non-elliptic element $g \in \Gamma$ is real if and only if its cyclic diagram $\mathcal{D}_g$ admits a literal reflection symmetry.* ▷ [143]

It is clear that $\hat{\tau}$ takes a conjugacy class to a conjugacy class. Furthermore, the image $\bar{\mathcal{C}} := \hat{\tau}(\mathcal{C})$ does not depend on τ; hence, we can speak about the *conjugate* conjugacy class $\bar{\mathcal{C}}$ and about *real* conjugacy classes, *i.e.*, those satisfying $\bar{\mathcal{C}} = \mathcal{C}$. If $\mathcal{C}$ is not elliptic, the cyclic diagram of $\bar{\mathcal{C}}$ is obtained from that of $\mathcal{C}$ by a reflection or, equivalently, reversing the order. Due to Proposition 10.40, a conjugacy class $\mathcal{C}$ is real if and only if some (equivalently, any) representative of $\mathcal{C}$ is real. As another consequence, if an element $g \in \Gamma$ is τ-real, then so is any element of $\mathcal{Z}(g)$ and the set of real structures stabilizing g is a principal homogeneous space over $\mathcal{Z}(g)$.

Extend $\hat{\tau}$ to the set of r-factorizations *via*

$$\hat{\tau}(\mathfrak{m}_1, \dots, \mathfrak{m}_r) := (\hat{\tau}(\mathfrak{m}_r), \dots, \hat{\tau}(\mathfrak{m}_1)).$$

(Observe the reversed order.) An r-factorization $\bar{\mathfrak{m}}$ is called *real* if $\hat{\tau}(\bar{\mathfrak{m}})$ is strongly Hurwitz equivalent to $\bar{\mathfrak{m}}$ for some real structure τ. If $\bar{\mathfrak{m}}$ is τ-real, so is $\mathfrak{m}_\infty$; the converse is not true, see below.

In the case of 2-factorizations, there are essentially two kinds of real structures. Indeed, replacing τ with an appropriate product $\tau\mathfrak{m}_\infty^m$, we can assume that either $\hat{\tau}(\bar{\mathfrak{m}}) = \bar{\mathfrak{m}} \uparrow \sigma_1$ or $\hat{\tau}(\bar{\mathfrak{m}}) = \bar{\mathfrak{m}}$. Hence, $\bar{\mathfrak{m}}$ is the monodromy of a real Lefschetz fibration over the disk $B \cong \bar{D}^2 \subset \mathbb{C}$ with two singular fibers; in the former case, the two fibers are real, in the latter case, they are c_B-conjugate.

It is clear that each of the 2-factorizations $\mathbb{X} = R \odot L^{-1}$ and $R^2 = R \odot R$ admits both kinds of real structures. In the hyperbolic case, real 2-factorizations are given by Corollary 10.36: they are all as in 10.36 (2). It is worth mentioning that the elements $g \in \Gamma$ as in Corollary 10.36 (3) and (4) are real and 2-factorizable, but they admit no *real* 2-factorizations. Such elements are used in Öztürk, Salepci [134] to construct examples of real contact 3-manifolds admitting no real Lefschetz filling.

As in the case of single elements, we can speak about the *conjugate* $\bar{\mathcal{W}} := \hat{\tau}(\mathcal{W})$ of a *weak* Hurwitz equivalence class $\mathcal{W}$: it does not depend on τ. As a consequence of the characterization above, a class $\mathcal{W}$ is *real*, *i.e.*, $\bar{\mathcal{W}} = \mathcal{W}$, if and only if some (equivalently, each) representative of $\mathcal{W}$ is real.

10.2.3 The transcendental lattice

The material of this section is inspired by the computation in Section 9.2.1. Here, we do *not* assume factorizations simple or Γ-valued. We give a formal purely algebraic definition of the transcendental lattice of a factorization (in a slightly more general setting), prove its Hurwitz invariance, and discuss a few very first properties, posing a number of open problems. In my opinion, the concepts introduced in this section deserve a thorough study, as a means of distinguishing Hurwitz equivalence classes.

Let $\mathcal{R}$ be a commutative ring. Recall that a *quadratic form* between two $\mathcal{R}$-modules $\mathcal{L}$, $\mathcal{V}$ is a function $q\colon \mathcal{L} \to \mathcal{V}$ such that $q(ru) = r^2 q(u)$ for all $r \in \mathcal{R}$ and $u \in \mathcal{L}$ and the difference $(u, v) \mapsto q(u+v) - q(u) - q(v)$ is a bilinear form. The *kernel* of q is the submodule

$$\ker q = \ker \mathcal{L} := \{u \in \mathcal{L} \mid q(v+u) = q(v) \text{ for all } v \in \mathcal{L}\}.$$

Fix a commutative ring $\mathcal{R}$, a pair of $\mathcal{R}$-modules $\mathcal{U}$ and $\mathcal{V}$, and a symplectic form $\bigwedge^2_{\mathcal{R}} \mathcal{U} \to \mathcal{V}$, $u \wedge v \mapsto u \cdot v$. Fix, further, a symplectic representation $G \to \operatorname{Sp}(\mathcal{U})$.

Definition 10.41. Given a G-valued factorization $\bar{\mathfrak{m}} = (\mathfrak{m}_1, \dots, \mathfrak{m}_r)$, define

- the $\mathcal{R}$-module $\mathcal{U} \otimes \bar{\mathfrak{m}} := \bigoplus_{i=1}^r \mathcal{U}$,
- the $\mathcal{R}$-linear map $\chi\colon \mathcal{U} \otimes \bar{\mathfrak{m}} \to \mathcal{U}$, $\bigoplus_i u_i \mapsto \sum_i u_i \uparrow (\mathfrak{m}_i - 1)$, *cf.* 9.23 (1), and
- the $\mathcal{R}$-quadratic map $q\colon \mathcal{U} \otimes \bar{\mathfrak{m}} \to \mathcal{V}$, *cf.* (9.28),

$$\bigoplus_i u_i \mapsto \sum_{i=1}^r u_i \cdot u_i \uparrow \mathfrak{m}_i - \sum_{1 \leqslant i < j \leqslant r} u_i \uparrow (\mathfrak{m}_i - 1) \cdot u_j \uparrow (\mathfrak{m}_j - 1).$$

Let $\mathcal{U}_{\bar{\mathfrak{m}}} := \operatorname{Ker} \chi$. The *transcendental lattice* of $\bar{\mathfrak{m}}$ is the quotient $\mathbf{T}(\bar{\mathfrak{m}}) := \mathcal{U}_{\bar{\mathfrak{m}}} / \ker$ equipped with the induced quadratic form $q\colon \mathbf{T}(\bar{\mathfrak{m}}) \to \mathcal{V}$.

Let $u_i, v_j \in \mathcal{U} \otimes \bar{\mathfrak{m}}$ be two 'basis' elements. A simple computation using the fact that each $\mathfrak{m}_j$ is a symplectic isometry shows that

$$q(u_i + v_j) - q(u_i) - q(v_j) = \chi(u_i) \cdot \chi(v_j) + \begin{cases} 2u_i \cdot v_j \uparrow (\mathfrak{m}_j - 1), & \text{if } i = j, \\ 0, & \text{if } i < j. \end{cases}$$

Corollary 10.42. *One has* $q(u+v) - q(u) - q(v) = \chi(u) \cdot \chi(v) \bmod 2\mathcal{V}$ *for any pair* $u, v \in \mathcal{U} \otimes \bar{\mathfrak{m}}$. ◁

Corollary 10.43. *If* $\mathcal{V}$ *is free of* 2*-torsion,* $q\colon \mathbf{T}(\bar{\mathfrak{m}}) \to \mathcal{V}$ *extends to a nondegenerate symmetric bilinear form* $\mathbf{T}(\bar{\mathfrak{m}}) \otimes \mathbf{T}(\bar{\mathfrak{m}}) \to \mathcal{V}$. ◁

Consider triples $(\mathcal{L}, \chi, q)$, where $\mathcal{L}$ is an $\mathcal{R}$-module, $\chi\colon \mathcal{L} \to \mathcal{U}$ is an $\mathcal{R}$-linear map, and $q\colon \mathcal{L} \to \mathcal{V}$ is a quadratic form. A *strong* (*weak*) *isomorphism* between two such triples $(\mathcal{L}_i, \chi_i, q_i)$, $i = 1, 2$, is an $\mathcal{R}$-isomorphism $\varphi\colon \mathcal{L}_1 \to \mathcal{L}_2$ such that $q_1 = q_2 \circ \varphi$ and $\chi_1 = \chi_2 \circ \varphi$ (respectively, $\chi_1 = (\chi_2 \circ \varphi) \uparrow g$ for a certain fixed element $g \in G$).

Theorem 10.44. *The triples* $(\mathcal{U} \otimes \bar{\mathfrak{m}}, \chi, q)$, $(\mathcal{U} \otimes \bar{\mathfrak{m}}', \chi', q')$ *corresponding to two strongly* (*weakly*) *equivalent factorizations* $\bar{\mathfrak{m}}$, $\bar{\mathfrak{m}}'$ *are strongly* (*respectively, weakly*) *isomorphic. In particular, the lattice* $\mathbf{T}(\bar{\mathfrak{m}})$ *is a weak Hurwitz invariant.*

Proof. If $\bar{\mathfrak{m}}' = g^{-1} \bar{\mathfrak{m}} g$, $g \in G$, the weak isomorphism is $\bigoplus_i u_i \mapsto \bigoplus_i u_i \uparrow g$.

Assume that $\bar{\mathfrak{m}} = \bar{\mathfrak{m}}' \uparrow \sigma_i$ and define the isomorphism $\varphi\colon \bigoplus_i u'_i \mapsto \bigoplus_i u_i$ *via* $u_i = u'_{i+1} \uparrow \mathfrak{m}_{i+1}^{-1}$, $u_{i+1} = u'_i + u_i \uparrow (\mathfrak{m}_i - 1)$. (The other coordinates do not change.)

Then $u_i \uparrow (\mathfrak{m}_i - 1) + u_{i+1} \uparrow (\mathfrak{m}_{i+1} - 1) = u'_i \uparrow (\mathfrak{m}'_i - 1) + u'_{i+1} \uparrow (\mathfrak{m}'_{i+1} - 1)$. Hence, $\chi' = \chi \circ \varphi$ and the essentially different terms in the expressions for q and q' are

$$u_i \cdot u_i \uparrow \mathfrak{m}_i + u_{i+1} \cdot u_{i+1} \uparrow \mathfrak{m}_{i+1} - u_i \uparrow (\mathfrak{m}_i - 1) \cdot u_{i+1} \uparrow (\mathfrak{m}_{i+1} - 1)$$

(and the corresponding primed version). Since $u_{i+1} - u_i \uparrow (\mathfrak{m}_i - 1) = u'_i$, this expression can be rewritten in the form

$$u'_{i+1} \uparrow \mathfrak{m}_{i+1}^{-1} \cdot u'_{i+1} \uparrow \mathfrak{m}_{i+1}^{-1}\mathfrak{m}_i + u'_i \cdot u_{i+1} \uparrow (\mathfrak{m}_{i+1} - 1).$$

Since $\mathfrak{m}_{i+1}$ is a symplectic isometry, the first term equals $u'_{i+1} \cdot u'_{i+1} \uparrow \mathfrak{m}'_{i+1}$. The second term, after substituting, multiplying out, and identifying $u'_i \cdot u_i \uparrow (\mathfrak{m}_i - 1) = u'_i \uparrow \mathfrak{m}'_i \cdot u'_{i+1} \uparrow (\mathfrak{m}'_{i+1} - 1)$, turns into $u'_i \cdot u'_i \uparrow \mathfrak{m}'_i - u'_i \uparrow (\mathfrak{m}'_i - 1) \cdot u'_{i+1} \uparrow (\mathfrak{m}'_{i+1} - 1)$. Hence, $q' = q \circ \varphi$, as stated. □

The discriminant form

Assume that $\mathcal{R} = \mathcal{V} = \mathbb{Z}$ and $\mathcal{U}$ is a free abelian group with a nonsingular symplectic form, *i.e.*, such that the associated homomorphism $\mathcal{U} \to \mathcal{U}^*$, $u \mapsto \{v \mapsto u \cdot v\}$, is an isomorphism. Let $G = \operatorname{Sp}(\mathcal{U})$ be the identity representation. Below, we introduce a few objects and state their properties; a common topological proof is given at the end of this subsection.

Definition 10.45. Given an element $g \in \operatorname{Sp}(\mathcal{U})$, the *local discriminant form* $\operatorname{discr}(g)$ is the group $\operatorname{discr}(g) := \operatorname{Tors}(\mathcal{U}/g)$ with the bilinear form

$$\operatorname{discr}(g) \otimes \operatorname{discr}(g) \to \mathbb{Q}/\mathbb{Z}, \quad u \otimes v \mapsto u \uparrow (g-1)^{-1} \cdot v \bmod \mathbb{Z}.$$

In Definition 10.45, the original symplectic form extends to $\bigwedge^2_{\mathbb{Q}}(\mathcal{U} \otimes \mathbb{Q}) \to \mathbb{Q}$ by linearity. If $\det(g-1) = 0$, the definition should be understood as follows: an element of $\operatorname{discr}(g)$ is represented by a vector $u \in \mathcal{U}$ such that $mu = x \uparrow (g-1)$ for some $m \in \mathbb{Z}$ and $x \in \mathcal{U}$, and we let $u \otimes v \mapsto \frac{1}{m} x \cdot v \bmod \mathbb{Z}$. For example, if $g = t_a^n$ is a power of a Dehn twist, $n \neq 0$, then $\operatorname{discr}(g) = \mathbb{Z}_{|n|}$ with $1 \cdot 1 = \frac{1}{n} \bmod \mathbb{Z}$.

Proposition 10.46. *Let* $g \in \operatorname{Sp}(\mathcal{U})$. *Then* $\operatorname{discr}(g)$ *is a well defined nondegenerate finite bilinear form. There is an isomorphism* $\operatorname{discr}(g^{-1}) = -\operatorname{discr}(g)$.

For an $\operatorname{Sp}(\mathcal{U})$-valued factorization $\bar{\mathfrak{m}} = (\mathfrak{m}_1, \ldots, \mathfrak{m}_r)$, define the *discriminant form* $\operatorname{discr}(\bar{\mathfrak{m}}) := \operatorname{discr}(\mathfrak{m}_\infty) \oplus \bigoplus_i \operatorname{discr}(\mathfrak{m}_i^{-1})$. If $\bar{\mathfrak{m}}$ is simple (*i.e.*, each factor $\mathfrak{m}_i$ is a Dehn twist), then $\operatorname{discr}(\bar{\mathfrak{m}}) = \operatorname{discr}(\mathfrak{m}_\infty)$. Let also $\mathcal{Q}(\bar{\mathfrak{m}}) := \operatorname{Tors}(\mathcal{U}/\operatorname{Im}\bar{\mathfrak{m}})$. The quotient projection induces a homomorphism $\operatorname{pr}\colon \operatorname{discr}(\bar{\mathfrak{m}}) \to \mathcal{Q}(\bar{\mathfrak{m}})$, which does *not* need to be onto. Due to Lemma 10.46, we can regard the adjoint of pr as a homomorphism $\operatorname{pr}^*\colon \operatorname{Hom}(\mathcal{Q}(\bar{\mathfrak{m}}), \mathbb{Q}/\mathbb{Z}) \to \operatorname{discr}(\bar{\mathfrak{m}})$.

Theorem 10.47. *Let* $\bar{\mathfrak{m}}$ *be an* $\mathrm{Sp}(\mathcal{U})$*-valued factorization. Then* $\operatorname{Im}\mathrm{pr}^* \subset \operatorname{discr}(\bar{\mathfrak{m}})$ *is an isotropic subgroup and there is an isomorphism* $\operatorname{discr}\mathbf{T}(\bar{\mathfrak{m}}) = (\operatorname{Im}\mathrm{pr}^*)^{\perp}/\operatorname{Im}\mathrm{pr}^*$.

For the next statement, consider the reduction of the symplectic form to the finite group $\mathcal{U}\otimes\mathbb{Z}_2$: it is a symmetric bilinear form with values in $\mathbb{Z}_2 \subset \mathbb{Q}/\mathbb{Z}$. Consider the set $\mathrm{Qu}(\mathcal{U}\otimes\mathbb{Z}_2)$ of quadratic extensions of this form; recall that $\mathrm{Qu}(\mathcal{U}\otimes\mathbb{Z}_2)$ is a principal homogeneous space over $\mathrm{Hom}(\mathcal{U},\mathbb{Z}_2)$, see Remark A.1. There is an obvious action of $\mathrm{Sp}(\mathcal{U})$ on $\mathrm{Qu}(\mathcal{U}\otimes\mathbb{Z}_2)$. For example, a Dehn twist t_a, $a\in\mathcal{U}$, acts *via* $q\uparrow t_a = q + l_a$, where $l_a\colon x \to (x\cdot a)[q(a)+1] \bmod 2$; it follows that a quadratic extension q is t_a-invariant if and only if $q(a) = 1 \bmod 2\mathbb{Z}$.

Theorem 10.48. *Let* $\bar{\mathfrak{m}}$ *be an* $\mathrm{Sp}(\mathcal{U})$*-valued factorization, and assume that the action of* $\operatorname{Im}\bar{\mathfrak{m}}$ *on* $\mathrm{Qu}(\mathcal{U}\otimes\mathbb{Z}_2)$ *has an invariant element. Then the lattice* $\mathbf{T}(\bar{\mathfrak{m}})$ *is even.*

Proof of Lemma 10.46 *and Theorems* 10.47 *and* 10.48. Since any two nonsingular symplectic forms of the same rank are isomorphic, we can represent $\mathcal{U}$ as $H_1(S)$, where S is a closed surface of genus $\frac{1}{2}\operatorname{rk}\mathcal{U}$. Then the assignment $\varphi \mapsto \varphi_*$ gives us an epimorphism $\mathrm{Map}^+(S) \twoheadrightarrow \mathrm{Sp}(\mathcal{U})$.

Given an element $g\in\mathrm{Sp}(\mathcal{U})$, represent it by a diffeomorphism $\tilde{g}\colon S\to S$ and consider the *mapping torus* $T_{\tilde{g}} := S\times I/(s,0)\sim(s\uparrow\tilde{g},1)$. It is an oriented closed 3-manifold, and Definition 10.45 is merely a geometric way to compute the linking coefficient form on $\operatorname{Tors}H_1(T_{\tilde{g}})$. Indeed, let u, v, x be circles in S representing certain classes $[u],[v],[x]\in\mathcal{U}$. If $m[u] = [x]\uparrow(g-1)$ for some $m\in\mathbb{Z}$, then mu is homologous to the boundary of the chain $(x\times I)/{\sim}$, and we intersect this chain with v in a fiber $S\times\{\epsilon\}$, $0<\epsilon\ll 1$.

The mapping torus $T_{\tilde{g}^{-1}}$ differs from $T_{\tilde{g}}$ by orientation only. Together with the previous observation, this fact proves Lemma 10.46. (The linking coefficient form is known to be symmetric and nondegenerate.)

For Theorem 10.47, lift each $\mathfrak{m}_i$ to a diffeomorphism $\tilde{\mathfrak{m}}_i\colon S\to S$ and represent $\bar{\mathfrak{m}}$ as the monodromy of an S-bundle $X^\circ \to \bar{D}^\circ_r$ over a punctured disc (or, more precisely, closed disk $\bar{D}^2$ with r disjoint open disks removed): the disk $\bar{D}^\circ_r$ retracts to a wedge of r circles, and the restriction of the bundle to the i-th circle is the mapping torus $T_{\tilde{\mathfrak{m}}_i}$.

The total space X° is a compact oriented 4-manifold, and the boundary ∂X° is the union of the mapping tori $T_{\tilde{\mathfrak{m}}_\infty}$ and $T_{\tilde{\mathfrak{m}}_i^{-1}}$, $i=1,\dots,r$. Acting as in the proof of Theorem 9.25, we can see that the homology group $H_2(X^\circ)$ is torsion free and $H_2(X^\circ)/[S] = \mathcal{U}_{\bar{\mathfrak{m}}}$, where $[S]$ is the class of a generic fiber. Hence, $\operatorname{discr}\mathbf{T}(\bar{\mathfrak{m}})$ is the torsion of the cokernel of the homomorphism

$$H_2(X^\circ) \to H_2(X^\circ,\partial X^\circ)/\operatorname{Tors} = \mathrm{Hom}(H_2(X^\circ),\mathbb{Z})$$

(Poincaré–Lefschetz duality and the universal coefficient formula). Furthermore, the homomorphism pr can be identified with the inclusion homomorphism

$$\operatorname{Tors}H_1(\partial X^\circ) \to \operatorname{Tors}H_1(X^\circ) = \mathcal{Q}(\bar{\mathfrak{m}}).$$

Now, in the exact sequence of pair

$$H_2(X^\circ) \longrightarrow H_2(X^\circ, \partial X^\circ) \xrightarrow{\partial} H_1(\partial X^\circ) \xrightarrow{i_*} H_1(X^\circ),$$

we have $\operatorname{Tors} H_2(X^\circ, \partial X^\circ) = \operatorname{Tors} H^2(X^\circ) = \operatorname{Hom}(\mathcal{Q}(\bar{\mathfrak{m}}), \mathbb{Q}/\mathbb{Z})$ (the universal coefficient formula) and the restriction $\partial : \operatorname{Tors} H_2(X^\circ, \partial X^\circ) \to \operatorname{Tors} H_1(\partial X^\circ)$ is identified with pr^*. Since $\operatorname{Im} \partial \subset \operatorname{Ker} i_*$, the image $\operatorname{Im} \mathrm{pr}^*$ is isotropic, and the last statement of Theorem 10.47 follows from the fact that $\operatorname{discr} \mathbf{T}(\bar{\mathfrak{m}})$ is nondegenerate.

Finally, for Theorem 10.48, recall that the set $\operatorname{Qu}(\mathcal{U} \otimes \mathbb{Z}_2) = \operatorname{Qu}(H_1(S; \mathbb{Z}_2))$ is canonically identified with the set of Spin-structures on S. An $\operatorname{Im} \bar{\mathfrak{m}}$-invariant element represents a Spin-structure invariant under all autodiffeomorphisms $\bar{\mathfrak{m}}_i$, $i = 1, \ldots, r$. This Spin-structure extends to the mapping tori and, hence, to X°. Thus, X° is a Spin-manifold and $w_2(X^\circ) = 0$. On the other hand, $w_2(X^\circ) = 0$ is a characteristic class of the intersection index form: one has $u^2 = u \circ w_2(X^\circ)$ for any homology class $u \in H_2(X^\circ; \mathbb{Z}_2) \supset H_2(X) \otimes \mathbb{Z}_2$. □

Problem 10.49. Is there a simple expression for the rank and signature of $\mathbf{T}(\bar{\mathfrak{m}})$, at least for simple factorizations? In all examples this lattice is positive definite.

The representation $\mathbb{B}_{2g+1} \to \mathrm{Sp}(H_1(S_g))$

The specialization of the Burau representation at $t = -1$ gives rise to a representation $\mathbb{B}_{2g+1} \to \mathrm{Sp}(H_1(S_g))$, where S_g is a closed surface of genus g. It can be regarded as the double covering of the punctured disk $\bar{D}^\circ_{2g+1}$ defined by the homomorphism $\deg \bmod 2\colon \pi_1(\bar{D}^\circ_{2g+1}) \to \mathbb{Z}_2$, with the punctures compactified and the single boundary component contracted to a point. (Alternatively, we can keep the boundary and consider the surface S°_g with one puncture; the inclusion $S^\circ_g \hookrightarrow S_g$ induces an isomorphism in the 1-homology.) In the basis $\mathbf{e}_1, \ldots, \mathbf{e}_{2g}$, the symplectic form (the intersection index form of S_g) is $\mathbf{e}_i \cdot \mathbf{e}_{i+1} = 1$ and $\mathbf{e}_i \cdot \mathbf{e}_j = 0$ for $|i - j| \neq 1$.

Example 10.50. Consider a simple sextic $D \subset \mathbb{P}^2$. Fix a generic point $P \in D$ and let L be the tangent to D at P. Blowing P up, we obtain a *pentagonal model* $C \subset \Sigma_1$ of D, which is proper over the complement Ω of a small regular neighborhood of the fiber at infinity L. Choosing a proper section over Ω, we can consider the braid monodromy of C, see Section 5.1.1, which gives rise to a $\mathbb{B}_5$-valued factorization $\bar{\mathfrak{m}}$ of the element $\mathfrak{m}_\infty := \Delta^2 \sigma_1 \sigma_2 \sigma_3 \sigma_4^2 \sigma_3 \sigma_2 \sigma_1 \in \mathbb{B}_5$. Clearly, the conjugacy classes of the factors of $\bar{\mathfrak{m}}$ are determined by the set of singularities $\mathbf{S}$ of D: there is one factor for each singular point $S \in \mathbf{S}$ (determined by the type of S) and a number of Artin generators (corresponding to the 'vertical' tangents to D) to match the degree.

As in the proof of Theorem 9.25 (see also [5]), one can easily show that $\mathbf{T}(\bar{\mathfrak{m}})$ is isomorphic to the transcendental lattice $\mathbf{T} := (\mathbf{S} \oplus \mathbb{Z}h)^\perp$ of the covering $K3$-surface, see page 214. There is a great deal of pairs (sometimes triples) of simple sextics sharing the same set of singularities but having non-isomorphic lattices $\mathbf{T}$, see [148].

Hence, we obtain $\mathbb{B}_5$-valued factorizations that differ by their transcendental lattices. (In many cases, even the discriminant forms $\operatorname{discr} \mathbf{T}(\bar{\mathfrak{m}})$ differ, *cf.* Theorem 10.47.) An explicit example, a pair of sextics with the set of singularities $\mathbf{A}_{10} \oplus \mathbf{A}_9$, is worked out in Arima, Shimada [5]. It is worth mentioning that the two sextics are conjugate over $\mathbb{Q}(\sqrt{5})$; hence, $\mathbf{T}$ is a topological, but not algebraic invariant.

If $g = 1$, the specialization of the Burau representation at $t = -1$ turns into the canonical projection $\mathbb{B}_3 \twoheadrightarrow \mathbb{B}_3/\Delta^2 = \tilde{\Gamma}$, see (2.39), used throughout the book.

Proposition 10.51. *For a $\tilde{\Gamma}$-valued factorization $\bar{\mathfrak{m}}$ (and the identity representation $\tilde{\Gamma} = \operatorname{Sp}(\mathcal{H})$), the lattice $\mathbf{T}(\bar{\mathfrak{m}})$ is even.*

Proof. There is a unique quadratic extension $q \in \operatorname{Qu}(\mathcal{H} \otimes \mathbb{Z}_2)$ taking value 1 mod $2\mathbb{Z}$ at each nontrivial element of $\mathcal{H} \otimes \mathbb{Z}_2$. This extension q is obviously invariant under the full group $\tilde{\Gamma}$, and the statement follows from Theorem 10.48. □

Example 10.52. Consider a simple sextic $D \subset \mathbb{P}^2$ with a distinguished triple singular point P, and let C be the proper model of D, see Section 8.1.2. The braid monodromy of C (with one of the distinguished singular fibers F_{I} taken for the fiber at infinity) is a factorization $\bar{\mathfrak{m}}$ of a certain element $\mathfrak{m}_\infty \in \mathbb{B}_3$, both $\mathfrak{m}_\infty$ and the conjugacy classes of the factors depending on the set of singularities of D only. As above, it is easy to show that $\mathbf{T}(\bar{\mathfrak{m}})$ is the transcendental lattice of the covering $K3$-surface, see page 214, and, using [148], we obtain a large number of pairs of factorizations that differ by $\mathbf{T}$. Some of them are discussed in Example 10.24 (the sets of singularities marked with a * in Table 10.1).

If $\bar{\mathfrak{m}}$ is a simple $\tilde{\Gamma}$-valued factorization, the construction of $\mathbf{T}(\bar{\mathfrak{m}})$ can be simplified. Let w_i be a primitive invariant vector of $\mathfrak{m}_i$, $i = 1, \ldots, r$. Then the subgroup $\mathcal{H}^0_{\bar{\mathfrak{m}}} := \{\bigoplus_i a_i w_i \mid a_i \in \mathbb{Z}\}$ is in the kernel of q, the quotient $\mathcal{H}'_{\bar{\mathfrak{m}}} := (\mathcal{H} \otimes \bar{\mathfrak{m}})/\mathcal{H}^0_{\bar{\mathfrak{m}}}$ is a free abelian group of rank r (the length of $\bar{\mathfrak{m}}$), and both q and χ descend to $\mathcal{H}'_{\bar{\mathfrak{m}}}$, taking the form $\chi(c_1, \ldots, c_r) = \sum_i c_i w_i$ and

$$q(c_1, \ldots, c_r) = -\sum_{i=1}^{r} c_i^2 - \sum_{1 \leqslant i < j \leqslant r} c_i c_j (w_i \cdot w_j),$$

cf. (9.33). (If $v_i \in \mathcal{H}$ is such that $v_i \cdot w_i = 1$, then $v_i \uparrow (\mathfrak{m}_i - 1) = -w_i$.) In addition, in order to obtain a more 'recognizable' lattice, the expression for q can be modified by adding any quadratic form vanishing on $\operatorname{Ker} \chi$. It is this approach that was used in the proof of Theorem 9.31 for Items 1 and 2.

In view of the uniqueness of a simple lift to $\tilde{\Gamma}$, the transcendental lattice can be defined for a simple Γ-valued factorization as well. In an attempt to construct a more robust invariant, define a *coloring* of length r as a function $\ell\colon \{1, \ldots, r\} \to \{\pm 1\}$. Given an r-factorization $\bar{\mathfrak{m}}$ and a coloring ℓ of length r, let $\mathbf{T}(\bar{\mathfrak{m}}, \ell)$ be the transcendental lattice of the $\tilde{\Gamma}$-valued factorization defined as follows: a factor $\mathfrak{m}_i = g^{-1} R g \in \Gamma$,

$i = 1, \ldots, r$, lifts to $g^{-1}\ell(i)Rg \in \tilde{\Gamma}$. Alternatively, the lift can be described as the one with the eigenvalues of sign $\ell(i)$, and in this form the concept can be extended to a wider class of factorizations, *e.g.*, those with unipotent factors. The following statement is straightforward.

Proposition 10.53. *Assume that two Γ-valued factorizations $\bar{\mathfrak{m}}$, $\bar{\mathfrak{m}}'$ of length r are weakly equivalent. Then there exists a permutation $\sigma \in \mathbb{S}_r$ such that, for any coloring ℓ of length r, the lattices $\mathbf{T}(\bar{\mathfrak{m}}, \ell)$ and $\mathbf{T}(\mathfrak{m}', \ell \circ \sigma)$ are isomorphic.* $\triangleleft$

Example 10.54. The lattices $\mathbf{T}(\mathfrak{m}, \ell)$ for the simple Γ-valued factorizations $\mathfrak{m}$ given by Theorem 10.20 and colorings ℓ taking at most one value (-1) are computed in Theorem 9.31; unfortunately, they are all isomorphic. At present, I do not know any examples of simple factorizations distinguished by the transcendental lattice.

Problem 10.55. Does Proposition 10.53 distinguish (some of) the simple Γ-valued factorizations given by Theorem 10.20? Does it distinguish 2-factorizations?

10.2.4 2-factorizations *via* matrices

In this section, we work in the matrix group $\tilde{\Gamma}$. Consider a weak Hurwitz equivalence class $\mathcal{W}$ of hyperbolic 2-factorizations. It has a representative $t_{\mathbf{a}} \odot t_{\mathbf{u}}$, where $\mathbf{u} := [s, n] \in \mathcal{H}$ is a primitive vector. We can assume that $n \geqslant 0$, and then $\mathcal{W}$ is hyperbolic if and only if $n \geqslant 2$. In terms of the word A in Theorem 10.27 (3), we have

$$\mathbf{u} = \pm(\mathbf{a} \uparrow h), \quad \text{where } h := L^{-1}\mathbb{Y}A^{\mathrm{t}}L, \tag{10.56}$$

see (10.28) and (10.26), so that $t_{\mathbf{u}} = h^{-1}Rh$. This representative $t_{\mathbf{a}} \odot t_{\mathbf{u}}$ is defined up to the global conjugation by a power of $R = t_{\mathbf{a}}$ and up to the Hurwitz move σ_1 followed by the global conjugation by Rh^{-1}. The former adds to s a multiple of n; hence, s can be regarded as an element of $\mathbb{Z}_n^*$ (the group of units of $\mathbb{Z}_n$). The latter operation replaces $t_{\mathbf{u}}$ with hRh^{-1}, taking s to $-s^{-1} \in \mathbb{Z}_n^*$. (In (10.56), we replace A^{t} with A.) Hence, we have the following characterization of 2-factorizations.

Theorem 10.57. *A weak equivalence class of hyperbolic 2-factorizations is determined by an integer $n \geqslant 2$ and a pair $s, -s^{-1} \in \mathbb{Z}_n^*$ of anti-reciprocal units.* $\triangleleft$

Given a pair (s, n), $s \in \mathbb{Z}_n^*$, denote by $\mathcal{W}_n(s)$ and $\mathcal{C}_n(s)$, respectively, the weak equivalence class of 2-factorizations given by Theorem 10.57 and the conjugacy class $[\![\mathfrak{m}_\infty]\!]$, $\bar{\mathfrak{m}} \in \mathcal{W}_n(s)$. There is a canonical map $\mathcal{W}_n(s) \to \mathcal{C}_n(s)$, $\bar{\mathfrak{m}} \mapsto \mathfrak{m}_\infty$; due to Theorem 10.30, unless $(s, n) = (1, 2)$ (*i.e.*, $\mathcal{C} = [\![-L^4]\!]$), the preimage under this map of an element $g \in \mathcal{C}_n(s)$ is a single strong equivalence class. Some simple properties of 2-factorizations are easily expressed in terms of (s, n), but the question of uniqueness remains open.

Proposition 10.58. *Let $s \in \mathbb{Z}_n^*$, $n \geqslant 2$. Then:*

1. *$\mathcal{C}_n(s)$ is a square, see Corollary 10.36 (1), if and only if $s^2+1=0$ in $\mathbb{Z}_n$;*
2. *one has $\bar{\mathcal{W}}_n(s) = \mathcal{W}_n(-s)$; hence, $\mathcal{W}_n(s)$ is real if and only if $s^2=1$ in $\mathbb{Z}_n^*$;*
3. *one has $\operatorname{trace} g = 2-n^2$ for any $g \in \mathcal{C}_n(s)$.*

Proof. The first statement follows from the description of hyperbolic 2-factorizations: the factorizations $\bar{\mathfrak{m}}$ and $\bar{\mathfrak{m}} \uparrow \sigma_1$ are conjugate if and only if $A = A^{\mathrm{t}}$ in 10.28, *i.e.*, $\mathfrak{m}_\infty$ is a square. For the second one, since the conjugate class $\bar{\mathcal{W}}$ does not depend on the choice of a real structure, we can take $\tau\colon \mathbf{a} \mapsto \mathbf{a}$, $\mathbf{b} \mapsto -\mathbf{b}$; then $s \mapsto -s$. Hence, $\mathcal{W}_n(s)$ is real if and only if $s=-s$ or $s=s^{-1}$ in $\mathbb{Z}_n^*$. In the former case, since s is a unit, we would have $2=0$ in $\mathbb{Z}_n$, *i.e.*, $n=2$. But then $s=1$ and still $s^2=1$. Finally, for the last statement, it suffices to compute $\operatorname{trace}(Rt_{\mathbf{u}}) = 2-n^2$ and use the fact that trace is a class function. □

Problem 10.59. Characterize pairs $s', s'' \in \mathbb{Z}_n^*$ such that $\mathcal{C}_n(s') = \mathcal{C}_n(s'')$.

Example 10.60. By induction, one can show that

$$(LR)^m = \begin{bmatrix} a_{2m+1} & a_{2m} \\ a_{2m} & a_{2m-1} \end{bmatrix}, \quad (RL)^m = \begin{bmatrix} a_{2m-1} & a_{2m} \\ a_{2m} & a_{2m+1} \end{bmatrix},$$

where a_i is the i-th Fibonacci number. Hence, the two 2-factorizations of V_m, see (10.31), correspond to $n = a_{2m+3}$ and $a_{2m+1}, a_{2m+2} \in \mathbb{Z}_n^*$. The other 'regular' sequence $W_{1/m}(\varnothing)$ is similarly related to the sequence b_i: $b_0=0$, $b_1=1$, and $b_{i+2} = 2b_{i+1}+b_i$ for $i \geqslant 0$: we have $n = 2b_{2m+1}$ and the two 2-factorizations correspond to the units $s' := b_{2m-1}+b_{2m}$ and $s'' := b_{2m}+b_{2m+1} = -s' \bmod n$.

Proposition 10.58 (3) gives us a necessary condition for the existence of a 2-factorization of an element $g \in \tilde{\Gamma}$: the difference $(2-\operatorname{trace} g)$ must be a perfect square. This condition is far from sufficient; for a counterexample, one can take the element $R^3LR^2 = (L^2RL^3)^{\mathrm{t}}$ of trace 7.

Another necessary condition is (10.6) with $r=2$ and $\mathcal{P}_1 = \mathcal{P}_2 = [\![R]\!]$ applied to a finite quotient of $\tilde{\Gamma}$. For example, one can consider the groups $SL(2,\mathbb{Z}_p) = \tilde{\Gamma}/\tilde{\Gamma}(p)$ for p prime; their irreducible characters are known see, *e.g.*, [122], and, for each prime p, the condition in (10.6) can be checked effectively in terms of Gauss sums.

Problem 10.61. Does an analogue of the Hasse principle hold for the 2-factorization problem, *i.e.*, do Proposition 10.58 (3) and condition (10.6) applied to all $SL(2,\mathbb{Z}_p)$ constitute a sufficient condition for the existence of a 2-factorization?

The transcendental lattice

As another example illustrating Section 10.2.3, consider the transcendental lattice $\mathbf{T}(\bar{\mathfrak{m}}, \ell)$ of a hyperbolic 2-factorization $\bar{\mathfrak{m}}$. Since $\operatorname{rk} \mathbf{T}(\bar{\mathfrak{m}}, \ell) \leqslant \operatorname{Card} \ell^{-1}(-1)$ in this case (in fact, the equality holds), this lattice is only interesting if $\ell \equiv -1$.

Regard s as an integer and consider a representative $R \odot t_{\mathbf{u}}$, $\mathbf{u} := s\mathbf{a} + n\mathbf{b} \in \mathcal{H}$, of the class $\mathcal{W}_n(s)$. Since $\ell \equiv -1$, we lift this factorization to $\bar{\mathfrak{m}}^- = (-R, -t_{\mathbf{u}})$ in $\tilde{\Gamma}$. Denote by $\mathbf{a}_i$, $\mathbf{b}_i$, $i = 1, 2$, the generators of the summand $\mathcal{H} \otimes \mathfrak{m}_i$ of $\mathcal{H} \otimes \bar{\mathfrak{m}}^-$. Then $\mathcal{H}_{\bar{\mathfrak{m}}^-}$ is a finite index subgroup of the free abelian group $\mathcal{H}'_{\bar{\mathfrak{m}}^-}$ generated by the rational vectors $\tilde{\mathbf{a}}, \tilde{\mathbf{b}} \in \mathcal{H}_{\bar{\mathfrak{m}}^-} \otimes \mathbb{Q}$:

$$\tilde{\mathbf{a}} := \mathbf{a}_1 - \frac{sn+2}{2}\mathbf{a}_2 - \frac{n^2}{2}\mathbf{b}_2, \quad \tilde{\mathbf{b}} := \mathbf{b}_1 + \frac{\tilde{s}s - 2}{4}\mathbf{a}_2 + \frac{\tilde{s}n - 4}{4}\mathbf{b}_2,$$

where $\tilde{s} := 2s - n$. In this basis, the quadratic form q is given by

$$q(x\tilde{\mathbf{a}} + y\tilde{\mathbf{b}}) = \frac{1}{4}(2nx + \tilde{s}y)^2 + y^2.$$

In particular, q is positive definite and, hence, $\mathbf{T}(\bar{\mathfrak{m}}^-) = \mathcal{H}_{\bar{\mathfrak{m}}^-}$.

Depending on the residue $n \bmod 4$, a vector $x\tilde{\mathbf{a}} + y\tilde{\mathbf{b}}$ belongs to the subgroup $\mathcal{H}_{\bar{\mathfrak{m}}^-} = \mathcal{H}'_{\bar{\mathfrak{m}}^-} \cap \mathcal{H} \otimes \bar{\mathfrak{m}}^-$ if and only if:

- $n = 1 \bmod 2$: $y = 2x \bmod 2$; hence, $\mathcal{H}_{\bar{\mathfrak{m}}^-}$ is generated by $2\tilde{\mathbf{a}}$ and $\tilde{\mathbf{a}} + 2\tilde{\mathbf{b}}$;
- $n = 2 \bmod 4$: $y = 0 \bmod 2$; hence, $\mathcal{H}_{\bar{\mathfrak{m}}^-}$ is generated by $\tilde{\mathbf{a}}$ and $\tilde{\mathbf{a}} + 2\tilde{\mathbf{b}}$;
- $n = 0 \bmod 4$: no restriction, and $\mathcal{H}_{\bar{\mathfrak{m}}^-}$ is generated by $\tilde{\mathbf{a}}$ and $\tilde{\mathbf{b}}$.

Note that $\mathcal{H}_{\bar{\mathfrak{m}}^-} \subset \mathcal{H}'_{\bar{\mathfrak{m}}^-}$ can be reinterpreted as the maximal subgroup on which q takes even integral values, *cf.* Proposition 10.51. (Note also that, if a quadratic form q takes even values, it automatically extends to an integral symmetric bilinear form.)

Let $\tilde{n} := n$ if n is odd and $\tilde{n} := n/2$ if n is even. In the basis for $\mathcal{H}_{\bar{\mathfrak{m}}^-}$ indicated above, the Gram matrix C of q is

$$4\begin{bmatrix} \tilde{n}^2 & s\tilde{n} \\ s\tilde{n} & s^2+1 \end{bmatrix} \quad \text{or} \quad \begin{bmatrix} 4\tilde{n}^2 & 2\tilde{n}(s-\tilde{n}) \\ 2\tilde{n}(s-\tilde{n}) & (s-\tilde{n})^2+1 \end{bmatrix} \tag{10.62}$$

for $\tilde{n}$ odd or even, respectively. Denote $d := \text{g.c.d.}(s^2+1, \tilde{n})$. Then we have:

- if $\tilde{n}$ is odd, then $\det C = 16\tilde{n}^2$ and $\text{g.c.d.}(C_{ij}) = 4d$;
- if $\tilde{n}$ is even, then $\det C = 4\tilde{n}^2$ and $\text{g.c.d.}(C_{ij}) = 2d$.

As a partial classification statement, we have the following proposition.

Proposition 10.63. *Let s, n, $\tilde{n}$ be as above, and assume that* $\text{g.c.d.}(s^2+1, \tilde{n}) = 1$ *(hence $\tilde{n}$ is odd). Then the transcendental lattices $\mathbf{T}(\bar{\mathfrak{m}}, -1)$ and $\mathbf{T}(\bar{\mathfrak{m}}', -1)$ of two 2-factorizations $\bar{\mathfrak{m}} \in \mathcal{W}_n(s)$ and $\bar{\mathfrak{m}}' \in \mathcal{W}_n(s')$ are isomorphic if and only if $s' = \pm s^{\pm 1}$ in $\mathbb{Z}_n^*$.*

Proof. Let C be the Gram matrix of $\mathbf{T}(\bar{\mathfrak{m}}, -1)$. Under the hypotheses, we have $\text{g.c.d.}(C_{ij}) = 4$. Hence, all values of the bilinear form can be divided by 4, and for the new lattice $\mathbf{T}'$ we have $\operatorname{discr} \mathbf{T}' \cong \mathbb{Z}_{\tilde{n}^2}$ as a group.

Due to (10.62), $\mathbf{T}'$ can be represented as the sublattice of $\mathbf{B}_2^+$ generated by $\tilde{n}\mathbf{v}_1$ and $s\mathbf{v}_1 + \mathbf{v}_2$. Since $\operatorname{discr} \mathbf{T}'$ is cyclic, it has a unique subgroup of order $\tilde{n}$; hence, $\mathbf{T}'$ has a *unique* finite index extension to a unimodular lattice, see Theorem A.4. Thus, the representation of $\mathbf{T}'$ as a sublattice of $\mathbf{B}_2^+$ generated by a specific pair of vectors is unique up to a power of $R \in SL(\mathbf{T}')$ (s is defined modulo $\tilde{n}$) and an automorphism of $\mathbf{B}_2^+$, followed by a linear transformation of $\mathbf{T}'$ bringing the generators of $\mathbf{T}'$ back to a pair of the form $\tilde{n}\mathbf{v}_1, s'\mathbf{v}_1 + \mathbf{v}_2 \in \mathbf{B}_2^+$. The orthogonal group $O(\mathbf{B}_2^+)$ is generated by the automorphisms $\mathbf{v}_2 \mapsto -\mathbf{v}_2$ and $\mathbf{v}_1 \leftrightarrow \mathbf{v}_2$, which can only transform s to $\pm s^{\pm 1} \bmod \tilde{n}$. If $n = 2\tilde{n}$ is even, observe, in addition, that $s, s' \in \mathbb{Z}_n^*$ must be odd and the congruence $s = s' \bmod \tilde{n}$ implies $s = s' \bmod n$. □

According to Proposition 10.63, under the assumption g.c.d.$(s^2 + 1, \tilde{n}) = 1$, the transcendental lattices corresponding to two classes $\mathcal{W} := \mathcal{W}_n(s)$ and $\mathcal{W}'$ are isomorphic if and only if $\mathcal{W}' = \mathcal{W}$ or $\mathcal{W}' = \bar{\mathcal{W}}$, see Proposition 10.58 (2).

Remark 10.64. Assume that $\tilde{n}$ is odd and let $\tilde{n} = \prod_{i=1}^r p_i^{d_i}$ be its prime factorization. Denote $q_i := p_i^{d_i}$, so that $\mathbb{Z}_{\tilde{n}} = \bigoplus_i \mathbb{Z}_{q_i}$, and let $1_i \in \mathbb{Z}_{q_i}$ be the multiplicative identity and $s_i := s1_i \in \mathbb{Z}_{q_i}$, $i = 1, \ldots, r$. Then g.c.d.$(s^2 + 1, \tilde{n}) > 1$ if and only if $s_i \bmod p_i$ is a square root of -1 in $\Bbbk_{p_i}$ for at least one $i = 1, \ldots, r$. It follows that most, in a sense, pairs (s, n) with $\tilde{n}$ odd satisfy the hypotheses of Proposition 10.63.

If g.c.d.$(s^2 + 1, \tilde{n}) = 1$ and $\mathfrak{m} \in \mathcal{W}_n(s)$, then

$$\operatorname{discr} \mathbf{T}(\mathfrak{m}, -1) \cong \operatorname{discr} \mathbf{D}_5^+ \oplus \operatorname{discr} \mathbf{D}_5^+ \oplus \mathbb{Z}_{\tilde{n}^2},$$

the quadratic form sending a generator of $\mathbb{Z}_{\tilde{n}^2}$ to $4(s^2 + 1)/\tilde{n}^2 \bmod 2\mathbb{Z}$ (a simple consequence of (10.62)). Since $\mathbb{Z}_{\tilde{n}^2}^*$ is the product of the cyclic groups $\mathbb{Z}_{q_i^2}^*$, $i = 1, \ldots, r$, the lattices $\mathbf{T}(\mathfrak{m}, -1)$ fall into at most 2^r genera, see Theorem A.2.

As another extremal case, assume that $s^2 + 1 = 0$ in $\mathbb{Z}_n$, *cf.* Proposition 10.58 (1). Then C is a multiple of a unimodular matrix and $\mathbf{T}$ is obtained from $\mathbf{B}_2^+$ (the only positive definite unimodular lattice of rank 2) by multiplying the form by l.c.m.$(n, 4)$. Hence, all such lattices (with n fixed) are isomorphic.

In general, any pair of lattices with the Gram matrices as in (10.62) can be compared using the canonical forms given by Theorem A.10.

10.3 Geometric applications

The geometric applications of Γ- and $\tilde{\Gamma}$-valued factorizations are extremal elliptic surfaces, see Section 10.3.1, ribbon curves, including trigonal M-curves, see Section 10.3.2, and real Lefschetz fibrations, see Section 10.3.3.

10.3.1 Extremal elliptic surfaces

In this section, we discuss the relation between isomorphisms and homeomorphisms of extremal elliptic surfaces.

Theorem 10.65. *The map $X \mapsto \operatorname{Im}\mathfrak{h}_X$ establishes a bijection between the set of isomorphism classes of extremal elliptic surfaces without exceptional singular fibers and the set of conjugacy classes of finite index free subgroups of $\tilde{\Gamma}$.*

Proof. Consider an extremal elliptic surface $X \to B$ without exceptional singular fibers and let $\tilde{H} := \operatorname{Im}\mathfrak{h}_X \subset \tilde{\Gamma}$ and $H := \tilde{H}/\Delta^2 \subset \Gamma$. Due to Proposition 5.69 and Theorem 5.62, we have $H = \operatorname{Stab}\mathfrak{S}$, where $\mathfrak{S}$ is the skeleton of X. Under the assumptions, $\mathfrak{S}$ is regular (hence H is free and of finite index), $\pi_1(B^\sharp) = \pi_1(\mathfrak{S}) = H$, and the map $\mathfrak{h}_X \colon H \twoheadrightarrow \tilde{H} \hookrightarrow \tilde{\Gamma}$ can be regarded as a section of the projection $\tilde{H} \to H$. Hence, $\tilde{H}$ is also free.

Conversely, given a free finite index subgroup $\tilde{H} \subset \tilde{\Gamma}$, the projection $\tilde{H} \twoheadrightarrow H := \tilde{H}/\Delta^2$ is an isomorphism. Thus, $\tilde{H}$ gives rise to a unique, up to isomorphism, extremal elliptic surface X: by definition, its skeleton is $\mathfrak{S} := \operatorname{Sk} H$, and its homological invariant $\mathfrak{h} \colon \pi_1(\mathfrak{S}) = H \to \tilde{H} \hookrightarrow \tilde{\Gamma}$ is the inverse of the projection $\tilde{H} \twoheadrightarrow H$. □

Corollary 10.66. *Two extremal elliptic surfaces without exceptional singular fibers are isomorphic if and only if they are related by a bi-oriented homeomorphism.* ◁

Corollary 10.67. *All surfaces given by Theorem* 9.30 *differ topologically.* ◁

Theorem 10.68. *Two extremal elliptic surfaces over a rational base $B \cong \mathbb{P}^1$ and without singular fibers of type* II* *are isomorphic if and only if they are related by a bi-oriented homeomorphism.*

Proof. The only if part is obvious. For the converse, observe that an extremal elliptic surface over $\mathbb{P}^1$ is determined by its skeleton $\mathfrak{S}$ and type specification tp, and both are recovered from the homological invariant $\mathfrak{h}$ regarded as a factorization of $1 \in \tilde{\Gamma}$. The skeleton is $\operatorname{Sk}(\operatorname{Im}\mathfrak{h})$ and its regions correspond to the factors of $\mathfrak{h}$ conjugate to $\pm R$, so that the sign determines the value of tp on each region. The values on the monovalent •- and ∘-vertices are 10 and 9, respectively, as the exceptional fibers of an extremal surface are of types IV* and III*. □

Remark 10.69. Theorem 10.68 obviously extends to all Jacobian elliptic surfaces ramified at maximal trigonal curves and without singular fibers of type IV or II*. In the proof, the fixed values $\mathrm{tp}(\bullet) = 10$ and $\mathrm{tp}(\circ) = 9$ should be replaced with those given by the corresponding factors of $\mathfrak{h}$.

Problem 10.70. Does Corollary 10.66 extend to surfaces with singular fibers of types III* and IV*? Do Corollary 10.66 and Theorem 10.68 extend to surfaces with singular fibers of type II*? (In the latter case, one should consider skeletons with bivalent •-vertices, which do not correspond directly to subgroups of Γ.)

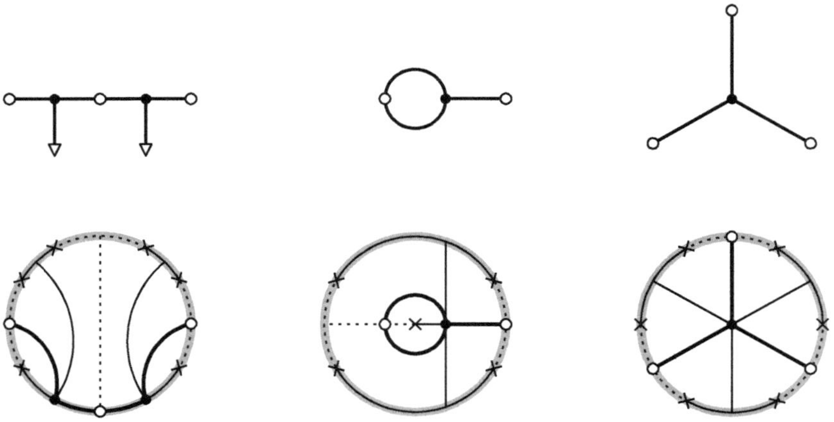

(a) I_2 (b) I_1 (c) II_3

Figure 10.6. Fragments of $\mathfrak{S}^+$ *vs.* ribbon blocks.

10.3.2 Ribbon curves *via* skeletons

Consider a skeleton $\mathfrak{S}$, not necessarily finite, denote by $\mathfrak{S}^+$ the skeleton obtained from $\mathfrak{S}$ by removing all edges $\circ\!\!-\!\!\!-\!\!\bullet$ incident to monovalent $\bullet$-vertices, and assume that $\mathfrak{S}^+$ can be cut at (some of) its bivalent $\circ$-vertices into a number of fragments shown in Figure 10.6, top. (Note that some of the monogons of $\mathfrak{S}^+$ may be preserved, as in Figure 10.6 (b), whereas some may be cut into fragments as in Figure 10.6 (c).)

Replace each fragment with a ribbon block and then with the corresponding dessin, as shown in Figure 10.6. The skeleton of each dessin is naturally identified with a subgraph of the original fragment, the $\circ$-vertices in the zigzags corresponding to the original monovalent $\circ$-vertices, and this identification extends to a unique, up to isotopy, embedding of the dessin to the minimal supporting surface $\operatorname{Supp}\mathfrak{S}^+$. Define a ribbon curve structure $\mathcal{S}$ performing a junction at each bivalent $\circ$-vertex of $\mathfrak{S}^+$ and selecting for self-junction each monovalent $\circ$-vertex *not* adjacent to a monovalent $\bullet$-vertex of the original skeleton $\mathfrak{S}$. Orientations for the junctions are chosen so that the resulting ribbon dessin $\mathfrak{D}$ is still embedded to $\operatorname{Supp}\mathfrak{S}^+$; in particular, $\mathcal{S}$ is oriented. We obtain a certain ribbon curve $C \subset \Sigma \to B$ whose base B is naturally directed: the distinguished orientation of $B_{\mathbb{R}} = \partial B^+$ is the one induced from the underlying surface B^+ of $\mathfrak{D}$ regarded as a subset of $\operatorname{Supp}\mathfrak{S}^+$.

Lemma 10.71. *Up to orientation preserving isomorphism, the oriented ribbon curve structure $\mathcal{S}$ as above is determined by the skeleton $\mathfrak{S}$ and the set of the monogonal regions of $\mathfrak{S}$ that are* not *cut into fragments as in Figure* 10.6 (c).

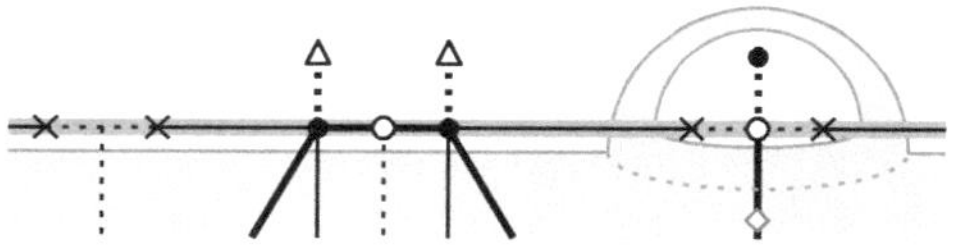

Figure 10.7. The monodromy regions $\bar{B}^+$ and $\bar{B}^\sharp$; extending $\operatorname{Sk}\mathfrak{D}$ to $\mathfrak{S}$.

Proof. The construction of S becomes unambiguous as soon as the skeleton $\mathfrak{S}^+$ is cut into fragments as in Figure 10.6. On the other hand, with one exception, the data as in the statement do determine the cut. The exception is the series of skeletons $\mathfrak{S}_k := \operatorname{Sk}\langle R^{2k}\rangle$, $k \in \mathbb{N}$: each $\mathfrak{S}_k$ looks like a circle bounding the $2k$-gonal region, to which $2k$ Farey branches are attached, all pointing outwards. In this case, there are two cuts, which differ by a shift along the circle. (The skeleton is to be cut at every other $\circ$-vertex.) Clearly, these cuts result in isomorphic ribbon curve structures. □

To distinguish the curves topologically, we consider the monodromy of a certain auxiliary curve (or rather fibered embedded topological surface) $C' \subset \Sigma' \to \bar{B}^+$ over an appropriate monodromy domain $\bar{B}^+$. Namely, let $\bar{B}^+ \subset B$ be the subset obtained from B^+ by removing a small regular neighborhood of the real part $B_{\mathbb{R}}$ and adding disjoint regular neighborhoods of the zigzags of C, see Figure 10.7. (According to the construction, zigzags of C appear at monovalent $\bullet$-vertices of $\mathfrak{S}$, and the dotted bold lines in the figure represent the edges of $\mathfrak{S}$ incident to such vertices; they are missing in $\mathfrak{S}^+$.) Let, further, $\bar{B}^\sharp$ be obtained from $\bar{B}^+$ by removing each zigzag (or a smaller regular neighborhood thereof, see Figure 10.7) and puncturing at each inner $\times$-vertex of $\mathfrak{D}$. Note that $\bar{B}^+$ can also be regarded as a subset of $\operatorname{Supp}\mathfrak{S}$, and the embedding can be chosen so that the 'missing' edges of $\mathfrak{S}$ are in the difference $\bar{B}^+ \smallsetminus \bar{B}^\sharp$, as shown in the figure. Then, choosing for reference an edge e of $\mathfrak{S}^+$ and using the fact that each inner $\times$-vertex is inside a monogonal region of each of the skeletons, see Figure 10.6, we obtain a canonical epimorphism $\pi_1(\bar{B}^+, e) \twoheadrightarrow \pi_1^{\mathrm{orb}}(\mathfrak{S}, e)$.

Denote by $p\colon \Sigma \to B$ the projection and rebuild C and Σ over $\bar{B}^+$ as follows: cut $p^{-1}(\bar{B}^+)$ over the grey dotted arcs shown in the figure and reattach the pieces *via* fiberwise diffeomorphisms inducing in each fiber the braid $\sigma_1\sigma_2\sigma_1$ with respect to a canonical basis over the edge of $\mathfrak{S}$ intersecting the cut. (For example, over each cut the fibration can be trivialized so that $F^\circ = \mathbb{C}$ and $F^\circ \cap C = \{0, \pm 1\}$, and the attaching diffeomorphism is $z \mapsto -z$.) The resulting 4-manifold and embedded surface fibered over $\bar{B}^+$ are denoted by Σ' and C', respectively. Now, as in Section 5.1.1, we can choose a proper section s of Σ' over $\bar{B}^+$ and use it to define the braid monodromy $\pi_1(\bar{B}^\sharp) \to \mathbb{B}_3$. Similar to Theorem 5.62, the Γ-valued reduction $\mathfrak{m}\colon \pi_1(\bar{B}^\sharp) \to \Gamma$ does not depend on the section; moreover, we have the following lemma.

Lemma 10.72. *The* Γ*-valued reduction* $\mathfrak{m}$ *splits via*

$$\pi_1(\bar{B}^+, e) \twoheadrightarrow \pi_1^{\mathrm{orb}}(\mathfrak{S}, e) \xrightarrow{\mathrm{val}} \Gamma,$$

where the left arrow is the epimorphism introduced above.

Proof. The statement follows mostly from Theorem 5.62, as the monodromy can be computed along chains in $\mathfrak{S}$. Extra attention is needed at the monovalent vertices of $\mathfrak{S}$, where the relation between $\mathfrak{S}$ and C is somewhat artificial. Each monovalent $\circ$-vertex v of $\mathfrak{S}$ corresponds to an odd hyperbolic component of C, and it is immediate that, in a canonical basis over the edge incident to v, the monodromy along such a component is an odd power of $\sigma_1\sigma_2\sigma_1$, *i.e.*, $\mathbb{Y} \in \Gamma$. Each monovalent $\bullet$-vertex v corresponds to a zigzag. Using Theorem 5.65, one can see that, in a canonical basis over the edge shown in Figure 10.7, the Γ-valued monodromy about the zigzag of the *original* curve C is $RL^{-1} = \mathbb{X}$. In the same basis, the monodromy of $\mathfrak{S}$ about v is $\mathbb{Y}\mathbb{X}\mathbb{Y}$. To compensate for the extra conjugation by $\mathbb{Y}$, we change C to C'. □

Theorem 10.73. *Consider two oriented ribbon curve structures* $\mathcal{S}_1$ *and* $\mathcal{S}_2$ *without ribbon blocks of type* II_1 *or* II_2*, and let* $C_i \subset \Sigma_i \to B_i$*,* $i = 1, 2$*, be corresponding ribbon curves; their bases* B_1*,* B_2 *are naturally directed. Then the following three statements are equivalent:*

1. *the structures* $\mathcal{S}_1$ *and* $\mathcal{S}_2$ *are related by an orientation preserving isomorphism;*
2. *the curves* C_1 *and* C_2 *are directedly deformation equivalent, i.e., related by a deformation and/or directed isomorphism of the ruled surfaces;*
3. *the curves* C_1 *and* C_2 *are related by a directed homeomorphism* $\Sigma_1 \to \Sigma_2$*.*

Proof. The statement is obvious for single block ribbon curve structures and, since blocks of type I_0 or II_0 cannot participate in a junction, we can assume that all blocks involved are of types I_1, I_2, or II_3. The implications (1) $\Longrightarrow$ (2) $\Longrightarrow$ (3) are also obvious, and it remains to show that (3) implies (1).

Any ribbon curve structure with all blocks of types I_1, I_2, or II_3 is obtained by the construction described at the beginning of this section: one can take for $\mathfrak{S}$ the skeleton of the corresponding ribbon dessin, extended by a pair of Farey branches at each odd real bold segment (block of type I_2) and an edge $\circ$—$\bullet$ at each zigzag, see Figure 10.7. Thus, we only need to show that the set of data ($\mathfrak{S}$, {uncut regions}) as in Lemma 10.71 can be recovered from the directed equivariant fiberwise topology of the ribbon curve. To this end, observe that the auxiliary curve $C' \subset \Sigma' \to \bar{B}^+$ is defined in purely topological terms, and hence so is its monodromy $\mathfrak{m}$. Pick a reference edge e and consider the image $H := \operatorname{Im}\mathfrak{m}$ and the H-conjugacy classes $r_i := [\![\mathfrak{m}(l_i)]\!]_H$, where l_i are some lassoes about the inner $\times$-vertices of $\mathfrak{D}$ (*i.q.* singular fibers of C). All r_i are classes of Dehn twists. The pair $(H, \{r_i\})$ is well defined up to conjugation in Γ. On the other hand, due to Lemma 10.72, we have $\mathfrak{S} \cong \operatorname{Sk} H$ and the classes r_i correspond to the uncut monogonal regions of $\mathfrak{S}$, see Lemma 1.23. □

Corollary 10.74 (of Theorems 10.73 and 4.69). *Two real trigonal M-curves are in the same (directed) deformation class if and only if they are related by a (directed) homeomorphism.* $\triangleleft$

Theorem 10.73 can be extended to non-oriented and even non-orientable ribbon curve structures; for the latter, one should consider the orientation double covering. We leave the precise statements to the reader.

Curves in Hirzebruch surfaces

One can easily see that the ribbon curve $C \subset \Sigma \to B$ obtained from a skeleton $\mathfrak{S}$ has rational base $B \cong \mathbb{P}^1$ if and only if $\mathfrak{S} = \mathrm{Sk}(\mathfrak{T}, \ell)$ is a pseudo-tree with the vertex function ℓ taking values in $\{0, \bullet, \vartriangle\}$ and all monogonal regions (loops) of $\mathfrak{S}$ are *not* cut, *i.e.*, each loop is converted to a ribbon block of type I_1, see Figure 10.6 (b).

Let k be an integer. Define a *k-flat necklace diagram* as a class of broken necklace diagrams modulo the equivalence relation generated by $\mathcal{N} \sim \mathcal{N}^*$ and $S_1 S_2 \ldots S_r \sim S_2 \ldots S_r S_1$ if k is even or $S_1 S_2 \ldots S_r \sim S_2 \ldots S_r S_1^*$ if k is odd. (Here S^* is the dual stone, see Table 3.3.) The *monodromy* of a k-flat necklace diagram $[\mathcal{N}]_k$ is the conjugacy class $\mathfrak{m}[\mathcal{N}]_k := [\![\mathfrak{m}(\mathcal{N})]\!]$ if k is even or $[\![\mathfrak{m}(\mathcal{N})\mathbb{Y}]\!]$ if k is odd.

The real part of a non-hyperbolic real trigonal curve C in a directed Hirzebruch surface Σ_k can be described by a k-flat necklace diagram. Indeed, C gives rise to a pair of opposite real Jacobian elliptic surfaces over $\mathbb{P}^1 \smallsetminus b$, where $b \in \mathbb{P}^1_{\mathbb{R}}$ is a point outside the ovals and zigzags of C, *cf.* page 98, and the diagram $[\mathcal{N}]_k$ in question is the class of the broken (at b) necklace diagrams $\mathcal{N}, \mathcal{N}^*$ of these surfaces, regarded as directed fibrations. (If k is odd, the extra twists S_1^* and $\mathbb{Y}$ in the definition are due to the fact that $(\Sigma_k)_{\mathbb{R}}$ is non-orientable and the fibrations do not extend through b.) The monodromy $\mathfrak{m}[\mathcal{N}]_k$ is merely the Γ-valued reduction of the braid monodromy of C evaluated at the boundary of $B^+ \smallsetminus U$, where U is a regular neighborhood of $\mathbb{P}^1_{\mathbb{R}}$, *cf.* the proof of Theorem 3.93 and Figure 3.5.

Lemma 10.75. *A k-flat necklace diagram $[\mathcal{N}]_k$ without arrow type stones is uniquely determined by the parity of k and the monodromy $\mathfrak{m}[\mathcal{N}]_k$.*

Proof. If k is even, we have $\mathfrak{m}[\mathcal{N}]_k = L^m$ for $\mathcal{N} = \square^m \sim \bigcirc^m$ and

$$\mathfrak{m}[\mathcal{N}]_k = (RL^{m_1-1}R)(RL^{n_1-1}R)\ldots(RL^{m_s-1}R)(RL^{n_s-1}R) \tag{10.76}$$

for $\mathcal{N} = \square^{m_1}\bigcirc^{n_1}\ldots\square^{m_s}\bigcirc^{n_s}$. (In the latter case, we can always break the diagram between two stones of distinct types.) If k is odd, then we have $\mathfrak{m}[\mathcal{N}]_k = RL^{m-1}R$ for $\mathcal{N} = \square^m \sim \bigcirc^m$ and

$$\mathfrak{m}[\mathcal{N}]_k = (RL^{m_1-1}R)(RL^{n_1-1}R)\ldots(RL^{m_s-1}R)$$

for $\mathcal{N} = \square^{m_1}\bigcirc^{n_1}\ldots\square^{m_s}$. $\square$

Corollary 10.77 (of the proof). *With the following exceptions*:

$$(\square\bigcirc)^n \mapsto R^{4n}, \quad (\bigcirc\square)^n \mapsto L^{-4n}, \quad \square^n \mapsto LR^{-n}L^{-1}, \quad \bigcirc^n \mapsto R^{-1}L^nR,$$

the total monodromy $\mathfrak{m}(S_1)\ldots\mathfrak{m}(S_r)$ *of a nonempty sequence* $S_1 \ldots S_r$ *of stones of types* $\square$ *or* $\bigcirc$ *is hyperbolic.* ◁

Definition 10.78. A positive word A in the alphabet $\{L, R\}$ is called *L-even* (*R-even*) if it can be obtained from a word in $\{l, R\}$ (respectively, $\{L, r\}$) by the substitution $l \mapsto L^2$ (respectively, $r \mapsto R^2$). A word is called *even* if it is both L- and R-even. A parabolic or hyperbolic element $g \in \Gamma$ is *R-even*, *L-even*, or *even* if the conjugacy class $[\![g]\!]$ admits a representative (2.16) with this property.

Corollary 10.79 (of the proof of Lemma 10.75). *The monodromy of a k-flat necklace diagram without arrow type stones is R-even.* ◁

A ribbon curve $C \subset \Sigma_k$ obtained from a pseudo-tree $\mathfrak{S} := \operatorname{Sk}(\mathfrak{T}, \ell)$ has no zigzags (the k-flat necklace diagram $[\mathcal{N}]_k$ of C has no arrow type stones) if and only if $\mathfrak{S}$ is regular, *i.e.*, the vertex function ℓ takes values in $\{0, \vartriangle\}$. In this case, we do not need to consider the modified curve C' and $\mathfrak{m}[\mathcal{N}]_k = \mathfrak{m}_\infty(\mathfrak{S})$, see Lemma 10.72; hence $\mathfrak{m}_\infty(\mathfrak{S})$ is R-even. A simple analysis of Proposition 10.22 shows that the converse also holds: a regular pseudo-tree $\mathfrak{S}$ corresponds to a ribbon curve (*i.e.*, $\mathfrak{S}$ can be cut into fragments as in Figure 10.6) if and only if $\mathfrak{m}_\infty(\mathfrak{S})$ is R-even.

***M*-curves** (see [64])

According to Theorem 4.69, any M-curve $C \subset \Sigma_k$ is obtained from a pseudo-tree $\mathfrak{S} := \operatorname{Sk}(\mathfrak{T}, \ell)$ with the vertex function ℓ taking exactly two values distinct from $\vartriangle$; conversely, any such curve is an M-curve. In this case, $\mathfrak{S}$ is a 'linear' pseudo-tree, *cf.* Figure 10.5, where one or both loops can be contracted to monovalent •-vertices. If an M-curve C has at least one zigzag, the directed deformation class of C is determined by its k-flat necklace diagram $[\mathcal{N}]_k$, see [60]. If C has no zigzags, it coincides with the auxiliary curve $C' \subset \Sigma'$ over $\bar{B}^+$ and, in view of Lemma 10.72, the Γ-valued monodromy $\mathfrak{m}\colon \pi_1(\bar{B}^\sharp) \to \Gamma$ of C factors through $\operatorname{val}\colon \pi_1(\mathfrak{S}) \to \Gamma$.

Theorem 10.80 (see [64]). *Two trigonal M-curves $C_i \subset \Sigma_{k_i}$, $i = 1, 2$, without zigzags are in the same directed deformation class (in particular, $k_1 = k_2$) if and only if their monodromy groups* $\operatorname{Im}\mathfrak{m}_i$ *are conjugate in Γ. A subgroup $H \subset \Gamma$ is the monodromy group of a trigonal M-curve without zigzags if and only if the skeleton* $\mathfrak{S} := \operatorname{Sk} H$ *is a regular pseudo-tree with two loops and even monodromy at infinity.*

Proof. The first statement follows from Theorem 4.69 (each M-curve is a ribbon curve with all blocks of type I), Theorem 10.73 (ribbon curves are classified by their skeletons), and Lemma 10.72 (in the absence of zigzags, the skeleton is $\operatorname{Sk}(\operatorname{Im}\mathfrak{m})$).

The second statement is given by Theorem 4.69 and the fact that the cyclic diagram of the monodromy at infinity $\mathfrak{m}_\infty$ of a regular pseudo-tree with two loops has a para-symmetry, *cf.* Theorem 10.38; hence, $\mathfrak{m}_\infty$ is R-even if and only if it is even. □

Corollary 10.81 (*cf.* Theorem 10.27). *A k-flat necklace diagram $[\mathcal{N}]_k$ without arrow type stones is realized by a trigonal M-curve if and only if $\mathfrak{m}[\mathcal{N}]_k \overset{c}{\sim} L^2AL^2A^t$ for some* even *positive word A in $\{L, R\}$.* ◁

Corollary 10.82 (*cf.* Theorem 10.32). *A diagram $[\mathcal{N}]_k$ as in Corollary* 10.81 *is shared by at most two non-equivalent trigonal M-curves. In other words, at most two equivalence classes of curves may share homeomorphic real parts.* ◁

Corollary 10.83 (*cf.* Theorem 10.32). *A diagram $[\mathcal{N}]_k$ as in Corollary* 10.81 *is shared by two non-equivalent trigonal M-curves if and only if $\mathfrak{m}[\mathcal{N}]_k \overset{c}{\sim} W_q(B)$ for some* even *positive word B in $\{L, R\}$.* ◁

Remark 10.84. The two curves given by (4.72) correspond to the 2-factorizations of $W_{1/3}(L^2R^2)$; their common flat necklace diagram is $(\bigcirc^5\square^5\bigcirc\square^3)^2$.

Example 10.85. In general, the same real part can be shared by a large number of curves. For example, all simple pseudo-trees with a fixed number $2k$ of $\bullet$-vertices give rise to $T(k)$ non-equivalent ribbon curves in Σ_{2k} with the same flat necklace diagram $\bigcirc^{5k-1}$, *cf.* Theorem 10.20.

10.3.3 Maximal Lefschetz fibrations are algebraic (see [64])

Let $\mathcal{N}$ be a broken necklace diagram and $w \geqslant 0$ an integer. A *w-pendant* on $\mathcal{N}$ is a *strong* Hurwitz equivalence class of w-factorizations $\bar{\mathfrak{m}}$ of the monodromy $\mathfrak{m}(\mathcal{N})$. The cyclic permutation $1 \mapsto 2 \mapsto \dots$ acts on pairs $(\mathcal{N}, \bar{\mathfrak{m}})$ *via*

$$\mathcal{N} = S_1 \dots S_r \mapsto S_2 \dots S_n S_1, \qquad \bar{\mathfrak{m}} \mapsto P_1^{-1}\bar{\mathfrak{m}}P_1,$$

where P_1 is the monodromy of S_1, and we define a *w-pendant necklace diagram* as an orbit of this action. The following statement refines Theorem 3.96, and its proof is literally the same: the w-pendant structure controls the extension of the fibration from the circle $B_\mathbb{R}$ to the disk B^+, see Figure 3.5 and the geometric interpretation of monodromy factorizations on page 290.

Theorem 10.86 (see [64]). *There is a canonical one-to-one correspondence between the set of isomorphism classes of directed fibrations $X \to B$ with $2r > 0$ real and $w \geqslant 0$ pairs of c_B-conjugate singular fibers and the set of w-pendant necklace diagrams of length r.* ▷

Example 10.87. According to Theorem 10.86, the same necklace diagram can be shared by a large number of directed fibration. Thus, Theorem 10.20 gives us an example of $T(k)$ non-isomorphic directed fibration with the same necklace diagram $\bigcirc^{5k-1}$ (or $\square^{5k-1}$) and $w := k + 1$ pairs of conjugate singular fibers.

Theorem 10.88 (see [64]). *Any maximal directed fibration is algebraic.*

This theorem is one of the principal applications of Theorem 10.27, and we outline its proof. In view of (3.94), a maximal directed fibration $p\colon X \to B \cong \mathbb{P}^1$ may have $w = 0$, 1, or 2 pairs of conjugate singular fibers, and the necklace diagram $\mathcal{N}$ of p has $(2 - w)$ arrow type stones. The case $w = 0$ is covered by Corollary 4.80.

Proof of Theorem 10.88: *the case* $w = 2$. The necklace diagram $\mathcal{N}$ of p has no arrow type stones and its monodromy is either conjugate to L^m, $m > 0$, see Corollary 10.77, or of the form (10.76). On the other hand, the cyclic diagram of $\mathfrak{m}(\mathcal{N})$ has a para-symmetry, see Theorems 10.86, 10.27, and 10.38; hence, $\mathfrak{m}(\mathcal{N})$ is even and the monodromy group of the 2-pendant of p is realized by an M-curve C, see Theorem 10.80. Then $\mathcal{N}$ is a representative of the flat necklace diagram of C and p is isomorphic to one of the two opposite real elliptic surfaces ramified at C. □

To complete the proof, we need an additional construction. Let $p\colon X \to B$ be a directed fibration, and let S be an arrow type stone of its necklace diagram, regarded as a segment in $B_{\mathbb{R}}$. Consider a c_B-invariant regular neighborhood U of S. We can cut off the preimage $p^{-1}(U)$ and reglue it back to form a pair F^+, F^- of conjugate singular fibers. (Algebraically, this procedure corresponds to deforming the fibration through a singular fiber of type $\tilde{\mathbf{A}}_0^{**}$.) If a real basepoint b is in U and l is a lasso at b about F^+ contained in U, then the monodromy $\mathfrak{m}(l)$ of the new fibration is P^{-1}, where P is the monodromy of S, see Table 3.3. Conversely, given a lasso l (at a real basepoint b outside the stones) about a singular fiber $F^+ \in B^+$ with $\mathfrak{m}(l) = R$ or L^{-1}, we can cut off and reglue the preimage of a c_B-invariant regular neighborhood of the union $l \cup c_B(l)$ (including the encompassed disks) to produce a new stone S of type $<$ or $>$, respectively, next to the basepoint b.

Lemma 10.89. *Let $p\colon X \to B$ be a real elliptic surface ramified over a trigonal M-curve $C \subset \Sigma_k$ without zigzags. Then, up to isotopy, there are exactly two lassoes l as above with $\mathfrak{m}(l) = R$ and exactly two lassoes with $\mathfrak{m}(l) = L^{-1}$. In all four cases, the modified directed fibration with an arrow type stone is algebraic.*

Proof. The four lassoes are contained in the two ribbon blocks of type I_1: there is one lasso at a basepoint in each of the four real solid segments contained in these blocks. The modified fibration can be realized by a real elliptic surface ramified at an M-curve obtained from C by replacing the corresponding block with one of type I_2, properly oriented. Hence, it is algebraic.

Consider one of the four lassoes l and let $b \in B_{\mathbb{R}}$ be its basepoint. Since the disk B^+ has only two punctures, any other lasso l' about the same singular fiber is homotopic to $\gamma l \gamma^{-1}$, where γ is a path (not necessarily minimal) along $B_{\mathbb{R}}$ connecting the basepoint b' of l' to b. Hence, $\mathfrak{m}(l') = P^{-1}\mathfrak{m}(l)P$ or $P\mathfrak{m}(l)P^{-1}$, where P is the monodromy of a contiguous sequence of stones starting right after b or, respectively, ending right before b. Assume that $\mathfrak{m}(l) = R$. From the classification of M-curves it follows that b has a neighborhood in $B_{\mathbb{R}}$ of the form —□—<—□—□— or —○—○—<—○—, *cf.* the row $b_* = 12$ in Table 3.4 (we represent b by the $<$ type stone corresponding to R), and Corollary 10.77 implies that P never commutes with R. Hence, the other lasso with monodromy R is about the other singular fiber, and a similar argument shows that there are no otherline such lassoes. If $\mathfrak{m}(l) = L^{-1}$, then b has a neighborhood —□—□—>—□— or —○—>—○—○— and the proof is literally the same. □

Proof of Theorem 10.88: *the case* $w = 1$. The necklace diagram $\mathcal{N}$ of p has a single arrow type stone. Converting this stone to a pair of conjugate singular fibers, we obtain a maximal directed fibration p', which is algebraic due to the first part of the proof. Then, Lemma 10.89 implies that, topologically, there are only four ways to recreate the arrow type stone, and all four ways result in algebraic directed fibrations. □

Alternative proof of Corollary 4.80. The necklace diagram $\mathcal{N}$ of p has two arrow type stones. Converting each to a pair of conjugate singular fibers, we obtain an algebraic fibration. Then, Lemma 10.89 still applies to show that the original fibration is also algebraic. □

Appendices

Appendix A

An algebraic complement

A.1 Integral lattices

An (*integral*) *lattice* if a finitely generated free abelian group L equipped with a symmetric bilinear form $\varphi\colon L \otimes L \to \mathbb{Z}$. If φ is understood, we use the notation $x^2 := \varphi(x, x)$ and $x \cdot y := \varphi(x, y)$. The *orthogonal complement* of a subgroup $N \subset L$ is the subgroup $N^\perp := \{x \in L \mid x \cdot y = 0 \text{ for all } y \in N\}$.

Occasionally, we will also consider *rational lattices*, where the bilinear form φ is allowed to take values in $\mathbb{Q}$.

The *kernel* of a lattice L is the subgroup $\ker L := \{x \in L \mid x \cdot y = 0 \text{ for all} y \in L\}$. (The notation ker is opposed to the kernel Ker of a homomorphism.) A lattice L is *nondegenerate* if $\ker L = 0$. The lattice $L/\ker := L/\ker L$ is always nondegenerate. The *determinant* $\det L$ of a lattice L is the determinant of its Gram matrix; since the transition matrix between any two integral bases is unimodular, $\det L$ is well defined. A lattice L is *unimodular* if $\det L = \pm 1$. A lattice L is *even* if $x^2 = 0 \bmod 2$ for all $x \in L$; otherwise, L is *odd*. A *characteristic class* of a unimodular lattice L is an element $w \in L$ such that $x^2 = x \cdot w \bmod 2$ for any $x \in L$; such an element exists and is unique modulo $2L$. The *signature* of a nondegenerate lattice L is the pair (σ_+, σ_-) of the inertia indices of the bilinear form.

When speaking about lattices, we assume all direct sums orthogonal. A lattice is called *irreducible* if it does not split into orthogonal direct sum of two nontrivial sublattices. A nondegenerate sublattice $N \subset L$ is an orthogonal summand if and only if the orthogonal projection $L \to N \otimes \mathbb{Q}$ is defined over $\mathbb{Z}$. If this is the case, N is also an orthogonal summand in any smaller sublattice $M \supset N$ contained in L. As a simple consequence, any unimodular lattice N is an orthogonal summand in any larger lattice $L \supset N$.

A *root* in a lattice L is a vector of square ± 1 or ± 2. A *root lattice* is a negative definite lattice generated by its roots.

A.1.1 Nikulin's theory of discriminant forms (see [125])

A *finite bilinear form* is a finite abelian group $\mathcal{L}$ equipped with a symmetric bilinear form $\varphi\colon \mathcal{L} \otimes \mathcal{L} \to \mathbb{Q}/\mathbb{Z}$. The *orthogonal complement* $\mathcal{N}^\perp$ and the *kernel* $\ker \mathcal{L} = \mathcal{L}^\perp$ are defined in the same way as for lattices, and $\mathcal{L}$ is called *nondegenerate* if $\ker \mathcal{L} = 0$. In this case $(\operatorname{Card} \mathcal{N})(\operatorname{Card} \mathcal{N}^\perp) = \operatorname{Card} \mathcal{L}$ for any subgroup $\mathcal{N} \subset \mathcal{L}$.

A *finite quadratic form* is a finite bilinear form $(\mathcal{L}, \varphi)$ equipped with a *quadratic extension* of φ, *i.e.*, a map $q\colon \mathcal{L} \to \mathbb{Q}/2\mathbb{Z}$ quadratic in the sense that

$$q(x+y) - q(x) - q(y) = 2\varphi(x,y), \quad q(x) = \varphi(x,x) \bmod \mathbb{Z},$$

where $2\colon \mathbb{Q}/\mathbb{Z} \to \mathbb{Q}/2\mathbb{Z}$ is the homomorphism $q \mapsto 2q$.

As in the case of lattices, we use the shortcuts $x^2 := q(x)$ and $x \cdot y := \varphi(x,y)$.

Remark A.1. Since $2\colon \mathbb{Q}/\mathbb{Z} \to \mathbb{Q}/2\mathbb{Z}$ is an isomorphism, the bilinear form φ is uniquely determined by q. On the other hand, the action $q \uparrow l := q + l$ turns the set $\operatorname{Qu}(\mathcal{L})$ of all quadratic extensions of a given bilinear form φ into a principal homogeneous space over $\operatorname{Hom}(\mathcal{L}, \mathbb{Z}_2)$. As a special case, if $\operatorname{Tors}_2 \mathcal{L} = 0$, any symmetric bilinear form has a unique quadratic extension.

If L is a nondegenerate lattice, the bilinear form extends to a $\mathbb{Q}$-valued bilinear form on $L \otimes \mathbb{Q}$ and the dual group $L^* := \operatorname{Hom}(L, \mathbb{Z})$ is naturally identified with the subgroup $\{x \in L \otimes \mathbb{Q} \mid x \cdot y \in \mathbb{Z} \text{ for all } y \in L\}$. Thus, L is a finite index subgroup of L^* and one can consider the quotient $\operatorname{discr} L := L^*/L$; it is called the *discriminant form* of L. This quotient inherits from $L \otimes \mathbb{Q}$ a nondegenerate symmetric bilinear form $[x] \otimes [y] \mapsto x \cdot y \bmod \mathbb{Z}$. If L is even, $\operatorname{discr} L$ inherits also a quadratic extension $q\colon [x] \mapsto x^2 \bmod 2\mathbb{Z}$. One has $\operatorname{Card}(\operatorname{discr} L) = |\det L|$.

For a degenerate lattice L, we let $\operatorname{discr} L := \operatorname{discr}(L/\ker)$.

Recall that two lattices are said to be in the same *genus* if they become isomorphic after tensoring by $\mathbb{R}$ and by p-adic integers for all primes p.

Theorem A.2 (Nikulin [125]). *The genus of a non-degenerate* even *lattice L is determined by its signature (σ_+, σ_-) and quadratic discriminant form* $\operatorname{discr} L$. ▷

Theorem A.3 (Nikulin [125]). *An* indefinite *non-degenerate even lattice L satisfying the condition* $\operatorname{rk} L \geqslant \ell(\operatorname{discr} L) + 2$ *is unique in its genus.* ▷

In general, each genus contains a finite number of isomorphism classes. Some other conditions sufficient for the uniqueness in the genus and a criterion of realizability of a given triple $(\sigma_+, \sigma_-, \mathcal{L})$ by a nondegenerate even lattice are also found in [125]. On the other hand, the classification and comparison of finite bilinear or quadratic forms is a relatively simple task, see, *e.g.*, [125] or [59] for a slightly different approach.

Lattice extensions

All lattices considered in this subsection are nondegenerate. An *extension* of a lattice N is a lattice L containing N as a sublattice, *i.e.*, such that the bilinear form on N is the restriction of that on L. An *isomorphism* between two extensions $L_i \supset N$, $i = 1, 2$, is an isometry $L_1 \to L_2$ identical on N.

Let $L \supset N$ be a finite index extension. Then, within $N \otimes \mathbb{Q}$, we have inclusions $N \subset L \subset L^* \subset N^*$. The *kernel* of the finite index extension $L \supset N$ is the subgroup

$\mathcal{K}_L := L/N \subset \operatorname{discr} N$. Since $\mathcal{K}_L^\perp = L^*/N$, the kernel is an *isotropic subgroup*, *i.e.*, $\mathcal{K}_L \subset \mathcal{K}_L^\perp$. If L (and hence N) is even, $\mathcal{K}_L$ is also q-isotropic, *i.e.*, the restriction to $\mathcal{K}_L$ of the quadratic form $q\colon \operatorname{discr} L \to \mathbb{Q}/2\mathbb{Z}$ is identically zero. Conversely, given an isotropic subgroup $\mathcal{K} \subset \operatorname{discr} N$, the pull-back of $\mathcal{K}$ under the quotient projection $N^* \to \operatorname{discr} N$ is an extension of N. Thus, we have the following statement.

Theorem A.4 (Nikulin [125]). *The map $L \mapsto \mathcal{K}_L := L/N \subset \operatorname{discr} N$ establishes a bijection between the set of isomorphism classes of (even) finite index extensions $L \supset N$ of a given lattice N and that of isotropic (respectively, q-isotropic) subgroups $\mathcal{K} \subset \operatorname{discr} N$. One has $\operatorname{discr} L = \mathcal{K}_L^\perp/\mathcal{K}_L$.* ◁

Another extreme case is that of a primitive extension $L \supset N$ to a unimodular lattice L. (An extension is *primitive* if $N \subset L$ is a primitive subgroup, *i.e.*, the quotient L/N is torsion free.) In this case, L is a finite index extension of $N \oplus N^\perp$ and one can speak about its kernel $\mathcal{K} \subset \operatorname{discr} N \oplus \operatorname{discr} N^\perp$. Since both N and $N^\perp$ are primitive in L, one has $\mathcal{K} \cap \operatorname{discr} N = \mathcal{K} \cap \operatorname{discr} N^\perp = 0$ and the projections of $\mathcal{K}$ to both summands are monomorphisms. On the other hand, L is unimodular if and only if $(\operatorname{Card} \mathcal{K})^2 = \operatorname{Card}(\operatorname{discr} N \oplus \operatorname{discr} N^\perp)$; in this case, both projections above are isomorphisms and $\mathcal{K}$ is the graph of a certain isomorphism $\varphi\colon \operatorname{discr} N \to \operatorname{discr} N^\perp$. Since the graph of an isomorphism φ is isotropic if and only if φ is an anti-isometry, we have the following theorem.

Theorem A.5 (Nikulin [125]). *An extension $L \supset N$ is primitive and unimodular if and only if the kernel of the finite index extension $L \supset N \oplus N^\perp$ is the graph of an anti-isometry $\varphi\colon \operatorname{discr} N \to \operatorname{discr} N^\perp$. In this case, the lattice L is even if and only if both N and $N^\perp$ are even and φ is a q-anti-isometry.* ◁

A.1.2 Definite lattices

The theory of definite integral lattices is a vast subject. Unlike the indefinite case, the genera of definite lattices tend to contain huge numbers of isomorphism classes, *cf.* Warning A.9 below. Referring to [31] and [147] for a detailed account of the modern state of affairs, we only cite a few facts, mainly those concerning root lattices.

Theorem A.6 (Eichler [67]; see also Kneser [94]). *A definite lattice L is the orthogonal direct sum of all its irreducible orthogonal summands.*

We emphasize that, according to Theorem A.6, the decomposition of a definite lattice L into direct sum of irreducible orthogonal summands is not just unique up to isomorphism: each irreducible orthogonal summand is 'rigid' in L and is present in any (in fact, the only) such decomposition.

We precede the proof of Theorem A.6 with a lemma.

Lemma A.7. *Let L be a definite lattice and $N_1, N_2 \subset L$ two orthogonal summands of L. Then the intersection $S := N_1 \cap N_2$ is an orthogonal summand in both N_1 and N_2 and the sublattices $N_i' := S^{\perp}_{N_i}$, $i = 1, 2$, are orthogonal to each other.*

Proof. Without loss of generality, we can assume that L is positive definite. Denote by p_i the orthogonal projection to N_i, $i = 1, 2$. Assume that there is a vector $v_1 \in N_1'$ not orthogonal to N_2'. Then the vector $v_2 := p_2(v_1)$ does not belong to S and one has $0 < v_2^2 < v_1^2$ and $v_2 \cdot v_1 = v_2^2 \neq 0$, hence v_2 is not orthogonal to N_1'. Let $v_3 := p_1(v_2)$, $v_4 := p_2(v_4)$, *etc.* Each time, the vector v_{2k+i} is in $N_i' \smallsetminus S$ and is not orthogonal to N_{3-i}', $i = 1, 2$. Hence, we obtain an infinite sequence of *integral* vectors of strictly decreasing positive length. This is a contradiction; hence $N_1' \perp N_2'$.

Since $N_1' \perp N_2'$, the orthogonal projection $p_1 \colon L \to N_1$ restricts to a map $N_2 \to S$ and hence S is an orthogonal summand of N_2. Similarly, S is also an orthogonal summand of N_1. □

Proof of Theorem A.6. Let N_1 and N_2 be two distinct irreducible orthogonal summands of L. Then, according to Lemma A.7, $N_1 \cap N_2 = 0$ and $N_1 \perp N_2$, *i.e.*, N_2 belongs to the orthogonal complement $N_1^{\perp}$ and is an orthogonal summand thereof. Proceeding by induction, one concludes that all irreducible orthogonal summands of L are orthogonal to each other and their direct sum equals L. □

Corollary A.8. *Let L_1, L_2, and M be three positive definite lattices, and assume that $L_1 \oplus M \cong L_2 \oplus M$. Then $L_1 \cong L_2$.* ◁

Warning A.9. The conclusion of Corollary A.8 does not hold if not all three lattices are assumed definite. For example, it is not difficult to show that, if L_1 and L_2 are unimodular of the same rank and signature and $M = \mathbb{Z}h_1 \oplus \mathbb{Z}h_2$ with $h_1^2 = -h_2^2 = 1$, then $L_1 \oplus M \cong L_2 \oplus M$. But L_1 and L_2 may differ; for example, one of them may be even and the other one, odd. In fact, the number of isomorphism classes of unimodular *definite* lattices of a given rank r grows exponentially in r, see, *e.g.*, [147]. There are more than eighty million isomorphism classes of even unimodular definite lattices of rank 32; according to [147], their classification has not been completed *yet*.

Theorem A.10 (Gauss [78]). *A positive definite lattice L of rank two has a unique canonical form in which the Gram matrix C satisfies $0 \leqslant 2C_{12} \leqslant C_{11} \leqslant C_{22}$.*

Proof. Let $\mathbf{u}, \mathbf{v}$ be a basis for L. The inequalities $C_{11} \leqslant C_{22}$ and $2|C_{12}| \leqslant C_{11}$ can be achieved by the basis changes $\mathbf{u} \leftrightarrow \mathbf{v}$ and $\mathbf{u} \mapsto \mathbf{u} \pm \mathbf{v}$, respectively. Since all entries are integers and C_{11} does not increase, this process converges.

Assume that C satisfies the inequalities as in the statement. Then the square of a vector $w := x\mathbf{u} + y\mathbf{v} \in L$ can be estimated as follows:

$$w^2 = C_{11}x^2 + 2C_{12}xy + C_{22}y^2 \geqslant C_{11}(x^2 + y^2 - |xy|) \geqslant \frac{1}{2}C_{11}(x^2 + y^2).$$

Since x, y are integers, we conclude that $\mathbf{u}$ is a shortest vector and, unless $C_{11} = C_{22}$, all shortest vectors are $\pm\mathbf{u}$. If $C_{11} = C_{22}$, another pair of shortest vectors is $\pm\mathbf{v}$ and, if also $2C_{12} = C_{11}$, there is yet another pair $\pm(\mathbf{u} - \mathbf{v})$. Using these observations, one easily concludes that the Gram matrix as above is unique. □

Root lattices

Given an integer $k \geqslant 1$, define the lattices $\mathbf{B}_k$ and $\tilde{\mathbf{B}}_k$ as the orthogonal direct sums

$$\mathbf{B}_k = \bigoplus_{i=1}^{k} \mathbb{Z}\mathbf{v}_i, \qquad \tilde{\mathbf{B}}_k = \bigoplus_{i=1}^{k} \mathbb{Z}\mathbf{v}_i \oplus \mathbb{Z}\mathbf{w},$$

where $\mathbf{v}_i^2 = -1$, $i = 1, \ldots, k$, and $\mathbf{w}^2 = 0$. These root lattices are related to the irreducible root systems (Dynkin diagrams) of type $\mathbf{B}_k$ and $\mathbf{C}_k$, see [27]. Modulo a technical detail (one should consider graphs with dotted edges, representing negative intersections), the following statement is a consequence of Proposition 1.10.

Theorem A.11 (see [27]). *Any irreducible root lattice is either* $\mathbf{B}_1$, *or* $\mathbf{A}_p$, $p \geqslant 1$, *or* $\mathbf{D}_q$, $q \geqslant 4$, *or* $\mathbf{E}_6$, $\mathbf{E}_7$, $\mathbf{E}_8$. ▷

Convention A.12. Since we mainly deal with singularities, we reserve the notation $\mathbf{A}$, $\mathbf{B}$, $\mathbf{D}$, and $\mathbf{E}$ for the *negative* definite root lattices. Their positive definite counterparts are denoted by $\mathbf{A}^+$, $\mathbf{B}^+$, $\mathbf{D}^+$, and $\mathbf{E}^+$, respectively.

We are interested in the classification, up to the left/right equivalence, of isometric embeddings $S \to T$ of *even* root lattices. Since in an even definite lattice a vector of square (-2) cannot decompose into two nontrivial orthogonal summands, it suffices to consider the case when the target T is irreducible. All statements cited below are known but difficult to find in the literature. They can be obtained by a rather tedious computation using the description of roots in T found in [27] and Nikulin's theory of lattice extensions, see Section A.1.1.

For an embedding $\varphi\colon S \hookrightarrow T$, denote $\ell(\varphi) := \ell(\operatorname{Tors} T/\varphi(S))$ and let $\mathbf{R}(\varphi)$ be the maximal root lattice contained in $\varphi(S)^\perp$. An embedding is *primitive* if $\ell(\varphi) = 0$, *i.e.*, if $\varphi(S)$ is a primitive subgroup of T.

Proposition A.13. *Let T be an irreducible root lattice. For another root lattice S, a primitive embedding $\varphi\colon S \hookrightarrow T$ (respectively, an embedding with $\ell(\varphi) \leqslant 1$) exists if and only if the Dynkin diagram S can be represented as an induced subgraph of the Dynkin diagram T (respectively, of the affine diagram $\tilde{T}$).* ▷

Proposition A.14. *An embedding* $\varphi\colon S \hookrightarrow \mathbf{A}_p$, $p \geqslant 1$, *is determined by the lattice S uniquely up to equivalence; any such embedding is primitive, see Proposition* A.13. ▷

Proposition A.15. *Any embedding* $\varphi\colon S \hookrightarrow \mathbf{D}_q$, $q \geqslant 4$, *of an* irreducible *root lattice* S *is primitive, see Proposition* A.13; *unless* $S \cong \mathbf{A}_3$ *and* $q > 4$, *it is uniquely determined by* S. *One has*:

- *if* $S \cong \mathbf{D}_p$, *then* $\mathbf{R}(\varphi) \cong \mathbf{D}_{q-p}$;
- *if* $S \cong \mathbf{A}_p$, $p \neq 1, 3$, *then* $\mathbf{R}(\varphi) \cong \mathbf{D}_{q-p-1}$;
- *if* $S \cong \mathbf{A}_1$, *then* $\mathbf{R}(\varphi) \cong \mathbf{D}_{q-2} \oplus \mathbf{A}_1$;
- *if* $S \cong \mathbf{A}_3$, *there are two embeddings, with* $\mathbf{R}(\varphi) \cong \mathbf{D}_{q-3}$ *or* $\mathbf{D}_{q-4}$. ▷

Roughly, Proposition A.15 extends to reducible lattices as follows: an embedding $\bigoplus \mathbf{A}_{p_i} \oplus \bigoplus \mathbf{D}_{q_j} \hookrightarrow \mathbf{D}_q$ is unique up to equivalence provided that some of the $\mathbf{A}_3$ summands of S are regarded as $\mathbf{D}_3$ and some of the pairs $2\mathbf{A}_1$ are regarded as $\mathbf{D}_2$, see Convention 1.11.

Proposition A.16. *An embedding* $\varphi\colon S \hookrightarrow \mathbf{E}_6$ *is uniquely determined by* S; *for any such embedding, one has* $\ell(\varphi) \leqslant 1$, *see Proposition* A.13. ▷

Proposition A.17. *With the exception of*

1. $S \cong \mathbf{A}_5 \oplus \mathbf{A}_1$, $\mathbf{A}_3 \oplus 2\mathbf{A}_1$, $4\mathbf{A}_1$ (*each admitting one primitive and one imprimitive embedding*);
2. $S \cong \mathbf{A}_5$, $\mathbf{A}_3 \oplus \mathbf{A}_1$, $3\mathbf{A}_1$ (*two primitive embeddings that differ by* $\varphi(S)^\perp$)

an embedding $\varphi\colon S \hookrightarrow \mathbf{E}_7$ *is uniquely determined by* S. *With the exception of* $S \cong 6\mathbf{A}_1$, $\mathbf{D}_4 \oplus 3\mathbf{A}_1$ ($\ell(\varphi) = 2$) *or* $7\mathbf{A}_1$ ($\ell(\varphi) = 3$), *one has* $\ell(\varphi) \leqslant 1$, *see Proposition* A.13. ▷

The imprimitive embeddings in Proposition A.17(1) factor through $\mathbf{A}_7 \hookrightarrow \mathbf{E}_7$. For S as in A.17(2) the two embeddings also differ by whether they extend to a primitive embedding $S \oplus \mathbf{A}_1 \hookrightarrow \mathbf{E}_7$.

Proposition A.18. *Ten root lattices* S *admit embeddings* $\varphi\colon S \hookrightarrow \mathbf{E}_8$ *with* $\ell(\varphi) > 1$:

- $S \cong 2\mathbf{D}_4$, $\mathbf{D}_6 \oplus 2\mathbf{A}_1$, $\mathbf{D}_4 \oplus 3\mathbf{A}_1$, $2\mathbf{A}_3 \oplus 2\mathbf{A}_1$, $\mathbf{A}_3 \oplus 4\mathbf{A}_1$, $4\mathbf{A}_2$, $6\mathbf{A}_1$: *one has* $\mathbf{E}_8/\varphi(S) \cong \mathbb{Z}_3 \oplus \mathbb{Z}_3$ *for* $S \cong 4\mathbf{A}_2$ *and* Tors $\mathbf{E}_8/\varphi(S) \cong \mathbb{Z}_2 \oplus \mathbb{Z}_2$ *otherwise*;
- $S \cong \mathbf{D}_4 \oplus 4\mathbf{A}_1$ *or* $7\mathbf{A}_1$: *one has* Tors $\mathbf{E}_8/\varphi(S) \cong \mathbb{Z}_2^3$;
- $S \cong 8\mathbf{A}_1$: *one has* $\mathbf{E}_8/\varphi(S) \cong \mathbb{Z}_2^4$.

For $S \cong \mathbf{A}_7$, $\mathbf{A}_5 \oplus \mathbf{A}_1$, $2\mathbf{A}_3$, $\mathbf{A}_3 \oplus 2\mathbf{A}_1$, *or* $4\mathbf{A}_1$, *there is one primitive and one imprimitive embedding*; *otherwise, an embedding is determined by* S. ▷

The imprimitive embeddings $S \hookrightarrow \mathbf{E}_8$ of $S \cong \mathbf{A}_7$, $\mathbf{A}_5 \oplus \mathbf{A}_1$, $2\mathbf{A}_3$, $\mathbf{A}_3 \oplus 2\mathbf{A}_1$, or $4\mathbf{A}_1$ factor through $\mathbf{A}_7 \hookrightarrow \mathbf{E}_7$.

A.2 Quotient groups

We use flexible self-explanatory notation for quotient groups. In addition to G/H for a normal subgroup $H \subset G$, we use $G/S := G/NC(S)$ for a subset $S \subset G$ and $G/\{r_1, \dots, r_k\}$ if $S = \{r_1, \dots, r_k\}$. For a one element set, $G/\{r\}$ is abbreviated to G/r. Thus, the precise meaning of the symbol $G/\cdot$ depends on the nature of the object following the slash. This notation is stretched even further in the next section.

Sometimes, the relations are written in the more human readable form $r_i = 1$ or $a_i = b_i$ if $r_i = a_i b_i^{-1}$. This convention applies to presentations of groups as well.

A.2.1 Zariski quotients

For this section, we fix a pair of groups A, G and a right group action $A \times G \to A$, *i.e.*, a homomorphism $G \to \operatorname{Aut} A$.

Definition A.19. The *relator subgroup* $\operatorname{Rel} H \subset A$ of a subset $H \subset G$ is the normal closure in A of the set $\{(\alpha \uparrow h)\alpha^{-1} \mid \alpha \in A,\, h \in H\}$. The *Zariski quotient* A/H is the quotient group $A/\operatorname{Rel} H$.

For a one element set $H = \{g\}$, we use the abbreviation $\operatorname{Rel} g := \operatorname{Rel} H$ and $A/g := A/H$. Describing a presentation of a group A/H or its further quotients, we often use the notation $g_i = \mathrm{id}$, $g_i \in H$ for the relations.

Lemma A.20. *Given a subgroup $K \subset A$, the set*

$$\operatorname{Zar}_G K := \{g \in G \mid (\alpha \uparrow g)\alpha^{-1} \in K \text{ for all } \alpha \in A\}$$

is a subgroup of G.

Proof. For $g, h \in G$ and $\alpha \in A$, one has

$$(\alpha \uparrow gh)\alpha^{-1} = (\beta \uparrow h)\beta^{-1} \cdot (\alpha \uparrow g)\alpha^{-1}, \qquad (\alpha \uparrow g^{-1})\alpha^{-1} = ((\gamma \uparrow g)\gamma^{-1})^{-1},$$

where $\beta := \alpha \uparrow g \in A$ and $\gamma := \alpha \uparrow g^{-1} \in A$. Hence, whenever g and h belong to $\operatorname{Zar}_G K$, so do gh and g^{-1}. □

Definition A.21. For a subgroup $K \subset A$, the subgroup $\operatorname{Zar}_G K \subset G$ introduced in Lemma A.20 is called the *Zariski kernel* of K. The *saturation* of a subset $H \subset G$ is the subgroup $\operatorname{Sat}_G H := \operatorname{Zar}_G(\operatorname{Rel} H)$. A subgroup $H \subset G$ is called *saturated* if $\operatorname{Sat}_G H = H$. (If G is understood, the subscript G is omitted in the notation.)

Corollary A.22. *For a subset $H \subset G$, one has $A/H = A/\operatorname{Sat} H$. In particular, in the Zariski quotient A/H one has the relations $\alpha = \alpha \uparrow h$ for any $\alpha \in A$ and any element h of the subgroup generated by H.* ◁

Lemma A.23. *For any subgroups $K \subset A$ and $H \subset G$ and an element $g \in G$ one has the identities* $\operatorname{Zar}(K \uparrow g) = g^{-1}(\operatorname{Zar} K)g$ *and* $\operatorname{Rel}(g^{-1}Hg) = (\operatorname{Rel} H) \uparrow g$.

Proof. Both statements follow from the obvious identity

$$((\alpha \uparrow h)\alpha^{-1}) \uparrow g = (\beta \uparrow g^{-1}hg)\beta^{-1},$$

where $h \in G$, $\alpha \in A$, and $\beta := \alpha \uparrow g \in A$. □

For a subgroup $K \subset A$, define the *set stabilizer* $G_K := \{g \in G \mid K \uparrow g = K\}$; it is obviously a subgroup of G.

Corollary A.24. *For any subgroups $K \subset A$ and $H \subset G$ one has $G_K \subset \mathcal{N}(\operatorname{Zar} K)$ and $\mathcal{N}(H) \subset G_{\operatorname{Rel} H}$. If $H \subset G$ is a saturated subgroup, then $\mathcal{N}(H) = G_{\operatorname{Rel} H}$.* ◁

Corollary A.25. *For a subgroup $H \subset G$, the restricted $\mathcal{N}(H)$-action on A factors to a well defined action on A/H.* ◁

Since $\operatorname{Rel} H$ is a normal subgroup, *i.e.*, $G_{\operatorname{Rel} H} \supset \operatorname{Inn} A$ (assuming that G is a subgroup of $\operatorname{Aut} A$), we also have the following consequence of Corollary A.24.

Corollary A.26. *Assume that $G \subset \operatorname{Aut} A$. Then the normalizer $\mathcal{N}(H)$ of any saturated subgroup $H \subset G$ contains $G \cap \operatorname{Inn} A$.* ◁

A.2.2 Auxiliary lemmas

Given two elements α, β of a group and a nonnegative integer m, introduce the notation

$$\{\alpha,\beta\}_m = \begin{cases} (\alpha\beta)^k(\beta\alpha)^{-k}, & \text{if } m = 2k \text{ is even,} \\ ((\alpha\beta)^k\alpha)((\beta\alpha)^k\beta)^{-1}, & \text{if } m = 2k+1 \text{ is odd.} \end{cases}$$

The relation $\{\alpha,\beta\}_m = 1$ is equivalent to $\sigma_1^m = \mathrm{id}$, where σ_1 is the generator of the braid group $\mathbb{B}_2 \cong \mathbb{Z}$ acting on the free group $\langle \alpha, \beta \rangle$. Hence,

$$\{\alpha,\beta\}_m = \{\alpha,\beta\}_n = 1 \quad \text{is equivalent to} \quad \{\alpha,\beta\}_{\text{g.c.d.}(m,n)} = 1. \tag{A.27}$$

The relation $\{\alpha,\beta\}_m = 1$ implies

$$[\alpha,(\alpha\beta)^m] = [\beta,(\alpha\beta)^m] = 1. \tag{A.28}$$

If $m = 2k$ is even, in (A.28) one can replace m with k, and any of the resulting relations is equivalent to $\{\alpha,\beta\}_m = 1$. For small m, the relation $\{\alpha,\beta\}_m = 1$ takes the following form:

- $m = 0$: tautology;
- $m = 1$: the identification $\alpha = \beta$;

- $m = 2$: the commutativity relation $[\alpha, \beta] = 1$;
- $m = 3$: the braid relation $\alpha\beta\alpha = \beta\alpha\beta$.

Next few simple group theoretic lemmas are used in the computation of the fundamental groups of algebraic curves; usually, they are employed to reduce an obviously infinite group to a potentially finite one, which can be analyzed with GAP.

Lemma A.29. *Let G be a group, and let $H \subset G$ be a normal subgroup such that the quotient projection $G \twoheadrightarrow G/[G,G]$ restricts to a monomorphism $H \to G/[G,G]$. Then the projection $G \to G/H$ restricts to an isomorphism $[G,G] = [G/H, G/H]$ of the commutants.*

Proof. In general, there is a short exact sequence

$$0 \to [G,G] \cap H \to [G,G] \to [G/H, G/H] \to 0.$$

Under the assumptions, one has $[G,G] \cap H = 0$ and the statement follows. □

We will usually apply Lemma A.29 to the cyclic subgroup $H = \langle g \rangle$ generated by a central element $g \in G$.

Corollary A.30. *Let G be a group and let $g \in G$ be a central element that projects to an infinite order element in the abelianization $G/[G,G]$. Then the quotient projection $G \to G/g$ restricts to an isomorphism $[G,G] = [G/g, G/g]$ of the commutants.* ◁

Now, let A and K be abelian groups and let

$$0 \longrightarrow K \longrightarrow G \longrightarrow A \longrightarrow 0$$

be a central extension. It is straightforward that the commutant map $G \times G \to G$, $(\alpha, \beta) \mapsto [\alpha, \beta]$, takes values in K and descends to a well defined symplectic form $[\,,]\colon \bigwedge^2 A \to K$. Clearly, G is abelian if and only if this form is identically zero. As any symplectic form on a cyclic group is trivial, we have the following corollary.

Corollary A.31. *Any central extension of a cyclic group is abelian.* ◁

A.2.3 Alexander module and dihedral quotients

Let G be a group equipped with an epimorphism $\deg\colon G \twoheadrightarrow \mathbb{Z}$. The *Alexander module* of G (or, more precisely, of deg) is the abelianization A_G of the kernel Ker deg, regarded as a module over the ring $\Lambda := \mathbb{Z}[t, t^{-1}]$ of integral Laurent polynomials *via* the action $t\colon \mathrm{A}_G \to \mathrm{A}_G$, $[h] \mapsto [\alpha h \alpha^{-1}]$, where $h \in \operatorname{Ker}\deg$ and $\alpha \in G$ is any element of degree 1. With a certain abuse of the language, we also apply this and subsequent definitions to groups of the form $\bar{G} := G/\rho$, where $\rho \in G$ is a central element of positive degree. Due to Corollary A.30, we have $\mathrm{A}_{\bar{G}} = \mathrm{A}_G$; furthermore, we can assert that this module is annihilated by $t^{\deg \rho} - 1$. In the applications, usually G is the affine fundamental group of an algebraic curve C, whereas $\bar{G}$ is its projective fundamental group. In this case, we use subscript C instead of G in the notation.

If the vector space $\mathrm{A}_G \otimes \mathbb{Q}$ is of finite dimension, one can speak about the characteristic polynomial of the operator $t\colon \mathrm{A}_G \otimes \mathbb{Q} \to \mathrm{A}_G \otimes \mathbb{Q}$; this polynomial $\Delta_G(t)$ is called the (classical) *Alexander polynomial* of G.

A generalized dihedral quotient $G \twoheadrightarrow \mathbb{D}[Q]$ is said to be *uniform* if it factors the homomorphism $(\deg \bmod 2)\colon G \twoheadrightarrow \mathbb{Z}_2$. If $G/[G,G] = \mathbb{Z}$, any dihedral quotient of G is uniform. The maximal uniform dihedral quotient of a group G is denoted by $\mathbb{D}[\mathcal{Q}_G]$.

We usually assume that G is given by a presentation $\mathbb{F}_n := \langle \alpha_1, \ldots, \alpha_n \rangle \twoheadrightarrow G$ so that $\deg\colon G \twoheadrightarrow \mathbb{Z}$ factors the map $\alpha_i \mapsto 1$, $i = 1, \ldots, n$. (Any finite presentation of G can be converted into this form using Nielsen transformations.) If this is the case, A_G is a quotient of the universal Alexander module A_n, see Definition 2.30; in particular, it is a finitely generated Λ-module.

Lemma A.32. *For a finitely generated group G, one has $\mathcal{Q}_G = \mathrm{A}_G/(t+1)$.*

Proof. By the Reidemeister–Schreier method, the abelianization K_G of the kernel $\operatorname{Ker}(\deg \bmod 2)$ is $\mathbb{Z}[\alpha_1^2] + \mathrm{A}_G/(t^2-1)$. In A_n, we have the identities

$$(t+1)(\mathbf{e}_i + \ldots + \mathbf{e}_1) = [\alpha_{i+1}^2] - [\alpha_1^2], \quad i = 1, \ldots, n-1.$$

By the definition of a dihedral group, $\mathcal{Q}_G$ is the quotient of K_G by $[\alpha_1^2]$ (so that $\mathbb{Z}_2$ is a semi-direct factor) and $(t+1)$ (so that $\mathbb{Z}_2$ acts *via* the multiplication by (-1)); it follows that $\mathcal{Q}_G = \mathrm{A}_G/(t+1)$. □

Corollary A.33 (of the proof). *Let K_G be the abelianization of the kernel of the map $(\deg \bmod 2)\colon G \twoheadrightarrow \mathbb{Z}_2$. Then $\mathcal{Q}_G$ is the quotient of K_G by the subgroup generated by the classes $[\alpha_i^2]$, $i = 1, \ldots, n$.* ◁

Lemma A.34 (*cf.* Zariski [168]). *If the abelianization $G/[G,G]$ is cyclic, the Alexander polynomial $\Delta_G(t)$ is not divisible by $\Phi_q(t)$ for any prime power $q = p^d$.*

Proof. Assume the contrary. Then the module $\mathrm{A}_G/(t^q - 1)$ is infinite and, hence, $K_p := \operatorname{Hom}(\mathrm{A}_G/(t^q-1), \mathbb{Z}_p) \neq 0$. The p-group $\mathbb{Z}_q$ acting on the p-group K_p has a fixed element $\varphi \neq 0$, and G factors to the semi-direct product $(\mathrm{A}_G/\operatorname{Ker}\varphi) \rtimes \mathbb{Z}$, which is in fact direct. Hence, $G/[G,G]$ is at least $\mathbb{Z}_p \times \mathbb{Z}$. □

Lemma A.35. *Assume that $\Phi_{2q}(t) \mid \Delta_G(t)$ for some odd prime power $q = p^d$. Then G has a uniform quotient $\mathbb{D}_{2p}$.*

Proof. As in the previous proof, $K_p := \operatorname{Hom}(\mathrm{A}/(t^q+1), \mathbb{Z}_p) \neq 0$. On the p-group K_p there is an action of $\mathbb{Z}_{2q} = \mathbb{Z}_2 \times \mathbb{Z}_q$; the first factor acts *via* (-1), and the second one has a fixed element $\varphi \neq 0$. Hence, G factors to the dihedral group $\mathbb{D}[\mathrm{A}_G/\operatorname{Ker}\varphi] = \mathbb{D}_{2p}$. □

Finally, assume that G is a Zariski quotient $\mathbb{F}_n/H$, where $H \subset \operatorname{Aut}\mathbb{F}_n$ is a subgroup whose elements preserve deg, thus descending to Λ-automorphisms of A_n. The following two statements are straightforward.

Lemma A.36. *Let $\bar{\mathrm{A}}_G = \mathrm{A}_n/H$. Then A_G is the quotient of $\bar{\mathrm{A}}_G$ by the submodule generated by $[(\alpha \uparrow h)\alpha^{-1}]$, $h \in H$, with any fixed element $\alpha \in \mathbb{F}_n$ of degree ± 1.* ◁

Lemma A.37. *The group $\mathcal{Q}_G$ is the quotient of $(\mathrm{A}_n/(t+1))/H$ by the subgroup generated by $[(\alpha \uparrow h)\alpha^{-1}]$, $h \in H$, with any fixed element $\alpha \in \mathbb{F}_n$ of odd degree.* ◁

Appendix B

Bigonal curves in Σ_d

For completeness and comparison, we give a brief account of some old results (see [41, 43]) concerning the classification, fundamental groups, and simplest geometric properties of bigonal (hyperelliptic) curves in Hirzebruch surfaces. Such curves are related to plane curves $D \subset \mathbb{P}^2$ with a singular point of multiplicity $(\deg D - 2)$. As an application, we discuss plane sextics with a quadruple point (including Oka's conjecture for such sextics) and plane quintics with a triple point.

Here, we deal exclusively with complex curves. For the general approach and a number of particular results concerning the classification of real bigonal curves in Σ_d (sometimes in the presence of additional sections), see [59].

B.1 Bigonal curves in Σ_d

A pair $(p, q) \neq (0, 0)$ of nonnegative integers is called *admissible* if either $p = q$ or the smallest of p, q is even. An *abstract formula* is a finite unordered sequence (multiset) $\{(p_i, q_i)\}$ of admissible pairs such that the difference $\sum_i q_i - 2\sum_i p_i$ is nonnegative and even, equipped with the following additional structure: if all integers q_i are even, then the pairs (p_i, q_i) with $p_i > q_i$ are partitioned into two subsets $\mathcal{P}_1$, $\mathcal{P}_2$. (If at least one of q_i is odd, we assume the trivial partition with $\mathcal{P}_2 = \varnothing$.)

Consider a *bigonal curve* $C \subset \Sigma_d$, not necessarily proper. In affine coordinates (x, t) such that $E = \{x = \infty\}$ and the fiber $\{t = \infty\}$ is nonsingular, the curve is given by an equation of the form

$$x^2 a(t) + x b(t) + c(t) = 0, \qquad \deg c = \deg b + d = \deg a + 2d \tag{B.1}$$

with the discriminant $\Delta(t) := b^2 - 4ac$ of degree $2 \deg a + 2d$. The singular fibers of C are the roots t_i of a or Δ, and to each such root t_i we assign the pair (p_i, q_i) of its multiplicities in a and Δ, respectively. This pair is admissible. The curve is reducible (splits into two sections C_1, C_2 of Σ_d) if and only if all integers q_i are even. A pair (p_i, q_i) with $p_i > q_i = 2k_i$ corresponds to a type $\mathbf{A}_{2k_i-1}$ singular point of C in E, the two branches of C intersecting E with the local multiplicities k_i and $p_i - k_i > k_i$. Thus, if C is reducible, we can partition such pairs into two subsets according to the component of C that has higher order of intersection with E at the corresponding point. The multiset $\{(p_i, q_i)\}$ partitioned as just described is an abstract formula; it is called the *formula* of C.

Theorem B.2. *The map $C \mapsto$ formula of C establishes a bijection between the set of equisingular deformation classes of bigonal curves $C \subset \Sigma_d$ and the set of abstract formulas $\{(p_i, q_i)\}$ with $\sum_i q_i - 2\sum_i p_i = 2d$.*

Proof. The statement is mostly obvious; we will only show that any abstract formula is realized by a curve and that the curves sharing the same formula form an irreducible family.

By a sequence of positive Nagata transformations we can convert a given curve C to its minimal proper model $C' \subset \Sigma_{d'}$. In terms of the formulas, we inductively apply a Nagata transformation at each pair (p_i, q_i) with $p_i > 0$. If $\min\{p_i, q_i\} \geqslant 2$, the pair is converted to $(p_i - 2, q_i - 2)$; otherwise, $(p_i, 0) \mapsto (p_i - 1, 0)$ and $(1,1) \mapsto (0,1)$. Note that these operations take an abstract formula to an abstract formula. (Here, we temporarily accept formulas with trivial pairs $(p_i, q_i) = (0,0)$.)

The topological type of a singular fiber t_i of a proper bigonal curve is determined by the multiplicity q_i of t_i in Δ. The space of such curves, even with *ordered* singular fibers (t_i, q_i), $i = 1, \ldots, r$, of given types, is irreducible: it is fibered over the space of ordered r-tuples of distinct points $t_i \subset \mathbb{P}^1$ (a Zariski open set in $(\mathbb{P}^1)^r$), and the fiber over an r-tuple (t_i) is the irreducible variety constituted by the curves given by polynomials of the form $x^2 + 2xb + \prod_{i=1}^r (t - t_i)^{q_i} + b^2$, where $b \in \mathbb{C}[t]$, $\deg b \leqslant d'$.

It remains to observe that the inverse Nagata transformations converting C' back to C are encoded in the formula of C, and these transformations can be performed continuously in a continuous family of curves. □

Consider a bigonal curve $C \subset \Sigma_d$ with formula $\mathcal{F}$ and define

$$s := \text{g.c.d.}\{q_i - p_i \mid q_i \geqslant p_i\}, \quad r_j := \text{g.c.d.}\{p_i - q_i \mid (p_i, q_i) \in \mathcal{P}_j\},\ j = 1, 2,$$

with the convention that $\text{g.c.d.}\,\varnothing = 0$.

Theorem B.3. *If $\mathcal{F}$ contains a pair (p,p) with p odd, then $\pi^{\text{aff}}(C) = \mathbb{Z}$. Otherwise,*

$$\pi^{\text{aff}}(C) = \langle \alpha_1, \alpha_2 \mid [\alpha_1^{r_1}, \alpha_2] = [\alpha_1, \alpha_2^{r_2}] = \{\alpha_1, \alpha_2\}_s = 1 \rangle,$$

where r_1, r_2, s are as above.

Remark B.4. Note that, if s is odd, the generators α_1, α_2 are conjugate and the first two relations are equivalent to $\alpha_1^r = \alpha_2^r$, where $r = \text{g.c.d.}(r_1, r_2)$.

Proof of Theorem B.3. The fundamental group is given by Theorem 5.50, and the braid monodromy and slopes are computed as explained in Section 5.2.2, where the third branch of the curve is disregarded by merely setting $\alpha_3 = 1$.

Table B.1. Groups π_1 as in Theorem B.3 with finite commutant $[\pi_1, \pi_1] \neq 0$.

(r_1, r_2, s)	$[\pi_1, \pi_1]$	$\operatorname{ord}[\pi_1, \pi_1]$
$(2, 2, 2k)$	$\mathbb{Z}_k$	k
$(2, 2, 2k+1)$	$\mathbb{Z}_s$	$2k+1$
$(2, r, 4),\ r \geqslant 2$	$\mathbb{Z}_r$	r
$(2,3,6), (3,3,4), (3,3,3)$	Q_8	8
$(2,3,8), (2,4,6), (3,4,4), (4,4,3)$	$SL(2, \Bbbk_3)$	24
$(2,3,10), (2,5,6), (3,3,5), (3,5,4), (5,5,3)$	$SL(2, \Bbbk_5)$	120

More precisely, locally, a singular fiber of C corresponding to an admissible pair $(p_i, q_i) \in \mathcal{F}$ results in the following relations:

1. $q_i \geqslant p_i = 2k$: (5.72) with $m = q_i - p_i$ and $n = k$;
2. $p_i \geqslant q_i = 2k$: (5.75) with $m = 0$ and $n = p_i - q_i$;
3. $p_i = q_i = 2k+1$: $\alpha_1 = \alpha_2$, see (5.74); hence, the group is cyclic.

(In Case 2, the remaining k Nagata transformations do not change the slope and we can replace (p_i, q_i) with $(p_i - q_i, 0)$.) Since the group $\mathbb{B}_2$ is cyclic, these relations have the same form in any geometric basis for $\mathbb{F}_2$, except that, if the curve is reducible, in Case 2 one needs to keep track of the component containing the blow-up centers, *cf.* Remark 5.76. Combining the like relations, we arrive at the presentation given by the theorem. □

Remark B.5. The projective fundamental group $\pi^{\text{proj}}(C)$ of a bigonal curve C is also given by Theorem 5.50, with the slopes computed as explained in Section 5.2.2. The relation at infinity has the form

$$\alpha_1^{m_1} = (\alpha_1\alpha_2)^n \quad \text{or} \quad \alpha_1^{m_1}\alpha_2^{m_2} = (\alpha_1\alpha_2)^n$$

for s odd or even, respectively, where $r_i \mid m_i$, $i = 1, 2$, and $s \mid 2n$.

Proposition B.6. *Let π_1 be a group as in Theorem* B.3, *and assume that $r_1 \leqslant r_2$ and $r_1 = r_2$ if s is odd, cf. Remark* B.4. *Then:*

- *π_1 is abelian if and only if $r_1 = 1$, $r_2 = 1$, or $s = 1$ or 2;*
- *otherwise, $[\pi_1, \pi_1]$ is finite if and only if the triple (r_1, r_2, s) is as in Table* B.1.

Proof. It is obvious that π_1 is abelian whenever $r_1 = 1$, $r_2 = 1$, or $s = 1, 2$.

If $r_1 = r_2 = 0$ and $s > 0$, then π_1 is the fundamental group of a torus link of type $(2, s)$ and $[\pi_1, \pi_1] = \mathbb{F}_{s-1}$ if s is odd or $\mathbb{F}_\infty$ if s is even.

According to Lemma A.29, there is an isomorphism $[\pi_1, \pi_1] = [G, G]$, where $G := \pi_1/\{\alpha_1^{r_1}, \alpha_2^{r_2}\}$.

If $s = 0$, then $G \cong \mathbb{Z}_{r_1} * \mathbb{Z}_{r_2}$ (with the convention $\mathbb{Z}_0 = \mathbb{Z}$) and its commutant is a free group of positive rank whenever both factors are nontrivial. Similarly, if

$r_1 = 0 < r_2$ (and hence $s = 2k$ is even), by Corollary A.30 we have $[G, G] = [\bar{G}, \bar{G}]$, where $\bar{G} := G/(\alpha_1\alpha_2)^k \cong \mathbb{Z}_{r_2} * \mathbb{Z}_k$.

Finally, assume that $r_1, r_2, s > 0$. Then G is the group

$$r_1[s]r_2 := \langle \alpha_1, \alpha_2 \mid \alpha_1^{r_1} = \alpha_2^{r_2} = \{\alpha_1, \alpha_2\}_s = 1 \rangle$$

and $G/[G, G]$ is finite. Hence, $[\pi_1, \pi_1]$ is finite if and only if so is G. According to [39], G is finite if and only if $(1/r_1) + (1/r_2) + (2/s) > 1$, and the commutants of these finite groups (see Table B.1) are easily computed using `GAP`. □

The Alexander module and maximal uniform dihedral quotient of a bigonal curve are easily computed from the presentation given by Theorem B.3, and we merely state the final result. It is remarkable that, although bigonal curves are 'simpler' than the trigonal ones, their metabelian invariants are not universally bounded.

Proposition B.7. *Let π_1 be a group as in Theorem* B.3, *and assume that $r_1 \leqslant r_2$ and $r_1 = r_2$ if s is odd, cf. Remark* B.4. *Let, further, $r :=$* g.c.d.(r_1, r_2). *Then:*

- *the Alexander module A_{π_1} is $\Lambda/(\tilde{\varphi}_s(-t), t^r - 1)$;*
- *the maximal uniform dihedral quotient of π_1 is $\mathbb{D}_{2m}$, where $m := s$ if r is even and $m :=$* g.c.d.$(s, 2)$ *if r is odd.* ◁

Since C is irreducible if and only if s is odd, we have a reducibility criterion similar to Proposition 6.2: *a bigonal curve C is reducible if and only if its affine fundamental group factors to $\mathbb{D}_4$.* There also is a relaxed version of *Oka's conjecture*, see Theorem 6.10. A bigonal curve C is said to be of *$(2, m)$-torus type*, $m \in \mathbb{N}$, if its equation (B.1) can be written in the form

$$[xu(t) + v(t)]^2 + w^m(t) = 0 \tag{B.8}$$

for some polynomials $u, v, w \in \mathbb{C}[t]$.

Theorem B.9. *A bigonal curve $C \subset \Sigma_d$ is of $(2, m)$-torus type, $m \geqslant 2$, if and only if its affine fundamental group has a geometric quotient $\langle \alpha_1, \alpha_2 \mid \{\alpha_i, \alpha_2\}_m = 1 \rangle$.*

Proof. Equation (B.1) of C is in the form (B.8) if and only if $a = u^2$ is a square, $a \mid \Delta$, and $\Delta/a = -4w^m$ is an m-th power. In terms of the formula $\mathcal{F}$ of C, this condition can be restated as follows: for each pair $(p_i, q_i) \in \mathcal{F}$, one has $q_i \geqslant p_i = 0 \bmod 2$ and $q_i - p_i = 0 \bmod m$. Finally, in terms of the presentation given by Theorem B.3, this condition is equivalent to $r_1 = r_2 = 0$ and $m \mid s$. □

Note that, unlike Theorems 6.10 and 7.33, neither the divisibility $\tilde{\varphi}_m(-t) \mid \Delta_C(t)$ nor the existence of a quotient $\pi^{\mathrm{aff}}(C) \twoheadrightarrow \mathbb{D}_{2m}$ (even for $m = 3$ and irreducible curve C) are sufficient for the curve to be of $(2, m)$-torus type.

Digression: monogonal curves

Let $C \subset \Sigma_d$ be a *monogonal curve, i.e.*, merely a section of Σ_d distinct from E. Define the *formula* of C as the multiset $\mathcal{F} := \{p_1, \dots, p_r\}$ of the local intersection indices at all points of intersection $C \cap E$. The following theorem is obvious and is stated for reference purposes only.

Theorem B.10. *The map $C \mapsto$ formula of C establishes a bijection between the set of equisingular deformation classes of monogonal curves $C \subset \Sigma_d$ and the set of finite multisets $\{p_i\}$ of positive integers. Any monogonal curve C is irreducible and its affine fundamental group is $\pi^{\mathrm{aff}}(C) \cong \mathbb{Z}$.* ◁

B.2 Plane quartics, quintics, and sextics

Let $D \subset \mathbb{P}^2$ be a plane curve of degree m with a distinguished singular point P of multiplicity $(m-2)$ and without linear components through P. Blowing P up, we obtain a Hirzebruch surface Σ_1; the proper transform $C \subset \Sigma_1$ of D is a bigonal curve, which is called the *bigonal model* of D. The formula $\mathcal{F} := \{(p_i, q_i)\}$ of C is subject to the conditions $\sum_i p_i = m - 2$ and $\sum_i q_i = 2m - 2$.

Similarly, given a plane curve $D \subset \mathbb{P}^2$ of degree m with a distinguished singular point P of multiplicity $(m-1)$ and without linear components through P, we can define its *monogonal model* $C \subset \Sigma_1$; its formula $\{p_i\}$ is subject to the condition $\sum_i p_i = m - 1$.

The next two theorems follow immediately from Theorems B.2, B.3, and B.10.

Theorem B.11 (probably Klein). *Up to equisingular deformation, an irreducible plane quartic $D \subset \mathbb{P}^2$ is determined by its set of singularities. The only such quartic with nonabelian fundamental group $\pi_1(\mathbb{P}^2 \smallsetminus D)$ has three cusps; for this curve, one has $\pi_1(\mathbb{P}^2 \smallsetminus D) \cong \mathbb{Z}_3 \rtimes \mathbb{Z}_4$.* ◁

Theorem B.12. *Up to equisingular deformation, an irreducible plane quintic $D \subset \mathbb{P}^2$ with a triple or quadruple singular point is determined by its set of singularities. The fundamental groups of all such quintics are abelian.* ◁

Theorem B.13. *Up to equisingular deformation, an irreducible plane sextic $D \subset \mathbb{P}^2$ with a quadruple or quintuple singular point is determined by its set of singularities. The following is a complete list of such sextics with nonabelian fundamental group $\pi_1 := \pi_1(\mathbb{P}^2 \smallsetminus D)$:*

- $\mathbf{W}_{12} \oplus 2\mathbf{A}_4$: *one has* $\pi_1 \cong \langle u, v \mid u^4 = v^4, \{u, v\}_5 = 1, v^4 = (uv)^5 \rangle$;
- $\mathbf{Y}^1_{1,1} \oplus 2\mathbf{A}_4$: *one has* $\pi_1 \cong \mathbb{D}_{10} \times \mathbb{Z}_3$;
- $\mathbf{Y}^1_{2,2} \oplus 2\mathbf{A}_2$, $\mathbf{Y}^1_{2,5} \oplus \mathbf{A}_2$, $\mathbf{Y}^1_{5,5}$, $\mathbf{W}_{1,0} \oplus 2\mathbf{A}_2$, $\mathbf{W}^{\#}_{1,3} \oplus \mathbf{A}_2$: *one has* $\pi_1 \cong \Gamma$.

A sextic D as above is of torus type if and only if $\pi_1(\mathbb{P}^2 \smallsetminus D)$ factors to $\mathbb{S}_3 = \mathbb{D}_6$ (the last item of the list).

Proof. The case of quintuple point P is covered by Theorem B.10. For a quadruple point P, the classification and the fundamental group are given by Theorems B.2 and B.3, respectively. Let $\mathcal{F} := \{(p_i, q_i)\}$ be the formula of the bigonal model C of D. According to Proposition B.6, for the group $\pi_1(\mathbb{P}^2 \smallsetminus D) = \pi^{\mathrm{proj}}(C)$ to be nonabelian one must have $p_i = 0$ or $p_i > 1$ for all i. Since also $\sum_i p_i = 4$, the nonzero entries p_i are either 2 and 2 or 4. Another necessary condition is that the integer $s :=$ g.c.d.$\{q_i - p_i \,|\, q_i > p_i\}$ must be odd and at least 3. Together with $\sum_i q_i = 10$, these restrictions leave the following seven formulas, listed together with the corresponding sets of singularities $\mathbf{S}$ of the original sextics:

- $\{(4,0),(0,5),(0,5)\}$: then $\mathbf{S} = \mathbf{W}_{12} \oplus 2\mathbf{A}_4$;
- $\{(2,0),(2,0),(0,5),(0,5)\}$: then $\mathbf{S} = \mathbf{Y}^1_{1,1} \oplus 2\mathbf{A}_4$;
- $\{(2,2),(2,2),(0,3),(0,3)\}$: then $\mathbf{S} = \mathbf{Y}^1_{2,2} \oplus 2\mathbf{A}_2$;
- $\{(2,5),(2,2),(0,3)\}$: then $\mathbf{S} = \mathbf{Y}^1_{2,5} \oplus \mathbf{A}_2$;
- $\{(2,5),(2,5)\}$: then $\mathbf{S} = \mathbf{Y}^1_{5,5}$;
- $\{(4,4),(0,3),(0,3)\}$: then $\mathbf{S} = \mathbf{W}_{1,0} \oplus 2\mathbf{A}_2$;
- $\{(4,7),(0,3)\}$: then $\mathbf{S} = \mathbf{W}^{\#}_{1,3} \oplus \mathbf{A}_2$.

For the last statement, note that both (B.1) and (B.8) can be regarded as affine equations of the original sextic D. Hence, D is of torus type if and only if C is of $(2,3)$-torus type. According to Theorem B.9, the latter condition is equivalent to the requirement that $\pi^{\mathrm{aff}}(C)$ should have $\mathbb{B}_3$ as a geometric quotient, which is indeed the case whenever $\pi_1(\mathbb{P}^2 \smallsetminus D) \cong \Gamma$, the last item in the statement. On the other hand, for a sextic D as in the first two items, the maximal dihedral quotient of $\pi_1(\mathbb{P}^2 \smallsetminus D)$ is $\mathbb{D}_{10}$, see Proposition B.7. □

Note that for $\mathbf{S} = \mathbf{W}_{12} \oplus 2\mathbf{A}_4$ the group is infinite, see Proposition B.6. So far this is the only known example of an irreducible sextics not of torus type with infinite fundamental group, *cf.* Conjecture 7.47.

Appendix C
Computer implementations

C.1 GAP implementations

All files found in this section can be downloaded from my web page at

http://www.fen.bilkent.edu.tr/~degt/dessins/GAP.zip.

C.1.1 Manipulating skeletons in GAP [76]

Listing C.1 contains a few common objects and functions related to the modular group Γ, braid group $\mathbb{B}_3$, and its Burau representation. This file is input by all other routines used in the book. Note that "S" $= \mathbb{B}_3$ and "Gamma" $= \Gamma$ are merely used as name spaces; we do not need the defining relations of these groups.

Listing C.1. A few common objects ("common.txt").

```
# The group "S" = B3, "Gamma" = Γ, and free group "G" = F3.
# The generators "x" and "y" are commonly used below to specify paths in skeletons
```

```
G := FreeGroup(3);
S := FreeGroup(2);        # We do not need the relations
Gamma := FreeGroup(["x", "y"]);   x := Gamma.1;   y := Gamma.2;
rho := G.1*G.2*G.3;       # Usually there is a relation "rho^n"
```

```
# The homomorphism B3 → Γ and the inverse expressions
```

```
XX := function(s1, s2) return s1*s2; end;
YY := function(s1, s2) return s1^-1*s2^-1*s1^-1; end;
sigma1 := function(x, y) return x^-1*y^-1; end;
sigma2 := function(x, y) return y^-1*x^-1; end;
```

```
# "WordPath(path)" converts "path" in the form "[0, -1, 1, 0, ...]", with 0 (or ±2) for Y
     and ±1 for X^±1 (for easy input) to a word in "x", "y" (an element of "Gamma")
# "Path(p)" returns path "p" as a braid (an element of "S")
```

```
WordPath := function(path)
 if not(path in Gamma) then
  path := AssocWordByLetterRep(FamilyObj(x), List(path,
             function(i) if i = 0 then return 2; else return i; fi; end));
 fi;
 return path;
end;
Path := p -> MappedWord(WordPath(p), [x, y], [XX(S.1, S.2), YY(S.1, S.2)]);
```

The Burau representation
"BurauPath(path, t[: normalized, transposed])" returns the image of "Path(path)" in the Burau representation. If "normalized" is set, the result is normalized to clear the denominators; if "transposed" is set, the result is transposed.

```
t := Indeterminate(Integers, "t");          # Use this to enter polynomials
if not IsBound(out) then out := 0; fi;      # Just bind the variable
OpenOut := function(name)                   # Create a stream for output
 out := OutputTextFile(name, false);  SetPrintFormattingStatus(out, false);
end;
Burau := [t -> [[-t, 0*t], [t^0, t^0]],     # Burau matrices
          t -> [[t^0, t], [0*t, -t]]];
BurauPath := function(path, t)
 local res, d;
 res := MappedWord(Path(path), [S.1, S.2], [Burau[1](t), Burau[2](t)]);
 if ValueOption("transposed") = true then res := TransposedMat(res); fi;
 if ValueOption("normalized") = true then
  d := Minimum(List(Flat(res), p -> CoefficientsOfLaurentPolynomial(p)[2]));
  res := t^(-3*QuoInt(d - 2, 3))*res;
 fi;
 return res;
end;
```

Next file, Listing C.2, contains most skeleton manipulating routines; in particular, it is used extensively in the proof of Theorem 6.16. The meaning and usage of most functions are explained in the listing. Skeletons are stored as pairs of permutations and can hold an unlimited amount of user defined data. The functionality can be extended by adding extra fields to the 'virtual method table' "get_".

Listing C.2. Skeleton manipulation in GAP ("skeleton.txt").

A skeleton is stored as a record, with fields computed on demand
Initialized are "Count" (number of edges) and permutations "X" = nx, "Y" = op
Other fields are accessed *via* "get(S, name)"; after first call, "S.name" will also do

A home grown VMT and virtual method dispatcher
For a field "name" let "get_.name := [f, ...]", where "f = function(S, arg)" and other members are 'field dependent' arguments: "f" receives "get_.name" in "arg"
Probably, there is a more civilized way to do this in GAP ...

```
get_ := rec(XY := [function(S, arg) return S.X*S.Y; end]);
get := function(S, name)
 if not IsBound(S.(name)) then
  S.(name) := get_.(name)[1](S, get_.(name));
 fi;
 return S.(name);
end;
```

This function creates a new skeleton

Usage: "`Skeleton([n, ]x, y{, name1, value1})`", where "`n`" is the number of edges, "`x`", "`y`" are two permutations (possibly as lists), and "`name1, value1`" *etc.* are optional user data to be stored

```
Skeleton := function(arg)
 local n, s, x, y, perm, res;
 perm := function(p)
  if IsPerm(p) then p := ListPerm(p); fi;
  p := PermList(p);
  if p <> fail then n := Maximum(n, LargestMovedPoint(p)); fi;
  return p;
 end;
 n := 0; s := 0;
 if arg[1] in Integers then n := arg[1]; s := 1; fi;
 x := perm(arg[s+1]); y := perm(arg[s+2]);
 if (x = fail) or (y = fail) then return fail; fi;
 res := rec(Count := n, X := x, Y := y);
 for n in [s+3, s+5..Length(arg) - 1] do res.(arg[n]) := arg[n+1]; od;
 return res;
end;
```

Here are more computable fields

"`XCounts`", "`YCounts`", "`XYCounts`": numbers of vertices/regions as lists (by sizes)

```
get_.XYCounts := [function(S, arg)
 local l, i, s;
 l := List(CycleStructurePerm(get(S, arg[2]))); s := 0;
 for i in [1..Length(l)] do
  if not IsBound(l[i]) then l[i] := 0; fi;  s := s + (i+1)*l[i];
 od;
 return Concatenation([S.Count - s], l);  # Include length 1 orbits as well
end, "XY"];
get_.XCounts := [get_.XYCounts[1], "X"];
get_.YCounts := [get_.XYCounts[1], "Y"];
```

"`XCount`", "`YCount`", "`XYCount`": total numbers of vertices/regions

```
get_.XYCount := [function(S, arg)
 return Sum(get(S, arg[2]));
end, "XYCounts"];
get_.XCount := [get_.XYCount[1], "XCounts"];
get_.YCount := [get_.XYCount[1], "YCounts"];
```

"`XOrbits`", "`YOrbits`", "`XYOrbits`": lists of orbits of $\mathbf{x}$, $\mathbf{y}$, $\mathbf{xy}$

```
get_.XYOrbits := [function(S, arg)
 return OrbitsDomain(Group(get(S, arg[2])), [1..S.Count]);
end, "XY"];
get_.XOrbits := [get_.XYOrbits[1], "X"];
get_.YOrbits := [get_.XYOrbits[1], "Y"];
```

"`Group`": the permutation group G generated by $\mathbf{x}$, $\mathbf{y}$
"`Stabilizer`": the stabilizer $H := \operatorname{stab}_G 1 \subset G$
"`Aut`" $= \operatorname{Aut} \mathfrak{S} = \mathcal{N}_G(H)/H$

```
get_.Group := [function(S, arg) return Group(S.X, S.Y); end];
get_.Stabilizer := [function(S, arg)
 return Stabilizer(get(S, "Group"), 1);
end];
get_.Aut := [function(S, arg)
 local H;
 H := get(S, "Stabilizer");
 return FactorGroup(Normalizer(get(S, "Group"), H), H);
end];
```

"Components": the list of connected components of $\mathfrak{S}$
"ComponentCount": the number of components
"IsConnected" returns "true" if $\mathfrak{S}$ is connected (silently assumed by other methods)

```
get_.Components := [function(S, arg)
 return OrbitsDomain(get(S, "Group"), [1..S.Count]);
end];
get_.ComponentCount := [function(S, arg)
 return Length(get(S, "Components"));
end];
get_.IsConnected := [function(S, arg)
 return get(S, "ComponentCount") = 1;
end];
```

"Genus": the genus of $\mathfrak{S}$
"Type": the 'minimal' type $\mathbf{t}$ of $\mathfrak{S}$
"IsRegular" returns "true" if $\mathfrak{S}$ is $\mathbf{t}$-regular

```
get_.Genus := [function(S, arg)
 return (2+S.Count-get(S, "XCount")-get(S, "YCount")-get(S, "XYCount"))/2;
end];
get_.Type := [function(S, arg)
 local lcm;
 lcm := l -> Lcm(Filtered([1..Length(l)], i -> l[i] > 0));
 return [lcm(get(S, "XCounts")), lcm(get(S, "YCounts"))];
end];
get_.IsRegular := [function(S, arg)
 local t;
 t := get(S, "Type");
 return (PositionProperty(get(S, "XCounts"), i -> i > 0) = t[1]) and
        (PositionProperty(get(S, "YCounts"), i -> i > 0) = t[2]);
end];
```

"Get" provides conventional access to the fields: *e.g.*, after "XY := Get('XY')" one can use "XY(S)" instead of "get(S, 'XY')"

```
Get := name -> S -> get(S, name);
```

```
Count := Get("Count"); XY := Get("XY");       # We cannot assign to "X"
Black := Get("XOrbits"); Blacks := Get("XCounts"); BlackNo := Get("XCount");
White := Get("YOrbits"); Whites := Get("YCounts"); WhiteNo := Get("YCount");
Reg := Get("XYOrbits"); Regs := Get("XYCounts"); RegNo := Get("XYCount");
Genus := Get("Genus"); Stab := Get("Stabilizer"); Aut := Get("Aut");
IsConnected := Get("IsConnected"); Components := Get("Components");
```

```
IsRegularSkeleton := Get("IsRegular"); Type := Get("Type");
# These functions are for printing
PartitionNotation := function(list)          # A list in the partition notation
 return NormalizedWhitespace(JoinStringsWithSeparator(List(
     [1..Length(list)],
      function(i)
       if list[i] = 0 then return ""; fi;
       return Concatenation(String(i), "^", String(list[i]));
      end), " "));
end;
RegStr := S -> PartitionNotation(Regs(S)); # Regions in partition notation
get_.info := [function(S, arg)              # General info on the skeleton
 return [S.Count, Whites(S)[1], Blacks(S)[1], RegStr(S), Size(Aut(S))];
end];
SkeletonInfo := Get("info");
```

Below are a few honest functions that do not store the results

"`stab(S, e)`": the stabilizer $\mathrm{stab}_G\, e$
"`Component(S, c)`": the c-th component (a sublist of "`[1..S.Count]`")
"`ExtractComponent(S, c)`" extracts the c-th component as a new skeleton (c being an integer or the actual component returned by "`Component(S, c)`")

```
stab := function(S, e) return Stabilizer(get(S, "Group"), e); end;
Component := function(S, c) return get(S, "Components")[c]; end;
ExtractComponent := function(S, c)
 local p;
 if IsInt(c) then c := Component(S, c); fi;
 p := MappingPermListList([1..Length(c)], c);
 return Skeleton(Length(c), p*RestrictedPerm(S.X, c)*p^-1,
     p*RestrictedPerm(S.Y, c)*p^-1, "parent", S, # original skeleton and inclusion
     "inc", l -> List(l, i -> c[i]));
end;
```

"`Prod(S1, S2)`" returns Cartesian product $\mathfrak{S}_1 \times \mathfrak{S}_2$
"`IsIsomorphic(S1, S2)`" returns "`true`" if $\mathfrak{S}_1 \cong \mathfrak{S}_2$ (assuming both connected)

```
# Cartesian product of two permutations
prod := function(p1, n1, p2, n2)
 local i, j, res;
 res := [1..n1*n2];
 for i in [1..n1] do for j in [1..n2] do
  res[(i - 1)*n2 + j] := (i^p1 - 1)*n2 + j^p2;
 od; od;
 return PermList(res);
end;
Prod := function(S1, S2)
 return Skeleton(S1.Count*S2.Count,
   prod(S1.X, S1.Count, S2.X, S2.Count),
   prod(S1.Y, S1.Count, S2.Y, S2.Count),
   "factor", [S1, S2],   # Store the factors
   "proj", [             # The canonical projections to the factors
       l -> List(l, i -> QuoInt(i - 1, S2.Count) + 1),
       l -> List(l, i -> RemInt(i - 1, S2.Count) + 1)]);
```

```
end;
IsIsomorphic := function(S1, S2)
 if S1.Count <> S2.Count then return false; fi;
 return S1.Count in List(get(Prod(S1, S2), "Components"), Length);
end;
```

Morphisms of skeletons are stored as lists: "List([1..S.Count], i -> f(i))"
"Morphisms(S1, S2)" lists all morphisms $\mathfrak{S}_1 \to \mathfrak{S}_2$
"IsMorphism(S1, S2, m)" returns "true" if "m" is a morphism $\mathfrak{S}_1 \to \mathfrak{S}_2$
"IsAutomorphism(S, a)" returns "true" if $a \in \operatorname{Aut}\mathfrak{S}$ ("a" may also be a permutation)

```
Morphisms := function(S1, S2)
 local S, f, p;
 f := function(l)
  p := PermList(S.proj[1](l))^-1;  l := S.proj[2](l);
  return List([1..S1.Count], i -> l[i^p]);
 end;
 if not IsInt(S1.Count/S2.Count) then return []; fi;
 S := Prod(S1, S2);
 return List(Filtered(get(S, "Components"), i -> Length(i) = S1.Count), f);
end;
IsMorphism := function(S1, S2, m)
 local i;
 for i in [1..S1.Count] do
  if (m[i^S1.X] <> m[i]^S2.X)or(m[i^S1.Y] <> m[i]^S2.Y) then
   return false;
  fi;
 od;
 return true;
end;
IsAutomorphism := function(S, a)
 if IsList(a) then a := PermList(a); fi;
 return (a*S.X = S.X*a) and (a*S.Y = S.Y*a);
end;
```

"MapOrbits(p)" converts a list of orbits to a map
"FactorSkeleton(S, m)": factor of $\mathfrak{S}$ by "m". Here, "m" is either a morphism (see above), or a permutation (possibly given by a list), or a list of orbits

```
MapOrbits := function(p)
 local i, j, res;
 res := List([1..Sum(p, Length)], i -> 0);
 for i in [1..Length(p)] do for j in p[i] do res[j] := i; od; od;
 return res;
end;
FactorSkeleton := function(S, p)
 local i, j, x, y, m, res;
 if IsList(p) and (Maximum(p) = S.Count) then p := PermList(p); fi;
 if IsPerm(p) then p := OrbitsDomain(Group(p), [1..S.Count]); fi;
 if IsList(p) and IsList(p[1]) then p := MapOrbits(p); fi;
 m := Maximum(p);  x := List([1..m], i -> 0);  y := List([1..m], i -> 0);
 for i in [1..Length(p)] do x[p[i]] := p[i^S.X]; y[p[i]] := p[i^S.Y]; od;
 res := Skeleton(m, x, y, "parent", S, "proj", l -> List(l, i -> p[i]));
```

```
 if IsMorphism(S, res, p) then return res; fi;
 return fail;
end;
```

Finally, in Listing C.3, we suggest a simple 'brute force' approach to the computation of the fundamental group $\pi_1 := \mathbb{F}_3/H$ (or, more precisely, the projective group $\pi^{\text{proj}} := \pi_1/\rho^{\text{dp}\,H}$) for a given subgroup $H \subset \mathbb{B}_3$. The function "Pi1" takes the extended skeleton Sk˜ H as a parameter and returns π^{proj} as a finitely presented group. The computation is extremely slow and resource consuming; however, it does succeed on most skeletons listed in Table 6.2.

Listing C.3. Computing $\pi_1(\mathfrak{S})$ ("SkPi1.txt").

```
Reread("skeleton.txt");     # (see Listing C.2)
Reread("braid.txt");        # (see Listing 2.5)
Reread("common.txt");       # (see Listing C.1)
```

"Pi1(S[: simplified])": a 'brute force' presentation for $\pi_1(\mathfrak{S})$, assuming $\mathfrak{S}$ the extended skeleton of a subgroup of $\mathbb{B}_3$. The "simplified" option tries to simplify the presentation obtained by applying "SimplifiedFpGroup"

```
Pi1 := function(S)
 local dp, hom, H, B; # Using the true braid group reduces the number of relations
 dp := get(S, "Type");  dp := Lcm(dp[1]*2, dp[2]*3)/6;
 B := FreeGroup(2);  B := B/[B.1*B.2*B.1/(B.2*B.1*B.2), (B.1*B.2)^(3*dp)];
 hom := GroupHomomorphismByImagesNC(B, get(S, "Group"), [B.1, B.2],
                          [sigma1(S.X, S.Y), sigma2(S.X, S.Y)]);
 H := PreImages(hom, get(S, "Stabilizer"));
 H := Union(List(GeneratorsOfGroup(H), # see "BraidRelations" in Listing 2.5
         g -> BraidRelations(UnderlyingElement(g), G, [G.2, G.3])));
 H := G/Union(H, [rho^dp]);
 if ValueOption("simplified") = true then H := SimplifiedFpGroup(H); fi;
 return H;
end;
```

C.1.2 Proof of Theorem 6.16

The computer aided part of the proof for the case $N \geqslant 7$, see Section 6.2.2, is implemented in Listing C.4, functions "Pass1" and "Pass2". The input for "Pass1" (Step 1, see page 195) is a hand tuned list of pairs $(N, \{g_1, \ldots, g_r\})$ stored in the file "Burau_start.txt"; the primes $p = 2$ and 3 are treated separately: for $p = 2$, we have a different value of $M = \mathrm{e}_2(N)$, for $p = 3$, a different list of admissible types, III_0 *vs.* $\mathrm{III}_\pm$. The output of "Pass1" is saved to "log/pass1.txt"; it is read by "Pass2" (Step 2, see page 195), which saves its output to "log/genus0.txt" in both human and GAP readable form. This file can be used for the further analysis of the Alexander modules (see, *e.g.*, "Burau2.txt").

In Step 2, in order to avoid repetition, we compute the G-orbit of the point $[\mathbf{u}_T^{\perp}(\xi)]$ in $\mathbb{P}(\bar{\mathbb{k}}^2)$. If, for another type T', the point $[\mathbf{u}_{T'}^{\perp}(\xi)]$ lies in this orbit, the group $H_{T'}(\xi)$ is not considered, as it is conjugate to $H_T(\xi)$. As another simplification, for $N \geqslant 11$ we try to improve Euler's inequality (6.26) first. Namely, for each pair $T \neq T'$ of types, we compute the resultants $\mathcal{R}_d \in \mathbb{k}_p$ of (6.25) and $\Phi_M(t)$. If $\mathcal{R}_d \neq 0$ for all $d = 0, \ldots, N-1$, the two types cannot appear within one trivial region. Thus, the set $\mathcal{T}$ of all admissible types splits into disjoint subsets $\mathcal{T}_1, \mathcal{T}_2, \ldots$ so that two types are in the same subset if and only if they can share a common trivial region. Hence, the *a priori* bound $k^{\mathrm{all}} \leqslant 21$ used in Section 6.2.2 can be improved to $k^{\mathrm{all}} \leqslant \max_i \sum_{T \in \mathcal{T}_i} c^T$, where $c^{\mathrm{I}} = c^{\mathrm{II}} = 5$, $c^{\mathrm{III}_\epsilon} = 4$, and $c^{\mathrm{IV}} = 3$. If N does not satisfy the improved inequality $N < 6 + k^{\mathrm{all}}$, the pair (p, ψ_ξ) is disregarded.

Listing C.4. Proof of Theorem 6.16 ("Burau.txt").

```
# Auxiliary functions and objects

Reread("SkPi1.txt");    # (see Listing C.3)
e := function(N, p)     # The e_p(N) function
 if (p = 2) or (N mod 4 = 0) then return N; fi;
 if N mod 2 = 0 then return N/2; fi;
 return N*2;
end;
v\* := function(v)      # Convert v to v⊥ if "transposed" option is specified
 if ValueOption("transposed") = true then return [v[2], -v[1]]; fi;
 return v;
end;
vectors := [    # Canonical generators, see (6.23)
  function(N, p, t) return v\*([t^0, 0*t]); end,    # type I
  function(N, p, t) return v\*([t^0, t+t^0]); end,  # type II
  function(N, p, t)                                 # type III+ or III0
   if (p = 3) and (N mod 3 <> 0) then return v\*([t^0, -t]); fi;
   if (p <> 3) and (N mod 3 = 0) then return v\*([-t^(e(N, p)/3), t]); fi;
   return fail;
  end,
  function(N, p, t)                                 # type III-
   if (p <> 3) and (N mod 3 = 0) then return v\*([-t, t^(e(N, p)/3)]); fi;
   return fail;
  end,
  function(N, p, t)                                 # type IV
   if e(N, p) mod 2 <> 0 then return v\*([t^((e(N, p) - 1)/2), t^0]); fi;
   return fail;
  end];
Vectors := function(N, p, t)    # All generators u_T as a list
 return List([1..Length(vectors)], i -> vectors[i](N, p, t));
end;
Region := function (N, p, t)    # Powers of xy (a region of width N)
 t := BurauPath((x*y)^-1, t);  return List([0..N-1], i -> t^i);
end;
Coeff := [5, 5, 4, 4, 3];       # Coefficients for k^all in Euler's formula
```

"`CreateSkeleton(N, p, psi, vector[: full])`" creates the skeleton $\mathfrak{S}$ of the universal subgroup $H \subset \mathbb{B}_3$ corresponding to the triple (N, p, ψ), starting from the generator "`vector`". Also stored are "`N`", "`p`", "`psi`", and the automorphism "`U`" such that $\mathfrak{S}/U$ is the skeleton of the universal subgroup of Bu_3. If "`full`" is specified, the extended skeleton $\mathrm{Sk}^{\sim} H$ is computed

```
CreateSkeleton := function(N, p, psi, vector)
 local pp, i, G, H, U, pd, root, vec, m, m1, m2, os, rt, hom, IsFound, proj;
 IsFound := function(v)
  if v = fail then return false; fi;
  for i in [0..pd-1] do if v^(m^i) in os.orbit then return true; fi; od;
  return false;
 end;
 proj := function()
  if U in G then return Image(hom, U); fi;
  for i in G do if (vec^U)^i = vec then
   rt := RightTransversal(G, H);
   return PermList(List(rt, g -> PositionCanonical(rt, i*g)));
  fi; od;
  return Image(hom, U^3);
 end;
 pp := p;  if p = 0 then p := 7; fi;      # A safe value
 pd := p^DegreeIndeterminate(psi, t);
 root := RootsOfUPol(GF(pd), PolynomialModP(psi, p))[1];
 U := root*IdentityMat(2);  m := Z(pd)*IdentityMat(2);
 m1 := BurauPath(x, root: transposed);
 m2 := BurauPath(y, root: transposed);
 G := Group(m1, m2);  vec := vectors[vector](N, p, root: transposed);
 os := OrbitStabilizer(G, vec);  H := os.stabilizer;
 if ValueOption("full") <> true then H := ClosureGroup(H, U^3); fi;
 hom := FactorCosetAction(G, H);
 return Skeleton(Index(G, H), Image(hom, m1), Image(hom, m2),
         "N", N, "p", pp, "psi", psi, "U", proj(), "vector", vector,
         "orbit", List(Vectors(N, p, root: transposed), IsFound),
         "expected", pd);
end;
# Data to be printed
DataFields := ["N", "p", "psi", "info", "id", "vector"];
UserData := function(S) return List(DataFields, i -> get(S, i)); end;
get_.N := [function(S, arg) return get(S.parent, arg[2]); end, "N"];
get_.p := [get_.N[1], "p"];  get_.psi := [get_.N[1], "psi"];
get_.id := [get_.N[1], "id"];  get_.vector := [get_.N[1], "vector"];
get_.orbit := [get_.N[1], "orbit"];
```

Step 1 of the algorithm (see page 195): call "`Pass1`".

"`tryPathList(N, p, list)`" does the main job: it tries a list of elements $g_1, \dots, g_r \in \Gamma$ given in "`list`", returning a list of pairs (p, ψ_ξ) that fail the test.
The input values "`N`" and "`p=0,2,3`" are used for computing M and detecting allowed types $T \neq 0$.

```
tryPathList := function(N, p, list)
 local mat, vec, reg, cp, m1, m2, v1, v2, i, zero, addp, found, d, f, r, s;
 addp := function(pp)   # Add to the list all common factors as integral polynomials
  if N mod pp = 0 then return; fi;
```

```
  for f in Set(Factors(Gcd(PolynomialModP(cp, pp), PolynomialModP(d, pp))))
   do
    AddSet(found, [pp, N, UnivariatePolynomial(Integers,
        List(CoefficientsOfUnivariatePolynomial(f), IntFFE), t)]);
   od;
 end;
 zero := function()      # Fails if R = 0; otherwise, calls "addp" on all prime factors of R
  d := DeterminantMatDestructive([(vec[v1]^mat[m1])^reg[i],
                                    vec[v2]^mat[m2]]);
  r := Resultant(cp, d, t);
  if p <> 0 then if r mod p = Zero(r) then addp(p); fi; return false; fi;
  if r = Zero(r) then return true; fi;
  for s in Set(Filtered(Factors(AbsInt(r)), s -> not(s in [1, 2, 3]))) do
   addp(s);
  od;
  return false;
 end;
 mat := List(list, m -> BurauPath(m, t: normalized)); # g_i ∈ Γ as Burau matrices
 vec := Filtered(Vectors(N, p, t), v -> v <> fail); # admissible types as vectors
 reg := Region(N, p, t: normalized);                # a trivial region
 found := Set([]);
 cp := Value(CyclotomicPolynomial(Integers, e(N, p)),
           [Indeterminate(Integers)], [t]);
 for m1 in [1..Length(mat)] do for m2 in [1..Length(mat)] do  # the main loop
  for v1 in [1..Length(vec)] do for v2 in [1..Length(vec)] do
   for i in [1..Length(reg)] do
    if (i <> 1) or (m1 <> m2) or (v1 <> v2) then
     if zero() then return fail; fi;    # if one of the resultants is zero, return "fail"
    fi;
   od;
  od; od;
 od; od;
 return found;
end;
```

"tryPathLists(N, ...)" calls "tryPathList" iteratedly on all primes $p = 0, 1, 2$ and all sets $\{g_1, \dots, g_r\}$ passed as arguments, computing intersection over all sets and union over all primes. The result is united with the master list passed as option "result=..." (if any).

```
tryPathLists := function(arg)
 local N, i, p, isfound, found, res, Res;
 N := arg[1];  Res := Set([]);
 Print(">>N=", N, ":\n");
 for p in [0, 2, 3] do             # loop over p = 0, 2, 3
  if (p <> 0) and (N mod p = 0) then continue; fi;
  found := Set([]); isfound := false;
  Print(" p=", p, ":\n");
  for i in [2..Length(arg)] do  # loop over all arguments
   if arg[i] = [] then continue; fi;  Print("  ", arg[i], ": ");
   res := tryPathList(N, p, arg[i]);
   if res = fail then Print("fail\007\n"); continue; fi;
   if isfound then IntersectSet(found, res); else found := res; fi;
```

```
   isfound := true;             # after one successful step, compute intersections
   Print("found=", Length(res), "/", Length(found), "\n");
   if found = [] then break; fi;
  od;
  if not isfound then Print("Failed\n"); return fail; fi;
  UniteSet(Res, found);
 od;
 Print("Total=", Length(Res), ": ", Set(List(Res, i -> i[1])), "\n");
 res := ValueOption("result");
 if IsSet(res) then UniteSet(res, Res); fi;
end;
```

"Pass1()" initializes all variables and calls "tryPathLists" on the hand tuned lists contained in "Burau_start.txt". For further processing, the pairs (p, ψ_ξ) with $N \leqslant 5$ (computed manually) from "Burau_fixed.txt" are also added to the result, which is saved to "log/pass1.txt". Output is logged to "log/pass1.log".

```
Pass1 := function()
 local Res, rc;
 Res := Set([]);
 OutputLogTo("log/pass1.log");                      # log output, just in case
 PushOptions(rec(result := Res));
 \[do\] := tryPathLists;  Read("Burau_start.txt");  # Step 1 of the algorithm
 \[do\] := function(N, p, psi, vec) AddSet(Res, [p, N, psi, [vec]]); end;
 Read("Burau_fixed.txt");                           # add pairs with N ⩽ 5
 PopOptions(); OutputLogTo();
 Print(">>Total found: ", Length(Res), "\n");
 OpenOut("log/pass1.txt");                          # save the result
 PrintTo(out, "\# Total found: ", Length(Res), "\n");
 for rc in Res do
  PrintTo(out, "\\[do\\](", rc[2], ", ", rc[1], ", ", rc[3]);
  if Length(rc) > 3 then PrintTo(out, ": vectors := ", rc[4]); fi;
  PrintTo(out, ");\n");
 od;
 CloseStream(out);
end;
```

Step 2 of the algorithm (see page 195): call "Pass2".

"tryEuler(N, p, psi)" checks if $N < 6 + k^{\text{all}}$, see (6.26), with an improved estimate on k^{all}

```
tryEuler := function(N, p, psi)
 local vec, reg, pd, root, v, u, r, test;
 pd := p^DegreeIndeterminate(psi, t);
 root := RootsOfUPol(GF(pd), PolynomialModP(psi, p))[1];
 vec := Vectors(N, p, root);  reg := Region(N, p, root);
 test := function(u) # return "Coeff[u]" if "u" and "v" can share a region, or 0 otherwise
  if vec[u] = fail then return 0; fi;
  for r in reg do
   if Determinant([v^r, vec[u]]) = Zero(root) then return Coeff[u]; fi;
  od;
  return 0;
 end;
```

```
 for v in vec do
  if v = fail then continue; fi;
  if N < 6 + Sum(List([1..5], test)) then return true; fi;
 od;
 return false;
end;
```

"tryGenus(N, p, psi)" checks a candidate (p, ψ_ξ) for genus zero

```
tryGenus := function(N, p, psi)
 local print, list, SS, g, found, i, prefix;
 print := function(S)
  g := Genus(S);
  if not IsBound(S.id) then S.id := fail; fi;
  Print(prefix, [N, p, psi], ", size=", S.Count, ", genus=", g,
        ", orbit=", get(S, "orbit"), "\n");
  S.info := SkeletonInfo(S);
  if (g = 0) and IsOutputTextStream(out) then               # save if g = 0
# Output format: [N, p, ψ_ξ, "SkeletonInfo", 𝔖 ≠ 𝔖/U, type T] # [types]
   PrintTo(out, "\\[do\\](", UserData(S), "); \# ", S.orbit, "\n");
  fi;
 end;
 found := List(Vectors(N, p, t), i -> i = fail);  prefix := "*";
 if (N > 10) and not tryEuler(N, p, psi) then             # "tryEuler" first
  Print(prefix, [N, p, psi], ", Euler <= 0\n");
  return;
 fi;
 list := ValueOption("vectors");  if list = fail then list := [1..5]; fi;
 for i in list do                                          # try all types T ≠ 0,
  if found[i] then continue; fi;                           # but avoid repetition
  SS := CreateSkeleton(N, p, psi, i);                      # compute 𝔖
  found := UnionBlist(found, SS.orbit);
  SS.id := SS.U <> One(SS.U);  print(SS);  prefix := " ";          # try 𝔖
  if SS.id and (g > 0) then print(FactorSkeleton(SS, SS.U)); fi; # try 𝔖/U
 od;
end;
```

"Pass2()" reads candidates (p, ψ_ξ) from "log/pass1.txt", tests them for genus zero, and saves those passed to "log/genus0.txt". Output is logged to "log/pass2.log".

```
Pass2 := function()
 \[do\] := tryGenus;
 OpenOut("log/genus0.txt");  OutputLogTo("log/pass2.log");
 Read("log/pass1.txt");
 OutputLogTo();  CloseStream(out);
end;
```

A few examples of the further processing of the final list of universal subgroups of genus zero can be found in "Burau2.txt". (This file was intended for private use and is not very well documented.) In particular, we compute double and triple intersections and inclusions of the subgroups, see Addenda 6.17 and 6.34, decide which skeletons are isomorphic, see Remark 6.18, and attempt a 'brute force' computation of the fundamental groups using the function "`Pi1`" in Listing C.3. Here, we show that, with the exception of Lines 11 and 12, the affine fundamental group of each proper universal curve C listed in Table 6.2 is minimal expected, *i.e.*, $\pi^{\text{aff}}(C) \cong \mathrm{A}_C \rtimes \mathbb{Z}$. (On Lines 11 and 12, my laptop runs out of memory.)

Appendix D

Definitions and notation

D.1 Common notation

In this section, we fix the terms and notation, more or less common, that are not introduced elsewhere in the book. For the reader's convenience, we also discuss a few concepts and facts that are used throughout the whole text but might be not very well known to non-specialists.

D.1.1 Groups and group actions

We fix the following notation for some classical groups:

- 0: the trivial (sub-)group, even if multiplicative;
- $\mathbb{Z}_n := \mathbb{Z}/n\mathbb{Z}$: the cyclic group of order n;
- $\mathbb{S}(X)$: the group of permutations of a set X;
- $\mathbb{S}_n := \mathbb{S}\{1, \ldots, n\}$: the symmetric group of degree n;
- $\mathbb{A}_n := \operatorname{Ker}[\mathbb{S}_n \to \mathbb{Z}_2]$: the alternating group of degree n;
- $\mathbb{D}[H] := H \rtimes \mathbb{Z}_2$: the generalized dihedral group; here H is an abelian group and the generator of $\mathbb{Z}_2$ acts on H *via* the multiplication by (-1);
- $\mathbb{D}_{2n} := \mathbb{D}[\mathbb{Z}_n]$: the classical dihedral group of order $2n$;
- $\mathbb{F}_n$: the free group on n generators (*not* a field with n elements, see $\Bbbk_p$);
- $Q_8 := \{\pm 1, \pm i, \pm j, \pm k\} \subset \mathbb{H}$: the quaternion group of order 8;
- $GL(n, \mathcal{R})$, SL, $PGL := GL/\mathcal{R}^*$, $PSL := SL/\mathcal{R}^*$: the general and special linear groups of rank n over a commutative ring $\mathcal{R}$ and their projective versions.

Given a subset $S \subset G$,

- $NC(S) := \bigcap_H H$, $H \supset S$ is a normal subgroup, is the *normal closure* of S,
- $\mathcal{Z}_G(S) := \{g \in G \mid g^{-1}hg = h \text{ for all } h \in S\}$ is the *centralizer* of S, and
- $\mathcal{N}_G(S) := \{g \in G \mid g^{-1}Sg = S\}$ is the *normalizer* of S.

The subscript G is omitted whenever the group G is understood.

The *commutator* of two elements $a, b \in G$ is $[a, b] := a^{-1}b^{-1}ab$. The *commutant* (derived subgroup) of a group G is the subgroup $[G, G]$ generated by all commutators. Occasionally, when describing groups with long composition series, we use the short notation $G' := [G, G]$; however, we do not reserve $'$ for the commutant, reintroducing it each time when it is used: normally, G' is just another group.

The notation $g \overset{c}{\sim} h$ means that two elements or subgroups are conjugate. The *conjugacy class* of an element $g \in G$ is denoted by $[\![g]\!] = [\![g]\!]_G$; the same notation applies to subgroups. A subgroup H is *subconjugate* to a subgroup F, $H \prec F$, if $H \subset g^{-1}Fg$ for some $g \in G$. This notion extends to conjugacy classes: if $H \prec F$, then $H \prec [\![F]\!]$ and $[\![H]\!] \prec [\![F]\!]$.

By default, all group actions are *right* and we use the notation $(x, g) \mapsto x \uparrow g$ (*cf.* ^ in GAP or the exponential notation in many algebra texts). This convention applies to matrix groups as well: matrices act on row vectors by the right multiplication. On the few occasions when a left group action is considered, it is denoted by $(g, x) \mapsto g \downarrow x$. Given a group action $X \times G \to X$, the *stabilizer* (isotropy subgroup) of an element $x \in X$ is $\operatorname{stab} x := \{g \in G \mid x \uparrow g = x\}$. If X is a group and G acts by homomorphisms, we use the term *subconjugate* and notation $H \prec F$ in a more general sense: $H \prec F$ if $H \subset F \uparrow g$ for some $g \in G$.

The group of automorphisms of a group G (acting on G from the right) is denoted by $\operatorname{Aut} G$; the group of inner automorphisms is $\operatorname{Inn} G \subset \operatorname{Aut} G$. There is a natural epimorphism $\operatorname{inn}\colon G \twoheadrightarrow \operatorname{Inn} G$, so that $h \uparrow (\operatorname{inn} g) = g^{-1}hg$. Often (especially if $\mathcal{Z}(G) = 0$, *e.g.*, if G is a free group), we identify $g \in G$ and its image $\operatorname{inn} g \in \operatorname{Aut} G$.

If G is an abelian group, then $\operatorname{Tors} G \subset G$ is its *torsion subgroup* and $\operatorname{Tors}_p G$ is the p-torsion subgroup, where p is a prime. The minimal number of generators of G is denoted by $\ell(G)$, and we use the shortcut $\ell_p(A) := \ell(\operatorname{Tors}_p A)$.

D.1.2 Topology and homotopy theory

My favorite reference book in algebraic topology is [74]; unfortunately, it is only available in Russian. Other standard references are [68, 155].

As usual, $H_i(\,\cdot\,)$ and $H^i(\,\cdot\,)$ stand for the (co-) homology of topological spaces, pairs, sheaves, *etc*. In the topological settings, the default coefficients are $\mathbb{Z}$. The Betti numbers are denoted by $b_i(\,\cdot\,) := \operatorname{rk} H_i(\,\cdot\,)$ or $b_i(\,\cdot\,; \Bbbk) := \dim_{\Bbbk} H_1(\,\cdot\,; \Bbbk)$ (where $\Bbbk$ is a field), and the (topological) Euler characteristic is the alternating sum $\chi(\,\cdot\,) := \sum_{i=0}^{\infty}(-1)^i b_i(\,\cdot\,)$.

In homotopy theory we fix the notation $I := [0, 1]$ for the unit segment and $D^2 := \{z \in \mathbb{C} \mid |z| < 1\}$ for the 'standard' unit disk. We use S^n for the n-sphere and $T^n := (S^1)^n$ for the n-torus.

The fixed point set of an auto-homeomorphism $c\colon X \to X$ is denoted by $\operatorname{Fix} c$.

The i-th Stiefel–Whitney class of a vector bundle L over X is $w_i(L) \in H^i(X; \mathbb{Z}_2)$, and the i-th Chern class of a complex vector bundle is $c_i(L) \in H^{2i}(X)$.

Intersection index and linking coefficients

For a compact manifold X of dimension n, the *Poincaré duality* isomorphism is

$$\mathrm{D} := [X] \cap\colon H^p(X, \partial X) \to H_{n-p}(X) \quad \text{or} \quad H^p(X) \to H_{n-p}(X, \partial X),$$

where the coefficients are $\mathbb{Z}$ if X is oriented or $\mathbb{Z}_2$ otherwise. The *intersection index* is the pairing $H_p(X) \otimes H_{n-p}(X, \partial X) \to \mathbb{Z}$, $a \otimes b \mapsto a \circ b := \langle a, \mathrm{D}^{-1} b\rangle$. If X is closed and $n = 2k$ is even, we have a bilinear form $H_k(X) \otimes H_k(X) \to \mathbb{Z}$, symmetric if k is even and skew-symmetric (but not necessarily symplectic) if k is odd.

Recall that, for any finite abelian group G, there is an isomorphism $\operatorname{Ext}(G, \mathbb{Z}) = \operatorname{Hom}(G, \mathbb{Q}/\mathbb{Z})$. Hence, for a closed oriented 3-manifold X, we have an isomorphism $\mathrm{D}\colon \operatorname{Tors} H^2(X) = \operatorname{Hom}(\operatorname{Tors} H_1(X), \mathbb{Q}/\mathbb{Z}) \to \operatorname{Tors} H_1(X)$, which gives rise to the *linking coefficient form* $\operatorname{lk}\colon \operatorname{Tors} H_1(X) \otimes \operatorname{Tors} H_1(X) \to \mathbb{Q}/\mathbb{Z}$. It is symmetric and nondegenerate. Geometrically, lk is computed as follows: represent the two torsion classes by disjoint cycles a, b, and let $ma = \partial x$ for an integer m and a 2-chain x, which can be assumed transversal to b. Then $\operatorname{lk}([a], [b]) = \frac{1}{m}(x \circ b) \bmod \mathbb{Z}$.

If X is a closed oriented 4-manifold, we always identify $H^2(X) = H_2(X)$ *via* Poincaré duality. The quotient $L := H_2(X)/\operatorname{Tors}$ is a unimodular integral lattice. If X has boundary, L is still a lattice and the dual group L^* can be identified with $H_2(X, \partial X)/\operatorname{Tors}$. The homomorphism $L \to L^*$ associated with the intersection index form (*i.e.*, the homomorphism $a \mapsto \{b \mapsto a \circ b\}$) can be identified with the relativization $H_2(X) \to H_2(X, \partial X)$, and the homomorphism $\partial\colon H_2(X, \partial X) \to H_1(\partial X)$ maps $\operatorname{discr} L$ to a subgroup of $\operatorname{Tors} H_1(\partial X)$, taking the bilinear form on $\operatorname{discr} L$ to the linking coefficient form.

Monodromy

Let $p\colon X \to B$ be a locally trivial fibration (in the topological, smooth, *PL*-, or similar category). Denote by $F_b := p^{-1}(b)$ the fiber over a point $b \in B$. For a loop $\gamma\colon I \to B$, the pull-back $\gamma^* p$ is a locally trivial fibration over the contractible space I, and a simple application of the obstruction theory shows that $\gamma^* p$ admits a unique, up to fiberwise isotopy in the appropriate category, trivialization. Any such trivialization gives rise to an automorphism $F_b = (\gamma^* p)^{-1}(0) \to (\gamma^* p)^{-1}(1) = F_b$, where $b = \gamma(0) = \gamma(1)$, which is defined up to isotopy and depends on the homotopy class of γ only. Thus, we obtain a homomorphism $\mathfrak{m}\colon \pi_1(B, b) \to \operatorname{Map}(F_b)$, where $\operatorname{Map}(F)$ is the appropriate *mapping class group*, *i.e.*, the group of autohomeomorphisms (autodiffeomorphisms, *etc.*) of F modulo isotopies (diffeotopies, *etc*). Recall that, according to our general convention, $\operatorname{Map}(F)$ acts on F *from the right*; otherwise, $\mathfrak{m}$ would be an anti-homomorphism.

For an even more formal definition, one can replace p with the associated principal $\operatorname{Map}(F_b)$-bundle, which is a Galois covering; then $\mathfrak{m}$ is the boundary homomorphism $\partial : \pi_1(B, b) \to \pi_0(\text{fiber})$ in the Serre exact sequence.

The homomorphism $\mathfrak{m}\colon \pi_1(B, b) \to \operatorname{Map}(F_b)$ is called the *monodromy* of p. More generally, one can fix a pair of points $b_1, b_2 \in B$ and speak about the monodromy as a map $\pi_1(B; b_1, b_2) \to \operatorname{Map}(F_{b_1}, F_{b_2})$, where $\pi_1(B; b_1, b_2)$ is the set of path homotopy classes of paths in B from b_1 to b_2 and $\operatorname{Map}(F_{b_1}, F_{b_2})$ is the set of morphisms from F_{b_1} to F_{b_2} modulo isotopies.

Often, the monodromy is composed with some natural action of $\operatorname{Map}(F_b)$. Thus, one can speak about the monodromy in the (co-)homology $H_n(F_b)$ or $H^n(F_b)$ (with any fixed coefficients) or monodromy in the homotopy groups $\pi_n(F_b)$. In the latter case, extra care is needed: unless F_b is n-simple, one should fix a section $s\colon B \to X$ of p, require that the trivializations used in the definition should preserve s, and speak about the monodromy in $\pi_n(F_b, s(b))$.

The Riemann–Hurwitz formula

In algebraic geometry, the classical *Riemann–Hurwitz formula* applies to a nonconstant morphism $\varphi\colon X \to B$ of connected compact curves and states that

$$2 - 2g(X) = (2 - 2g(B))\deg\varphi - \sum_{x\in X}(r_x - 1),$$

where r_x is the ramification index of φ at a point $x \in X$. Since all but finitely many ramification indices are equal to 1, the sum is actually finite. In topology, this formula applies to any ramified covering between two surfaces, not necessarily compact, and computes the Euler characteristic $\chi(X)$ (which is equal to $2-2g(X)$ if X is compact, connected, and without boundary) in terms of that of B. In this form, the Riemann–Hurwitz formula generalizes to a much wider class of maps; however, it is usually easier to prove the formula in each particular case than to come up with the 'most general' statement. For example, if $\varphi\colon X \to B$ is a double covering ramified at $D \subset B$, the formula would state that $\chi(X) = 2\chi(B) - \chi(D)$, and the proof may refer to the Smith exact sequence.

The most general statement that I know is due to Viro [162], where the Riemann–Hurwitz formula takes the form of Fubini's theorem: for a 'sufficiently good' continuous map $p\colon X \to B$, one has

$$\chi(X) = \int_B \chi(F_b)\, d\chi,$$

where $F_b := p^{-1}(b)$ is the fiber over $b \in B$ and the integration is with respect to the Euler characteristic. Certainly, χ is not a σ-additive measure and the integral is not quite well defined; however, it works! (A few interesting applications are also discussed in [162].) The proof of this formula depends on the meaning of 'sufficiently good' in the statement: it may be as simple as mere cell counting or as technical as a reference to the Leray spectral sequence.

D.1.3 Algebraic geometry

All algebraic varieties are over $\mathbb{C}$. We also consider complex analytic manifolds that are not necessarily algebraic. References are [16, 86, 88].

As usual, $\mathcal{O}_X$ denotes the structure sheaf of an algebraic/analytic variety X; if X is smooth, then K_X is its canonical sheaf/divisor.

The cohomology of a sheaf $\mathcal{S}$ over X are denoted by $H^i(X; \mathcal{S})$. To avoid confusion with the modular group, we use $H^0(X; \mathcal{S})$ for the group/space of global sections of $\mathcal{S}$.

Given a complex $\mathcal{S}^*$ of sheaves, $\mathbf{H}^i(X;\mathcal{S}^*)$ is its hypercohomology. For an analytic map $p\colon X \to Y$ and a sheaf $\mathcal{S}$ on X, we denote by $p_*\mathcal{S}$ its direct image and by $R^i p_*\mathcal{S}$ its higher direct images (the right derived functors of p_*).

The notation G_X stands for the constant sheaf on X with the stalk G (a given abelian group). The *exponential exact sequence* is the short exact sequence

$$0 \longrightarrow \mathbb{Z}_X \xrightarrow{2\pi i} \mathcal{O}_X \xrightarrow{\exp} \mathcal{O}_X^* \longrightarrow 0$$

of sheaves and the associated cohomology exact sequence.

The *Picard group* is the group $\operatorname{Pic} X := H^1(X;\mathcal{O}_X^*)$ of classes of line bundles. For a smooth compact variety X, we identify line bundles, invertible sheaves (the sheafs of germs of sections), and classes of linear equivalence of divisors (divisors of zeroes and poles of meromorphic sections). There is a homomorphism $\operatorname{Pic} X \to H^2(X)$ (the connecting homomorphism in the exponential exact sequence); the image of a divisor D (line bundle $\mathcal{L}$) is denoted by $[D] = c_1(\mathcal{L})$. This image is indeed Poincaré dual to the fundamental class of D regarded as a (singular) manifold.

If X is a smooth compact surface, the *Néron–Severi lattice* $\mathit{NS}(X)$ is the image of $\operatorname{Pic} X$ in $H^2(X) = H_2(X)$, *i.e.*, the group of homology classes realized by divisors. It follows from the exponential exact sequence that $H_2(X)/\mathit{NS}(X)$ is torsion free.

Singularities

Recall that an isolated singular point O of (a germ of) a curve C in a surface S is called *simple* if it is of type $\mathbf{A}_p$, $p \geqslant 1$, $\mathbf{D}_q$, $q \geqslant 4$, $\mathbf{E}_6$, $\mathbf{E}_7$, $\mathbf{E}_8$. For more details and the remarkable properties of these singularities, we refer to [66]. The types of simple singularities are in a one-to-one correspondence with those of simply laced Dynkin diagrams, see Section 1.1.3, or, equivalently, of irreducible root lattices. This correspondence has a geometric meaning: denoting by X the minimal resolution of singularities of the double covering of S branched at C, the lattice $\mathbf{T}$ corresponding to O is the sublattice in $H_2(X)$ spanned by the exceptional divisors over O. If C has several simple singular points, the exceptional divisors over all these points span the orthogonal direct sum of the root lattices corresponding to the individual points.

Convention D.1. We always identify a combinatorial set of simple singularities and the corresponding negative root lattice $\mathbf{S}$; in view of Theorem A.6, the irreducible summands (*i.e.*, the types of the individual singular points) are uniquely recovered from $\mathbf{S}$. For this reason, we use the orthogonal summation symbol $\oplus$ when speaking about sets of singularities, extending this notation to non-simple singularities as well.

Given (a germ of) an analytic curve $C \subset S$ at an isolated singular point O, for any ball D_ϵ about O of sufficiently small radius ϵ the pair $(D_\epsilon, D_\epsilon \cap C)$ is homeomorphic to the cone over $(\partial D_\epsilon, \partial D_\epsilon \cap C)$, see [116]. Any ball D_ϵ with this property is called a *Milnor ball* for C at the origin.

The complement $D_\epsilon \smallsetminus C$ fibers over the punctured disk, the map being essentially given by $(x,y) \mapsto f(x,y)$, where $f(x,y)=0$ is the equation of C in local coordinates (x,y). A typical fiber F of this projection is called a *Milnor fiber*, and the rank $\mu(O) := \operatorname{rk} H_1(F)$ is called the *Milnor number* of C at O. The *total Milnor number* of a reduced curve C in a compact surface S is $\mu(C) := \sum \mu(O)$, the summation running over all singular points O of C.

Curves on surfaces

The *geometric genus* $p_g(C)$ of an irreducible compact curve C is the genus of its normalization. The *arithmetic genus* of a curve C in a compact smooth surface X is the number $p_a(C)$ given by the *adjunction formula*

$$p_a(C) = 1 + \frac{1}{2}(C^2 + C \circ K_X).$$

Intuitively, $p_a(C)$ is the genus of a nonsingular curve in the linear system $|C|$. Always $p_g(C) \leqslant p_a(C)$; if C is irreducible, then $p_g(C) = p_a(C)$ if and only if C is nonsingular. In general, to each isolated singular point O of C we can assign the number $\delta(O) := \frac{1}{2}(\mu + r - 1)$, where μ is the Milnor number and r is the number of analytic branches at O. Then we have the following *genus formula*: if $C \subset X$ is irreducible and with isolated singularities only, then

$$p_g(C) = p_a(C) - \sum_O \delta(O),$$

the summation running over all singular points of C. For simple singularities, we have $\delta(\mathbf{A}_p) = [\frac{1}{2}(p+1)]$, $\delta(\mathbf{D}_q) = [\frac{1}{2}(q+2)]$, $\delta(\mathbf{E}_6) = 3$, $\delta(\mathbf{E}_7) = \delta(\mathbf{E}_8) = 4$.

D.1.4 Miscellaneous notation

In agreement with the commonly accepted abuse of notation, we use the subscript $_*$ and superscript * for all kinds of covariant and, respectively, contravariant functors, in the hope that the precise meaning of this notation is always clear from the nature of the objects involved. In particular, $V^* := \operatorname{Hom}(V, \mathcal{R})$ is the dual module (over a commutative ring $\mathcal{R}$) and $\{\mathbf{v}_i^*\}$ is the basis dual to $\{\mathbf{v}_i\}$. As an exception from this rule, the transposed matrix is denoted by M^{t}.

The cardinality of a set X is denoted by $\operatorname{Card} X$.

We use the common notation $\mathbb{Z}$, $\mathbb{Q}$, $\mathbb{R}$, $\mathbb{C}$, and $\mathbb{H}$ for, respectively, integers, rational, real, and complex numbers, and quaternions. *Natural numbers* are *positive* integers: $\mathbb{N} := \{n \in \mathbb{Z} \mid n > 0\}$, and $\bar{\mathbb{N}} := \mathbb{N} \cup \{\infty\}$.

If q is a prime power, $\Bbbk_q$ is the (only) field with q elements. For uniformity, this notation extends to $\Bbbk_0 := \mathbb{Q}$. The ring $\mathbb{Z}/n\mathbb{Z}$ is denoted by $\mathbb{Z}_n$, and $\mathbb{Z}_n^*$ is its group of units (invertible elements).

The greatest common divisor and least common multiple of a subset $S \subset \mathbb{Z}$ are denoted by g.c.d. S and l.c.m. S, respectively. By convention, g.c.d. $\varnothing = 0$.

The cyclotomic polynomial of order n is $\Phi_n(t) \in \mathbb{Z}[t]$. Even more often we use the polynomial $\tilde{\varphi}_n(t) := (t^n - 1)/(t - 1) \in \mathbb{Z}[t]$, $n \in \mathbb{Z}$.

D.2 Index of notation

Symbols

Greek

Bibliography

[1] A. Adem, J. Leida and Y. Ruan, *Orbifolds and stringy topology*, Cambridge Tracts in Mathematics 171, Cambridge University Press, Cambridge, 2007.

[2] A. Akyol, Classical Zariski pairs, *J. Knot Theory Ramifications*, to appear.

[3] ______, *Classical Zariski pairs with nodes*, Master's thesis, Bilkent University, 2007.

[4] Z. Arad, J. Stavi and M. Herzog, *Powers and products of conjugacy classes in groups*, Products of conjugacy classes in groups, Lecture Notes in Math. 1112, Springer, Berlin, 1985, pp. 6–51.

[5] K.-i. Arima and I. Shimada, Zariski-van Kampen method and transcendental lattices of certain singular $K3$ surfaces, *Tokyo J. Math.* **32** (2009), 201–227.

[6] V. I. Arnol′d, S. M. Guseĭn-Zade and A. N. Varchenko, *Singularities of differentiable maps. Vol. I*, Monographs in Mathematics 82, Birkhäuser Boston Inc., Boston, MA, 1985, The classification of critical points, caustics and wave fronts, Translated from the Russian by Ian Porteous and Mark Reynolds.

[7] E. Artal, J. Carmona, J. I. Cogolludo and H.-O. Tokunaga, Sextics with singular points in special position, *J. Knot Theory Ramifications* **10** (2001), 547–578.

[8] E. Artal Bartolo, J. Carmona Ruber and J. I. Cogolludo Agustín, *On sextic curves with big Milnor number*, Trends in singularities, Trends Math., Birkhäuser, Basel, 2002, pp. 1–29.

[9] ______, Braid monodromy and topology of plane curves, *Duke Math. J.* **118** (2003), 261–278.

[10] ______, Effective invariants of braid monodromy, *Trans. Amer. Math. Soc.* **359** (2007), 165–183 (electronic).

[11] E. Artal Bartolo, J. I. Cogolludo and H.-o. Tokunaga, Pencils and infinite dihedral covers of $\mathbb{P}^2$, *Proc. Amer. Math. Soc.* **136** (2008), 21–29 (electronic).

[12] E. Artal Bartolo and J. I. Cogolludo-Agustín, On the connection between fundamental groups and pencils with multiple fibers, *J. Singul.* **2** (2010), 1–18.

[13] E. Artal Bartolo and H.-o. Tokunaga, Zariski k-plets of rational curve arrangements and dihedral covers, *Topology Appl.* **142** (2004), 227–233.

[14] E. Artin, Theory of braids, *Ann. of Math. (2)* **48** (1947), 101–126.

[15] V. G. Bardakov, The structure of a group of conjugating automorphisms, *Algebra Logika* **42** (2003), 515–541, 636, English translation: Algebra Logic 42 (2003), no. 5, 287–303.

[16] W. P. Barth, K. Hulek, C. A. M. Peters and A. Van de Ven, *Compact complex surfaces*, second ed, Ergebnisse der Mathematik und ihrer Grenzgebiete. 3. Folge. A Series of Modern Surveys in Mathematics [Results in Mathematics and Related Areas. 3rd Series. A Series of Modern Surveys in Mathematics] 4, Springer-Verlag, Berlin, 2004.

[17] A. Beauville, Application aux espaces de modules, *Astérisque* **126** (1985), 141–152, Geometry of $K3$ surfaces: moduli and periods (Palaiseau, 1981/1982).

[18] G. V. Belyĭ, Galois extensions of a maximal cyclotomic field, *Izv. Akad. Nauk SSSR Ser. Mat.* **43** (1979), 267–276, 479.

[19] I. Berstein and A. L. Edmonds, On the classification of generic branched coverings of surfaces, *Illinois J. Math.* **28** (1984), 64–82.

[20] F. Beukers and H. Montanus, *Explicit calculation of elliptic fibrations of $K3$-surfaces and their Belyi-maps*, Number theory and polynomials, London Math. Soc. Lecture Note Ser. 352, Cambridge Univ. Press, Cambridge, 2008, pp. 33–51.

[21] S. Bigelow, The Burau representation is not faithful for $n = 5$, *Geom. Topol.* **3** (1999), 397–404 (electronic).

[22] B. Birch, *Noncongruence subgroups, covers and drawings*, The Grothendieck theory of dessins d'enfants (Luminy, 1993), London Math. Soc. Lecture Note Ser. 200, Cambridge Univ. Press, Cambridge, 1994, pp. 25–46.

[23] J. S. Birman and T. E. Brendle, *Braids: a survey*, Handbook of knot theory, Elsevier B. V., Amsterdam, 2005, pp. 19–103.

[24] S. J. Blank, *Extending immersions and regular homotopies in codimension 1*, Ph.D. thesis, Brandeis University, Waltham, Mass., 1967.

[25] F. Bogomolov and Y. Tschinkel, *Monodromy of elliptic surfaces*, Galois groups and fundamental groups, Math. Sci. Res. Inst. Publ. 41, Cambridge Univ. Press, Cambridge, 2003, pp. 167–181.

[26] F. Bohnenblust, The algebraical braid group, *Ann. of Math. (2)* **48** (1947), 127–136.

[27] N. Bourbaki, *Lie groups and Lie algebras. Chapters 4–6*, Elements of Mathematics (Berlin), Springer-Verlag, Berlin, 2002, Translated from the 1968 French original by Andrew Pressley.

[28] E. Brieskorn, Über die Auflösung gewisser Singularitäten von holomorphen Abbildungen, *Math. Ann.* **166** (1966), 76–102.

[29] ______, Die Auflösung der rationalen Singularitäten holomorpher Abbildungen, *Math. Ann.* **178** (1968), 255–270.

[30] W. Burau, Über Zopfgruppen und gleichsinnig verdrillte Verkettungen, *Abh. Math. Sem. Hamburg* **11** (1936), 179–186.

[31] J. W. S. Cassels, *Rational quadratic forms*, London Mathematical Society Monographs 13, Academic Press Inc. [Harcourt Brace Jovanovich Publishers], London, 1978.

[32] F. Catanese, M. Lönne and B. Wajnryb, Moduli spaces of surfaces and monodromy invariants, in: *Proceedings of the Gökova Geometry-Topology Conference 2009*, pp. 58–98, Int. Press, Somerville, MA, 2010.

[33] O. Chisini, *Courbes de diramation des plans multiples et tresses algébriques*, Deuxième Colloque de Géométrie Algébrique, Liège, 1952, Georges Thone, Liège, 1952, pp. 11–27.

[34] O. Chisini, Una suggestiva rapresentazione reale per le curve algebriche piane, *Rend. Ist. Lombardo II* **66** (1933), 1141–1155.

[35] J. I. Cogolludo, Fundamental group for some cuspidal curves, *Bull. London Math. Soc.* **31** (1999), 136–142.

[36] J. H. Conway and R. K. Guy, *The book of numbers*, Copernicus, New York, 1996.

[37] D. A. Cox and W. R. Parry, Torsion in elliptic curves over $k(t)$, *Compositio Math.* **41** (1980), 337–354.

[38] ______, Genera of congruence subgroups in Q-quaternion algebras, *J. Reine Angew. Math.* **351** (1984), 66–112.

[39] H. S. M. Coxeter and W. O. J. Moser, *Generators and relations for discrete groups*, fourth ed, Ergebnisse der Mathematik und ihrer Grenzgebiete [Results in Mathematics and Related Areas] 14, Springer-Verlag, Berlin, 1980.

[40] C. J. Cummins and S. Pauli, Congruence subgroups of $\mathrm{PSL}(2, \mathbb{Z})$ of genus less than or equal to 24, *Experiment. Math.* **12** (2003), 243–255.

[41] A. Degtyarev, Isotopic classification of complex plane projective curves of degree 5, *Algebra i Analiz* **1** (1989), 78–101, English translation: Leningrad Math. J. 1 (1990), no. 4, 881–904.

[42] ______, Alexander polynomial of a curve of degree six, *J. Knot Theory Ramifications* **3** (1994), 439–454.

[43] ______, Quintics in CP^2 with nonabelian fundamental group, *Algebra i Analiz* **11** (1999), 130–151, English translation: St. Petersburg Math. J. 11 (2000), no. 5, 809–826.

[44] ______, Fundamental groups of symmetric sextics, *J. Math. Kyoto Univ.* **48** (2008), 765–792.

[45] ______, Oka's conjecture on irreducible plane sextics, *J. Lond. Math. Soc. (2)* **78** (2008), 329–351.

[46] ______, On deformations of singular plane sextics, *J. Algebraic Geom.* **17** (2008), 101–135.

[47] ______, Stable symmetries of plane sextics, *Geom. Dedicata* **137** (2008), 199–218.

[48] ______, Fundamental groups of symmetric sextics. II, *Proc. Lond. Math. Soc. (3)* **99** (2009), 353–385.

[49] ______, Irreducible plane sextics with large fundamental groups, *J. Math. Soc. Japan* **61** (2009), 1131–1169.

[50] ______, Oka's conjecture on irreducible plane sextics. II, *J. Knot Theory Ramifications* **18** (2009), 1065–1080.

[51] ———, *On irreducible sextics with non-abelian fundamental group*, Singularities—Niigata–Toyama 2007, Adv. Stud. Pure Math. 56, Math. Soc. Japan, Tokyo, 2009, pp. 65–91.

[52] ———, Zariski k-plets via dessins d'enfants, *Comment. Math. Helv.* **84** (2009), 639–671.

[53] ———, Classical Zariski pairs, *J. Singul.* **2** (2010), 51–55.

[54] ———, Plane sextics via dessins d'enfants, *Geom. Topol.* **14** (2010), 393–433.

[55] ———, Plane sextics with a type E_8 singular point, *Tohoku Math. J. (2)* **62** (2010), 329–355.

[56] ———, The fundamental group of a generalized trigonal curve, *Osaka J. Math.* **48** (2011), 749–782.

[57] ———, Plane sextics with a type-$\mathbf{E}_6$ singular point, *Michigan Math. J.* **60** (2011), 243–269.

[58] ———, *Topology of plane algebraic curves: the algebraic approach*, Topology of algebraic varieties and singularities, Contemp. Math. 538, Amer. Math. Soc., Providence, RI, 2011, pp. 137–161.

[59] A. Degtyarev, I. Itenberg and V. Kharlamov, *Real Enriques surfaces*, Lecture Notes in Mathematics 1746, Springer-Verlag, Berlin, 2000.

[60] ———, On deformation types of real elliptic surfaces, *Amer. J. Math.* **130** (2008), 1561–1627.

[61] A. Degtyarev, I. Itenberg and V. Zvonilov, Real trigonal curves and real elliptic surfaces of type I, *J. Reine Angew. Math.*, to appear, `arXiv:1102.3414`.

[62] A. Degtyarev and V. Kharlamov, Topological properties of real algebraic varieties: Rokhlin's way, *Uspekhi Mat. Nauk* **55** (2000), 129–212, English translation: Russian Math. Surveys 55 (2000), no. 4, 735–814.

[63] A. Degtyarev and M. Oka, *A plane sextic with finite fundamental group*, Singularities —Niigata–Toyama 2007, Adv. Stud. Pure Math. 56, Math. Soc. Japan, Tokyo, 2009, pp. 93–108.

[64] A. Degtyarev and N. Salepci, Products of pairs of Dehn twists and maximal real Lefschetz fibrations, *Nagoya Math. J.*, to appear, `arXiv:1110.4093`.

[65] A. Dimca, *Singularities and topology of hypersurfaces*, Universitext, Springer-Verlag, New York, 1992.

[66] A. H. Durfee, Fifteen characterizations of rational double points and simple critical points, *Enseign. Math. (2)* **25** (1979), 131–163.

[67] M. Eichler, *Quadratische Formen und orthogonale Gruppen*, Die Grundlehren der mathematischen Wissenschaften in Einzeldarstellungen mit besonderer Berücksichtigung der Anwendungsgebiete. Band LXIII, Springer-Verlag, Berlin, 1952.

[68] S. Eilenberg and N. Steenrod, *Foundations of algebraic topology*, Princeton University Press, Princeton, New Jersey, 1952.

[69] C. Eyral and M. Oka, On the fundamental groups of the complements of plane singular sextics, *J. Math. Soc. Japan* **57** (2005), 37–54.

[70] ———, Fundamental groups of the complements of certain plane non-tame torus sextics, *Topology Appl.* **153** (2006), 1705–1721.

[71] ———, Alexander-equivalent Zariski pairs of irreducible sextics, *J. Topol.* **2** (2009), 423–441.

[72] ———, On the geometry of certain irreducible non-torus plane sextics, *Kodai Math. J.* **32** (2009), 404–419.

[73] ———, *A proof of a conjecture of Degtyarev on non-torus plane sextics*, Singularities —Niigata–Toyama 2007, Adv. Stud. Pure Math. 56, Math. Soc. Japan, Tokyo, 2009, pp. 109–131.

[74] A. T. Fomenko and D. B. Fuks, *Kurs gomotopicheskoi topologii*, "Nauka", Moscow, 1989, With an English summary.

[75] R. Friedman and J. W. Morgan, *Smooth four-manifolds and complex surfaces*, Ergebnisse der Mathematik und ihrer Grenzgebiete (3) [Results in Mathematics and Related Areas (3)] 27, Springer-Verlag, Berlin, 1994.

[76] The GAP Group, *GAP – Groups, Algorithms, and Programming, Version 4.4.12*, 2008.

[77] F. A. Garside, The braid group and other groups, *Quart. J. Math. Oxford Ser. (2)* **20** (1969), 235–254.

[78] C. F. Gauss, *Disquisitiones arithmeticae*, Springer-Verlag, New York, 1986, Translated and with a preface by Arthur A. Clarke, Revised by William C. Waterhouse, Cornelius Greither and A. W. Grootendorst and with a preface by Waterhouse.

[79] I. P. Goulden and D. M. Jackson, The KP hierarchy, branched covers, and triangulations, *Adv. Math.* **219** (2008), 932–951.

[80] H. Grauert, Über Modifikationen und exzeptionelle analytische Mengen, *Math. Ann.* **146** (1962), 331–368.

[81] P. Griffiths and J. Harris, *Principles of algebraic geometry*, Wiley Classics Library, John Wiley & Sons Inc., New York, 1994, Reprint of the 1978 original.

[82] J. L. Gross and T. W. Tucker, *Topological graph theory*, Wiley-Interscience Series in Discrete Mathematics and Optimization, John Wiley & Sons Inc., New York, 1987, A Wiley-Interscience Publication.

[83] A. Grothendieck, *Esquisse d'un Programme*, unpublished, 1984.

[84] A. Haefliger and Q. N. Du, Appendice: une présentation du groupe fondamental d'une orbifold, *Astérisque* **116** (1984), 98–107, Transversal structure of foliations (Toulouse, 1982).

[85] A. Haefliger, *Orbi-espaces*, Sur les groupes hyperboliques d'après Mikhael Gromov (Bern, 1988), Progr. Math. 83, Birkhäuser Boston, Boston, MA, 1990, pp. 203–213.

[86] R. Hartshorne, *Algebraic geometry*, Springer-Verlag, New York, 1977, Graduate Texts in Mathematics, No. 52.

[87] F. Hirzebruch, *The signature of ramified coverings*, Global Analysis (Papers in Honor of K. Kodaira), Univ. Tokyo Press, Tokyo, 1969, pp. 253–265.

[88] F. Hirzebruch, *Topological methods in algebraic geometry*, Classics in Mathematics, Springer-Verlag, Berlin, 1995, Translated from the German and Appendix One by R. L. E. Schwarzenberger, With a preface to the third English edition by the author and Schwarzenberger, Appendix Two by A. Borel, Reprint of the 1978 edition.

[89] A. Hurwitz, Ueber Riemann'sche Flächen mit gegebenen Verzweigungspunkten, *Math. Ann.* **39** (1891), 1–60.

[90] I. M. Isaacs, *Algebra*, Brooks/Cole Publishing Co., Pacific Grove, CA, 1994, A graduate course.

[91] I. O. Kalinin, Cohomological characteristics of real projective hypersurfaces, *Algebra i Analiz* **3** (1991), 91–110.

[92] C. Kassel and V. Turaev, *Braid groups*, Graduate Texts in Mathematics 247, Springer, New York, 2008, With the graphical assistance of Olivier Dodane.

[93] F. Klein, Ueber die Transformation elfter Ordnung der elliptischen Functionen, *Math. Ann.* **15** (1879), 533–555.

[94] M. Kneser, Zur Theorie der Kristallgitter, *Math. Ann.* **127** (1954), 105–106.

[95] K. Kodaira, On compact analytic surfaces. II, III, *Ann. of Math. (2) 77 (1963), 563–626; ibid.* **78** (1963), 1–40.

[96] T. Koshy, *Catalan numbers with applications*, Oxford University Press, Oxford, 2009.

[97] V. S. Kulikov, Surjectivity of the period mapping for $K3$ surfaces, *Uspehi Mat. Nauk* **32** (1977), 257–258.

[98] ______, Hurwitz curves, *Uspekhi Mat. Nauk* **62** (2007), 3–86, English translation: Russian Math. Surveys 62 (2007), no. 6, 1043–1119.

[99] V. S. Kulikov and V. M. Kharlamov, On braid monodromy factorizations, *Izv. Ross. Akad. Nauk Ser. Mat.* **67** (2003), 79–118, English translation: Izv. Math. 67 (2003), no. 3, 499–534.

[100] ______, Surfaces with DIF $\neq$ DEF real structures, *Izv. Ross. Akad. Nauk Ser. Mat.* **70** (2006), 135–174, English translation: Izv. Math. 70 (2006), no. 4, 769–807.

[101] V. S. Kulikov, D. Oru and V. Shevchishin, Regular homotopy of Hurwitz curves, *Izv. Ross. Akad. Nauk Ser. Mat.* **68** (2004), 91–114, English translation: Izv. Math. 68 (2004), no. 3, 521–542.

[102] R. S. Kulkarni, An arithmetic-geometric method in the study of the subgroups of the modular group, *Amer. J. Math.* **113** (1991), 1053–1133.

[103] S. Lang, *Introduction to modular forms*, Grundlehren der Mathematischen Wissenschaften [Fundamental Principles of Mathematical Sciences] 222, Springer-Verlag, Berlin, 1995, With appendixes by D. Zagier and Walter Feit, Corrected reprint of the 1976 original.

[104] H. Larcher, The cusp amplitudes of the congruence subgroups of the classical modular group, *Illinois J. Math.* **26** (1982), 164–172.

[105] A. Libgober, Alexander polynomial of plane algebraic curves and cyclic multiple planes, *Duke Math. J.* **49** (1982), 833–851.

[106] ———, *Alexander invariants of plane algebraic curves*, Singularities, Part 2 (Arcata, Calif., 1981), Proc. Sympos. Pure Math. 40, Amer. Math. Soc., Providence, RI, 1983, pp. 135–143.

[107] ———, *Alexander modules of plane algebraic curves*, Low-dimensional topology (San Francisco, Calif., 1981), Contemp. Math. 20, Amer. Math. Soc., Providence, R.I., 1983, pp. 231–247.

[108] ———, Invariants of plane algebraic curves via representations of the braid groups, *Invent. Math.* **95** (1989), 25–30.

[109] ———, *Lectures on topology of complements and fundamental groups*, Singularity theory, World Sci. Publ., Hackensack, NJ, 2007, pp. 71–137.

[110] ———, *Problems in topology of the complements to plane singular curves*, Singularities in geometry and topology, World Sci. Publ., Hackensack, NJ, 2007, pp. 370–387.

[111] F. Loeser and M. Vaquié, Le polynôme d'Alexander d'une courbe plane projective, *Topology* **29** (1990), 163–173.

[112] D. D. Long and M. Paton, The Burau representation is not faithful for $n \geqslant 6$, *Topology* **32** (1993), 439–447.

[113] E. Looijenga, The complement of the bifurcation variety of a simple singularity, *Invent. Math.* **23** (1974), 105–116.

[114] W. Magnus, A. Karrass and D. Solitar, *Combinatorial group theory: Presentations of groups in terms of generators and relations*, Interscience Publishers [John Wiley & Sons, Inc.], New York-London-Sydney, 1966.

[115] W. Magnus and R. Moufang, Max Dehn zum Gedächtnis, *Math. Ann.* **127** (1954), 215–227.

[116] J. Milnor, *Singular points of complex hypersurfaces*, Annals of Mathematics Studies, No. 61, Princeton University Press, Princeton, N.J., 1968.

[117] R. Miranda and D. R. Morrison, The number of embeddings of integral quadratic forms. I, *Proc. Japan Acad. Ser. A Math. Sci.* **61** (1985), 317–320.

[118] ———, The number of embeddings of integral quadratic forms. II, *Proc. Japan Acad. Ser. A Math. Sci.* **62** (1986), 29–32.

[119] B. Moishezon, *The arithmetic of braids and a statement of Chisini*, Geometric topology (Haifa, 1992), Contemp. Math. 164, Amer. Math. Soc., Providence, RI, 1994, pp. 151–175.

[120] B. G. Moishezon, *Stable branch curves and braid monodromies*, Algebraic geometry (Chicago, Ill., 1980), Lecture Notes in Math. 862, Springer, Berlin, 1981, pp. 107–192.

[121] B. Moishezon, *Complex surfaces and connected sums of complex projective planes*, Lecture Notes in Mathematics, Vol. 603, Springer-Verlag, Berlin, 1977, With an appendix by R. Livne.

[122] M. A. Naĭmark and A. I. Štern, *Theory of group representations*, Grundlehren der Mathematischen Wissenschaften [Fundamental Principles of Mathematical Sciences] 246, Springer-Verlag, New York, 1982, Translated from the Russian by Elizabeth Hewitt, Translation edited by Edwin Hewitt.

[123] S. M. Natanzon, Spaces of moduli of real curves, *Trudy Moskov. Mat. Obshch.* **37** (1978), 219–253, 270.

[124] ______, Topology of 2-dimensional coverings and meromorphic functions on real and complex algebraic curves, *Selecta Math. Soviet.* **12** (1993), 251–291, Selected translations.

[125] V. V. Nikulin, Integer symmetric bilinear forms and some of their geometric applications, *Izv. Akad. Nauk SSSR Ser. Mat.* **43** (1979), 111–177, 238, English translation: Math USSR-Izv. 14 (1979), no. 1, 103–167 (1980).

[126] M. Nori, On certain elliptic surfaces with maximal Picard number, *Topology* **24** (1985), 175–186.

[127] M. Oka, Some plane curves whose complements have non-abelian fundamental groups, *Math. Ann.* **218** (1975), 55–65.

[128] ______, *A survey on Alexander polynomials of plane curves*, Singularités Franco-Japonaises, Sémin. Congr. 10, Soc. Math. France, Paris, 2005, pp. 209–232.

[129] M. Oka and D. T. Pho, Classification of sextics of torus type, *Tokyo J. Math.* **25** (2002), 399–433.

[130] ______, *Fundamental group of sextics of torus type*, Trends in singularities, Trends Math., Birkhäuser, Basel, 2002, pp. 151–180.

[131] S. Y. Orevkov, Riemann existence theorem and construction of real algebraic curves, *Ann. Fac. Sci. Toulouse Math. (6)* **12** (2003), 517–531.

[132] ______, *On braid monodromy monoid*, Talk at MSRI, 2004.

[133] ______, Quasipositivity problem for 3-braids, *Turkish J. Math.* **28** (2004), 89–93.

[134] F. Öztürk and N. Salepci, *Real open books and real contact structures*, to appear, `arXiv:1202.5928`.

[135] U. Persson, *Double sextics and singular K-3 surfaces*, Algebraic geometry, Sitges (Barcelona), 1983, Lecture Notes in Math. 1124, Springer, Berlin, 1985, pp. 262–328.

[136] I. I. Pjateckiĭ-Šapiro and I. R. Šafarevič, Torelli's theorem for algebraic surfaces of type K3, *Izv. Akad. Nauk SSSR Ser. Mat.* **35** (1971), 530–572, English translation: Math. USSR-Izv. 5, 547–588.

[137] V. Poenaru, *Extension des immersions en codimension 1 (d'après Samuel Blank)*, Séminaire Bourbaki, 20e année, 1967/68, pp. 1–33.

[138] A. N. Protopopov, Homeomorphisms of branched coverings of a two-dimensional sphere, *Dokl. Akad. Nauk SSSR* **290** (1986), 792–795.

[139] ______, Topological classification of branched coverings of a two-dimensional sphere, *Zap. Nauchn. Sem. Leningrad. Otdel. Mat. Inst. Steklov. (LOMI)* **167** (1988), 135–156, 192, English translation: J. Soviet Math. 52 (1990), no. 1, 2832–2846.

[140] B. Saint-Donat, Projective models of $K-3$ surfaces, *Amer. J. Math.* **96** (1974), 602–639.

[141] N. Salepci, Classification of totally real elliptic Lefschetz fibrations via necklace diagrams, *J. Knot Theory Ramifications*, to appear, `arXiv:1104.1794`.

[142] ______, *Real Lefschetz fibrations*, Ph.D. thesis, Université Louis Pasteur et IRMA, 2007.

[143] ______, Real elements in the mapping class group of T^2, *Topology Appl.* **157** (2010), 2580–2590.

[144] W. K. Seiler, *Moduln elliptischer Flächen mit Schnitt. (Dissertation).*, Ph.D. thesis, Fakultät für Mathematik der Universität Karlsruhe. 130 S., 1982.

[145] ______, Global moduli for elliptic surfaces with a section, *Compositio Math.* **62** (1987), 169–185.

[146] C. Series, *On coding geodesics with continued fractions*, Ergodic theory (Sem., Les Plans-sur-Bex, 1980) (French), Monograph. Enseign. Math. 29, Univ. Genève, Geneva, 1981, pp. 67–76.

[147] J.-P. Serre, *A course in arithmetic*, Springer-Verlag, New York, 1973, Translated from the French, Graduate Texts in Mathematics, No. 7.

[148] I. Shimada, *On the connected components of the moduli of polarized $K3$ surfaces*, `http://www.math.sci.hiroshima-u.ac.jp/~shimada/K3.html` (to appear), 2007.

[149] ______, Lattice Zariski k-ples of plane sextic curves and Z-splitting curves for double plane sextics, *Michigan Math. J.* **59** (2010), 621–665.

[150] G. Shimura, *Introduction to the arithmetic theory of automorphic functions*, Publications of the Mathematical Society of Japan 11, Princeton University Press, Princeton, NJ, 1994, Reprint of the 1971 original, Kanô Memorial Lectures, 1.

[151] T. Shioda, On elliptic modular surfaces, *J. Math. Soc. Japan* **24** (1972), 20–59.

[152] ______, On the Mordell-Weil lattices, *Comment. Math. Univ. St. Paul.* **39** (1990), 211–240.

[153] R. Silhol, *Real algebraic surfaces*, Lecture Notes in Mathematics 1392, Springer-Verlag, Berlin, 1989.

[154] V. Snyder, A. H. Black, A. B. Coble, L. A. Dye, A. Emch, S. Lefschetz, F. R. Sharpe and C. H. Sisam, *Selected topics in algebraic geometry*, second ed, Chelsea Publishing Co., New York, 1970.

[155] E. H. Spanier, *Algebraic topology*, Springer-Verlag, New York, 1981, Corrected reprint.

[156] C. C. Squier, The Burau representation is unitary, *Proc. Amer. Math. Soc.* **90** (1984), 199–202.

[157] J. G. Thompson, *A finiteness theorem for subgroups of* $\mathrm{PSL}(2, \mathbf{R})$ *which are commensurable with* $\mathrm{PSL}(2, \mathbf{Z})$, The Santa Cruz Conference on Finite Groups (Univ. California, Santa Cruz, Calif., 1979), Proc. Sympos. Pure Math. 37, Amer. Math. Soc., Providence, R.I., 1980, pp. 533–555.

[158] H.-o. Tokunaga, Dihedral coverings of algebraic surfaces and their application, *Trans. Amer. Math. Soc.* **352** (2000), 4007–4017.

[159] ———, Dihedral covers and an elementary arithmetic on elliptic surfaces, *J. Math. Kyoto Univ.* **44** (2004), 255–270.

[160] T. Urabe, *Combinations of rational singularities on plane sextic curves with the sum of Milnor numbers less than sixteen*, Singularities (Warsaw, 1985), Banach Center Publ. 20, PWN, Warsaw, 1988, pp. 429–456.

[161] E. R. van Kampen, On the fundamental group of an algebraic curve, *Amer. J. Math.* **55** (1933), 255–260.

[162] O. Y. Viro, *Some integral calculus based on Euler characteristic*, Topology and geometry—Rohlin Seminar, Lecture Notes in Math. 1346, Springer, Berlin, 1988, pp. 127–138.

[163] J. H. C. Whitehead, Combinatorial homotopy. I, *Bull. Amer. Math. Soc.* **55** (1949), 213–245.

[164] ———, Combinatorial homotopy. II, *Bull. Amer. Math. Soc.* **55** (1949), 453–496.

[165] B. Wu and J.-G. Yang, Some new Zariski pairs of sextic curves, *Tohoku Math. J.*, to appear.

[166] J.-G. Yang, Sextic curves with simple singularities, *Tohoku Math. J. (2)* **48** (1996), 203–227.

[167] O. Zariski, On the Problem of Existence of Algebraic Functions of Two Variables Possessing a Given Branch Curve, *Amer. J. Math.* **51** (1929), 305–328.

[168] ———, On the irregularity of cyclic multiple planes, *Ann. of Math. (2)* **32** (1931), 485–511.

[169] ———, The Topological Discriminant Group of a Riemann Surface of Genus p, *Amer. J. Math.* **59** (1937), 335–358.

Index of figures

Index of tables

Index

Bei Fragen zur Produktsicherheit wenden Sie sich bitte an:
If you have any questions regarding product safety,
please contact:

Walter de Gruyter GmbH
Genthiner Straße 13
10785 Berlin
productsafety@degruyterbrill.com